Introduction to Robotics

Introduction to Robotics

Arthur J. Critchlow
California State University, Sacramento

Macmillan Publishing Company
New York
Collier Macmillan Publishers
London

Macmillan Publishing Company
866 Third Avenue, New York, New York 10022

Collier Macmillan Canada, Inc.

Library of Congress Cataloging in Publication Data

Critchlow, Arthur J.
    Introduction to Robotics.

    Including bibliographies and index.
    1. Robotics. I. Title.

TJ211.C69   1985   6.29.8'92   84–17107
ISBN 0–02–325590–0

Printing: 1 2 3 4 5 6 7 8   Year: 5 6 7 8 9 0 1 2

ISBN 0-02-325590-0

Dedication

This book is dedicated to Charles Rosen and Joseph F. Engelberger, whose faith, perseverance, and leadership over a 20—year period made modern robots a reality.

Preface

Objectives of This Book

This book is an attempt to include, in one volume, a useful summary of the many complex technologies that go into a modern robot. Our definition of a robot requires that it be able to operate independently of human control and be capable of movement, either on a mobile base or by the use of a manipulator arm. This definition covers, therefore, industrial robots and the growing category of home robots, office robots, and mobile vehicles or combinations thereof.

This book is entitled, <u>Introduction to Robotics</u> because it is just that. It was written as an introduction to this subject for engineering and computer science students, technology students, and practicing engineers who want to obtain some knowledge of the growing and esoteric field of robotics. It was initiated because the author was assigned to teach a course in robotics and could find no text that covered all the basic information needed. There was one book that discussed the applications of industrial robots, another that was a useful assembly of articles from various periodicals, and an excellent text on the mathematics of kinematics and control. But nothing available was suitable for an introductory course on robotics at the junior or senior level. Since this book was initiated in early 1982 other books have been written, but I believe that this is the most complete one available on the subject. Due to the rapid growth of the robotics field, it is expected that the basic ideas discussed here will last but that many technical details will quickly become obsolete as improved methods and technologies come into being.

It is assumed that all readers have some knowledge of dynamics and statics or at least are familiar with elementary physical concepts such as Newton's laws. Force, mass, acceleration, velocity, centrifugal force, and gravitational attraction are of some importance to robots, as they are to humans. Other subjects touched on are described and defined as necessary before use. In addition, references to more detailed descriptions of each major subject area are provided.

Students coming into robotics courses taught by the author have often had gaps in knowledge that made some robotics concepts difficult to grasp. Mechanical engineers, in many cases, did not understand electronic circuits, logic, or control theory. Electronic engineers, well versed in those subjects, were not familiar with backlash or the static properties of cantilever beams. Except for a few computer science experts, none were familiar with computer vision and terms such as <u>pixel</u>, <u>gray level</u>, and <u>edge detection</u>. And computer science people were unacquainted with most of the mechanical and electrical engineering details discussed. Other important areas—industrial engineering, production engineering, CAD/CAM, simulation, and so on were only of peripheral concern to the average engineer.

It is my conviction that we each need to know enough about the other person's special field to be able to identify the important concepts. If a large percentage of those who read this book learn to identify the concepts discussed, the author's purpose will have been served.

Of course, if many people become interested in and excited by the fascinating field of robotics, even greater joy will result.

I hope you enjoy this book as much as I have enjoyed writing it.

Now to discuss the organization of this text:

Chapter 1 provides an overview of robots: historical background, definitions, a discussion of why robots are valuable, and a preliminary description of how robots work. It also includes a brief description of market, social, and economic factors affecting robots and a prediction of the changes to be expected in the near future.

The early develoment of robots is described in Chapter 2. Some major industrial and technical developments are described to provide a background for understanding how the modern robot became a practical and viable entity.

Robot technology is described in Chapters 3, 4, 5, and 6, which provide a summary of the mechanical, electrical, electronic, sensing, and control problems that must be solved to make a robot work and adapt it to a particular application. Computer control and interfacing, an important and increasingly complex part of this technology, is reviewed in Chapter 7. Robot software, including programming, is described in Chapter 8. The software area is expected to be a major determinant of robot system capability in the next few years. It should receive much more attention than is now being accorded it. Programming of a robot is expected to be one of the most difficult and demanding areas of development.

Until recently, there were few applications of robot vision. Now it is the most exciting growth area in robotics. Hardware and software for vision are described in detail in Chapter 9, although complete coverage of these important subjects was not possible in the space available. It is hoped that the background provided in Chapter 9 will provide the interested reader with sufficient information to understand the many references to these subjects listed at the end of the chapter. Also, there is a considerable amount of information on the capabilities and limitations of current vision systems and the progress being made in improving vision capability.

Major applications of robots in manufacturing are discussed in detail in Chapter 10, with emphasis on new developments. The use of vision in new applications should be of interest to many readers.

One of the perils of writing a book covering a broad field is that no two users want the same information. This book is especially open to criticism, since it not only covers a broad field but also tries to serve the needs of many types of users. It is hoped, and expected, that people with backgrounds in mechanical, electrical, and industrial engineering will find the book valuable in providing a sound background in new knowledge areas. In addition, the book should be of help to those who wish to improve their understanding of robotic vision, software, and control.

Feedback from users of this book is encouraged. There are bound to be many errors, some minor and some major. In addition, some topics are not included in enough detail, whereas others are too detailed for some users. Please feel free to send me your comments and criticisms. Between us, we can make the second edition much better.

Suggested Uses for This Book

1. Reference Source

The broad coverage of robotic engineering and technology makes this a useful reference on the operation of robots and robotic systems for the practicing engineer, researcher, or robot technologist. A thorough index and a comprehensive glossary have been supplied to make it easy to find and understand the information presented. In addition, there is an extensive list of references to the widely scattered literature in the many fields that are combined to make robotics possible. Appendixes are supplied to describe some terminology and mathematical concepts that may be new to some readers.

2. Educational Use

There is enough information in this book to cover a two-semester course in introduction to robotics, robotics engineering, or robotics technology. There is also sufficient information for a one-semester course in robot software at the senior computer science or engineering level. With appropriately selected subject matter, the book can support as many as three quarters in one of these subjects. A suggested selection of topics for each type of course is outlined in the following table.

Each chapter is largely self-contained. Basic concepts are introduced at the beginning of the chapter and are elaborated on as the discussion proceeds. Use of the first half of each chapter is a practical way to introduce each subject area. The more difficult concepts may be omitted for an introductory course and followed up in an advanced course.

This book may be used as a textbook for several categories of students based on their educational level and the goals of the course. Possible assignments for different students are indicated by the letter X in the indicated column. The letter R indicates that material previously covered by the students, possibly in other courses, should be reviewed. One- and two-semester courses, and a three-quarter course, are possible, as designated by the course code.

CODE A: Engineering Technology Majors—One-Semester Course
CODE B: Mechanical Engineering Majors—Two- Semester Course
CODE C: Electrical (Electronic) Engineering Majors—Two-Semester Course
CODE D: Computer Science Majors—One Semester Course
CODE E: Engineering Majors—Three-Quarter Course

An accelerated course for graduate students with a previous background could be completed in one semester if only advanced sections are covered.

	A	B		C		D	E		
COURSE CODE (Semester or Quarter)	1	1	2	1	2	1	1	2	3

Chapter 1 Overview

	A 1	B 1	B 2	C 1	C 2	D 1	E 1	E 2	E 3
1.1 Historical Perspective		X		X		X	X		
1.2 Definitions and Classifications of Robots	X	X	R	X	R	X	X	R	
1.3 Why Are Robots Used?	X	X	R	X	R	X	X	R	
1.4 Applications of Robots by Type	X	X	R	X	R	X	X	R	
1.5 Components of a Robot	X	X	R	X	R	X	X	R	
1.6 Robot Systems			X		X	X		X	
1.7 Present Status			X		X			X	
1.8 Future Trends			X		X			X	
References		X	X	X	X	X	X	X	
Exercises		X	X	X	X	X	X	X	

Chapter 2 Early Development of the Modern Robot

	A 1	B 1	B 2	C 1	C 2	D 1	E 1	E 2	E 3
2.1 Development Timetable		X		X		X			
2.2 Industrial Development	X	X		X			X		
2.3 Technical Development		X		X		X	X		
References		X		X			X		
Exercises		X		X			X		
								X	

	COURSE CODE (Semester or Quarter)								
	A	B		C		D	E		
	1	1	2	1	2	1	1	2	3
9.6 Preprocessing		X		X	X				X
9.7 Edge Detection		X		X	X				X
9.8 Recognizing Objects		X		X	X				X
9.9 Scene Understanding		X		X	X				X
9.10 Interfacing and Control	X	R	X	R	X				X
9.11 Examples Of Vision Systems		X	R	X	R	X			X
References		X	X	X	X	X			X
Exercises		X	X	X	X	X			X

Chapter 10 Applications of Robots

	A	B		C		D	E		
	1	1	2	1	2	1	1	2	3
10.1 Machine Loading and Unloading	X	X	R	X	R	X	X		R
10.2 Material Handling	X	X	R	X	R	X	X		R
10.3 Fabrication	X	X	R	X	R	X	X		R
10.4 Spray Painting and Finishing	X	X	R	X	R	X	X		R
10.5 Spot Welding	X	X	R	X	R	X	X		R
10.6 Arc Welding	X	X	R	X	R	X	X		R
10.7 Assembly Applications	X	X	R	X	R	X	X		R
10.8 Inspection And Test	X	X	R	X	R	X	X		R
10.9 Auxiliary Equipment	X	X	R	X	R	X	X		R
10.10 Optimizing Robotic Systems		X		X					X
10.11 The Future Factory		X		X					X
10.12 Management Considerations		X		X					X
References	X	X	X	X	X	X	X	X	X
Exercises	X	X	X	X	X	X	X	X	X

It is important that the exercises and references be used appropriately by the instructor. Much of the knowledge of robotics is widely scattered in the literature. In the second semester and in advanced topics, it is desirable that the instructor assign reading from references available at his or her facility. Exercises are in the order of the topics covered in the chapters and should be assigned to increase student understanding of text material.

Arthur J. Chritchlow
Sacramento, California

Acknowledgments

I owe a great debt to the many scientists, engineers, and others who developed the basic theory and technology of robotics and made them available through published papers. This cooperative effort is the foundation of research and development throughout the world.

In particular, I would like to thank Dr. David Nitzan at SRI International, Dr. Thomas Binford at Stanford University and Dr. Raj Reddy at Carnegie-Mellon University for making available to me the reports from their respective institutions. I have also used extensively the many publications from the Massachusetts Institute of Technology.

Dr. Fred Blackwell of the California State University in Sacramento, (CSUS) taught me much about computer vision, pattern recognition, and programming theory. He also contributed most of the software portion of Chapter 9 and reviewed portions of several chapters. Dr. Mohammed Zand at CSUS reviewed Chapter 6 and made several valuable comments. Bill Kerth, Jr., and Rob Kerth of Adaptive Technologies, Inc., also reviewed parts of the manuscript and made useful comments.

I would also like to thank the reviewers who reviewed this manuscript, in its entirety, and gave me much needed guidance. I have endeavored to mold the text to meet their needs without deviating excessively from my own goals. Finally, the people at Macmillan deserve many thanks for converting the original manuscript into a completed text.

Contents

Chapter 4 Drive Methods _____ 88

Overview

This chapter is a summary and overview of the field of robotics. It is also a guide to the remainder of the book. After each summary there is a reference to the chapter or section of the book that deals with that subject in more detail.

1.1 Historical Perspective

Humans have been fascinated since prehistoric times with extra-ordinary beings, mechanical men, and other creatures, that, to them, were fantasies. Priests of old built the first mechanical arms. Ancient Egyptians built statues of gods operated by priests who pretended to be acting under direct inspiration of the god represented by the statue. There seems to be no clear understanding of the reason for this interest in machines. Perhaps they were only meant to impress the faithful or coerce the unfaithful, which were very likely results. When a mammoth god, breathing smoke and flames, waved a hand at primitive people and commanded them to obey, it is under-standable that a deep impression was made. This interest in mechanical men, robots, and other exotic creatures has continued to the present day.

During the heyday of the Greek civilization many centuries later, there were several creators of hydraulically operated statues. Heron of Alexandria built simple mechanisms to illustrate the then new science of hydraulics. It was not his intention to duplicate human life. These models were designed as teaching aids or design exercises. Most of them were stored in temples and apparently served to inspire and entertain the worshipers.

1.1.1
Early Automata

Early automata, built during the seventeenth and eighteenth centuries in Europe, were truly mechanical wonders. Starting about 350 years ago, ingenious inventors designed and fabricated some strange devices. Medieval clocks mounted on top of churches and cathedrals generally had a life-sized figure of a man, angel, or devil that used a mace to strike the hours on a bell. These were refined as time went by and became more intricate. Williams [23-25] and Penniman [15] provide details on these developments.*

1. Bird Organ

A marvel of the period was the bird organ, invented about 1770 (Figure 1.1). An arrangement of cams and levers driven by a clockwork mechanism controlled the operation of the bird's wings, head, and beak. It also operated valves and pistons to generate

*References appear at the end of each chapter. Reference identifiers are given in brackets [ ].

Figure 1.1 Bird organ (from [23]).

Source: "Antique Mechanical Computers—Part I" by J.M. Williams appearing in the July, 1978 issue of BYTE magazine. Copyright © 1978 by McGraw-Hill, Inc. New York, NY 10020.

assorted bird whistles. In operation the bird moved its head and wings realistically while it whistled a tune, opening and closing its beak as it sang. These bird organs were made in large quantities and became household items in well-to-do homes.

The mechanisms used in the bird organs would be classified today as Sequence Machines, as defined in Section 1.2.2. Figure 1.2 shows the bird organ mechanism. A main spring drove a gear train that operated a bellows to compress air into a wind box. Another gear train, operating off the same spring, drove a shaft on which cams were mounted. As the cams turned, they operated the drive and control mechanisms. Push wires, riding on cam followers, drove the bird's wings, tail, head, and beak. Other push rods opened and closed valves that allowed air to flow from the wind box to the various whistles.

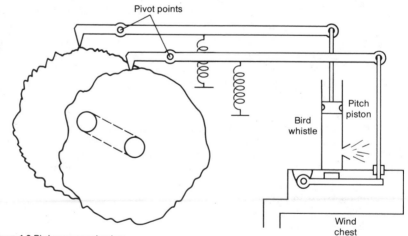

Figure 1.2 Bird organ mechanism.

Source: "18th & 19th Century Mechanical Marvels, Part II", by J.M. Williams appearing in th August, 1978 issue of BYTE magazine. Copyright © 1978 by McGraw-Hill, Inc., New York, NY 10020.

Note that the intricately carved cams that provided the bird's motions are equivalent to a kind of "read-only memory" (ROM) plus a mechanical driver. Electronic equivalents of this might be a ROM unit plus some transistor drivers that control small electrical motors.

More elaborate devices had stacks of cams that could be operated selectively, so that the devices gave the impression of lifelike motion that appeared to be nonrepetitive. An example of the complexity attainable was the automated flute player.

2. Automated Flute Player

   Clothed in a musician's clothes, the automated flute player held a flute to its lips, blew air across the flute, and manipulated its fingers to cover the control ports of the flute. An intricate carved drum, looking something like the drum of a music box, controlled a large number of motions through cam actions. The flute player was made by Jacques de Vaucanson in Paris and demonstrated in 1738. Details of the sophisticated flute player mechanism are shown in Figure 1.3.

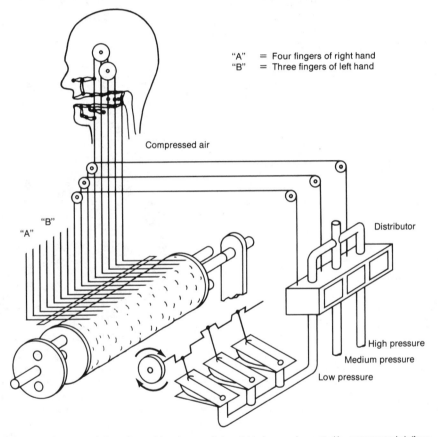

"A" = Four fingers of right hand
"B" = Three fingers of left hand

Figure 1.3 Vaucanson's flute player, driven by a music box-type drum and operated by compressed air (from [23]).

Source: "18th & 19th Century Mechanical Marvels, Part II", by J.M. Williams appearing in the August, 1978 issue of BYTE magazine. Copyright © 1978 by McGraw-Hill, Inc., New York, NY 10020.

   At this time there were no machines for metal fabrication, so that all of the intricate parts had to be made by hand. An example of the problems encountered by inventors was Charles Babbage's failure to make workable his early mechanical computer, the analytic engine, over a hundred years later. Vaucanson was clearly a clever inventor and a versatile mechanic. One wonders what he and Babbage could have done together.

   People of the period traveled great distances to see these marvels. It was possible for a genius like Vaucanson to make a good living building and exhibiting these devices. Kings and emperors were pleased when presented with a new mechanical art object that appeared to emulate a living creature. These clockwork automata were the technological frontier of the period. For these reasons, some of the most talented inventors chose to work with automata.

3. Maillardet's Automaton

Perhaps the most complicated and versatile of the clockwork automata was the incredible writing and drawing automaton built in London about 1805 by Henri Maillardet (Williams [24], Penniman [15]). It had an unusually large memory and very precise movements. A ship drawing by the Maillardet automaton shows a fully rigged three-masted schooner with three decks of ports and all the necessary lines and details. About 5 minutes are required by the automaton to make the drawing. Several items are in its repertoire and can be selected. Another example is a five-line poem, in French script, beautifully drawn by pen.

Figure 1.4 Maillardet's writing and drawing automaton, stripped of garments (from Penniman [15]).

Courtesy of The Franklin Institute Science Museum, Philadelphia, PA.

Maillardet's automaton even identifies itself. It was damaged in a fire and its identity was lost. After restoration by a patient staff machinist at the Franklin Institute in Philadelphia, where it had finally landed, its true identity was revealed. When it was working again and wrote out the French poem, it ended by writing "Ecrit par L'Automaton de Maillardet" ("Written by Maillardet's Automaton") after the poem. This automaton, operated by an unusually complicated series of cams, rods, levers, and shafts, is now on exhibit at the Franklin Institute.

Essential properties of all of these automata were:

1. They were built for display and entertainment. People came to marvel and told their friends of the great automata they had seen. Public interest was in the display; people were not interested in the method of operation or the significance of these simulations of life.
2. There was no feedback or sensory capability in any of the automata. Complicated motions were carried out sequentially, but there was no response to anything in the environment.

When a machine is required to maintain a specified output despite various internal and external disturbances, some kind of control system is required. Internal and external sensors must supply information on which the control system can operate.

A writing automaton was designed to write on a piece of paper. However, without sensory feedback, it could just as well wave its hand in the air if there was no paper or table to write on. At a minimum, then, it would be desirable to have some way of knowing that there was paper and a table under the pen. This information could be provided by using a pressure switch of some kind. Any small push-button switch could be made to close when the automaton's hand contacted the paper when the hand was in writing position--that is, if there was a source of electricity available and some way to sense the flow of the current.

Although the ancient Romans and Greeks apparently knew the general principles of feedback, they were not used much until the last century. While mechanical feedback is feasible, most devices use electrical and electronic feedback in control systems. The Sperry gyropilot developed for use in World War I was one of the earliest examples of a feedback device. In this case, a gyroscope supplied the sensory information that allowed the gyropilot to control the direction of a ship.

In later chapters, we will see how sensors and feedback make complex, adaptive, and perhaps even intelligent robots possible.

1.1.2
Robots and
Robotics

Even though the Greeks, Romans, and early Europeans had robot-like machines, it was left to a Czechoslovakian playwright, Karel Capek, to provide us with the name <u>robot</u>. In the Czech language, the word "robota" meant a worker providing compulsory service. In 1921, Capek wrote a play called <u>R.U.R.</u> for "Rossum's Universal Robots." In the play, robots were humanoid creations developed by Rossum and his son to be used as servants for humanity.

Isaac Asimov, one of the best of the science fiction writers, is credited with the first use of the word <u>robotics</u> to describe the science of dealing with robots. His stories about robots anticipated problems that have not yet arisen but are realistically possible. His "Three Laws of Robotics" are worth consideration by the modern designer and user of robots. These laws are:

1. A robot must not harm a human being or, through inaction, allow one to come to harm.
2. A robot must always obey human beings unless that is in conflict with the First Law.
3. A robot must protect itself from harm unless that is in conflict with the first or second laws.

1.1.3
Robot Relatives

Some devices that use the same technologies as robots but are not considered robots are those whose control comes from humans rather than from internal capability. Some related devices are now discussed.

1. Manipulators

Mechanically controlled manipulators were used and are still used in hazardous areas. Manipulator arms have clamps or pincers on the end that can be controlled by direct mechanical linkage from a remote point. Usually they are used through a wall that protects the operator from the hazard. Viewing is done through a protective window. These machines were used extensively in the early days of nuclear and atomic experimentation for handling dangerous materials. The protective window was made of heavily leaded glass several inches thick to absorb radiation.

Initially, manipulators were driven and controlled by human operators, whose strength and dexterity limited their use. It took considerable skill to operate the remote "hand" (clamps or pincers) by squeezing and twisting the control linkage. Later, power servo amplifiers were used to control electric motors in order to increase the lifting and handling capability of the manipulators.

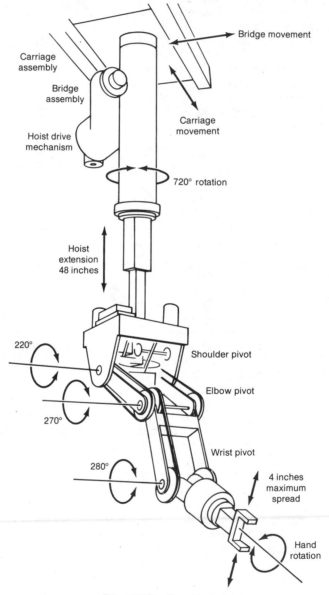

Figure 1.5 The Case Manipulator

As long as there was a direct mechanical linkage, the operator could sense by touch the effect of his actions. It was equivalent to holding an object with a pair of pliers, which any mechanic does with ease. When power servos were supplied, it became possible to exert great force without realizing it, sometimes with disastrous consequences. Therefore, the servo-equipped manipulators were supplied with a force feedback that used an electric or hydraulic motor to supply the operator with a sense of touch. There were two servo systems, one to drive the manipulator and one to provide the operator with feedback. A power-driven manipulator is shown in Figure 1.7. This manipulator, at the Case Institute of Technology, was used for research in computer control of a manipulator.

The "robot arm" on the U.S. space shuttle is a manipulator by our definition, since it is controlled by human operators through a computer control link.

2. Teleoperators

Teleoperators were the next step after manipulators. The term teleoperator was coined in about 1966 by Edwin Johnson, then at the Space Nuclear Agency, to describe servo-controlled manipulators that were not directly connected.

Teleoperators are operated by humans who provide remote control by using joysticks, wheels, control keyboards, or exoskeleton controls. They are used for dangerous work such as undersea activity, working with radioactive material, or working in space. Another name for these devices is telecherics or remote hands. Control can be provided by radio link, optical link, or even satellite communications. Vision is provided by remote television cameras. Touch feedback is still provided by the feedback servo system, but it is now transmitted over radio links.

Figure 1.6 Propelled anthropomorphic manipulator (SAM).

NASA HQ. Used by permission.

The self-propelled anthropomorphic manipulator (SAM) was developed under Johnson's direction in 1969 for the U.S. Atomic Energy Commission to handle radioactive reactor materials. A photograph of SAM is given in Figure 1.8.

3. Exoskeletons

More recently, the exoskeleton approach has been used for control. The operator wears a skeleton framework that models the manipulator's effectors. This is a lightweight structure that can be strapped to human limbs and used to sense limb motions and convert them to electrical signals. Each joint of the exoskeleton, corresponding to a human joint, has a potentiometer or encoder that measures the amount of motion and translates it into a signal that can be transmitted to a remote device.

As the operator moves his arm and hand, the remote effectors are driven by servo motors to follow this movement with great precision and controllable power. Manipulators have been built that can pick up an egg, pour a liquid from a flask into a test tube, and still have the strength to lift heavy objects. Such manipulators may have several fingers and a thumb on the hand.

Figure 1.7 Control station, SAM system, illustrating the use of exoskeleton control for a teleoperator.

NASA HQ. Used by permission.

Several manipulators are controlled by exoskeletons. The SAM vehicle with two manipulators is controlled by the exoskeleton shown in Figure 1.9. This vehicle and the arms was developed for the U.S. Atomic Energy Commission in 1969.

4. Prostheses and Orthoses

Prostheses are artificial replacements for human arms and legs. They often have servo motors, gears, and so on, as a robot would, but are controlled by myoelectric sensors that pick up human nerve impulses and translate them, through a computer, into commands to the artificial limbs. The simplest ones are artificial arms that are powered by electric motors or hydraulic drives. Some users have become very skilled in their use. A type of prosthesis called an orthosis works like an exoskeleton to move the arm on which it is mounted. Orthoses are used by people who are paralyzed. Some prostheses and orthoses can be controlled by voice commands.

1.2 Definitions and Classifications of Robots

Several definitions of robots exist:

1. The popular conception is of a mechanical man capable of carrying out tasks that a human might do and displaying some capability for intelligence. Robots are often portrayed in movies as strong and stupid. More recent movies such as Star Wars have shown friendly, capable robots such as C3PO and R2D2 who worked with their human counterparts in fighting the Empire.
2. Users and manufacturers of industrial robots have banded together to form the Robot Institute of America. They have a suitably precise definition: "A robot is a reprogrammable, multifunctional manipulator designed to move material, parts, tools or specialized devices through variable programmed motions for the performance of a variety of tasks."

Other useful ways of defining robots are illustrated in the following sections, where robots are defined by type of control, by capability level, and by mechanical configuration. Each of these definitions is useful for different purposes. Only general definitions are given here. We will discuss each type in more detail in later chapters.

1.2.1
Type of Control

1. Point-to-Point Robots

Point-to-point robots are able to move from one specified point to another but cannot stop at arbitrary points not previously designated. They are the simplest and least expensive type of robot. Stopping points are often just mechanical stops that must be adjusted for each new operation. Point-to-point robots driven by servos are often controlled by potentiometers set to stop the robot arm at a specified point.

2. Continuous-Path Robots

Continuous-path robots are able to stop at any specified number of points along a path. However, if no stop is specified, they may not stay on a straight line or constant curved path between specified points. Every point must be stored separately in the memory of the robot.

3. Controlled-Path (Computed Trajectory) Robots

Control equipment on controlled-path robots can generate straight lines, circles, interpolated curves, and other paths with high accuracy. Paths can be specified in geometric or algebraic terms in some of these robots. Good accuracy can be obtained at any point along the path. Only the start and finish coordinates and the path definition are required for control.

4. Servo versus Non-Servo Robots

Servo-controlled robots have some means for sensing their position and "feeding back" the sensed position to the means of control in such a way that the control can cause a particular path to be followed. Nonservo robots have no way of determining whether or not they have reached a specified location.

Controlled-path robots all have servo capability, so that they can correct their path constantly to carry out a specified motion.

1.2.2
Capability Level

Control capability and flexibility are additional criteria for evaluating and defining robots. Five types of machines are so classified.

1. Sequence-Controlled Machines

Sequence-controlled machines are able to go through a specified sequence of actions according to preset instructions. A household washing machine is of this type since it can fill, wash, rinse, and spin-dry clothes. Just as the washing machine uses a preset electrical timer, some simple robots are sequenced by mechanical or electrical means. Robots can be adjusted to move at different times and in different sequences. But once adjusted, they must follow the same sequence until they are physically readjusted. They are robots because they have manipulator arms that are being controlled by the sequence controller, and they can follow a continuous path rather than just move from one point to another.

2. Playback Machines

Playback machines can be "taught" to carry out a series of movements. A recording device, such as a magnetic disk, magnetic tape, or random access memory, is used to record a series of points or steps by recording the coordinate information from the position sensors. At each point, the robot coordinates along three axes are recorded. After the complete path is recorded, the robot can be directed to "play back" the path followed and to perform whatever task it has been taught.

Teaching is done by a human operator who guides the robot along the desired path. Every point along the path and every action taken must be controlled by the human operator. Methods for doing this teaching are described in Chapter 6.

3. Controlled-Path (Computed Trajectory) Robots

Some robots can be programmed to follow a certain path between particular points. No teaching is required. The user can specify points and the way in which these points are to be used in computing the path. Interpolating between points or establishing a spline curve can be done, for example. These robots are intermediate between the playback machines that require teaching and the adaptive robots that have sensors. The Japanese name for this type of robot is Numerically Controlled Machine because it is somewhat similar to numerically controlled machine tools.

4. Adaptive Robots

Adaptive robots have computer control and sensory feedback, so that they can react to their environment. Most of these robots have controlled path capability and will attempt to carry out a task, but can modify their path and actions as they go. For example, a welding robot can follow a weld path even though it is different from the prescribed path. In this case, a vision sensor gathers information that allows the control computer to adjust the path to weld along the actual path rather than the previously described path.

5. Intelligent Robots

Intelligent robots are the highest level of robot, and there is some argument about whether they now exist. In addition to the ability to sense their environment and modify their actions to fit, they must have a knowledge base and models of their environment. A robot with an expert system as defined by Artificial Intelligence researchers, a full set of sensors, a large memory, and a way to model its environment might be called intelligent by this definition. We will discuss this fascinating subject later in some detail.

**1.2.3
Configuration**

Robots have one advantage over humans. They can operate their arms in several different configurations corresponding to different coordinate systems.

1. Cartesian Coordinates

Cartesian coordinates are the familiar X, Y, Z coordinates used in computation. Using these coordinates, each part of the robot's arm slides at right angles to the previous part. Robots can reach any part of a volume bounded by the length of the separate sections of the arm. Cartesian coordinates, also called "rectangular coordinates," are illustrated in Figure 1.8.

2. Cylindrical Coordinates

The volume covered by the arm is cylindrical in form. A main support rotates around a base bearing and carries a two-section arm. One section of the arm can move horizontally and one section can move vertically, so the end of the arm sweeps out the volume between two cylinders. Normally, the main support cannot rotate through a complete circle because of mechanical interference. (See Figure 1.9.)

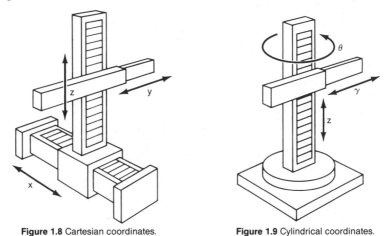

Figure 1.8 Cartesian coordinates.   **Figure 1.9** Cylindrical coordinates.

3. Polar Coordinates

A section of the arm is capable of linear motion in and out. It is supported by two other sections, one that rotates around the base and one that rotates about an axis perpendicular to the vertical through the base. This arm can reach almost anywhere in a volume bounded by an outer and an inner hemisphere. The radii of the two hemispheres correspond to the maximum and minimum extensions of the linear section, respectively. (See Figure 1.10.)

4. Revolute Coordinates

The arm that uses revolute coordinates has three rotary or revolute joints. A section or link of the arm mounted on the base joint can rotate around the base and carry two other sections that move like pieces of a hinge relative to each other. The rotary joints can be mounted either vertically or horizontally. (See Figure 1.11.)

Most of the first industrial robots made in the United States, starting about 1965, were made in either the cylindrical or polar coordinate (also called spherical) configuration. Although the revolute configuration was simpler to build and had simple rotary joints that could be easily sealed, it was not used. It was found that control of the revolute coordinate configuration was too difficult without the use of a computer. The advantages and disadvantages of each configuration are analyzed in detail in Section 3.1.

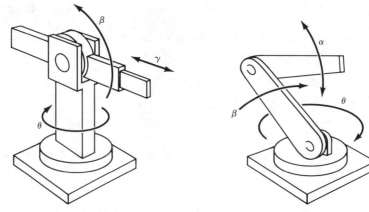

Figure 1.10 Polar coordinates.   **Figure 1.11** Revolute coordinates.

**1.2.4
Mobile versus
Fixed Robots**

Until recently, nearly all robots were fixed in position. They were mounted on a rigid base and bolted to the floor so that they could withstand the forces and torques applied when the arm lifted or moved something. There were several reasons for keeping robots fixed. One was economic; it was much cheaper to build a fixed robot. Another important reason was the control problem; additional movement meant more complex controls.

Now, with the availability of low-cost computers, the control problem is much easier to solve. Some manufacturers are now providing mobility to their robots. We can identify the following categories of robots:

1. Fixed Robots on a Pedestal Base
   This is the standard robot mounting.
2. Fixed Robots Supported from Above
   This robot is often used where floor space is critical.
3. Fixed Robots on a Track

There are now many robots that ride back and forth along a track so that their reach can be extended, as shown in Figure 1.12.

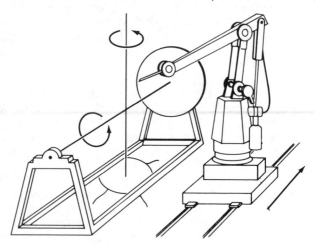

Figure 1.12 Tracked robot.

Welding and painting robots are the most common tracked robots. An additional problem exists when a robot is moved along a track: The power supply, air hoses, hydraulic hoses, electrical power wiring, and control wiring must be carried along with the robot. The gantry robot in Figure 1.13 has the same problem; it is solved by supporting the hoses and wiring in a moving chain connected to the robot. One is required on each axis.

4. Fixed Robots on an Overhead Gantry

As shown in Figure 1.13, the overhead-gantry robot is a large and versatile device. By moving the track overhead and supplying a movable cross bridge riding on a trolley, we have a robot that can travel 10 to 20 feet in two horizontal directions. This is essentially a rectangular coordinate robot with a polar coordinate arm mounted

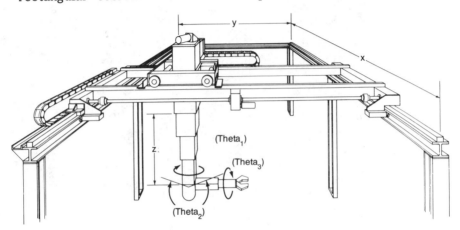

Figure 1.13 Overhead gantry robot.

Source: GCA Corporation. Used by permission.

from it. While these robots are expensive, they are able to carry out tasks that would be much more difficult if done in other ways. This robot can carry loads weighing up to 2,500 pounds.

5. Mobile—Wheeled Vehicle

Most mobile robots are mounted on wheels. There are no production versions of industrial robots mounted on wheels, even though there is a need for mobility on the factory floor. However, there are several hobby robots on wheels. These robots are being sold as household assistants and monitors. (See Section 1.8.2.) It would be relatively easy to mount a robot on a computer-controlled vehicle similar to a fork lift truck, with the robot's power coming from the heavy-duty truck batteries. However, no such machine is yet available.

6. Mobile—Tracked Vehicle

There is now a robot arm mounted on a tracked vehicle like a small tank or personnel carrier. However, it may not be an autonomous robot but really a teleoperator like SAM, which was discussed in Section 1.1.3. This field is wide open for both military and civilian use.

7. Mobile—with Four to Six Legs

Several vehicles are in the development stage, that have four to six legs so that the vehicle can walk in difficult terrain. These vehicles have been proposed for military and industrial use. One mobile robot, dubbed Odex I, the first functionoid, by its developer, Odetics, Inc., has been demonstrated before robot conferences. It can climb up on the back of a pickup truck and climb down again. It then picks up the truck by lifting it with a post supported on the frame of the robot. However, the demonstrations to date were remotely controlled by radio, so this device is really a teleoperator. Figures

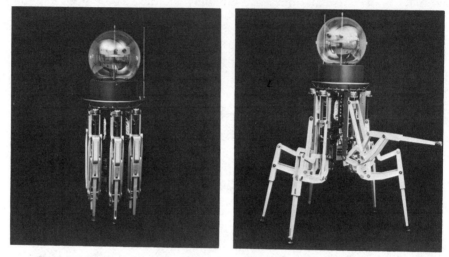

Figure 1.14 ODEX 1 in a narrow profile.   **Figure 1.15** ODEX 1 in a wide-articulated profile.

Courtesy of Odetics, Inc. of Anaheim, California. Used by permission.

1.14 and 1.15 show the ODEX 1 in two poses. Autonomous control of a walking robot is difficult. It has been accomplished and demonstrated by the Ohio State University with a hexapod, or six-legged device. This device is described in Orin [14]. It is shown in Figure 1.16.

Figure 1.16 The OSU hexapod experimental vehicle (from *ROBOTICS RESEARCH, Courtesy, MIT Press. Used by permission*).

8. Mobile—with Two Legs

Apparently, it will be some time before humans will have to worry about direct competition from robots. At this time, there is no published research on the development of two-legged robots. There are some difficult problems to be solved before such robots are feasible. Probably the biggest problem is providing balance so that the robot can stand upright while carrying out tasks. Every motion in a two-legged robot must be balanced by some other action. If we pick up an object with one hand, the body must bend back to offset the force; otherwise we will fall over forward. This balancing act requires a large amount of computation. No program is now available for this purpose.

Another problem for a two-legged robot is the power supply. Existing batteries would not run a 200-pound robot for more than a few hours at full activity. Needed is a power source as effective as the human source. Perhaps beamed power or a small nuclear power source could be supplied.

1.3 Why Are Robots Used?

In this section, we explain why robots are being used more and more in industrial and commercial applications. At the present time, the number of robots in use is increasing at the rate of about 35% per year. Sales volume is also increasing each year at about the same rate. This activity is discussed in detail in Section 1.7.

Some of the reasons for the increased use of robots are discussed in the following sections.

**1.3.1
Reduced
Production Cost**

Manufacturers are finding that the actual cost of making a product can be reduced by the use of robots. There are several reasons for this reduction.

1. The cost of a robot amortized over several years is often less than $5 per hour compared to average labor costs of $15 to $20 per hour when fringe benefits are included. Fringe benefits in many industries are 30 to 50% of the base salary. These fringe benefits cover such items as Social Security, workmen's compensation, vacations, holidays, sick leave, medical and dental benefits, and retirement pay. Robots get none of those benefits.
2. Robots work 98% of the time at their assigned task. Humans take coffee breaks, lunch breaks, and other time off for personal reasons. A standard industrial allowance for production workers is the Personal and Fatigue Allowance, which depends on the type of work. It is usually about 15 to 20%.
3. Robots produce a higher percentage of good parts or assemblies than human workers do because they repeat the same procedure every time and do not make parts incorrectly due to fatigue or lack of attention.

Frequently, robot systems can repay their entire capital, installation, and training costs within less than 12 months. These figures typically include the costs of auxiliary equipment used with the robot.

**1.3.2
Increased
Productivity**

1. Robots can work much faster at some tasks than human workers can. For example, a robot arc welder can maintain an average rate of 30 inches per minute on straight-line welds. A human welder, under the same conditions, can average less than 10 inches per minute.
2. In another application, two paint spray robots on an automobile assembly line can paint a complete car body in 90 seconds, inside and out, with two coats of paint, (the wet-on-wet coating method now in use). In the GM plant in Michigan, two GMF (General

Motors-Fanuc) robots achieve this rate and work 20 hours per day. Human painters can't compete with this rate or the quality of work performed. Even the best painters may take 15 to 30 minutes to do this job.

3. Increased productivity means that more work is completed on schedule and that schedules can be maintained. Utilization of space and equipment is improved, resulting in savings due to lower capital investment.

**1.3.3
Improved
Product Quality**

1. Accuracy of positioning is much greater in robots than in humans. Current robots with a 3-foot reach can achieve accuracies of 0.008 inch and repeatability of 0.004 inch. In a welding test, a robot produced a weld that didn't require grinding afterward and produced parts to better tolerance than any human welder could produce.
2. Speed of operation is another advantage in producing high-quality parts. Again, in welding very thin pieces, it is desirable to move quickly over the seam to be welded and to complete the weld before the pieces distort due to the heat of welding. The controlled accuracy and speed of the robot make possible some welds that were difficult to perform before.
3. Another example of improved quality is in die casting, where the casting cycle must be strictly adhered to in order to produce good parts. Humans can't adhere to a strict timing cycle for many hours in sequence; robots can. This control results in more good parts and improved die life when robots are used.

**1.3.4
Operation in
Hazardous and
Hostile
Environments**

1. Loading and unloading of hot forging presses was one of the early applications of robots. White-hot ingots must be held in place while a powered forging hammer hits them with massive, multiton strokes. Formerly two men, with long tongs, held the ingots in place during the forging operation. Now, one robot, holding the ingot with a steel end effector, positions the part accurately for forging. Higher operation rates result, workers are not exposed to flying sparks of hot metal, and product quality is improved.
2. Some painting is done with toxic paints that are extremely hazardous to the painter's health. Men were required to work completely covered with hoods and sealed garments, with an air supply piped into the hoods. Work under these conditions was hot and tiring. Robots were taught to do this work using the "teach" boxes described in Chapter 6, and men were freed from working under these conditions. In addition, the production rate and the quality of work were improved.

**1.3.5
Improved
Management
Control**

Computer-controlled robots can carry out preprogrammed procedures with great accuracy. In addition, they can record accurately what is being done. This information is then available and can be used to improve scheduling, planning, and monitoring operations in industrial plants.

**1.3.6
Integrated Systems**

Possibly every manager has dreamed of having fully competent workers who always do as they are told. Some managers are beginning to fulfill this dream by use of integrated systems of robots and other factory equipment. Work can be precisely scheduled, materials moved under computer control to robot work cells, and work performed as planned, all as part of an integrated system controlled by a host computer.

1.3.7
Meeting OSHA's
Standards

The Occupational Safety and Health Administration (OSHA) has regulations governing the exposure of workers to various hazards. For example, a worker may not put his hands under a punch press or other vertical press when there may be danger involved. Robots can be used to handle materials safely under these presses.

1.3.8
Decreased Reaction
and Debugging
Time

Robot flexibility means that some tasks can be modified quickly and made to run where the use of fixed automation would cause serious delays. It is also easier to reprogram or retrain a robot if an error is found in the operation than to modify a fixed automation system.

1.3.9
Longer Useful Life

Obsolescence can be reduced and system life extended when robots are used, since it is possible to change end effectors on a robot and reprogram it for a different task. Fixed automation must often be scrapped because it is cheaper to redesign than to modify the old equipment.

1.4 Applications of Robots By Type

In Section 1.3 we discussed some advantages of robots and illustrated them with applications. In this section, we will consider what kind of work can be done with specific classes of robots.

1.4.1
Sequence–
Controlled
Machines

Simple, unintelligent robots, as described in Section 1.2.2, have been used for more than 20 years to do routine handling tasks in factories. Typical tasks for these robots are:

1. Parts handling. This operation consists of picking up parts at a known fixed location and feeding them, one by one, to a machine. Pick-and-place robots are used that may be controlled by cam action, electrical sequencing relays, or hard-wired circuits. (Hard-wired means that the connections are permanently soldered or crimped together so that they cannot be altered by electrical control.)
2. Assembly. Simple robots have been used to some extent in basic assembly tasks. Since they are unable to recognize and orient parts, they act only as pick-and-place machines to place parts for other operations.

1.4.2
Playback Machines

1. Machine loading and unloading. Production equipment such as stamping presses, forge presses, die-casting machines, injection molding machines, and many types of metal-cutting machines are loaded and unloaded by robots that can be "trained" to go through a fixed series of operations. Many of these robots have control boxes that are used to lead the robot through a series of operations that the robot records in its memory. Then the operator can command the robot to repeat the operations indefinitely. Training can be done at slow speed, but the robot can do production operations at full speed. Synchronization of the robot with the operating cycle of the machine tools is provided.
2. Spray painting. A spray paint nozzel is attached to the robot's arm. The arm is programmed to move through a sequence of continuous path motions to carry out the painting operation. Training may be used for spray painting. Since paint spray and especially undercoating is hazardous to health and unpleasant to work around, workers in industry have accepted these robots willingly.

3. Welding. Spot welding is the largest application of robots in the automobile industry. More than 500 robots are working in this area alone. Heavy spot welding guns are carried by the robots to weld body frames and chassises together. Robots do not get tired and therefore are able to maintain high positional accuracy at all times, so that body accuracy and quality are improved.

1.4.3
Adaptive Robots

Adaptive robots are able to sense their environment and modify their actions in response to the information sensed. Computer control and built-in programs are necessary to make these actions possible.

Development of adaptive computer-controlled robots for industrial use is based on the joining of two technologies:

1. Programmable industrial robots with performance and costs acceptable to industry.
2. Artificial intelligence research in pattern recognition, visual sensing, tactile sensing, and robot control systems.

Adaptive robots can be applied in new areas. These uses are expected to proliferate in the next decade. Some of these areas are now discussed.

1. Assembly of Small Products

Development of vision systems made possible the use of robots in complex assembly tasks. Typical units to be assembled were water pumps, alternators, and similar small, in-line asemblies. Assembly of units of these types is now common, and several robot assembly systems are in use. Robot vision is used to identify parts, either in feeders or coming down conveyor belts, and guide the robot arm in picking up the part. Then the part is moved into place, over the product being assembled, and placed in position. Sometimes simple fixtures are used to hold the base while additional parts are placed on it.

The robot picks up bolts or large screws from feeders, aligns them in the proper holes, and screws them in place. When the assembly is complete, it is placed on an outgoing conveyor or on a pallet of similar assemblies. An important capability required in a robot system of this type is parallel processing. As parts move in on a conveyor, the robot must track them with its vision system and then move its arm at the same speed that the conveyor is moving. When the speeds of the conveyor and the robot hand are synchronized, the robot arm can pick up the part from the conveyor.

Tactile and proximity sensors used for adaptive robots are described in Chapter 5. Vision systems are described in Chapter 9. Complete assembly systems are described in Chapter 10.

2. Inspection Applications

Automobile bodies coming along the assembly line after spot welding are being inspected by robot vision systems. In some cases, the "eye" can be carried in the "hand" of the robot, so that the robot can make close-up measurements while standing alongside the assembly line. An important advantage of robot inspection is the ability to make an immediate digitized record of each inspection, and to store it in memory and finally in the associated disk storage unit.

3. Adaptive Welding

Both vision and tactile sensing have been used to guide robots in following a seam for welding. Often, the parts to be welded are not cut precisely or cannot be aligned exactly. In these cases, the robot can use a laser beam to scan ahead of the weld in order to locate the weld seam. The sensor information is fed back to the control computer and used to cause the robot arm, carrying the welding gun in its gripper, to follow along the actual seam path. Probes are also used to sense the weld seam by touch. Another technique is the use of the arc of the weld itself to determine the shape of the weld seam in the through-the-arc welding method.

Some of those techniques and the equipment required are discussed in Chapter 10.

1.4.4
Intelligent Robots
At present there are no intelligent robots as defined in Section 1.2.2. Several research laboratories are working on various aspects of the problem, however. Some writers use the term to describe what we have called <u>adaptive robots</u>. We have chosen to differentiate between the two capabilities because of the great potential difference in the final product.

1.5 Components of a Robot

It is now time to describe some of the elements and components of a robot to help explain how they work. A simple drawing of a complete robot is given in Figure 1.17. A complete robot doing a drilling and riveting operation is shown in Figure 1.18. It carries drilling and riveting equipment in its end effector.

Externally, a robot is observed to have a base support, an arm, and an end effector or gripper. The end effector carries a tool or part that is often called the <u>workpiece</u>. Inside the arm or the base support are the drives or motors that cause the arm to move. Some type of control device or computer is used to give instructions to the drives. Sensors are mounted on the robot for two different purposes: to sense the actions and status of the robot itself, or to sense the world external to the robot.

1.5.1
Links and Joints
The individual sections of the robot arm between the joints are called <u>links</u> by analogy to the links in a chain. Joints, as discussed previously, can be either rotating or sliding (or some combination of the two). Six joints are shown in Figure 1.19; in this case, all of them are rotary. Names given to the joints in industry are nonstandard, but those given here are the ones most used.

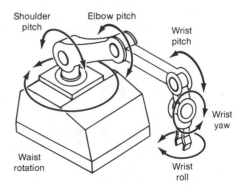

Figure 1.17 Axes or degrees of freedom in a revolute or articulated robot. The wrist has three axes: pitch, roll, and yaw.

Source: Tooling and Production Magazine.

Links are rigid members that support the loads carried by the robot. Usually they are hollow to reduce weight and provide space for gearing, electrical wiring, air hoses, and control wiring. A development that is expected to reduce the weight of robot arms is the use of graphite fiber composites in making the links of a robot arm. The Graco Corporation now makes a sleek, lightweight robot arm for spray painting out of these graphite fiber composites. It is smaller and lighter than the steel or aluminum arms but has the stiffness required for robot use.

Joints must be precisely made to maintain accurately the position of the arm. Ball bearings are used to allow joints to move smoothly and accurately.

1.5.2
Wrist

Wrists are mounted on the end of the arm (as might be expected) and may have two or three joints. They are usually made as a unit. It is desirable that wrist pitch, yaw, and roll occur about the same center rather than about separate centers, as seen in Figure 1.19. Control and dexterity are both improved by this concentric mounting.

There are a number of ingenious designs that make concentric mounting possible. Wrist design is an important selling point for robots, because wrist flexibility makes it possible for the robot to position the end effector to operate effectively in tight places.

1.5.3
End Effectors

In robot terminology, the end of the wrist is the end of the arm. A mounting plate is provided on the wrist so that different end effectors can be used. Typical end effectors are grippers, claws, welding guns, vacuum pickups, and many other devices. End effectors are discussed further in Section 3.2.

1.5.4
Drives and
Drive Mechanism

Electric motors, hydraulic motors, hydraulic cylinders, pneumatic motors, and pneumatic motors are used to move the links around or along the joints. Gears, ball screws, harmonic drives, and various kinds of linkages are used to convert the drive motion to the desired direction and speed.

Improved gearing, bearings, and ball screws have extended the life and reliability of robots. Perhaps the greatest recent improvement is the harmonic drive. Harmonic drives make it possible to provide speed reductions of 100 to 1 or more with high accuracy and in a compact volume. Backlash in gearing has been a major factor in the accuracy of robot drive mechanisms. With harmonic drives, backlash is reduced to very small amounts; some manufacturers advertise "zero backlash." (See Chapter 3.)

Early robots, such as the Unimate, used hydraulic cylinders or hydraulic rotary motors for drives. Two factors have been important in substituting electric motors for hydraulic drives in most medium-sized robots:

1. Rare earth magnets have provided greatly increased magnetic fields for motors in the same volume, thereby reducing the size of motors and increasing their efficiency.
2. More important is the development of high-power transistors for motor control. Electric motor drives for a two-horsepower motor required 8 to 10 power transistors 10 years ago, with resulting problems in balancing the load, cooling, and mounting. Today, one transistor can be used to control the same amount of power.

Another by-product of the solid-state revolution is the use of brushless direct current (DC) motors instead of commutated DC motors. Commutators and the brushes used to make electrical contact with them are sources of wear, which shortens the life of electrical motors and causes electrical noise through arcing of the contacts. Brushless motors, although more expensive, have removed these sources of wear and reduced reliability.

A new development is the use of reversible Alternating Current (AC) servo motors, which have improved reliability compared to DC motors.

Drives are described in Chapter 4. Drive mechanisms and the mechanical relationships of the robot are discussed in Section 3.1.4.

1.5.5 Controls

By controls we mean all of the equipment used to direct and sequence the arms, wrists, and end effectors. A simple control may be no more than a timing circuit that opens and closes air valves. Mechanical stops may be used to limit movement, as in a pick-and-place robot. More complex controls may be stepper switches, computer logic devices, or a complete microprocessor controlling the servo motors that drive the robot links.

Microcomputers have completely changed the control methods for robots. Many robots today have a microcomputer for every joint plus an overall computer to control the entire arm and interface to the external world. Vision and sensor systems used with the robot have one or several microcomputers or even minicomputers as part of their control system. Several robot developers are considering the use of multiple computers working in parallel to solve the equations of motion for the robot arm.

Controls are defined and described in Chapter 6. Control is so important, however, that Chapters 7 and 8 are devoted to a description of the hardware and software required to control robots and robot systems.

Figure 1.18 Drilling and riveting operations at Sikorsky Aircraft using an ASEA IRb–60 robot.
Courtesy of Robotics Today Magazine and ASEA Robotics, Inc. Used by permission.

1.5.6 Sensors

Position sensors are necessary to measure joint position on all but the simplest sequence machines. Playback machines record the position information from joint sensors and store it for future use. The stored position information for each joint is used to control the joint drive when the robot is used to carry out a prerecorded task.

Sensors are also used to measure force, torque, proximity, temperature, and many other factors in the more complex robots. Within the last 5 years, several new sensors have become available on the robot market. A wrist sensor, with three axes of force sensing and three axes of torque sensing, is available for use in assembly operations. Touch sensors, capable of recognizing small parts from their tactile patterns, have been built in the laboratory and are now available commercially. These developments are described in Chapter 5.

Computer vision has made possible a great increase in the capability of robots. Robots can now see objects and their surroundings, and can modify their operation accordingly so that complex tasks can be performed. Vision equipment and programming (hardware and software, respectively) are described in Chapter 9.

1.5.7
Interfaces

Interfaces are the robot's connections to the external world for all purposes. Electrical signals, in particular, must be received from auxiliary equipment, computers, and external sensors. Often, the robot is required to signal completion of a task or motion so that some other action can take place. It may be necessary for a numerically controlled lathe to open and allow the robot end effector to place a part in the lathe. Signals from the lathe are required to tell the robot that the lathe is open and has stopped to allow placement of a part in the lathe. After the robot has placed a part in the lathe, it must open its gripper to release the part and notify the lathe that the part has been released.

Interface standards, such as the RS-232 standard, are described in Chapter 7. The RS-170 standard, for television scanning and vision use, is described in Chapter 9.

1.6 Robot Systems

Robots must become part of a robot system before they can be useful to industry or other users. This section discusses some of the parameters of systems, the determination of system goals, and requirements, and the necessary components of a useful system.

1.6.1
What Is a System?

The Institute of Electrical and Electronics Engineers (IEEE) defines a system as "An integrated whole even though composed of diverse interacting structures or subjunctures." Another view is that a system is an entity with a structure and a set of boundaries defined to serve some purpose. Any defined collection of parts or subsystems working together may be considered a system. Examples of systems that fit this definition are the United States of America, the General Motors Corporation, the human body, and an adaptive robot assembly station.

Systems typically include subsystems to provide specific functions. In robot systems they may be control, vision, or conveyor subsystems. Subsystems are systems in their own right even though they are part of a larger system. Subsystems can be parts of more than one system in a system hierarchy.

1.6.2
Systems Goals and Requirements

In planning a robot system, the first step is to establish the goals to be achieved by the system. Then, by analysis of the goals, a set of requirements can be determined that the system must meet in order to achieve the desired goals.

Some of the functions that must be considered in specifying system requirements are now presented (Evans [8]).

1. Application Environment

What are the working conditions for the robot? Are they hot, wet, or dusty? Does the robot have freedom to move its arms, or is it constrained by other equipment or by moving vehicles in the area?

2. Range of Motion Required

Possible ranges of arm motion to consider are less than 1 foot, greater than 1 foot but less than 4 feet, greater than 4 but less than 10 feet, greater than 10 feet. Is mobility of the whole robot required, or just arm motions?

3. Speed Required

How fast are the arms, wrists, grippers, or other parts of the robot required to move? Both linear motion in feet per second, and rotary motion in degrees per second must be considered. Ranges to be considered might be the following:

1. Slow speed—less than 1 foot per second or 60 degrees per second.
2. Medium speed—1 foot per second to 5 feet per second or 60 to 180 degrees—per second.
3. High speed—more than 5 feet per second or 360 degrees per second.

4. Type of Control Required

1. Simple force—measure force along a single axis.
2. Complex force—measure force along two or more axes.
3. High positional accuracy—Tolerances less than 0.010 inch.
4. Precision positioning—Tolerances less than 0.001 inch.
5. Sensor—directed control—monitoring by vision or touch or force sensor.

5. Sensory Requirements

1. Proximity—noncontact detection of objects.
2. Touch—determine the presence or absence of a part.
3. Simple vision—detect edges, holes, corners, and so on.
4. Complex vision—recognize shapes and objects.

6. Interaction with Other Equipment

In some cases, a robot arm will have to pick up objects from a moving conveyor, which requires that the robot arm be synchronized with the conveyor before picking up the object. In other cases, the robot will have to wait for the completion of a machine cycle by another machine, such as a forge or punch press, before carrying out its own operation. A means must be provided to control the sequence of operations by passing signals of some kind between the robot arm and other machines. Use of two or more robot arms on the task may require their activity to be synchronized by communication signals.

Many activities in industry require very high production rates on a yearly basis. In these cases, it is desirable to make fully automated machines that produce one type of product in high quantities. A well-known example is the automobile engine, which is produced by a production line of automatic casting, machining, boring, and honing machines coordinated to do this one task efficiently. A rule of thumb used in the industry is that full automation is justified when more than a million items can be made on the same machine or group of machines.

Most of the products in industry are made in quantities of much less than a million per year. These products are usually made in job lots of a few hundred to a few thousand at a time. It is not economical to make large quantities of an item and then be forced to store them until they can be used. When products are made in small to medium quantities, it is likely that robots can be used effectively if other conditions are met. Operations requiring repetition of a few simple steps, such as machine loading and unloading, were the first to be assigned to robots.

Machine loading and unloading provide a useful example of the analysis required in determining requirements. It is apparent, after a little consideration of the problems, that loading and unloading are quite different operations. In unloading a machine, the location or part is known accurately because it has been positioned by or for the machine operation. In loading, the part may be in a storage bin or on a conveyor and difficult to find, orient, and pick up. Humans have no trouble picking items from a bin or a conveyor; simple robots can't do it. Only recently have robots with vision been able to handle such tasks, and they are still somewhat expensive. One immediate answer comes to mind; keep the part oriented as it comes out of the first manufacturing process. This is an excellent solution, but it may require several changes.

1.6.3 Classification of Robot Systems

The simplest robot system must have three basic items: the robot itself, a workpiece (part or object to be handled), and a place to work.

We can elaborate on each of these ideas. Several robots could be used; they could handle several different objects; and they could work in a complex environment with conveyors to supply parts, machines to be supplied with parts, and conveyors to take away parts. It might be desirable to have auxiliary vision systems, communication with other parts of a factory, and so on. We will identify three classes of systems as a background for discussion in later chapters.

1. Simple System

A pick-and-place robot that picks up a part from a fixed location, moves it to a new location, and sets it down is one of the simplest systems we can conceive. It has only the robot, a gripper on the robot to grasp the part, a pickup site, and a placement site. Control is done by adjusting cams or setting switches on the robot; there are no requirements for external communication.

2. Complex System

When two or more entities—robots, vision systems, conveyors, feeders, and so on—must work together, the control system is complex. There is a need for passage of information back and forth between entities. Usually this procedure requires the use of a computer to coordinate the information and control the sequencing of operations. An example of a complex system is the CONSIGHT system described in Chapter 9, which includes a robot, a vision system, a conveyor, and two light sources, all controlled by one computer.

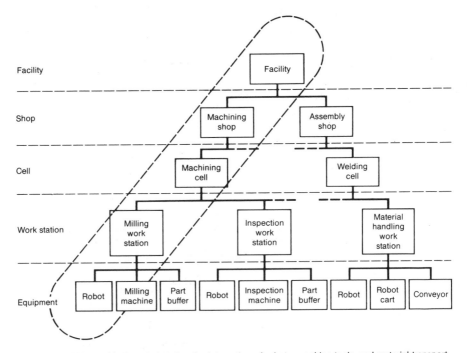

Figure 1.19 A hierarchical control system for integration of robots, machine tools, and material transport facilities into an automated factory. See Figure 1.22 for details enclosed in the dotted line.

3. Integrated System

As robotic systems become more important in the production activities of a business or industrial complex, it becomes necessary to coordinate them with greater portions of the overall production system. Systems that are connected to other production functions and operate with them in a coordinated and controlled way will be called integrated systems. Typically the overall system will have a hierarchy of computers controlling local, departmental, and higher-level functions. This organization is shown in Figure 1.19 with more detail in Figure 1.20.

Hierarchical control of integrated factory automation systems will join together the areas of Computer-Aided Design (CAD), computer-Aided Manufacturing (CAM), Factory Data Processing (FDP), and eventually the material handling, shipping, receiving, and other functions in a corporation.

Integrated systems now in use have automated material-handling equipment that brings parts to a robot, feeders to supply parts for assembly by the robot, and a hierarchical structure of control. An example of an integrated system is the Westinghouse APAS System described in Chapter 10. Other integrated systems are also described in Chapter 10 for welding and assembly functions.

Communications, large data bases, office functions, and all other corporate functions will become part of future integrated systems.

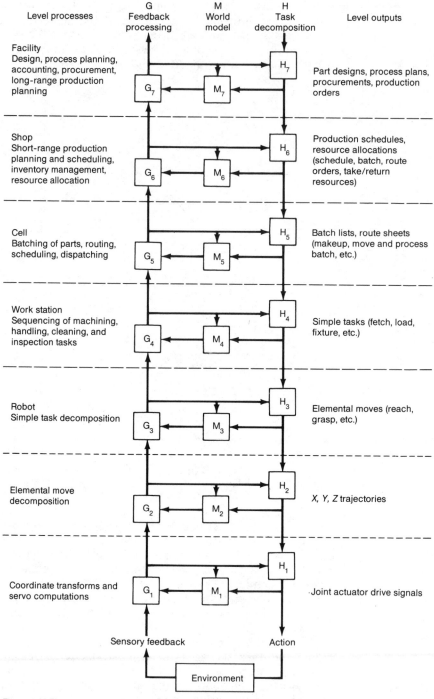

Figure 1.20 The computational hierarchy for a robot in a machining workstation. This hierarchy corresponds to the chain of command enclosed in dotted lines in Figure 1.21 (from Albus [1]).

1.7 Present Status

Robots are in use throughout the world for a multitude of tasks. In this section, we will discuss the past, present, and future markets for robots and the way these markets are divided among manufacturers, countries, and applications.

Table 1.1 lists the distribution of industrial robots by country for sample years. This list does not include manual manipulators and fixed-sequence robots, which are not considered robots in the United States. These values are estimates from different sources and should be considered as guides only. (See Conigliaro [5], Geverter [9], Schreiber [18], and Warnecke [20].) Estimates are not available for all countries for 1985 and 1990, but a growth rate of 35% is as accurate as any for those countries.

Table 1.1 Industrial Robots in Use, by Country

	1979	1982	1985	1990
Japan	14,000	21,000	90,000	160,000
United States	3,255	6,800	20,000	125,000
West Germany*	850	3,500		
Sweden	600			
Italy	500			
Poland	360			
France	200			
Britain	185	439		
Norway	170			
U.S.S.R.	25			50,000 (goal)
Belgium	13			

*Warnecke [20] reports a total of 3,500 robots in Germany. However, 59.8% of these are point-to-point control units, some of which may not be considered robots by our definition.

Total sales of robots in the United States are given in Table 1.2.

Table 1.2 Total U.S. Robot Market

	Total Number Installed	Annual Number Sold	Annual Revenue ($ millions)
1975	600	200	12
1980	3,000	2,000	100
1985	20,000	13,000	600
1990	125,000	50,000	2,000

Source: Conigliaro [5].

Other estimates for 1990 range from $1.5 to $2.5 billion. It appears highly probable that sales will become large unless there is a significant change in the economic or political climate of this country. Even higher sales are possible if there is general acceptance of robots and a movement to a shorter work week and earlier retirement.

There are over 50 manufacturers of robots in the United States today and over 100 manufacturers worldwide. Japanese manufacturers have sold the greatest total number of robots, with the U.S. second.

Industry forecasts by Prudential-Bache Securities (Conigliaro

[5]) reported in March 1983 are given in Table 1.3 for the years 1980 to 1983. Prior to 1980, two U.S. companies, Unimation and Cincinnati Milacron, accounted for more than 75% of total robot sales. Next in sales was Prab Robots, Inc., which had purchased the Versatran division of AMF and merged it with the Prab Conveyor Company.

Table 1.3 Manufacturers in the Robotics Market: Percentge of Market Share

	1980	1981	1982	1983
Unimation	44.4%	44.0%	32.9%	17.4%
Cincinnati Milacron	32.2	32.3	16.7	13.6
DeVilbiss	5.5	4.2	12.4	8.7
ASEA, Inc.	2.8	5.8	5.0	5.5
Prab Rotobs, Inc.	6.1	5.3	6.5	6.5
Copperweld Robotics	3.3	2.3	1.0	1.4
	94.3%	93.9%	74.5%	53.0%
New Venture Companies	2.3	2.5	8.2	14.1
Large Companies (IBM, etc.)			5.3	17.7
Others	3.4	3.6	12.0	15.2
	100.0%	100.0%	100.0%	100.0%

Source: Prudential-Bache Securities, 1983.

DeVilbiss, whose major business was painting equipment, had licensed the Trallfa robot from Norway and then became a subsidiary of Champion Spark Plug. The Trallfa hydraulic robot was the leader in spray painting in Europe, and many were sold in the United States. ASEA, a Swedish company, introduced a well-designed electrical servo robot with six degrees of freedom and sold many of them in the United States.

These statistics may seem a little drab on first viewing, but they would become dramatic if you worked for or had stock in one of these companies. A shakeout is expected in the robot industry within the next 5 years; the present 50 U.S. companies will be reduced to about 25 survivors that will be able to make a profit in a very competitive industry. Major companies with tremendous resources have entered the robotics field, so we can expect great changes in market structure.

Large companies that now supply robots and robot systems are IBM, General Electric, General Motors-Fanuc (GMF), Westinghouse, and Bendix. Allegheny, Textron, and United Technologies have entered the field but are less active.

Westinghouse purchased Unimate in 1983 to add to its own capability. The effect was to broaden the Westinghouse line, add Unimate's system capability, and reinforce it with the financial strength of Westinghouse. GMF is a jointly owned company of General Motors and the Fanuc Company of Japan. It has developed an excellent spray-painting robot that surpasses all competitive equipment at this time.

Computers, programming, and systems integration are of major importance in the new complex robotic systems. Old-line companies that can't provide this capability will have difficulty competing unless they pick a specialty and do it well.

New companies have come into the field using start-up venture capital. These companies are characterized by high-technology capability and strength in the use of computers and the system approach. Typical companies are Adaptive Technologies, Inc., Advanced Robotics Corporation, American Robots Corporation, Automatix, Inc., Control Automation, Intelledex, Machine Intelligence, Nova Robotics and U.S. Robots.

1.7.3 Breakdown of Robots by Application

Table 1.4 Usage of Robots in the United States, by Type of Application

Application	Period	
	Through 1981 (%)	Estimated 1990 (%)
Spot welding	35–45	3–5
Materials handling	25–30	30–35
Assembly	10	35–40
Paint Spraying	8–12	5
Arc Welding	5–8	15–20
Other	8–10	7–10

Source: Conigliaro [4], p. 6.

These estimates were made in 1981 based on the business situation at that time. Information was gathered from many sources: manufacturers, users, business reports, and various robotic organizations. We expect the general trend in these estimates to be correct, but technical changes and fluctuations in the business climate may cause substantial changes in values to occur.

Spot welding accounts for the largest percentage of current robot use, primarily because robots have been adopted by the automotive industry, which uses more than 60% of currently operating robots. This usage is nearing saturation; nearly all good applications of robots in spot welding are being done. Not much increase in usage is expected, therefore.

Assembly is expected to become the largest single application in 1990. This growth is due to the availability of improved computer vision systems and improved control techniques for robots.

Arc welding is just beginning to come into use. Since there are more than 800,000 welding operators in this country, there is a huge potential for this application. New technology in sensors and control is making robot arc welding feasible and practical for any part or assembly made in lots of 100 or more.

1.8 Future Trends

1.8.1 Social, Political, and Economic Factors

What effect will robots have on society? How will they change the ways of government? Will there be an improvement in production, providing more goods and services? These are the questions that come to mind when discussing the effects of robots. There are answers to only a few of them.

1. Effect on Employment

The Upjohn Report is a major research study of the impact of robots on the state of Michigan and the United States (Schreiber [18]). It was commissioned by the Michigan Occupational Coordinating Committee because of the unemployment in the Michigan automotive industry and was carried out by the W.E. Upjohn Institute for Employment Research. The U.S. robot population is estimated to be between 50,000 to 100,000 by 1990, with 15,000 to 25,000 employed in the auto industry, 7,000 to 12,000 in Michigan, and the remainder elsewhere. At an estimated two jobs displaced per robot, the total job displacement will be only 100,000 to 200,000 in the United States which is about 1 to 2% of the total U.S. employment expected at that time. This doesn't tell the whole story, however. Operatives and laborers in the auto industry will suffer a displacement rate of 5.1 to 8.6%, and 15 to 20% of welders and 30 to 40% of painters will be displaced. This is not a general problem but an important specific effect.

According to the Upjohn Report, more than two-thirds of the employees involved in robotics will be professional, technical, administrative, sales, or clerical workers. More than half of these jobs will require a minimum of 2 years of college. It is estimated that there will be a nationwide demand for 4,600 to 9,300 engineers in robotics by 1990. There is some doubt that this demand will be met because of the competition for engineers in other areas. This shortage may limit the growth of the robotics industry.

About 13,000 to 26,000 technicians will be needed to service the robots. In Michigan, most of these technicians will come from the retrained auto workers, but in other states there will be some new jobs.

2. Worker Reaction

If robots do the demeaning, dirty, and dangerous tasks in industry, they will be welcomed by most workers. Robots can reduce the work week and improve the quality of life if they are introduced correctly and used widely. Unions have accepted robots in many factory applications where working conditions are undesirable.

Union spokesmen have stated repeatedly that they are in favor of the introduction of robots if the workers share in the benefits that accrue. Thomas L. Weakley [21] discussed the United Auto Workers' (UAW) stand at the Robots IV Conference in October 1979 and pointed out that a standard clause in major contracts contained these terms:

"The improvement factor provided herein recognizes the principle that a continuing improvement in the standard of living of employees depends upon technological progress, better tools, methods, processes and equipment and a cooperative attitude on the part of all parties in such progress. It further recognizes the principle that to produce more with the same amount of human effort is a sound economic and social objective."

There has been no change in this policy. In addition to the benefits of increased production, the average worker is very willing to escape the monotonous and hazardous jobs in industry.

Another factor is the realization that the United States is competing with other countries for markets, especially in the auto industry, and that high-cost labor and inefficiency can prevent it from being competitive in world markets. It is better that some people be unemployed than that all auto workers be unemployed because of sales lost to the Japanese or the West Germans.

3. Political and Economic Factors

Clearly, political leaders and government administrators will listen to the workers and try to find a way to meet their needs. But they must also consider the demands of the rest of the population. As yet, the U.S. government has not formulated a suitable plan for meeting the conflicting requirements of the various groups.

Economic factors are constantly under consideration by modern industrial managers. Most of them are trying to maximize the immediate profit rather than the long-term profit of their companies. Promotions and bonuses depend, to a great extent, on annual profits rather than on the development of improved products and better production methods for the long term.

Until recently, if a production machine did not pay off in 2 or 3 years, the company's officers and controllers were not willing to buy it. Neglect of productivity had become a way of life. As a result, manufacturing costs increased and quality decreased because there was insufficient money allocated to produce high-quality goods efficiently. The recession of 1979-1983 demonstrated to many managers that they were doing something wrong. In the automobile industry, the Japanese cars took an increasing share of the business. The lesson was learned the hard way, but it now appears that the present managers are willing to pay for increased productivity and quality. Large numbers of robots are now being used in industry to provide the benefits of lower cost, higher quality, and increased productivity that are needed to be competitive.

Economic advantages of robots will be discussed in later chapters. Important as political and social factors are, they are not the subject of this book; the reader is referred to the voluminous literature on these subjects. For our purposes, it is sufficient to note that there are critical social and political problems that must be solved before the full benefits of robots can be provided to our society.

1.8.2
New Applications

1. Australian Sheep Shearing

One of the most interesting new applications of robots is in sheep shearing (Wong [26]). This development has resulted from research sponsored by the Australian Wool Corporation. There is a major market for this application. Australia earns $1.7 billion per year in exports from 135 million sheep. Tens of thousands of Australians work in the sheep industry.

An electrical sheep clipper is carried by the robot arm and follows the contours of the sheep's body. Sensors on the clipper determine the distance to the skin of the sheep within an accuracy of 0.005 inch. Servos on the robot are fast enough to move the clipper out of the way when the animal moves or breathes so that the sheep is not damaged.

Successful shearing has been demonstrated on the back, sides, and belly of the sheeep, but more work is required on shearing of the legs and head. Studies have been made of all facets of the system, including the handling of wet wool, ways of securing the sheep during shearing, the resulting quality of the wool, and so on. Several sensory techniques have been explored: ultrasonics, infrared scanning, and microwave scanning, among others.

Commercial feasibility and acceptance of the shearing method by animal protectionists are still to come, although it currently appears that these problems will be solved. Since there has been a shortage of people willing to become sheep shearers, robot usage may become necessary to protect the Australian wool business.

2. Agricultural Applications

Robots with vision and sensing, those we have called adaptive robots, are potentially capable of handling many agricultural tasks. Fruit picking, asparagus harvesting, potato digging, and similar activities are being studied.

The General Electric Company is investigating the use of a robot mounted on a tractor for fruit harvesting (Manimalethu [13]). Vision is considered essential for picking only those fruits that match the picture in the computer's memory without damaging the branches and leaves. (This implies the use of color vision to identify ripe fruit and to aid in separating fruits and branches.) A sense of touch in the picking hand is needed to enable the robot to exert only the required pressure so that the fruit is not damaged. Preliminary analysis indicates that it is technically and economically feasible to build a fruit-picking robotic system. In addition, such a system may help solve the social problem of the influx of alien workers who work for a period and then go on welfare or require other social services.

Another possibility is the harvesting of asparagus. In California at present, it is difficult to find sufficient workers for this task. As a result, there has been a considerable decrease in asparagus planted and a consequent increase in the cost of asparagus. Current harvesting costs range from 10 to 18 cents per pound, depending on the season.

Harvesting is stoop labor; the worker must bend over and walk along the row of asparagus, cutting only the stalks above a certain height and placing them carefully in a container dragged along by his side. Frequently, the temperature in the fields reaches 90 to 100°F, and workers can work only in the morning hours.

Preliminary analysis by the author indicates that a robotic system mounted on a vehicle is economically and technically feasible if the robot works 20 hours per day, 100 days per year. One of the problems in agricultural applications is the short harvesting season. This means that the use of the robot must be justified over a period of about 100 days per year. There are some possibilities of overlapping harvest periods, however, since crops mature at different times in different areas.

3. Household and Hobby Robots

Several manufacturers are now providing robots designed to do useful tasks in the home or for hobby and entertainment use. Three of these, the HERO 1 robot, the ANDROBOT, and the GENUS, will be discussed.

HERO 1, made by the Heath Company, is available both in kit form and as a fully assembled robot (Figure 1.21). HERO was designed as a teaching aid and a hobby robot. It can move around a room under computer control and has a jointed arm that can pick up small objects. It is depicted as walking a dog, bringing a drink to a guest, and performing similar tasks. It sells for $1,500 in kit form and $2500 fully assembled (see Figure 1.21).

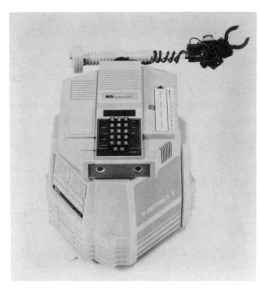

Figure 1.21 HERO 1 robot.

Courtesy of Heath Company. Used by permission.

Figure 1.22 Androbot BOB.

Courtesy of ANDROBOT, Inc. Used by permission.

There is a two-volume elementary text on robotics available for HERO. The novice user can learn some of the fundamentals of robotics by reading the text and experimenting with the robot, which stands about 24 inches tall and has an 18-inch arm driven by stepper motors. HERO has an ultrasonic ranging sensor, a separate sound detector and a simple voice synthesizer.

Androbot, Inc., was backed by Nolan Bushnell, a cofounder of the Atari Company. The ANDROBOT comes in two models: the Topo and the Brains on Board (BOB) model. BOB also has computer control and speech synthesis capability, and is mounted on wheels for mobility. It has three on-board 16-bit 8088 microprocessors and sells for $2,495. Apparently there is no arm on this first model (Figure 1.22). The Androbots, Topo, and BOB are 36.5 inches high by 24 inches wide.

GENUS, by the Robotics International Corporation, can vacuum floors and take charge of home security, among its other capabilities. It will sell for $10,000 to $12,000. This robot may be useful to those who really dislike vacuuming. Perhaps it would be useful in hotels and other public places.

These may be the prototypes of future household robots, but at present they are suitable primarily for hobby and educational purposes. One market research firm, Future Computing, Inc., estimates that sales of these robots will reach 500,000 to 1,000,000 annually by 1990.

4. Other Applications

Many other robot applications are feasible or are being studied. Some of these are:

1. Biomedical Testing. Moving trays and dishes into and out of ovens and other equipment. This type of manipulation is tedious and subject to error.
2. Underseas Exploration. Submersibles have used primarily manipulators. With vision, the need for a person aboard could be eliminated at considerable savings.

3. Firefighting. Robots with legs would be valuable in dangerous situations and in rescue work.
4. Mining. Coal mines are especially dangerous places. Some mines have only 4 feet of clearance between the floor and the ceiling, so it is difficult for men to do necessary roof bolting and other work. Robots or manipulators could be useful in these applications.
5. Textile Handling. Handling textiles and feeding them to machines for processing is being studied in the United Kingdom (Kemp [12]).

1.8.3
Long–Range Goals

1. Integrated Systems

Integrated systems are a definite part of the long-range goals of the robotics industry. Such systems are being put together today, so that by 1990 we can expect complete factorywide systems to be operational. Such systems will do design on CAD, fabricate parts with CAM, and assemble the parts in work cells operated by multiple robots with the necessary auxiliary equipment.

2. Intelligent Systems

Systems that contain a model of their environment and have expert systems capability and a knowledge base are being actively investigated in the research laboratories (Buchanan [3]). Expert systems can be applied to automated manufacturing in the near future by using the same techniques used for geological exploration and medical diagnosis.

3. Self-Reproducing Systems

In principle, at least, it is possible for a robotic factory to build another robotic factory. Dr. John von Neumann, one of this century's great mathematicians, proved that this operation was logically possible. This is still a very speculative idea, however, and no serious work is being done to demonstrate its feasibility.

Exercises

1. The bird organ mechanism in Figure 1.2 shows the cams required to operate the voice mechanism. Estimate roughly how many binary bits of information would be required in the cams used to operate the bird's beak in a realistic manner. Assume that the beak moves up and down 1 inch at a velocity of 1 inch per second.
2. Section 1.2.2 classifies robots by capability level. Can you think of a robot operation that is not covered by these definitions? Would a different set of terms be better?
3. Four configurations of robots are defined in Section 1.2.3. Which of these configurations would be best for a robot placing objects in an oven for curing? Why do you think so?
4. What are the advantages of a six-legged robot compared to a four-legged robot? Check your answer by referring to the references given.
5. Robots have numerous advantages, as noted in Section 1.3. Why, then, haven't more robots been put to work in U.S. industry?
6. What single development, in your opinion, has been most influential in extending the usefulness and acceptance of robots in the United States? Why do you think so? (Some reading of the references may help in this analysis.)

7. What are some important items to be considered in deciding whether or not to use robots? Which ones would be most important in a company making washing machines?
8. Extrapolate the data given in Tables 1.2 and 1.3 to determine the expected dollar value of sales for Unimation in 1984.

References

[1] J. S. Albus, C. R. McLean, A. J. Barbera, and M. L. Fitzgerald, "Hierarchical Control for Robots in an Automated Factory," Robots 7, 1983, pp. 13-29 to 13-43.

[2] J. T. Beckett, "A Computer-Aided Control Technique for a Remote Manipulator," Digital Systems Laboratory, Engineering Division, Case Institute of Technology, Cleveland, Ohio, 1967.

[3] B. G. Buchanan, "Partial Bibliography of Work on Expert Systems," Stanford University, Heuristic Programming Project, Report No. HPP-82-30, November 1982.

[4] L. Conigliaro, "Robotics Presentation, Institutional Investors Conference: May 28, 1981," Bache Robotics Newsletter 81-249, Bache Halsey Stuart Shields, Inc., New York, Oct. 28, 1981.

[5] L. Conigliaro, Prudential-Bache Securities Newsletter, March 1983.

[6] J. F. Engelberger, "Robotics in Practice," 1980, AMACOM, a Divison of the American Management Association, New York, 1980.

[7] H. A. Ernst, "MH-1, A Computer Operated Mechanical Hand." Ph.D. Thesis, Massachusetts Institute of Technology, 1961.

[8] J. M. Evans, Jr., J. S. Albus and A. J. Barbera, eds: "NBS/RIA Robotics Research Workshop," Williamsburg, Va. Published by the Institute for Computer Science and Technology, National Bureau of Standards, Washington, D.C., 1977.

[9] W. B. Gevarter, "An Overview of Artificial Intelligence and Robotics," NBSIR 82-2479, U.S. Department of Commerce, National Bureau of Standards, Washington, D.C., 1982.

[10] "Industrial Robots and Controls," RLB1182, GCA Corporation, Napiersville, Ill.

[11] E. G. Johnsen, "The Background of SAM, A Self-Propelled Anthropomorphic Manipulator. Space Nuclear Propulsion Office, U.S. Atomic Energy Commission, Washington, D.C., 1969.

[12] D. R. Kemp, G. E. Taylor, P. M. Taylor, and A. Pugh, "A Sensory Gripper for Handling Textiles," Robots 7, 1983, pp. 18-23 to 18-33.

[13] A. Manimalethu, "Agricultural Robot Application," Robots 7, 1983, pp. 10-76 to 10-87.

[14] D. E. Orin, "Supervisory Control of a Multilegged Robot," Journal of Robotics Research, Vol. 1, No. 1, Spring 1982, pp. 79-91.

[15] C. F. Penniman, "Philadelphia's 179 Year Old Android," Byte, August 1978, p. 90.

[16] J. C. Quinlan, "Robot Wrists and Grippers," Tooling and Production, January 1983, pp. 85-88.

[17] M. H. Raibert and I. E. Sutherland, "Machines That Walk," Scientific American, vol. 248, January 1983, pp. 44-53.

[18] R. A. Schreiber, "New Perspectives: The Upjohn Report," Robotics Today, April 1983, pp. 61-62.

[19] W. R. Tanner, ed., Industrial Robots, 2nd ed., Vol. 1, Fundamentals, Vol. 2, Applications. Robotics International of SME, Dearborn, Mich., 1981.

[20] H. J. Warnecke et al. "Simulation of Multi-Machine Service by Industrial Robots," Robots 7, 1983, pp. 2-10 to 2-22.

[21] T. L. Weakley, Robots 4 Conference, October 1979, Detroit, Mich.

[22] J. L. Wilf, "The Great Japanese Robot Show," Robotics Age, Vol. 3, Nov/Dec 1981, pp. 30-36.

[23] J. M. Williams, "Antique Mechanical Computers--Part 1: Early Automata," Byte, July 1978, p. 48.

[24] J. M. Williams, "Antique Mechanical Computers--Part 2: 18th and 19th Century Mechanical Marvels," Byte, August 1978, p. 96.

[25] J. M. Williams, "Antique Mechanical Computers--Part 3: The Torres Chess Automaton," Byte, September 1978, p. 82.

[26] P. C. Wong and P. R. W. Hudson, "The Australian Robotic Sheep Shearing Research and Development Program," Robots 7, 1983, pp. 10-56 to 10-63.

Chapter 2

Early Development of the Modern Robot

This chapter summarizes two important areas: (1) Industrial robot developments in capability and application, and (2) Technical development of advanced robot capabilities including sensors and vision. Emphasis is on the work done prior to 1978 on servo-controlled robots and associated sensors.

In each section, we will review some of the historical background of the technologies involved and then present the evolutionary steps that brought us to the present situation.

To set this information in perspective, a timetable is used to present some of the important developments. Each event is identified by type, and a reference to more information is given.

2.1 Development Time Table

In the following timetable there are four columns; the date of occurrence, the event category, a description of the event, and a reference to more information. Category identifications are shortened to "Ind." for Industrial or Commercial activity, "Tech." for Technical acitivity in a university or technical institute, and "Theor." for the theoretical developments in mathematics and analysis techniques. Dates are given in chronological order to show the relationships between events.

Date	Category	Description of the Event	Reference
1955	Theor.	Denavit and Hartenberg developed homogeneous transformations (D-H matrices)	Denavit and Hartenberg [7]
1961	Ind.	U.S. Patent 2,988,237 by George Devol on "Programmed Article "Transfer" (Basis for Unimate Robots)	EngPre [11]
1961	Ind.	First Unimate Installed; Used to tend a die-casting machine	EngPre [11]
1961	Tech.	MH-1, Mechanical Hand with sensors, developed by Ernst at MIT.	Ernst [12]
1963	Ind.	Versatran, a cylindrical coordinate robot, became commercially available.	Sutherland [45]
1965	Theor.	L.G. Roberts applied homogeneous matrices to robots	Roberts [38]

37

Date	Category	Description of the Event	Reference
1968	Tech.	Shakey robot developed at Stanford Research Institute; Computer-controlled mobile robot with vision	Munson [28]
1969	Tech.	Stanford Arm developed at Stanford University by V.C. Sheinman and associates.	Scheinman [43]
1969	Theor.	Robot vision demonstrated at SRI for mobile robot guidance	Duda and Hart [8]
1970	Tech.	Adaptive robot with vision developed by ETL in Japan	Mainichi [48]
1971	Ind.	Japan Industrial Robot Association (JIRA) formed	EngPre [11]
1972	Theor.	R.P. Paul developed trajectory calculation using D-H matrices	Paul [34]
1972	Theor.	D.E. Whitney developed coordinated motion control for manipulators.	Whitney [49]
1975	Ind.	Robot Institute of America formed	EngPre [11]
1975	Ind.	Unimate Corporation showed its first profit	EngPre [11]
1976	Tech.	Programmed assembly accomplished by a robot at SRI.	Nitzan and Rosen [33]
1978	Ind.	Machine Intelligence Company formed by C. Rosen and associates; first commercial robot vision system produced	Saveriano [42]

Many other important events occurred to make robots a reality, of course, but this timetable lists some of the key events that had great influence on the present robots. Events since 1978 are discussed further in the chapters on technical developments and applications.

2.2 Industrial Development

J.F. Engelberger, President of the Unimation Corporation, quotes a Department of Defense report (1967) (see Engelberger [9]) that "an innovation will gain acceptance only when there is a conjunction of three elements: (1) a recognized need, (2) competent people with relevant technological ideas and (3) financial support." Even though the need was recognized in the late 1950s and competent people were working on industrial robots in the early 1960s, it was several years before financial support was sufficient. Some courageous investors backed Engelberger and other developers in making robots, but finding buyers was an uphill battle.

Manufacturers, the group most likely to purchase robots, are inherently conservative; they are not willing to take risks that might close down their factories, incite union members to strike, or cause social problems. In addition, they want to be sure that any new equipment can be financially justified. The financial argument is frequently used even when other considerations are important.

Economic justification for any capital equipment investment is usually based on one of three types of analysis:

1. Payback period. This is the number of years required to return the capital investment through savings in operating costs.
2. Return on investment. This is the effective profit made on the capital investment when all costs are included and the equipment is depreciated over several years using good accounting procedures. Return on investment should be at least 25% per year in today's market. Risky investments may require higher rates of return to be acceptable.
3. Present value of future earnings. This analysis takes into account the cost of money invested. A robot system may require investment today but may not be operational for a year and take 2 or 3 years or longer to achieve a return. It is clear that money received 3 years from now is worth much less than money received today because the money received today could be earning compound interest for 3 years.

Even though analyses of this type were favorable in the early 1960s, it was difficult to convince most manufacturers to take the financial and other risks and to buy robots for their plants. Many articles have been written on the problems of selling robots. These problems are discussed by Engelberger [10], Tanner [47], Saveriano [42], and others.

**2.2.1
Unimation
Corporation**

In 1958, Joseph F. Engelberger, credited with being the "Father of Industrial Robots," started Unimation, the first commercial company making robots. The name Unimation is a contraction of the terms UNIversal autoMATION because of the belief that the robot was a universal tool that could be used for many kinds of tasks. Early development of this robot is described in Saveriano [41].

The first Unimate was a five-axis, hydraulically operated manipulator. Its design was based on a patent by George Devol, who worked with Engelberger on the design. It was first installed in the Trenton, New Jersey, plant of General Motors to run a die-casting machine. Arm control was provided by a special-purpose computer that could remember the sequence of operations required to do the job. Information was stored in a magnetic drum memory. Discrete solid-state digital control components were used. A teach control was part of this first robot. With the teach control, the robot could be moved in any desired direction. Memory was limited to about 180 steps, but was sufficient to do many tasks.

By 1964, similar models were doing spot welding and other tasks. Minor modifications were made in the robot from time to time. Product brochures in 1972 still quoted 180 steps as the memory capability, but there were provisions for augmenting this memory if desired. Also, a tape cassette could be used to copy the sequence information from the magnetic drum so that the same robot could be used to do several different tasks by selecting the proper tape cassette.

Performance Specifications for the Unimate in a 1972 brochure were:

Ranges and Velocities

Maximum radial velocity   30 inches per second
Maximum radial extension   3.5 feet

Maximum vertical velocity	50 inches per second
Maximum vertical stroke	3 to 90.5 inches
Maximum rotational velocity	110 degrees per second
Maximum rotation	220 degrees
Maximum wrist bend velocity	110 degrees per second
Maximum wrist bend	220 degrees
Maximum wrist swivel velocity	110 degrees
Maximum wrist swivel	180 degrees

Positioning accuracy: ± 0.05 inch

Maximum load: 25 lbs. at normal operating speed; 75 lbs. at reduced speed.

Design life: 40,000 operating hours

Maintenance: Routine 2,500-hour maintenance check recommended

Maximum clearance dimensions: 5 x 4 x 4.5 feet high

Power requirements: 220/440 volts, 3 phases, 60 cycles, 11.5 kVA

Weight: 3,500 pounds

Gripper clamping force: Adjustable to 300 lbs. at the end of 4-inch fingers

At this time, there were several Unimate models, the 2000, 2100, and 4000 series. Maximum reach was an awesome 10 feet in the 2100 and 4000 series, and the 4000 series could handle loads up to 350 lbs.

All of these machines were available in two to six articulations or degrees of freedom. The first Unimate had five degrees of freedom, but there were a few applications in which six degrees of freedom were required. Figure 2.1 shows the six axes or articulations available in the Unimate models. This is an example of the spherical coordinate (polar) robot defined in Part 3 of Section 1.2.3. A photograph of one of these models is shown in Figure 2.2.

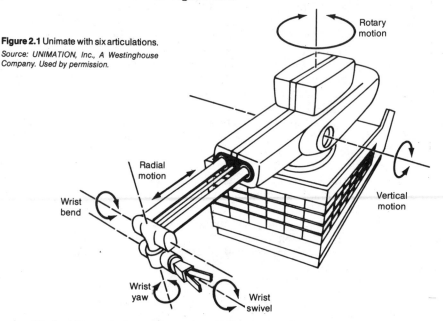

Figure 2.1 Unimate with six articulations.
Source: UNIMATION, Inc., A Westinghouse Company. Used by permission.

Hydraulic robots such as the Unimate have several advantages and disadvantages as compared to other robot types. These factors are discussed and tabulated in Chapter 3.

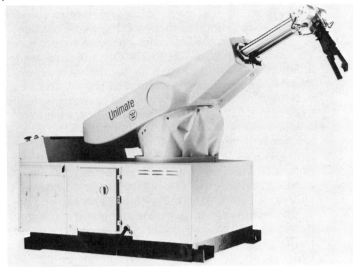

Figure 2.2 Unimate robot with controls and power supply.
Courtesy of UNIMATION, Inc., A Westinghouse Company. Used by permission.

It is noteworthy that the Unimation Company now makes both the Unimate series of robots and the PUMA robots, which are driven by electrical servo motors. There is a place for both. PUMA robots were introduced by Unimate in the late 1970s, and so belong to the recent period and not to the early development period, which is the subject of this chapter. However, the PUMA design was derived from the Stanford arm built by Vic Sheinman, as discussed in Section 2.3.2.

Unimate robots were and are still used for many different applications. Some of them are listed below:

1. Die Casting. Unload one or two die-casting machines. Quench parts in water or oil tank, trim parts in a trimming die, and insert them in a loading fixture.
2. Forging. Hold parts in a drop forge, upsetter, roll forge, or presses.
3. Stamping Pressing. Load and unload presses. Perform press-to-press transfer.
4. Welding. Perform spot welding, press welding, and arc welding.
5. Injection and Compression Molding. Unload molds, trim, insert, palletize, and package parts.
6. Investment Casting. Wax tree processing, dipping, manipulating, and transferring molds done routinely. Increased productivity and better quality control resulted.
7. Machine Tool Operations. Loading, unloading, palletizing, transfer between machine tools, and so on.
8. Spray Coat Applications. Mold releases, undercoats, finish coats, highlighting, and frit and sealant application.
9. Material Transfers. Many types of materials were transferred by robots by 1972—automotive assemblies and automotive parts, glass, textiles, ordnance, and so on. Robots worked in appliance manufacturing, molded products, heat treating, paper products, and plating operations. Many different types of grippers were used.

2.2.2
Versatran

AMF, Inc. started developing a robot in 1958 that was designed to relieve men of some of the tedious menial jobs in industry. The machine was called VERSATRAN from the words <u>VERSA</u>tile and

TRANSfer. It was used for transfer of materials between machines, machine loading, and many of the same applications listed in Section 2.2.1. Commercial activities on these robots commenced about 1963.

The standard VERSATRAN unit was made up of a single arm mounted in a column that, in turn, was mounted on a base. Rotation around the base, vertical movement of the arm on the column, and extension of the arm along a radius were possible. These movements gave the VERSATRAN the capability to move in a cylindrical coordinate system. Mounted on the end of the arm was a wrist unit and gripper that could move to grasp an object.

Hydraulic power was used to drive all motions of the device. Figure 2.3 illustrates the mechanism of the Versatran mechanical handling unit. Both the horizontal and vertical drives were hydraulic cylinders. Potentiometers ("pots") were used to measure the position of the cylinder and generate an electrical signal proportional to the movement of each cylinder. Electromagnetic servo valves controlled the flow of oil to the cylinders to determine the position to which the cylinder would move. A rotary hydraulic motor drove the base rotation mechanism. Solenoid valves controlled the flow of oil to the wrist and gripper mechanisms.

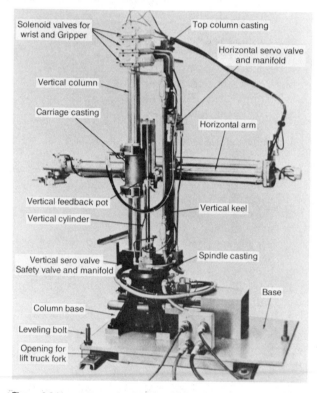

Figure 2.3 Versatran mechanical handling unit with "skins" removed to reveal the working parts. *Source: Prab Robots, Inc.*

Figure 2.4 shows a complete Versatran industrial robot system. Electrical power from the factory supply, usually at 220, or 440 volts, 60 cycles, 3 phases, was used to drive the hydraulic power pack. An electric motor drove the powerful hydraulic pump. The pump stored oil under high pressure in an accumulator tank, which was part of the power pack. Water was used to cool the high-pressure oil and was pumped from the factory water supply. Hydraulic lines carried oil to the mechanical handling unit to supply power for arm movement.

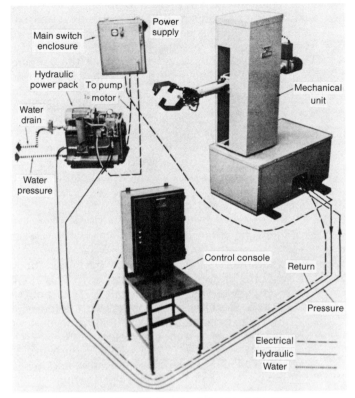

Figure 2.4 Complete Versatran industrial robot system.

Source: Prab Robots, Inc.

Use of high-pressure oil was a major problem in the early days of hydraulic systems. Hoses and lines carrying the oil were prone to leaking and would spray oil, at 2000 psi, over a large area. Another early problem was contamination of the oil with fine metal particles, which would prevent the high-precision servo valves from closing completely. Most of these problems have been eliminated by the use of greatly improved materials and designs.

An application of the Versatran can be seen in Figure 2.5. A vacuum pickup is used to hold a piece of flat glass during a transfer operation. Note the auxiliary hose coming down from overhead to supply the vacuum for the holding operation. A simple sensor is used;

Figure 2.5 Versatran robot with a seek-and-find sensor on the gripper transfers a flat glass from a stack through various processing steps.

Source: Prab Robots, Inc.

it is mechanically actuated when the vacuum cups contact the glass and cause the vacuum control valve to open so that the pickup can hold the part. This sensor reduces the air flow required in the vacuum system, since the vacuum is applied only when necessary. To release the part, a solenoid valve can shut off the vacuum supply.

Both continuous-path control and point-to-point control were available. Programs could be stored on magnetic tape decks in digital form by 1972. Earlier systems used adjustable potentiometers for setting position. Load-handling capability and operating speeds were similar to those provided by the Unimate.

Applications of the Versatran were similar to those of the Unimate as previously described. However, because of the use of a cylindrical coordinate system, the Versatran had some advantages in certain applications in which it was desirable to position the load accurately in a vertical direction.

2.2.3
Other Industrial
Developments in
the United States

1. Cincinnati Milacron

Cincinnati Milacron was the first U.S. robot manufacturer to use the revolute or jointed configuration. This configuration, called the " T^3 robot," was first operational in 1973 and had the advantages of being more flexible than robots with sliding joints. It could also position its gripper more flexibly due to a new wrist design. The first Cincinnati Milacron robot was hydraulic, so it had the same speed characteristics as the Unimate and Versatran. The T^3 was one of the largest early robots available; it had a working volume of 1,000 cubic feet and could reach 154 inches vertically.

These robots were used especially on moving production lines because of the development of the Controlled-Path System, an improved computer-controlled system that could track a moving conveyor. A block diagram of the control system is given in Figure 2.6.

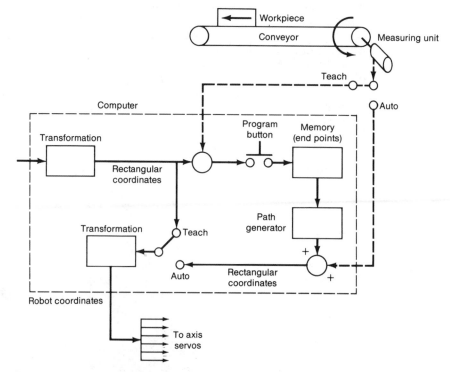

Figure 2.6 Controlled-path tracking and control system by Cincinnati Milacron.   *Source: Cincinnati Milacron.*

The position of the conveyor was measured by a sensor and sent through the computer, so that the computer could control the tip of the robot arm to track the conveyor location while performing its assigned task.

Major applications of this system were spot welding of automobile bodies, transfer of automobile bumpers, and machine tool loading. An article by B. L. Dawson [6] describes these and other applications.

2.2.4 Foreign Industrial Developments

Three foreign robot developments were important during this period.

1. ASEA

The ASEA was a five-axis robot with electrical drive motors. It was used at SAAB-SCANIA for spot welding of automobile rear floors. Detecting equipment was used on the electrodes to detect electrode wear. A control signal was generated to move the robot arm in order to compensate for this wear.

A new application of the ASEA IRb-6 in 1978 was the measurement of car body points in three dimensions. Points were measured to a repeatability of 0.5 millimeter at 120 points on 40 cars per hour. Four robots, each measuring 30 points on each car body, were used to maintain this inspection rate (Ford [15]).

2. TRALFFA

The TRALFFA robot was developed by the TRALFFA Nils Underhaug A/S company in Norway and was widely used by 1975 as a spray-painting robot. It was therefore one of the first continuous-path robots made, and led the world in spray-painting applications. There are over 1,000 of these robots now in use for spray painting, coating, and similar finishing tasks. In 1975 it was redesigned to be more accurate and was used in Automatic Arc Welding.

The TRALFFA was a hydraulic robot with either five or six degrees of freedom and was apparently the first industrial robot to use the revolute coordinate system for production robots. It could sweep 3,150 millimeters (124 inches) horizontally and 2,040 millimeters vertically, with a radial range from 835 to 1810 millimeters.

The DeVilbiss Company in the United States has licensed the Tralffa robot for spray painting and welding. It is now being sold as the DeVilbiss-Tralffa robot in this country.

3. Japanese Industrial Robots

Kawasaki Heavy Industries took out a license from Unimation, Inc., in 1968. This is believed to be the first servo-controlled industrial robot in Japan. Most Japanese robots were of the point-to-point type rather than the continuous path type. Japanese manufacturers had a labor shortage in the 1970s and used the simpler robots for machine feeding and transfer. There were only a few applications of servo robots.

In 1971 the Japanese Industrial Robot Association (JIRA) was formed. With government encouragement, the Japanese now lead the world in the total number of robots in use. However, their technology is at about the same level as, or slightly behind, U.S. technology.

2.3 Technical Development

A series of development projects occurred in the 1960-1978 period that provided the technical background required for successful use of robots. Major development efforts at the Stanford Research Institute (SRI), Stanford University, and Massachusetts Institute of Technology (MIT) were carried out and are discussed in this section. Other efforts were smaller and less productive, and thus will be discussed only briefly.

Much of this work was supported by the National Science Foundation, which deserves great credit for the way in which it aided this work without rigidly controlling its direction. Other work was supported by the Advanced Research Projects Agency (ARPA) of the Department of Defense and by the National Aeronautics and Space Administration (NASA). [ARPA has been renamed the Defense Advanced Research Projects Agency (DARPA) and will also be referred to by that name.] Some work, which appeared to have military value, was also supported by other parts of the Department of Defense. In addition, a few farsighted industrial organizations supported development work, primarily at SRI.

**2.3.1
Stanford
Research Institute**

Developments at all of the major development centers were reported in technical memoranda and periodicals. In addition, there were many visits and conferences among organizations, so that technical improvements were shared widely. Because of its emphasis on industrial applications, however, the work at SRI was more directly applicable to industrial robots.

1. SRI Intelligent Automation Program

SRI, located in Menlo Park, California, carried out the SRI Intelligent Automation Program in the 1966-1971 period (Munson [28]). The project's goal was to explore the basic functions demanded in the computer control of mobile machines operating in complex environments.

For the first time in one system, a vision system, a large computer system, computer-controlled servo systems, and advanced artificial intelligence ideas were combined. Computer control was maintained over two radio links: a narrow-band link for direct control and a wide-band link for transmitting the vision information from the mobile vehicle TV to the computer system.

Many sensors were used in the system: the TV scanner vision, a range-finding optical sensor, and "cat whisker"-type touch sensors to detect contact by the vehicle bumper with objects in its environment. Several motors were used: drive motors for the wheels and separate motors for the pan and tilt on the TV and range-finder systems. This information was continuously telemetered to the computer to be used for control of the mobile vehicle.

Mechanical design of the mobile vehicle did not provide a rigid structure, so that the upper part of the vehicle carrying the control equipment and vision scanners would vibrate slightly as it moved. From this attribute, it earned the name SHAKEY even after some design changes dampened down the vibration. Figure 2.7 is a photograph of Shakey showing the various components. (It is not clear where the spelling shakey rather than shaky first arose.)

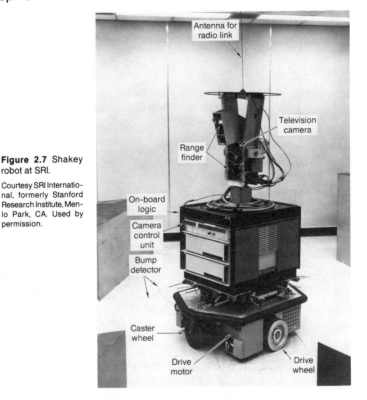

Figure 2.7 Shakey robot at SRI.

Courtesy SRI International, formerly Stanford Research Institute, Menlo Park, CA. Used by permission.

Shakey became famous in artificial intelligence circles and advanced development laboratories for its characteristics. Over a 5-year development period, Shakey successfully demonstrated that vision associated with computer control could enable a mobile vehicle to accomplish some complex tasks. The vehicle could steer down the center of a hall by visually sensing the edge of the hall and could manuever around objects in its path. It also performed feats that were impressive to all observers. An operator could give a command on the teletype keyboard that required sophisticated reasoning, and Shakey would carry it out. A particularly impressive example was the following.

Given the command "Push the cube off the platform," Shakey would locate the cube visually by scanning, recognize that the cube was on a platform about a foot high, realize that it couldn't get up on the platform directly, and work out a procedure to carry out its task. First, it searched the room to find a movable ramp of the right height, selecting it from among several pyramids, blocks, and so on. Then it pushed the ramp around so that the high end of the taper contacted the platform, with the low end facing out from the platform. Next, it rode up the ramp on to the platform and (triumphantly?) pushed the cube off the platform. This was not a prestored program but was built up by use of the predicate calculus implemented in a LISP program on a DEC PDP-10 computer.

Although much was learned from the automaton program, it became clear that it would not produce a directly usable industrial device. Therefore, the program was reoriented in 1971. Dr. Charles Rosen, director of the project, described the way in which this decision was implemented in a published interview that reveals how difficult it was to obtain support and run a development project at that time (Saveriano [41]).

2. Industrial Robot with Vision

The new project was designed to equip an existing industrial robot with the best available vision system and demonstrate its use in industrial applications. SRI's management supported the project, and two industrial clients assisted. A Unimate robot was purchased and equipped with the necessary sensors and vision equipment. Computer control of the vision system and the robot made it possible to carry out complex tasks.

Many of the applications developed and demonstrated by SRI depended on the development of the sophisticated vision system called the SRI Vision Module. The system was made possible initially by carefully determining what features were required for industrial use and supplying only these features. This procedure kept the system simple and relatively inexpensive. These features and the way they were implemented are described in detail in Section 9.11.1.

Other sensors were also developed, such as the wrist sensor described in Section 5.3.5 and various tactile sensors, to feed information back from the robot to the computer to allow accurate positioning of the robot arm and grippers in response to its environment. The robot as developed falls within our definition of an adaptive robot (Section 1.2.2).

Technical development by SRI has spread throughout the robotics industry and is therefore discussed in many sections of this text, so that a detailed listing of other early developments will be omitted.

A series of reports to the National Science Foundation cover the 1972–1980 period (Rosen et al. [39], Nitzan. (Dr. David Nitzan became the director in 1979). Some developments during that period were:

1. Interactive vision system
2. World and joint coordinates analysis for manipulators
3. End effector and sensor development
4. System control hardware: joystick control, multimode servo control, and voice control
5. Design of system software
6. Denavit-Hartenberg matrices for the Stanford arm, the six-joint Unimate arm, and the five-joint Unimate arm
7. Force-controlled part mating
8. Bolting with visual feedback
9. Visual tracking algorithms
10. Robot Programming Language (RPL)
11. Technology transfer to industrial affiliates
12. Communication between computers for distributed processing
13. Recognizing and locating partially visible objects
14. Inspection of printed circuit boards for part integrity

2.3.2
Stanford University
A broad program in robotics and computer vision was started in the Stanford Artificial Intelligence Laboratory (SAIL) in 1965 by Professor John McCarthy, assisted by Les Earnest and Raj Reddy. Many of the leaders of today's robotics efforts worked at SAIL and have moved on to head programs at other universities and several industrial establishments.

Much of this work was done by graduate students and was published as doctoral theses. Therefore, there was an emphasis on research and development leading to a fundamental understanding of the phenomena involved. Although the laboratory is currently more active than ever, we will discuss here only the research and development work done from 1965 to 1980 (Binford [3], Eighth Report).

Hardware, software, and experimental development was carried out. A hydraulic arm was built in 1968. This arm was extremely fast and powerful, and also somewhat dangerous. It was protected from people, or vice versa, by a steel and glass cage. Since some of the early hydraulic hoses and joints leaked occasionally, the glass protected the remainder of the laboratory from being sprayed with oil at 2,000 pounds pressure per square inch. As a result of the knowledge gained, Vic Sheinman, in 1970 designed and built the Stanford Arm, which used electrical drive motors. Several copies of the Stanford Arm are still in use in research laboratories and it has become a standard for comparison. (See, for example, Section 6.7 and Figure 6.26).

Richard Paul created the WAVE program in 1971. This was the first system designed as a general-purpose robot programming language. The philosophy of WAVE was that robot motions could be preprogrammed so that only small deviations from the trajectory would be needed during operation. It pioneered some important capabilities in robot programming languages: description of end effector position by Cartesian coordinates, coordination of joint motions to move smoothly through corner points of the path while maintaining velocities and accelerations, and specification of compliance in Cartesian coordinates. These developments are evaluated in a recent article on programming (Lozano-Perez [21]). The VAL program used on the Unimation PUMA robot is based on WAVE. Also, as a result of the WAVE experience, development of the AL language was begun (Finkel et al. [14]).

AL is a sophisticated language on which development is still continuing. Synchronized control of two robot arms was possible in the early version of AL. It also included data types and linear algebra for coordinate transforms and attachment between objects. AL was used for control of the Stanford arm and the PUMA arms for assembly. AL is described in Section 8.5.2.

An important development by Paul was the use of homogeneous matrices for robot arm control (Paul [34, 35]). This approach has become fundamental to robot control applications and is described further in Section 6.6.2.

Force and touch sensing were used on the Stanford arm along with vision to assemble an automobile water pump from 10 components. Vision research carried out by Binford, Agin, and Nevatia led to recognition of complex objects.

Other work done during this period was on a wrist force sensor and force control based on approximation in joint coordinates.

Errors and tolerancing problems in manufacturing and assembly were analyzed by Grossman [19]. A number of other analyses were done by other SAIL researchers. The research work at SAIL during the 1965-1980 period is reported in a series of reports by T. O. Binford [2, 3].

**2.3.3
Massachusetts
Institute of
Technology (MIT)**

Marvin Minsky and John McCarthy initiated the robotics work at MIT. One of the early successes of their program was the Mechanical Hand 1, or MH-1, developed by H. A. Ernst [12], as suggested by Claude Shannon and Marvin Minsky. This was the first computer-controlled manipulator and hand with sensors. Several problems in servo control, computer control, and sensors had to be solved to make this hand a reality. The hand and arm together had 35 degrees of freedom; it was necessary to simplify the system before it could be made to work.

Ernst also developed an interpretive programming language with three types of statements: symbolic statements that were concerned with outside-in control--the sense organs for touch; the explicit statements, which influence the positioning operations; and the transfer statements. Even by today's standards, it was a well-designed system. Speed control, move control, and various logical statements made the language easy to use. Interrupt control was provided in the hardware and accepted by the software so that sense information could be acted on immediately. Searching for an object was possible by programming the hand to move until the sensors indicated a desired logical pattern. As Ernst pointed out, however, the initial language, which he called MHI, had several disadvantages. Flexibility of programming was poor; a new program was required for each new application, and there was no memory for the real world (no world model). Also, there was no way to set up subroutines or write them automatically. Some of these problems were handled somewhat better by another language called THI, which Ernst developed. This experience provided a basis for many later developments at MIT.

MIT researchers have contributed substantially to the fields of Artificial Intelligence, robot motion planning, sensor development, task planning, and compliant motion control. Much of this work has been gathered together and published in Brady et al. [4], along with the work of other researchers. The reader is referred to this collection of papers for much of the work done at MIT. The work of Hollerbach [20], for example, is an important contribution to the formulation of manipulator dynamics.

Tomas Lozano-Perez has done a variety of work in task planning, (Brady et al. [4]), assembly systems, and obstacle avoidance (Lozano-Perez and Wesley [22]. He and P. Winston developed the LAMA language for automatic mechanical assembly [23]. Marr and Nishihara [24] developed methods for the representation and recognition of three-dimensional shapes. Mason [26] reports on compliance and force control for computer-controlled manipulators. Silver [44] developed the "Little Robot System." It was controlled by a language called MINI, which was an extension of the LISP language developed by McCarthy while at MIT. A principal feature of this system was the availability of a force-sensing wrist with six degrees of freedom, the most sensitive wrist developed up to that time.

Work at MIT was on advanced Artifical Intelligence concepts and theoretical analysis rather than the more directed and focused research done at SRI and Stanford University. As a result, it is difficult to pull together a record of all of the contributions made.

2.3.4
Other Early
Developments

During the early development period as we have defined it (1965-1978), there was little activity on robotics in most university and industrial laboratories. There was some development related to robotics but no concentrated effort, as there was at SRI, Stanford, and, in some respects, MIT.

Some examples of the work done are:

1. Draper Laboratories developed the force vector concept in 1972 so that a servo controller could guide the assembly of parts.
2. The Remote Center Compliance robot wrist was developed by the Draper Laboratory in 1976 (see Section 5.3.7). It demonstrated a new way to insert pins into holes and perform assembly operations.

3. The HARPY speech-understanding system was completed by Raj Reddy at Carnegie-Mellon University in 1976.
4. Robert McGhee, at Ohio State University, developed a hexapod walking vehicle in 1977. (This vehicle is discussed in Section 1.2.4.)

2.3.5
U.S. Government

U.S. government agencies contributed both to the technical development of robotics and to its financial support. In addition, much of the technology that made robotics possible came from the research and development resulting from the space program and military discoveries.

1. Financial Support

The National Science Foundation has supported the work at SRI, Stanford, Purdue, Carnegie-Mellon (CMU), University of Rhode Island, MIT, University of Florida, University of Mass., University of Rochester, University of California at Berkeley, UCLA, Ohio State University, and Draper Laboratories.

The Defense Advanced Research Projects Administration (DARPA) has supported work at MIT, Stanford, CMU, and Ohio State.

The Office of Navy Research (ONR) has supported work at MIT, CMU, SRI, and the University of Rochester. They are interested in robot arms for undersea work and related sensing capabilities.

The National Aeronautics and Space Administration (NASA) has supported work at Stanford and in-house work at JPL, which is run by Cal-Tech under NASA contract. Work is also supported at Langley Research Center and the Marshall Space Flight Center, primarily for space use.

Nearly all of these grants are relatively small. Usually they range from $50,000 to $500,000 per year. Projects supported have been manipulator design, kinematics, vision systems, tactile sensing, programming development, and applications to manufacturing.

In addition, some state governments such as those of Texas, Michigan, and Florida, have provided financial support.

2. Technical Guidance and In-House Research

The National Bureau of Standards (NBS) is developing interface standards, performance standards, and programming language standards for robots. These standards will be essential for the factory of the future so that robots can work together and communicate with other equipment and the data processing functions in the factory.

NBS has also performed a valuable function in determining the needs of robot users through various symposia and analysis of robot characteristics (Evans [13]). It maintains surveillance over the robot industry and periodically reports on the new developments and trends (Gevarter [16]).

J. S. Albus [1] developed a new approach to manipulator control that is based on a learning model similar to the operation of the human cerebellum. Control methods are attempted based on a first approach and modified to take advantage of the results found. Modifications of the control pattern are stored in memory each time a successful action is carried out and used as the basis of the next control attempt. Studies of robot manufacturing systems are continuing.

2.3.6
Japanese Research

Although there has been a great deal of research in Japan, it has not influenced U.S. development to a great extent and thus will be mentioned only briefly. An early development by the Electric-Technical Laboratory (ETL) in Tokyo of the Ministry of International Trade and Industry (MITI) was claimed as a "feat far surpassing the present international standard in robot engineering" by the Mainichi Daily News, an English-language paper published in Tokyo (Mainichi [48]). This was a "thinking" robot that had a visual system, a multijoint manipulator, and an advanced control system. It was capable of picking up small cubical blocks from a table and stacking them. It could distinguish nine colors and correct its own errors. By our definition, it was an adaptive robot.

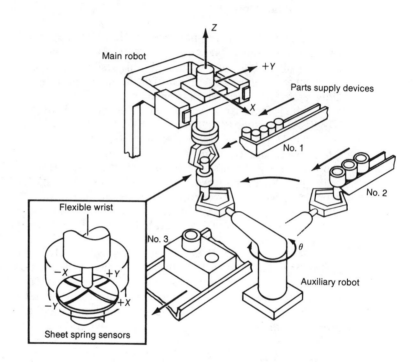

Figure 2.8 Construction of the HI-T-Hand Expert-2.

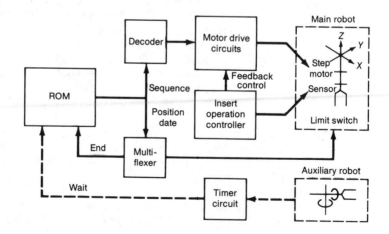

Figure 2.9 Schematic diagram of a controller.

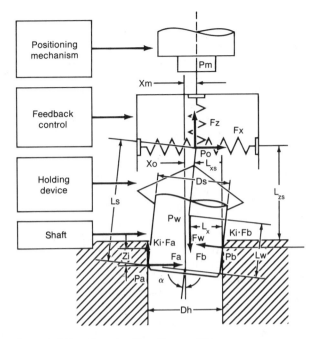

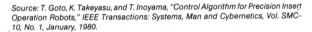

Figure 2.10 Two-dimensional insertion operation.

Source: T. Goto, K. Takeyasu, and T. Inoyama, "Control Algorithm for Precision Insert Operation Robots," IEEE Transactions: Systems, Man and Cybernetics, Vol. SMC-10, No. 1, January, 1980.

Development of tactile sensors has been an important activity in Japan (Goto [18]). Some of this work was done in developing the HI-T-HAND, which was one of the earliest robots capable of precision insertion operations for assembly (Takeyasu [46]). Force feedback was used to control the hand in assembling parts of an electric motor. This work was done at Hitachi, Ltd., one of the foremost robot manufacturers in Japan. A recent description of the HI-T-HAND is given in Goto et al. [17]. Figures 2.8, 2.9, and 2.10 are abstracted from Goto et al. [17]. This work was completed in 1978, and thus belongs in the early development period.

Exercises

The following exercises are designed to help you review what you have learned from this chapter.

1. What was the first robot to have adaptive capability as defined in Section 1.2.2? What made it an adaptive robot?
2. The first Unimate was installed in 1961 for what application?
3. What are the three commonly used justifications for capital equipment purchases?
4. Name and describe three early applications of the hydraulic robots made by Unimate and Versatran.
5. What was the positioning accuracy of the 1972 Unimate robot?
6. Which foreign robot was used as a continuous path spray painter?
7. What sensors were used on the Shakey robot at SRI? How was it controlled?
8. Read Section 9.11.1 and discuss the advantages and disadvantages of the SRI Vision Module. What approach was used to make it work more reliably?

9. What were the advantages of the WAVE program developed at Stanford by Richard Paul?
10. Describe some of the contributions made by MIT to the theoretical development of robots. Read one of the references given in Section 2.3.3 and discuss the meaning and importance of the work done.
11. What were some of the contributions of the NBS to the development of robots? Read at least one of the references given in Section 2.3.5 and discuss the work described.

References

[1] J.S. Albus, "A New Approach to Manipulator Control: The Cerebellar Model Articulation Controller (CMAC)," Journal of Dynamic Systems, Measurement and Control, vol. 97, 1975, pp. 220-227. Also "Data Storage in the Cerebellar Model Articulation Controller (CMAC)," Journal of Dynamic Systems, Measurement and Control, Vol. 97, 1975, pp. 228-233.

[2] T. O. Binford et al., "Exploratory Study of Computer Integrated Assembly Systems," Artificial Intelligence Laboratory, First Report, Stanford University, Stanford, Calif., 1974.

[3] Same as [2], Second through Eighth Reports.

[4] M. Brady et al. eds.: Robot Motion: Planning and Control. MIT Press, Cambridge, Mass., 1982.

[5] Conference Proceedings, 13th International Symposium on Industrial Robots and Robots 7, April 17-21, 1983, Chicago, Ill. Robotics International of SME, Dearborn, Mich., 1983.

[6] B. L. Dawson, "Moving Line Applications with a Computer Controlled Robot," (Cincinnati Milacron), Robots 2 Conference, October 1977. Reprinted in [47] Vol. 1, pp. 294-308.

[7] J. Denavit and R. S. Hartenberg, "A Kinematic Notation for Lower Pair Mechanisms Based on Matrices," Journal of Applied Mechanics, ASME, June 1955, pp. 215-221.

[8] R. O. Duda and P. E. Hart, "Experiments in Scene Analysis," Stanford Research Institute, Stanford, Calif. Published in [36], pp. 113-117.

[9] J. F. Engelberger, "Economic and Sociological Impact of Industrial Robots." Published in [36], 1970, pp. 7-12.

[10] J. F. Engelberger, Robotics in Practice. AMACOM, a division of the American Management Association, New York, 1980.

[11] J. F. Engelberger [10], Preface, with much historical information.

[12] H. A. Ernst, "MH-1, A Computer-Operated Mechanical Hand." Ph.D. thesis, Massachusetts Institute of Technology, Cambridge, Mass., 1961.

[13] J. M. Evans, Jr., J. S. Albus, and A. J. Barbera, eds.: "NBS/RIA Robotics Research Workshop," Williamsburg, Va., 1977, Published by the Institute for Computer Science and Technology, National Bureau of Standards, Washington, D.C.

[14] R. Finkel, R. H. Taylor, R. C. Bolles, R. Paul, and J. A. Feldman, "AL, A Programming System for Automation," Memo AIM-243, Artificial Intelligence Laboratory, Stanford University, Stanford, Calif., 1974.

[15] B. Ford, "Industrial Applications for Electrically Driven Robots." Published in [47], Vol.   pp. 286-293.

[16] W. B. Gevarter, "An Overview of Artificial Intelligence and Robotics," NBSIR 82-2479, U.S. Department of Commerce, National Bureau of Standards, Washington, D.C., March 1982.

[17] T. Goto, K. Takeyasu, and T. Inoyama, "Control Algorithm for Precision Insert Operation Robots," IEEE Transactions of Systems, Man and Cybernetics, vol. SMC-10, no. 1, January 1980, pp. 19-25.

[18] T. Goto, K. Takeyasu, T. Inoyama, and R. Shimomura, "Compact Packaging by Robot with Tactile Sensors", Proceedings of the Second International Symposium on Industrial Robots, May 1972, pp. 149-159.

[19] D. D. Grossman, "Monte Carlo Simulation of Tolerancing in Discrete Parts Manufacturing and Assembly," Memo AIM-280, Artificial Intelligence Laboratory, Stanford University, Stanford, Calif., May 1976.

[20] J. M. Hollerbach, "A Recursive Lagrangian Formulation of Manipulator Dynamics and a Comparative Study of Dynamics Formulation Complexity," IEEE Transactions of Systems, Man and Cybernetics, vol. SMC-10, no. 11, November 1980, pp. 730-736.

[21] T. Lozano-Perez, "Robot Programming", Proceedings of the IEEE, vol. 71, No. 7, July 1983, pp. 821-841.

[22] T. Lozano-Perez and M. A. Wesley, "An Algorithm for Planning Collision-free Paths Among Polyhedral Obstacles," Comm. ACM vol. 22, October 1979, pp. 560-570.

[23] T. Lozano-Perez and P. H. Winston, "LAMA: A Language for Automatic Mechanical Assembly," Fifth International Joint Conference on Artificial Intelligence, Massachusetts Institute of Technology, Cambridge, Mass. August 1977, pp. 710-716.

[24] D. Marr and H. K. Nishihara, "Representation and Recognition of the Spatial Organization of 3-D Shapes," Artifical Intelligence Laboratory, Massachusetts Institute of Technology, Memo AIM-416, May 1977.

[25] M. T. Mason, "Compliance and Force Control for Computer-Controlled Manipulators," Artificial Intelligence Laboratory, Massachusetts Institute of Technology, memo AIM-515, April 1979.

[26] M. T. Mason, "Compliance and Force Control for Computer-Manipulators," IEEE Transactions in Systems Man and Cybernetics, vol. SMC-11, no. 6, 1981, pp. 418-432.

[27] V. Milenkovic and B. Huant, "Kinematics of Major Robot Linkage." In [5], 1983, pp. 16-31 to 16-47, vol. 2.

[28] J. H. Munson, "The SRI Intelligent Automation Program", Stanford Research Institute, Stanford, Calif. Published in [36], pp. 113-117.

[29] J. L. Nevins, D. E. Whitney, and A. E. Woodin, "A Scientific Approach to the Design of Computer Controlled Manipulators", R-837, Draper Laboratory, Cambridge, Mass., November 1974.

[30] J. L. Nevins, D. E. Whitney, and S. N. Simunovic, "System Architecture for Assembly Machines," R-764, Draper Laboratory, Cambridge, Mass., November 1973.

[31] D. Nitzan, S. Barnard, R. Bolles, R. Cain, J. Hill, W. Park, and R. Smith, "Machine Intelligence Research Applied to Industrial Automation," Tenth Report, SRI International, November 1980.

[32] D. Nitzan et al.: Same as [31], Eleventh Report, January 1982.

[33] D. Nitzan and C. A. Rosen, "Programmable Industrial Automation," IEEE Transactions on Computers, vol. c-25, December 1976, pp. 1259-1270.

[34] R. P. Paul, "Modeling, Trajectory Calculation and Servoing of a Computer-Controlled Arm," Artificial Intelligence Laboratory,

Stanford University, AIM 177, 1972.

[35] R. P. Paul, Robot Manipulators: Mathematics, Programming and Control. MIT Press, Cambridge, Mass., 1981.

[36] Proceedings of the First National Symposium on Industrial Robots, April 2-3, IIT Research Institute, Chicago, Ill. 1970.

[37] D. R. Reddy and S. Rubin, "Representation of Three-Dimensional Objects," CMU-CS-78-113. Department of Computer Science, Carnegie-Mellon University, April 1978.

[38] L. G. Roberts, "Homogeneous Matrix Representation and Manipulation of N-Dimensional Constructs." Lincoln Laboratory, Massachusetts Institute of Technology, Document No. MS1045, 1965. (Referenced by [35]).

[39] C. Rosen, D. Nitzan, G. Agin, A. Bavarsky, G. Gleason, J. Hill, D. McGhie, and W. Park, "Machine Intelligence Research Applied to Industrial Automation," SRI International, Second through Eighth Reports, 1974-1978. Reports not issued every year.

[40] B. Roth, "Evaluation of Manipulators from a Kinematic Viewpoint," NBS Special Publication 459. U.S. Department of Commerce/National Bureau of Standards, Performance Evaluation of Programmable Robots and Manipulators, October 1976.

[41] J. W. Saveriano, "An Interview with Joseph Engelberger," Robotics Age, vol. 3, no. 1, January to February 1981, pp. 10-23.

[42] J. W. Saveriano, "Interview with Charles Rosen, Robotics Age, vol. 3, no. 3, May-June 1981, pp. 16-26.

[43] V. C. Sheinman, "Design of a Computer-Controlled Manipulator," Artificial Intelligence Laboratory, Stanford University, Stanford, Calif., memo AIM 92, June 1969.

[44] D. Silver, "The Little Robot System," Artificial Intelligence Laboratory, Massachusetts Institute of Technology, memo AIM-273, January 1973.

[45] J. M. Sutherland, "Robot Applications." In [36], pp. 13-37.

[46] K. Takeyasu, T. Goto, and T. Inoyama, "Precision Insertion Control Robot and its Application," Trans. ASME, vol. 98, no. 4, November 1976, pp. 1313-1318.

[47] W. R. Tanner, ed.: Industrial Robots, 2nd ed., Vol. 1, Fundamentals, Vol. 2, Applications. Robotics International of SME, Society of Manufacturing Engineers, Dearborn, Mich., 1981.

[48] "'Thinking' Robot Developed Here," Mainichi Daily News, Tokyo, Japan, October 15, 1970, p.

[49] D. E. Whitney, "Use of Resolved Rate to Generate Torque Histories for Arm Control," Draper Laboratory, Cambridge, Mass., memo MAT-54, July 1972.

[50] D. E. Whitney, "The Mathematics of Coordinated Control of Prosthetic Arms and Manipulators," Journal of Dynamic systems, Measurement and Control, December 1972, pp. 303-309. Reprinted in [4].

[51] M. W. Wichman, "The Use of Optical Feedback in Computer Control of an Arm," Artifical Intelligence Laboratory, Stanford University, Stanford, Calif., AIM 56, 1967.

Chapter 3

Mechanical Considerations

This chapter will provide background on the mechanical considerations involved in selecting and using a robot. Basic mechanical principles will be discussed to illustrate some of the important items to be considered. For those with no mechanical engineering background, the terms in section 3.5.1 will be helpful. In most cases, the terms will be defined either specifically or by context before being used.

Coordinate systems will be defined first, including the resulting work envelope. Robot motions—degrees of freedom, arm and body motions and wrist motions—will then be discussed. Drive mechanisms—linear drives, rotary drives and wrist mechanisms—will also be discussed, followed by a section on the end effectors or hands that do the actual work performed by a robot.

Other sections in this chapter describe the important characteristics that must be considered in determining the required specifications for a robot arm such as accuracy, repeatability, reliability, maintainability, and life expectancy. Some items are defined and discussed that affect accuracy and repeatability, such as gravitational effects, acceleration forces, backlash, thermal effects and friction. These terms will be familiar to mechanical engineers but may be new to those who are more familiar with other disciplines.

Because mobility of robots is of growing importance, we will review some of the current development efforts on mobile robots and discuss some of the problems to be considered. An interesting mobile robot with three-pronged supports for wheels is being developed in Japan at the University of Tokyo. It is shown in Figure 3.1 and described in Section 3.6.

Figure 3.1 Mooty mobile robot for maintenance.

Courtesy Masahuru Takno, University of Tokyo, Tokyo, Japan. And SRI International, formerly the Stanford Research Institute, Menlo Park, CA. Used by permission.

3.1 Physical Configurations

3.1.1
Coordinate
Systems

As mentioned in Chapter 1, there are several commonly used configurations of robots. These are:

1. Cartesian coordinate configuration—Three directions x, y, and z, are specified. Coordinate directions are <u>orthogonal</u> (at right angles) to each other. Also, the coordinate axes are set up so that a clockwise rotation from the x axis toward the y axis would advance a right-handed screw in the positive z direction. These coordinates are shown in Figure 3.2(a).

2. Cylindrical coordinate configuration—Movement of the robot arm will describe the surface of a cylinder if the radius of the arm is fixed. The coordinates are R, Θ, and z. R is the radius of the arm, Θ is the angular position, and z is the position in the vertical direction, as shown in Figure 3.2(b).

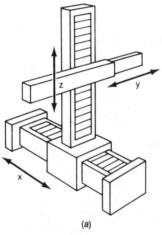 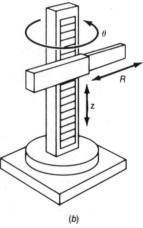

(a)   (b)

Figure 3.2 (a) Cartesian coordinates.   (b) Cylindrical coordinates.

3. Polar coordinate configuration. The surface described is a hemisphere of radius R. Coordinates are R, Θ, and ϕ. Θ is the rotation around a vertical through the base support of the arm. ϕ is the angle from the horizontal through 180 degrees vertically.

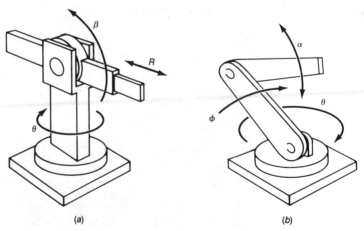

(a)   (b)

Figure 3.3 (a) Polar coordinates.   (b) Revolute coordinates.

4. Revolute coordinates. All movements are in angular form. Three angles, Θ, α, and α are specified. Θ is rotation around the vertical through the base, α is the angle between the horizontal through the base and the first arm member, and α is the angle of the second arm member in reference to the first as in Figure 3.3.b. The arm can reach most portions of a spherical volume. The actual shape of the volume traversed depends on the relative lengths of the first and second arm members.

Each of these coordinate systems has advantages and disadvantages, as discussed in the following paragraphs and summarized in Table 3.1.

3.1.2
Work Envelopes

Work envelopes enclose all the points in the surrounding space that can be reached by the robot arm or the mounting point for the end affector or tool. The area reachable by the end effector itself is not considered part of the work envelope. All interaction with other machines, parts, or processes must take place within this volume of space.

Figures 3.4, 3.5, 3.6, and 3.7 show the work envelopes for rectangular, cylindrical, polar, and revolute coordinates, respectively.

1. Rectangular Coordinate Robots

Rectangular coordinate robots can extend the work envelope to large areas and can be made modular, since it is not difficult to replace one linear drive with another. One company, the Mobot Corporation in San Diego, specializes in modular robots and can quickly build up specialized robots from their stock of modular parts. Several other companies also provide modular systems.

Rectangular robots are useful in providing an overhead carrier for working over large floor areas, much as a traveling crane might be used. In this case, the overhead structure carries a robot arm and wrist with considerable flexibility. A potential problem in dusty or corrosive areas is the need for some means to protect the sliding members of the rectangular structure. An advantage of rectangular robots is the ability to do straight-line insertions into furnaces and other equipment. Also, they are the easiest to program. It is not difficult for most users to visualize three-dimensional rectangular motion.

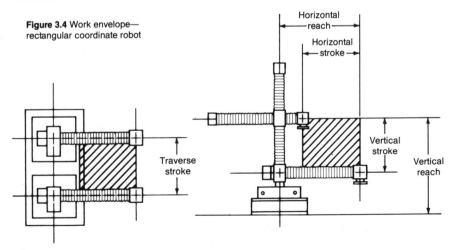

Figure 3.4 Work envelope—rectangular coordinate robot

2. Cylindrical Coordinate Robots

Cylindrical coordinate robots have a vertical lift axis carrying an arm that can move in and out radially and around the vertical axis. The resulting work envelope is enclosed by two partial cylinders, one inside the other. Because of mechanical limitations on rotary motion, the arm cannot reach all parts of the volume enclosed by the two cylinders. Also, it is not possible for the tip of the arm to reach floor level, since it is limited by the base of the vertical lift. Those areas that cannot be reached will be referred to as <u>dead space</u>. They can be a serious limitation to all robots, depending on their configuration, and should be carefuly considered in robot system design.

Where straight-line motion is required to place parts in a cavity or to remove parts from a lathe or molding machine, the cylindrical coordinate robot is valuable. It does not require complex programming, and the path of the arm tip is easy to visualize. In corrosive atmospheres, the sliding members must be protected, however. Note that the radial arm also extends backward from the center post. Therefore, a clearance volume must be made available when using a cylindrical robot. It is even possible for the clearance volume to overlap the work envelope, with potentially disastrous results.

Prab Versatran robots are made in the cylindrical configuration, and a large number of them have been sold.

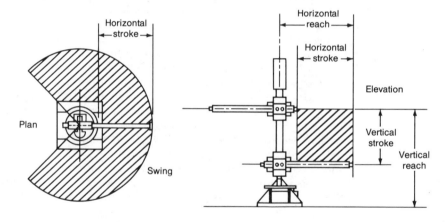

Figure 3.5 Work envelope—cylindrical coordinate robot.

3. Spherical (Polar) Coordinate Robots

Spherical or polar coordinate robots have an extension arm that pivots around a base and also in a vertical plane around a horizontal pivot point in the base support, as shown Figure 3.6. The Unimate® robots, made by the Unimation Corporation, used this configuration exclusively for many years, but now use other configurations as well.

The extension arm sweeps out the volume between two partial spheres. As in the cylindrical robot, there is a space inside the inner sphere that cannot be reached. It is possible to reach the floor if the horizontal stroke is made long enough. It is noteworthy that the interior of a tunnel inclined up or down can be reached more easily with this configuration than with any other, although the need for this capability may not be very large.

4. Revolute (Jointed) Coordinate Robots

Both terms, <u>revolute</u> and <u>jointed</u>, are used by many different robot engineers, manufacturers, and users. We will use both terms but prefer <u>revolute</u> because it alone describes the important idea that all joints are rotating in this configuration.

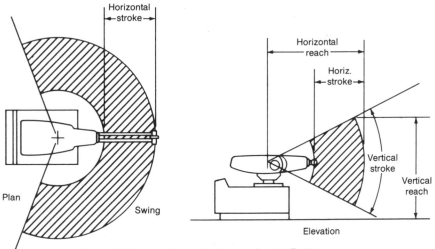

Figure 3.6 Work envelope—spherical or polar coordinates.

A major advantage of this configuration is the ability to reach any part of the work volume. The major disadvantage is the difficulty of visualizing and controlling the three links. The same point in space can be reached in many ways, so it is difficult to decide on the best way. Also, the path that the robot arm will take to reach a particular point is in question. This is not a problem with other configurations.

It is almost necessary to use a microcomputer or minicomputer to control the revolute arm because of the complexity of control. In the early days of robots, the cost of computers for this purpose was so large that the spherical coordinate configuration was chosen for this and other reasons.

Most of the new robots developed in the last few years have been revolute configurations because of the flexibility of the configuration

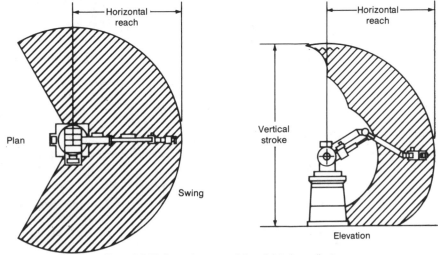

Figure 3.7 Work envelope—revolute or jointed coordinates.

and the fact that all joints can be sealed easily. Hydraulic robots of this type are the DeVilbiss Trallfa, the Cincinatti Milacron, and the Nordson.   Electrical robots in this configuration are the Unimate PUMA and the ASEA.

Table 3.1 Robot Configurations, Advantages, and Disadvantages

Advantages	Disadvantages
CARTESIAN COORDINATES x, y, z —Base travel, reach, and elevation (Also called "Rectangular Coordinates")	
Moves in three linear directions and is thus easy to visualize.	Requires large volume to operate in, even though whole space is not used.
Easy computation.	Largest surface area required of all configurations.
Most rigid structure for given length, since it is supported at both ends.	Exposed guiding surfaces require covering or boots in dusty or corrosive atmosphere.
CYLINDRICAL COORDINATES Θ, R, Z—BASE ROTATION, REACH, AND ELEVATION	
Easy to visualize and compute.	Restricted volume of access. Cannot reach cylindrical volume near vertical support or floor.
Linear drives well suited to use of hydraulic drives. Therefore, can provide great power.	Both radius drive and vertical drive are exposed. Difficult to seal from dust and liquids.
Good access into cavities and machine openings.	Possible to have rear clearance volume overlap and interfere with work envelope.
SPHERICAL COORDINATES Θ, R, ϕ—BASE ROTATION, REACH, ELEVATION ANGLE	
Covers a large volume from one central support.	Complex coordinates difficult to visualize and control.
Two rotary drives can be easily sealed.	One linear drive. May be sealing and protection problems.
Covers a large volume.	Restricted volume coverage.
REVOLUTE COORDINATES Θ, ϕ, α—BASE ROTATION, ELEVATION ANGLE, AND REACH ANGLE (Also called "Jointed Coordinates")	
All joints are rotary. Maximum flexibility, since any location in the total volume can be reached.	Require more expensive drives when using hydraulics, but well suited to electric motor drives.
Joints can be completely sealed. Useful in dusty or corrosive locations, or under water.	Visualization and control are most difficult in this configuration.
	Restricted volume coverage.

3.2 Robot Motions

3.2.1
Degrees of
Freedom

Six degrees of freedom are necessary to emulate the motion of a human arm and wrist. The human upper arm, moving at the shoulder joint, has two degrees of freedom since it can rotate up and down and forward and backward in two angular directions. The lower arm, moving at the elbow joint, is the third degree of freedom. Wrist rotation is the fourth, wrist movement up and down is the fifth, and wrist movement left and right is the sixth. Operation of the fingers and thumb provides other degrees of freedom.

Degrees of freedom in a robot arm depend on the configuration but can approximate the same flexibility as the human arm. The motions available are now discussed.

3.2.2
Arm and Body
Motions

1. Vertical Traverse. Up-and-down motion of the arm. In the Z axis direction, this may be caused by moving the whole robot body vertically. A hydraulic piston or vertical lift may be another way to obtain this motion.
2. Radial Traverse. Extension and retraction of the arm allows the effective length of the arm to be changed. In cylindrical coordinates, the diameter of the cylindrical surface being assessed is determined by the arm length.
3. Rotational Traverse. Rotation about the vertical axis determines the angular position that can be reached.

3.2.3
Wrist Motions

1. Wrist swivel. Rotation of the wrist about the axis of the arm. Some robots are limited to less than 360 degrees. Others allow several rotations to occur, limited by the number of wraps of a control cable up to as much as 7,200 degrees or 20 turns. Continuous rotation is possible in a robot wrist by providing rotating contacts for electrical control cabling. This arrangement is useful for operating a screwdriver and similar tools.
2. Wrist Bend. Up-and-down movement of the wrist. This movement is also known as pitch, a term borrowed from aircraft terminology. Rotation and pitch together could produce yaw as described next but it is sometimes desirable to provide for yaw as well.
3. Wrist Yaw. Right or left angular movement of the wrist. Wrist motions are usually provided at separate joints, rather than at a flexible, multipurpose joint, as in the human wrist.

3.3 Drive Mechanisms

3.3.1
Linear Drives

Examples of linear drives are the x, y and z drives of rectangular coordinate systems, the vertical and horizontal positioner of cylindrical systems, and the radial drive on spherical systems. Linear motion may be generated directly by a hydraulic or pneumatic piston in a cylinder or by conversion of rotary motion to linear motion through rack and pinion gearing, lead screws, worm gears, or ball screws.

1. Rack and Pinion

A rack is a long metal bar that has gear teeth cut across its width. A pinion gear is an auxiliary smaller, round gear placed so that its teeth can mesh with the teeth of the rack. Usually the rack is fixed in place. As the pinion gear rotates, it moves an attached carriage linearly along the rack. The rotary motion of the pinion gear is converted to linear motion of the carriage, as shown in Figure 3.8. Support for the carriage is provided by a guide rod or guide way.

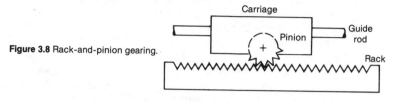

Figure 3.8 Rack-and-pinion gearing.

2. Lead Screws

Lead screw drives consist of a precise screw rotating to drive the nut along a track.   In high-precision equipment such as numerically controlled machine tools, it is common to use preloaded lead screws that are accurate to one ten-thousandth of an inch under the right conditions.   Long lead screw drives must be large enough to be rigid, but then have considerable friction and large inertia, so that they are seldom used in robots.

3. Ball Screws

Ball screws are frequently used in robots because they have very low friction and are capable of rapid response.   Ball screws are lead screws in principle, but a large number of ball bearings, riding in the screw of the gear, carry the load and reduce the friction to a low value. It is possible, for example, to turn a ball screw from the output end, which is difficult to do with lead screws because of the large friction in such gears.

As illustrated in Figure 3.9, the balls of a ball screw feed out of a carrier tube into the polished and ground screw thread, rotate two or three turns, and then feed back into the carrier. Energy transmission efficiency of 90% is achieved, so that power requirements are minimized and smaller lighter drive train components can be used. Rolling friction has been substituted for sliding friction. This change also reduces starting friction and essentially eliminates chatter. Life expectancy is also increased because of the low friction. Preloading can be done by using two nut assemblies back to back to reduce backlash to a low value.

Figure 3.9 Ball screw drive assembly, cut away to show the ball bearing carrier.

Courtesy Warner Electric Brake and Clutch Company, South Beloit, Illinois. Used by permission.

4. Hydraulic Cylinder Drives

Linear hydraulic drives have a close-fitting piston inside a polished cylinder. Oil under pressure is supplied at one end of the cylinder and drives the piston toward the other end. Movement is controlled by adjusting the amount of oil and the oil pressure on each end of the piston. An example of a hydraulic drive is provided in Figure 4.5.

Many of the early robots used hydraulic pistons controlled by servo valves for linear motions. Hydraulic pistons are powerful, simple, and inexpensive. Although high-quality servo values are expensive, there was a net saving for linear motion because there was no need to convert rotary motion to linear motion. Unimate®robots made by the Unimation Corporation are among the most widely used robots today. These robots use a linear hydraulic cylinder for the radial drive. Prab Versatran robots use linear hydraulic cylinders for both the vertical and radial drives of their cylindrical coordinate robot.

5. Pneumatic Cylinder Drives

Pneumatic cylinders operate the same way as hydraulic cylinders except that air is used. Airflow is usually controlled by solenoid valves. Since air is compressible, the force obtainable from pneumatic cylinders is limited.

3.3.2
Rotary Drives

Most electric motors and servo drive motors generate rotary motion directly, but often at a lower torque and higher rotational velocity than is desired. It is necessary, therefore, to use some kind of gear train, belt drive, or other mechanism to convert the available high speed to lower speed with greater torque. There are also some cases in which linear hydraulic cylinders or pneumatic cylinders are used as a power source, so that the linear motion must be converted to rotary motion. As discussed in Section 3.5.2, these conversions must be accomplished efficiently and without impairing the desirable characteristics of the system, especially positioning speed, accuracy and reliability. Several options are available for consideration.

1. Gear Trains

A gear train is made up of two or more gears used to change the angular velocity, torque, force, and displacement of the output relative to the input. For simplicity, we will consider a system with two gears, one on the input shaft and one on the output shaft, as shown in Figure 3.10. The following terms will be defined to assist us in analyzing this relationship.

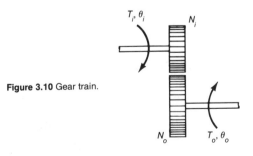

Figure 3.10 Gear train.

T_i = input torque
T_o = output torque
N_i = number of teeth on input gear
N_o = number of teeth on output gear

R_i = radius of gear circle, input gear
R_o = radius of gear circle, output gear
Θ_i = angular movement of input gear
Θ_o = angular movement of output gear
w_i = angular velocity of input gear
w_o = angular velocity of output gear
J_i = inertia of input drive
J_o = inertia of output drive

We will neglect the inertia and friction of the gears to simplify the analysis. It is assumed that there is no loss of energy in the operation.

Then we have the following relationships between the input drive and the output, as can be verified from study of Figure 3.10.

Since there is no loss, the total work on one shaft, the angular distance multiplied by the torque, must be equal to the total work on the other shaft, so $T_i \Theta_i = T_o \Theta_o$

The total rotational distance traveled by the meshed gears must be the same, so $R_i \Theta_i = R_o \Theta_o$

Also, since the gears mesh together, the number of teeth on each gear must be proportional to their radii, so $N_i / R_1 = N_o / R_o$

or $N_o / R_1 = N_i / R_o$

By similar reasoning, it can be shown that $N_i / w_1 = N_o / w_o$

or $\dfrac{N_i}{N_o} = \dfrac{w_o}{w_i}$

Therefore, the values seen at the output shaft as a function of the input and the mechanical parameters are

Torque: $T_o = \dfrac{N_o}{N_i} T_i$

Angular Displacement: $\Theta_o = \dfrac{N_i}{N_o} \Theta_i$

Angular Velocity: $w_o = \dfrac{N_i}{N_o} w_i$

If we wish to know the total effective inertia at the drive motor, we find

Inertia: $J_e = (N_i / N_o)^2 J_o + J_i$

where J_o is the total inertia on the output and J_i is the total inertia of the input drive.

Usually, the motor speed is several times larger than the desired output shaft rotational velocity, but the torque from the motor is not sufficient to drive the load directly. With a small gear on the input shaft and a large gear on the output shaft, we can improve both of these characteristics.

As an example, take $N_i = 10$ and $N_o = 100$. Then the output angular velocity $w_o = 0.1 w_i$. Also, the output torque T_o, has increaed by a factor of 10, so that both requirements are filled at once. In addition, the effective inertia seen by the drive motor due to the output shaft inertia has been decreased by a factor of 100 due to the squaring of the gear ratio. This means that the effective response time of the drive motor is greatly decreased, thus improving the ease of control by the servo system.

A detailed analysis of a servo system using gear drives is given in Section 6.7.

A problem with gear trains is the error due to backlash between gears. This backlash can cause direct positioning errors, as discussed in Section 3.5.3, but can also lead to servo mechanism instability unless some means of offsetting it is applied.

Either spur gears or bevel gears may be used in gear trains. Some ingenious wrist mechanisms use bevel gears to allow all three wrist motions to occur about the same center. In general, however, bevel gears are less efficient and more subject to wear than spur gears.

2. Timing Belt Drives

Timing belts are similar to fan belts and other drive belts, except that they have teeth molded into the belt and ride on timing belt pulleys that have matching gear teeth. These are essentially flexible gears in operation. They have the advantage of flexibility and low cost. Also, timing belts can be used where it is difficult to maintain alignment between shafts, since they will work satisfactorily over misalignments of as much as half an inch or so if the belt is long enough, so that the angular error is not too large. In servo systems in which the output shaft position is measured by some type of encoder, the timing belt on the input drive may be outside the servo loop and have little effect on the accuracy and repeatability of the system; repeatability of 50 thousandths of an inch is not difficult to achieve. In addition, of course, timing belts are much less expensive and less difficult to construct than gear trains. Often, gear trains and timing belt drives will be used together as a matter of convenience.

3. Harmonic Drives

Although harmonic drives have been around for a long time, they have not been used extensively until recently. Now they promise to solve some of the problems of robots in an efficient manner.

As seen in Figure 3.11, the harmonic drive is made up of three main parts: the circular spline, the wave generator, and the flexspline. In operation the rigid circular spline is mounted in place, while the flexspline, with its teeth outside, rotates in the inside teeth of the circular spline. The flexspline has two teeth fewer than the circular spline, so that on each rotation it gains the equivalent amount of rotational angle. Bearings on the elliptical wave generator support the flexspline, while the wave generator drives it in rotation and also causes it to flex. Only a few teeth are engaged, at the ends of the oval shape assumed by the flex spline while rotating, so there is freedom for the flexspline to rotate the equivalent of two teeth relative to the circular spline during each full revolution of the wave generator.

If the rigid, circular spline has 100 teeth and the flex spline has 2 teeth fewer, the flexspline will rotate only once for each 50 rotations of the wave generator, providing a gear ratio of 1:50 in a compact space. Because of the large number of teeth meshed at one time, the harmonic gear has high torque capacity. Although any two of the units may act as input and output, it is usual to mount the wave generator on the input shaft and the flexspline on the output shaft to obtain a large gear ratio reduction.

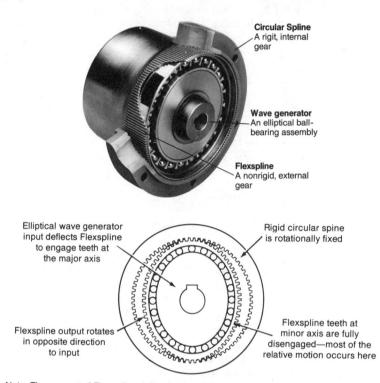

Circular Spline
A rigit, internal gear

Wave generator
An elliptical ball-bearing assembly

Flexspline
A nonrigid, external gear

Elliptical wave generator input deflects Flexspline to engage teeth at the major axis

Rigid circular spine is rotationally fixed

Flexspline output rotates in opposite direction to input

Flexspline teeth at minor axis are fully disengaged—most of the relative motion occurs here

Note: The amount of Flexspline deflection has been exaggerated in the diagram in order to demonstrate the principle. Actual deflection is much smaller than shown and is well within the material fatigue limits. Deflection is, therefore, not a factor in the life expectancy of the gearing.

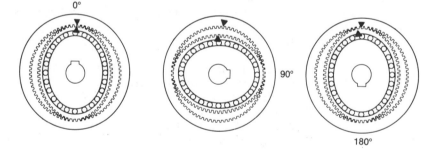

0°

90°

180°

Figure 3.11 Harmonic drive mechanism showing the method of operation.

Courtesy Emhart Machinery Group, Harmonic Drive Division. Used by permission.

Backlash is nearly zero in a harmonic drive because of the natural preloading, the many teeth in contact, and the smooth operation of the teeth. There is a potential source of angular error in the "windup" of the flexspline due to loading. Any shaft can exhibit windup, which is the rotational deflection of a shaft due to loading. Since the flexspline is not as rigid as a spur gear, it has some windup error. The windup is nonlinear at low torques. This characteristic is discussed in the manufacturer's brochure (Harmonic [7]). For most applications, it is not a major problem.

Luh et al. [7] describe the use of a harmonic drive in a modified Stanford manipulator in which strain gages provide torque sensing for feedback control. Backlash nonlinearity in the control loop and the effect of friction were reduced to very low values by the use of phase-lead compensating networks in the control loop.

4. Rotary Hydraulic Motors

Another way to obtain rotary motion is to drive the joint directly with a rotary hydraulic motor. These motors are discussed in Section 4.2.

5. Direct-Drive Torque Motors

A relatively new development is the availability of electrical torque motors that are capable of directly driving the joints of a robot. Excellent positional repeatability and compact packaging were obtained in an arm built by the Robotics Laboratory at Carnegie-Mellon Institute (Asada et al. [2]). This arm is discussed further in Chapter 6.

3.3.3
Wrist Mechanisms   Wrist mechanisms are designed to be mounted on the end of a robot arm or manipulator and support the hand or gripper or other end effector mechanism. They may be driven by electric motors, hydraulic drives, or pneumatic drives. Usually they provide all three of the degrees of freedom described in Section 3.2.3—rotation or swivel, bend or pitch, and yaw or side-to-side motion.

It is desirable that wrist mechanisms be compact and powerful, and that all motions take place about a common center. These are tough design requirements, and much design work has been done to make these features possible. Several designs are discussed by Quinlan [10] and Rosheim [11].

Several approaches to wrist design are available. We will discuss a few of them here.

1. Linkage Drives. One way to keep the wrist mechanism compact is to drive the wrist through linkages rotated remotely, as is done in the ASEA wrist show in Figure 3.12. Electric motors are mounted in the base of the robot or in a shoulder link, and the rotational motion is transmitted through push-pull links that rotate a disk, at a joint, on which additional links are mounted to transmit the motion further.

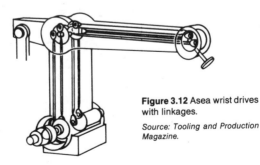

Figure 3.12 Asea wrist drives with linkages.

Source: Tooling and Production Magazine.

2. Torque Tubes. Drive shafts can be used to transmit power to the wrist with bevel gears or other drives in order to convert the rotary motion of the shaft to rotary motion at the wrist. The three-axis wrist of the Bendix ML-360 robot is driven by torque tubes and spiral bevel gearsets, as illustrated in Figure 3.13.

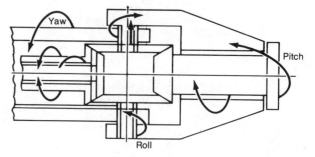

Figure 3.13 Bendix wrist driven by torque tubes and spiral bevel gearsets.
Source: Robotics Today.

A more complex wrist mechanism is the Cincinatti Milacron three-roll electric wrist shown in Figure 3.14. (Quinlan [10]. As reported by Rosheim [11], this wrist is driven by a triordinate drive shaft system, as shown in Figure 3.15. The cross section of Figure 3.16 shows the detail of the concentric torque tubes and the bevel gearing used to provide great flexibility in a very compact wrist. Further designs and patents are discussed in Rosheim [11].

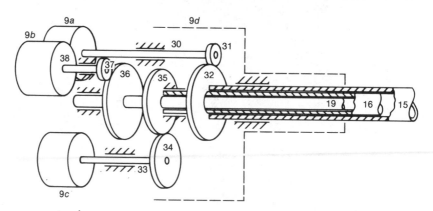

Figure 3.14 Cincinnati Milacron three-roll wrist.

Courtesy of Cincinnati Milacron. Used by permission.

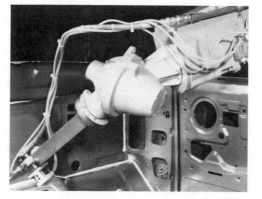

Figure 3.15 Cincinnati Milacron three-roll wrist with triordinate drives. (from a patent drawing)

3. Hydraulic Wrists. Compact hydraulic wrists can be made by placing simple actuators at the wrist itself and controlling fluid flow from a common manifold or distribution point. Small electro-magnetically controlled valves are used for fluid control. A good example of this type of wrist is the Hyd-ro-wrist made by the Bird-Johnson Company.

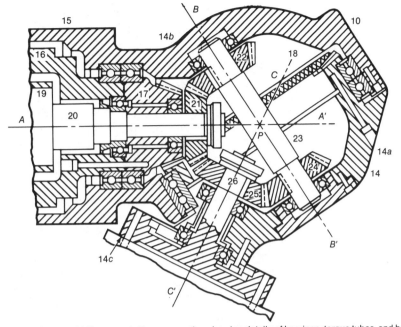

Figure 3.16 Cincinnati Milacron wrist in cross section showing details of bearings, torque tubes, and bevel gears.

3.3.4 Linear versus Rotary Drives

Before inexpensive computers were available, the computation required for controlling rotary motions was a major expense that made linear drives a good solution at that time. DC electric motors were a good power source for rotary drives, except for the major expense of the servo power amplifiers required for precise control. In 1970, for example, there were no reliable, high-power transistors. It was necessary to use many power transistors in parallel to drive a large electric servo motor.

Today, there has been a considerable change in the economics of electric motor drives and control. Power transistors are available with high wattage capacity, so only a few are needed to drive a large electric motor. Also, microcomputers have become so inexpensive that their cost is no longer an important factor. Some robots have a microprocessor for each joint or degree of freedom, for example.

These factors have caused a number of manufacturers to select rotary joints for their new robot designs. Older designs are still constrained by the use of linear drives. However, there are many cases in which the linear drives are preferable. Linear pneumatic cylinders are the cheapest drives of all, and where they can be applied they are chosen. New developments in control have made possible greater accuracy in pneumatic drives. Some manufacturers are using pneumatics, therefore, for light loads where low positioning accuracy can be tolerated, such as in pick-and-place systems.

3.3.5 Brakes

On many robot manipulators, it is necessary to have a brake on each joint to hold it in place whenever it is turned off or to protect it and its surroundings in case of power failure. High-quality gearing, harmonic drives, ball screw drives, and so on have so little friction that they will not support a load when the drives are not operating. When power is off, these joints would collapse unless some external device is used, such as a brake, clamp or external stop.

Brakes are usually operated in a fail-safe mode. Power must be applied to release the brake; otherwise the joint is held in place. This arrangement provides protection when power is turned off at the cost of continuous power use during operation to keep the brake released. If desired, of course, power need be applied only long enough to latch the brake off and hold it with a release pin. Then the only power required is that used to hold the release pin in place.

Positioning accuracy of the brake while holding the joint is often important. The amount of movement due to slight slippage of the brake can be reduced by coupling the brake to the drive side of the system, so that a large movement on the drive side is reduced by the gear ratio of the drive to a smaller movement in the joint itself. One application of this technique is to provide greater accuracy under load. Even if the robot arm is positioned accurately while unloaded, the drive servo motors might not be able to hold that position while loaded, so the brakes may be applied to hold the arm in place more accurately. An application for this might be in drilling, grinding, or other uses, where the major joints are locked in place while the end effector moves.

Caliper-type brakes, such as those used on automobile wheels, are used because they can stand repeated operation and can be made compact. A disk of rigid high-friction material is mounted on the joint. The calipers with pads are mounted on the frame, if possible, or on the larger link, so that the inertia effects are reduced as much as possible.

3.4 End Effectors

End effectors are the pickup devices, grippers, hands, and tools mounted on the end of the robot arm. These are the devices that perform the actual task. This general class of devices is also called end-of-arm tooling (EOAT), but we will use the simpler term, end effectors. End effectors are often designed for a particular task, but can also be made as all-purpose hands to do many types of tasks.

Robot end effectors are much better than human hands in dealing with heavy objects, corrosive substances, hot objects, or sharp and dangerous objects. They are not as good at handling complex shapes and fragile items. Also, since the current hands do not have good tactile sensing capability, they cannot do the many complex tasks that humans do by touch alone. However, new sensors, as described in Chapter 6, can provide useful tactile capability for many applications.

Examples of end effectors are given in Figures 3.17, 3.18, and 3.19. Engelberger [6] describes several other end effectors.

**3.4.1
Grasping and
Holding Techniques**

Many different techniques are used to provide the grasping or holding function in a robot end effector. Obvious techniques are:

1. Vacuum cups
2. Electromagnets
3. Clamps or mechanical grippers
4. Scoops, ladles or cups
5. Hooks
6. Hands with three or more fingers
7. Adhesives or strips of sticky tape

**3.4.2
Qualifications for
End Effectors**

1. Grippers

Since many different tasks are performed by robots, it is not surprising that there are many different requirements for gripper end effectors. Some of these requirements are:

1. Parts or items must be grasped and held without damage.
2. Parts must be positioned firmly or rigidly while being operated on.
3. Hands or grippers must accomodate parts of differing sizes or even of varying size.
4. Self-aligning jaws are required to ensure that the load stays centered in the jaws.
5. Grippers or end effectors must not damage the part being handled.
6. Jaws or grippers must make contact at a minimum of two points to ensure that the part doesn't rotate while being positioned.

Some of these characteristics are illustrated in Figure 3.17.

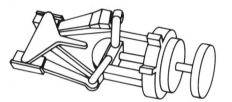

(a) Standard hand. A large variety of custom-made fingers can be attached to this all-purpose hand. Air- or solenoid-operated linkages provide the necessary clamping pressure to hold the item tightly. Parts of moderate weight can be handled reliably by this hand.

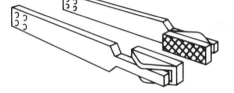

(b) Self-aligning fingers. Pivoted pads for fingers tilt to align with the part. This arrangement works well for flat-sided parts or those that have only a small angular difference between the sides. They can be mounted on the standard hand with bolts or screws.

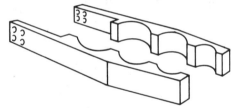

(c) Multiple-cavity fingers. These fingers can handle three sizes of parts. By positioning the robot hand, the cavity which best fits the parts can be selected.

Figure 3.17 Examples of mechanical grippers.

Source: Robotics in Practice by J.F. Engelberger, page 45. © 1980 by Joseph L. Engelberger. Published in the U.S. by AMACON a division of American Management Associations, New York. All rights reserved.

The clamping force required to hold parts must be adjusted to ensure that the part does not slip loose and yet is not damaged by the tight grip. Parts may be supported because the clamp has closed around an edge or protuberance on the part, or the part may be held by friction. Usually the part is held by the friction force, which is composed of pressure against the part due to the clamping force (the "normal" force of engineering dynamics) and a coefficient of friction due to the roughness or resilience of the contacting surfaces.

2. Friction Holding

If a part is to be held reliably, the friction force must be greater than the total of all the forces being applied—gravitational force, centrifugal force, and so on. Friction force is defined as:

$$F_f = \mu F_c$$

where F_c is the clamping force and u is the coefficient of friction. In a typical case, to hold a load of 50 pounds, we would need a clamping force of 100 and a coefficient of friction of $\mu = 0.5$ as a minimum. Actually, it would be desirable to have a factor of safety of at least two, so a clamping force of 200 pounds would be required.

Most robots use only one arm to handle objects, while humans use two arms when necessary. When large objects are to be handled, the human may pick up the object with one hand at each end. A robot will use one end effector that has been designed to handle a specific class of objects. The cost of end effectors is usually a small part of the total cost of the robotic system, so that specialized end effectors make good sense.

Usually the robot arm has a tool holder at its end to simplify the attachment of end effectors. This holder could be a <u>bayonet pin</u> mount or a plate with mounting holes. Bayonet pin mounts have a long, pointed shaft that fits into a support sleeve and a shorter pin that fits into another alignment hole to prevent the end effector from rotating. Both pins are secured in place by clamp rings or other means to prevent the end effector from being released accidentally. These mounts are similar to those used in quick-change machine tools. Mounts of this type allow automatic changing of end effectors. Solenoids can be used to release the clamping ring or pin used to hold the end effector or tool in place.

3. Vacuum Handling

Large sheets of metal or plastic can be picked up by using multiple vacuum cups on the end effector. Each cup is attached to the vacuum pump, so that the pressure differential inside and outside the cup is maintained. It is easy to maintain a differential of 10 pounds per square inch with a small vacuum pump.

Calculation of the vacuum cup area required for a desired grasping force is performed simply. As an illustration, assume that a horizontal metal plate weighing 100 pounds is to be picked up. Then the vacuum cup area required on the end effector is calculated from the formula

$$A = \frac{W}{(P_o - P_i)N}$$

where N =   number of vacuum cups on the end effector.
  W =   weight of the plate
  P_o =   outside pressure (standard atmospheric pressure), and
  P_i =   inside pressure.

When $N = 4$, $W = 100$ pounds, $P_o = 14.7$ psi, and $P_i = 4.7$ psi, we find that the required area $A = 2.5$ square inches. To provide a factor of safety of two, we would need 5 square inches per cup. For a circular cup, a radius of 1.26 inches is required.

If the metal sheet is to be turned vertically, the coefficient of friction between the cup and the metal must be considered. Sliding will occur between the vacuum cup and the metal unless the friction force is greater than the weight of the object. This force is calculated, as before, by multiplying the pressure differential due to the vacuum by the area of the cups and the coefficient of friction. Metal plates may be oily or dusty, so the friction coefficient may range from 0.1 to 0.7 depending on the situation. Vacuum cups with an area of A, will be able to pick up a sheet weighing W pounds when the coefficient of friction is μ and the pressure differential is p. The following equation may be used to calculate the weight-handling capability of the vacuum pickup. A factor of safety, s, has been included in the equation:

$$W = A\mu ps$$

Weight is in pounds when *A* is in square inches and *p* is in pounds per square inch.

A typical multiple vacuum cup pickup is shown in Figure 3.18.

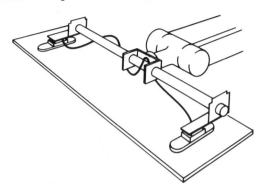

Figure 3.18 Multiple vacuum cup pickup.

Source: Robotics in Practice by J.F. Engelberger, page 52. © *1980 by Joseph L. Engleberger. Published in the U.S. by AMACON a division of American Management Associations, New York. All rights reserved.*

Vacuum pickup is essentially useful for lightweight, thin materials that would otherwise be difficult to handle. It is superior to magnetic pickup even for ferromagnetic materials. It is necessary to consider the surface conditions, however. Oil or dust may make the surface low in friction.

4. Specialized End-of-Arm Tooling

A large variety of specialized toolings may be mounted on a robot wrist. Stud-welding heads and a heating torch are the examples shown in Figure 3.19. Others in common use are arc welding guns, spray paint guns, grinders, drills, and so on.

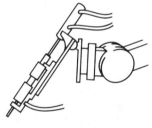

(a) Stud-Welding Head
Studs are fed to the head from a tubular feeder suspended overhead. Except for the mounting bracket, this could be the same tool used for manual operations.

(b) Heating Torch
A useful task for an industrial robot is to bake out foundry molds. Fuel can be saved because the robot holds the flame where it is most needed as long as it is needed. Bakeout is faster than it would be in a gas oven.

Figure 3.19 Tools mounted on robot wrists.

Source: Robotics in Practice by J.F. Engelberger, page 55. © *1980 by Joseph L. Engelberger. Published in the U.S. by AMACON a divison of American Management Associations, New York. All rights reserved.*

3.5 Determining Specifications

Now that we have described the mechanical portions of a robot, we can consider the characteristics that must be specified to ensure that the robot can carry out its assigned tasks. In this section, some of those important characteristics are defined and examined. Accuracy and repeatability are studied in more detail, using our knowledge of dynamics, statics, and physics to consider the most important factors that determine these characteristics.

3.5.1
Definitions of
Mechanical Terms

Several terms are used here which require definition to ensure that we are all attaching the same meaning to them. Many other terms are defined in the Glossary (Appendix C).

1. <u>Accuracy</u>. The absolute measurement of a quantity in relation to a reference base. Positioning accuracy is defined as the difference between the position desired and the position actually achieved. In most industrial robot work, it is given in inches or millimeters. Both positive and negative errors are specified.

2. <u>Cycle Time</u>. the time, in seconds, for a robot arm to carry out a specific operation or task. Typical cycle time specifications are full extension of an arm, rotation from extreme left to extreme right, or the time required to carry out a complete task such as picking up an object from a pallet and placing it in a machine tool.

3. <u>Joint</u>. A mechanism that allows relative movement between parts of a robot arm. Sliding joints allow links (separate arm sections) to move in a linear relationship. These joints are also called "prismatic" because the cross section of the joint is considered to be a generalized prism. Rotary joints allow only angular motion between links. Some joints are designed to allow both rotary and linear motion, however, they can always be considered as two separate joints for analysis.

4. <u>Link</u>. One portion of a robot arm separated by either linear or rotational joints. It is analogous to the human lower arm, upper arm, and so on.

5. <u>Reliability</u>. A measure of the length of time a robot will operate without failure. Mean time between failures (MTBF), stated in hours, is commonly used to indicate the reliability of a device. Another related measure, is mean time to repair (MTTR), also stated in hours. We want MTBF to be a maximum and MTTR to be a minimum, of course.

6. <u>Repeatability</u>. A measure of the difference between successive movements to the same commanded position. Repeatability in many robot arms is very good. Even though the actual position is in error the robot may move to the previous position with only a small difference. Note that the factors causing inaccuracies may not have much effect on repeatability. Deflection due to gravity will cause the same error every time, but the error will be repeatable. Backlash may vary, of course, since it is affected by friction and load effects.

7. <u>Resolution</u>. A measure of the size of the steps between positions. It is usually stated in steps per inch or per centimeter. It has the same meaning for mechanical positioning as optical resolution does in optics, where it may be stated in lines per millimeter. Digital devices, which determine position by discrete steps, always have a finite number of possible steps. Analog devices, in theory, could have an infinite number of steps. Accuracy and resolution are not necessarily related. A device could move to a commanded position with perfect accuracy, but if the resolution was low, it might not be in the position required.

8. <u>Speed</u>. The time rate of change in position. It is often given as a substitute for linear velocity or angular velocity. (See item 9)

9. <u>Velocity</u>. A measure of the speed and direction of motion. It is a vector quantity, which means that both the speed and direction must be stated. Linear velocity is measured in inches per second, or in millimeters per second; angular velocity is measured in degrees per second or radians per second. Direction of motion may be stated in any coordinate system or in terms of angular relationships.

Other mechanical terms are defined in the Glossary.

3.5.2
Design
Considerations

Desirable operational characteristics for a robot must be designed in. Therefore all of the necessary characteristics must be considered as part of the overall system. Clearly, for example, we could attain very high rigidity and accuracy by making all links massive and all joints highly precise, but at the cost of higher cost and slower operation.

1. Operating Cycle

A robot is not very useful if it cannot carry out a task in a reasonable amount of time. Usually a robot must compete with an alternative—a human worker or another type of machine. It must therefore operate rapidly enough to be cost effective compared to the alternative. In practice, this means that a complete operating cycle must take place in a few seconds.

Cycle time can be decreased in several ways. Increased drive power can be used, the mass and moment of inertia of joints and links can be reduced, or the means of control can be improved to perform the operation more efficiently, with less lost motion. All of these approaches are commonly used in design.

Mechanical analysis and simulation can be used to determine the critical parameters, and those parameters can be optimized. As discussed in Chapter 4, there are several types of drive motors available, some with high inertia and some with lower inertia. Whenever possible, the motors with lower inertia and lower weight are chosen.

Commonly, the arm links of a robot are tapered so that they are strong enough to provide the desired rigidity but do not have unnecessary stiffness. One manufacturer, Graco, has built a robot arm from graphite fibers to attain maximum stiffness with minimum weight.

Control methods to optimize the operating cycle are discussed in Chapter 6. Computer control to select the optimum trajectory has been refined so that much greater operating efficiency can be attained. In addition, new mathematical methods have reduced the computation required to determine a trajectory and perform a task.

2. Accuracy and Repeatability

As discussed in Section 3.5.3, several factors affect accuracy. It is important to consider these factors during the early stages of design to ensure that an optimum design is obtained. In addition, great care must be taken in fabrication to ensure that parts are machined to the required dimensions and that the materials selected are suitable for the purpose. There is no substitute for care in design and fabrication. In addition, the individual items must be carefully inspected and tested to be certain that they meet the design specifications.

Dimensional stability in mechanical parts should be considered. Metals can bend and creep during fabrication. Although they are initially machined to the correct dimension, there are changes due to stress release that can cause errors in the finished parts. These errors can be detected and eliminated by good shop practice and careful heat treatment.

3. Life Expectancy, Reliability, and Maintainability

Robots are expected to have a useful life of at least 40,000 working hours. Well-designed robots are expected to have an MTBF of at least 400 hours and an MTTR of no more than 8 hours (Engelberger [6]). In practice, over the last 10-year period, the Unimate® robots, for example, have demonstrated 98% up-time. That is, during the time when they are expected to be available, they have been available and ready to work 98% of the time. Up-time does not include the scheduled periods for preventive maintenance.

One means for decreasing MTTR is to modularize the design so that parts are easily replaceable and a minimum number of parts is required. Then, with a small stock of spare units, it is possible to do maintenance quickly by replacing the unit. In this way, the skill required for maintenance is reduced to the ability to replace the component. Rebuilding of components can be done at the factory by skilled personnel. Yearly maintenance costs are expected to be about 10% of the acquisition costs of a robot and associated equipment when the robot is working two shifts or 4,000 hours per year. This important subject is discussed in considerable detail in Chapter 5 of Engelberger [6].

3.5.3 Mechanical Positioning Accuracy and Repeatability Factors

Six major items affect positioning accuracy and repeatability in a robot:

1. The force of gravity acting on the arm members and load of the robot causes downward deflection of the arm and the support system.
2. Acceleration forces may act in any direction. Noticeable horizontal and vertical deflections occur when heavy loads are being accelerated.
3. Drive gears and belt drives often have noticeable amounts of slack that can cause positioning errors.
4. Thermal effects can expand or contract the links of the robot arm. In large robots, this effect can be of considerable magnitude.
5. Even bearing "play" can be significant when very high positioning accuracies are desired.
6. Windup can be important when long rotary members are used in the drive system and twist under load.

Errors due to sensor and control errors are considered in Chapters 4, 5, and 6 and will not be covered in this chapter.

The following sections provide a simplified analysis of the major factors affecting accuracy and repeatability.

1. Gravitational effects

To approximate the effect of gravity on a robot arm, we consider a steel cantilever beam of length, L, width, B, and height, H, fixed at one end and with a force, P, applied at the free end due to gravitational force on the load. Deflection D of this beam due to the force of gravity is given by equation (1).

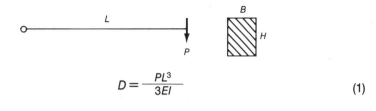

$$D = \frac{PL^3}{3EI} \qquad (1)$$

where E = Young's Modulus, 30.0×10^6 pounds per square inch for steel and $I = BH^3/12$ is the moment of inertia for a rectangular beam of width, B, and height, H taken about a horizontal line through its center. All dimensions are in inches, and the force is in pounds. (We are neglecting the weight of the beam in this simple example, which is used only to illustrate the error due to gravitational effects.)

We will use the following values in equation (1) to ascertain the magnitude of the deflection of a robot arm under a load:

P = 100 lbs.   Load on the end of the arm.
L = 60 inches   Length of the arm
B = 4 inches   Width of the arm link.
H = 6 inches   Depth or vertical dimension of the arm.

Inserting these values in equation (1), we find that $D = 0.0033$ inch, which is an acceptable value for the deflection. Note that this is a repeatable deflection obtained whenever the same conditions apply. In this case, the inaccuracy due to deflection is 0.0033 inch, but the repeatability error may be only a few ten-thousandths of an inch if the same load is applied again. Since an arm made up of multiple links and joints has an effective horizontal length that is changing, we can expect the error due to gravitational effects to vary as well. Note that the deflection is a function of the cube of the depth of the arm. Where we had a deflection of 0.0033 inch for H = 6 inches, the deflection would increase to 0.0111 inch with H = 4 inches. This deflection error may not be acceptable.

2. Acceleration Effects

In a cylindrical coordinate system, for example, an object being carried around the vertical with an angular velocity of w is subject to a radial force (assuming a point mass)

$$F_r = mrw^2 \qquad (2)$$

where $m = W/G$ is the mass of the object being carried,
W = weight of the object, in lbs.,
G = the gravitaitonal attraction, 32.2 feet per square second,
r = the radius of the path of the object's center of mass,
w = the angular velocity of the end of the arm in radians, and F_r
is given in pounds when the radius is in feet.

Angular deflection, D_a due to angular acceleration a_a is horizontal and of magnitude: $$D_a = (mra_aL^3)/(3EI) \qquad (3)$$

where $I = (B^3H)/12$ for the horizontal moment of inertia when B is the width of the beam, H is the depth, and other terms are as previously defined. Since the radius and the length of the arm are the same, we can replace Equation (3) with $D_a = (ma_aL^4)/(3EI)$.

In normal operation this deflection is neglibile, but it becomes important in some types of high-speed operation where the end of the robot arm starts and stops quickly. The resulting whipping motion takes some time to settle to zero position error and may not be zero before the arm is moved again, thus introducing a position error.

When a robot arm is designed, this effect must be taken into account because of the side force due to acceleration on the components of the arm itself.

3. Backlash Error

Backlash due to gearing or belt drives is another source of error in attempting to get high positioning accuracy and repeatability.

Gears do not mesh perfectly but have spaces between them, as shown in Figure 3.20. Note that the input member has spaces, D, between each side of the tooth and the output member. It is therefore necessary for the input member to move distance D before it can cause the output member to move. Any error at this point is multiplied by the gear ratio between the input gears and the final position of the robot arm or other member. Chains, belts, and other drives have the same kind of potential error. Some of this error can be taken out of the system by using very precise gearing or anti-backlash gearing, which is spring loaded to hold the input gear against one side of the driven gear. This solution is limited to the available strength of spring restraints.

Figure 3.20
Model of backlash.

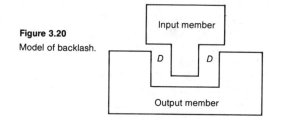

Backlash, with good gearing, can be held to less than 0.010 inch, but it becomes a difficult problem when positioning accuracies of 0.001 inch are desired.

In addition to the positioning errors caused by backlash, it is a potential source of oscillation due to the introduction of phase errors in the servo system.

4. Thermal Effects

Several robot manufacturers quote positioning accuracy and/or repeatability specifications of 0.001 or 0.002 inch. It is important to understand the conditions under which these specifications are valid. It is easy to calculate the amount of variation in the fully extended arm as a function of temperature using the equation:

$$V = Lc(T_2 - T_1)$$

where L is the length of the arm, c is the coefficient of linear expansion per degree, T_2 is the final temperature, and T_1 is the initial temperature.

As an example, we can use the following values in this equation:

L = 60 inches
c = 6.5 X 10^{-6} per $^{\circ}$F, for steel.
T_1 = 60°F
T_2 = 80°F

Therefore

$$V = 60 \text{ X } 6.5 \text{ X } 10^{-6} \text{ X } (80 - 60)$$
$$= 7.8 \text{ X } 10^{-3} = 0.0078 \text{ inch}$$

This is enough movement to cause a significant error. There may be some compensation due to other thermal expansions, but the net effect is large enough to be worth careful analysis.

5. Bearing Play

Bearing play is usually an insignificant source of error. Good bearings can hold tolerances of 0.0001 inch. In ordinary industrial robots, therefore, we can neglect this source of error. However, when a linkage revolves about a bearing and has a long lever arm, the possibility of bearing play should be considered.

6. Windup

Windup is the angular twisting, under load, of rotary drives or shafts. When windup is outside the servo loop, it must be kept small. Inside the servo loop, it is of less concern. The amount of windup can be calculated from

$$\Theta = \frac{32LT}{3.1416 \, D^4 G}$$

where L = length of the rotating member, T = torque, in inches per pounds, D = diameter, in inches, G = shearing modulus of elasticity (psi), and Θ = amount of twist in radians.

The shearing modulus of elasticity, G, is 12.5 X 10^6 psi for steel.

3.6 Mobility

As discussed in Section 1.2.4, several types of mobile robots are being developed; robots with wheels, tracks, and legs have been made. In addition, there are the robots mounted on tracks and overhead rails, which are in daily use in industry.

In this section, we will discuss freely moving robots capable of moving around a factory floor on wheels or some modification of wheels. There appears to be no immediate need, in the factory environment, for vehicles with legs. There is a need for mobility for robots to allow them to move from one machine to another or to move around an object being welded, painted, or fabricated. Instead of bringing the work to the robot, it would be desirable to move the robot to the work in many cases. Also, a robot may be able to do more work if it is able to move from one place to another instead of being forced to stay in one location at all times. When not busy on one task, it can move to another task, just as a human worker would.

To move around objects and from one job to another, the robot must be able to relocate itself in respect to an object. It must also be able to navigate around obstacles in its path when moving. Computer vision, as described in Chapter 9, is one way to provide these capabilities. In Chapter 2, the Shakey robot was described; it had a vision system and could navigate around obstacles. In this section, we will discuss only the mechanical considerations of mobility and leave the sensory, navigation, and other problems to other chapters.

A mobile vehicle capable of carrying a robot must be able to support the weight of the robot and to ·maintain stability as the robot moves around, carries a load, or works on a task. This means that the vehicle must either be counterbalanced against the moments and forces that might occur or be heavy and rigid enough to withstand the forces and torques that might be applied. Since it is much simpler to make the mobile vehicle heavy and rigid than it is to compute and supply the necessary counterbalancing forces, the former is the method most commonly used to provide robot mobility.

Maneuverability in all directions is desirable for a robot. Use of four wheels, two fixed and two steerable, as in an automobile, restricts the flexibility of the robot's path. Therefore, in mobile vehicles designed for robot use, attention has been concentrated on specialized, highly maneuverable wheels and wheel suspension systems.

**3.6.1
Unimations–
Stanford
Mobile
Vehicle
Project**

A survey of alternative designs has been reported by Brian Carlisle [4]. He lists the following alternatives:

1. Three Wheel Design with Rollers. This design, which has been patented, uses three wheels mounted on the points of an equilateral triangle. Each wheel is made up of rollers, as shown in Figure 3.21, that can roll freely. The vehicle can move forward by rotating the wheels or is free to move sideways on the rollers. It is a flexible design that allows the vehicle to move freely in any direction by driving two or three wheels at one time. All wheels are supplied with drive motors. A problem with this design is the reduced stability against tipping. Also, the stability effect varies with the wheel positions with respect to the load. There is also some wobbling due to the change in the point of roller contact as the wheel shifts from one roller to another.

Figure 3.21 Three wheels with rollers (patent no. 4,237,990).

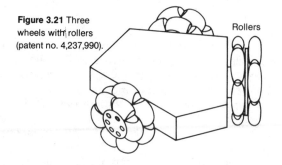

Rollers

2. Four-Wheel Design with Rollers. Four wheels provide greater stability but require the use of a wheel suspension system to ensure that all four wheels are in contact with the ground. It is also more expensive because it requires four drive motors and more complex control.

Other designs discussed by Carlisle [4] do not meet the requirements for stability, omnidirectional motion, and controllability that are the goals of the project. Therefore, the Unimation-Stanford group redesigned the three-wheel vehicle to reduce the gap between rollers, minimize wheel thickness, and improve stability. A unique feature of their design is the use of both long and short rollers, with the long rollers tapered at each end to fit within the concave ends of the short rollers so that the gap between rollers is very small. Softer rubber was used on the rollers so that there was sufficient compliance to allow transfer between rollers without wobbling.

**3.6.2
Moravec's
CMU Rover**

Hans Moravec at CMU has worked on the problem of mobile robot vehicles for several years (Moravec [9]). His current design uses an omnidirectional steering system with three pairs of driven wheels,

Figure 3.22 Rover wheelbase.
Courtesy Carnegie-Mellon University Robotics Institute. Used by permission.

as shown in Figures 3.22 and 3.23. With this design, the rover's trajectory will be an arc about the point in the floor plane where lines through the axles of all three wheels intersect at the center of the arc.

A unique feature of the CMU Rover's design is its ability to go through a door by moving in a straight line and simultaneously rotating

Shaft encoder

Drive motor

Shaft encoder

Steering motor

Differential gear

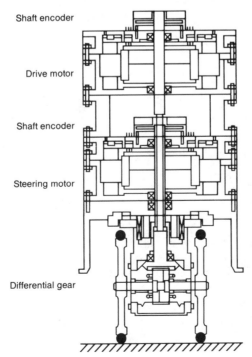

Figure 3.23 Rover wheel assembly.

Source: H.P. Moravac, "The Stanford Cart and the CMU Rover," from Proceedings of the IEEE, Volume 71, Number 7, July 1983, page 880, Figure 11. © 1983 IEEE. Used by permission.

about the vertical axis. This design allows a single arm mounted on the Rover to open the door and hold it open while the Rover glides through. This maneuver is controllable by a complex computer system mounted in the Rover. There is a 6805 computer on each wheel of the three pairs of wheels (six in all) plus a 68000 computer directly controlling the wheels. Several other computers are used for sensors, visions, and overall control of the Rover.

3.6.3
The TO-ROVER

Tokano [13] reports on the TO-ROVER mobile robot, which is used for the mooty maintenance robot system. This robot is being developed at the University of Tokyo for use in nuclear power plants for automatic inspections and repairs. It has a sensory system and a control system plus a unique wheel design so that it can climb stairs or roll along the floor with almost equal ease. Figure 3.1 shows the Mooty robot climbing stairs. Figure 3.24 shows more detail of the wheel mechanism design. Note that one set of wheels is wider than the other. This allows the mechanism to climb stairs and fold, as shown in Figure 3.25. Steering is done only by controlling the difference in velocity between the left and right wheels.

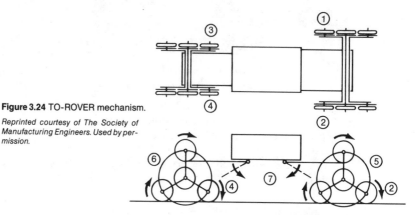

Figure 3.24 TO-ROVER mechanism.

Reprinted courtesy of The Society of Manufacturing Engineers. Used by permission.

Figure 3.25 Prototype TO-ROVER.

Reprinted courtesy of The Society of Manufacturing Engineers. Used by permission.

These different solutions to the mobile robot problem are amazingly similar. One or more of them should be available commercially within the next 5 years.

Exercises

These exercises test your knowledge of the information covered in Chapter 3. Some of the more difficult questions may require you to refer to a text from your local library, as noted.

1. Name and describe the four most commonly used robot coordinate systems.
2. How is a robot manipulator work envelope defined?
3. What are the major advantages of the rectangular coordinate robot?
4. Revolute coordinate robots are being used increasingly today. What disadvantage caused other configurations to be used instead in the early robots?
5. What type of interference can occur in a cylindrical coordinate robot that could be potentially dangerous?
6. What is the lowest friction linear drive mechanism used in robots? How is this low friction attained?
7. If a gear train has a ratio of drive gear to driven gear of n, what is the effective moment of inertia of the driven mechanism as reflected in the drive mechanism?
8. If a positioning accuracy of 50 thousandths of an inch is required, what is the least expensive drive that can be used?
9. A harmonic drive has a circular spline with 80 teeth, and the flex spine has 2 teeth fewer. What is the effective gear ratio of the drive?
10. What is the drive mechanism that has the lowest backlash?
11. How does the Cincinatti Milacron three-roll wrist achieve its small size and concentric operation?
12. In what mode are brakes used on robots? Why are they used in this way, and how is this mode of operation achieved?
13. Can a brake be used to improve the effective accuracy of a robot?
14. Name five types of end effectors for robots. What is the advantage of each?
15. What is the maximum holding force that can be exerted by a vacuum cup 2 inches in diameter?
16. Define accuracy and repeatability. What is the significant difference between them?
17. In what ways can the operating cycle time of a robot arm be decreased?
18. What is the meaning of MTBF? What value is typical for today's robots?
19. In your opinion, what are the two largest mechanical causes of inaccuracy in robots?
20. If you had a steel robot arm with a known repeatability of 0.001 inch at constant temperature and the arm was 40 inches long when fully extended, over what temperature range could you expect to achieve a repeatability of 0.003 inch due to all causes? Use Farenheit temperatures. ($c = 6.5 \times 10^{-6}$ per °F for steel.)
21. A load of 20 pounds is placed on the end of a robot arm. Assume it is a point mass. If the arm is 25 inches long when fully extended , what vertical accuracy could be attained? Assume the arm has a solid cross section 3 inches high by 3 inches wide and is made of aluminum. Young's modulus for aluminum is 10×10^{7} pounds per square inch. If you include the weight of the arm, what answer would you get? (The density of aluminum is 175 lbs. per cubic foot.)

22. What wheel configuration is used for the Unimate–Stanford mobile robot? Why was this configuration chosen?
23. What three wrist articulations does a robot require? Name and briefly describe them.
24. Name five important qualifications for robot end effectors and discuss the reasons why they are important.
25. Classify the following industrial robots by configuration. That is, do they operate in revolute, polar, cylindrical, or rectangular coordinates? See the manufacturer's brochures or Tanner, [12], vol. 1, p. 21.

Robot Name	Coordinate System
Asea IRb-60	
Unimate® (all)	
Unimation Puma (all)	
Cincinatti Milacron T^3	
DeVilbiss TR-3000	

27. Following is a table of specifications for four industrial robots. Fill in the table to indicate the approximate values for the specifications given. These specifications will be found in manufacturer's brochures or in Tanner [12]. (In case of conflict, use the most recently published values.)

	Auto-Place	Puma 250	Unimate 2000B	Asea IRb-60
Repeatability (inches)				
Load Capacity (lbs.)				
Power Required (kilowatts)				
Type of Drive				
Cost ($000's)				

References

[1] "A Look at the New Bendix Robots," Robotics Today, August 1982, pp. 49-50.
[2] H. Asada, T. Kanade, and I. Takeyama, "Control of a Direct Drive Arm, CMU-RI-TR-82-4, The Robotics Institute, Carnegie-Mellon University, March 9, 1982, pp. 1-15 plus figures.
[3] Bulletin B-58, Beaver Precision Products, Inc., Warner Electric Brake and Clutch Company, Troy, Mich., 1970.

[4] B. Carlisle, "An Omni-Directional Mobile Robot," unpublished paper, presented at the workshop, "The Future of Robotics Research and Development," Palo Alto, Calif. Feb. 10-11, 1983. Sponsored by the University of Calif. at Berkeley, Davis, Los Angeles, and San Diego. Coordinated by R. C. Dorf, University of California at Davis.

[5] Conference Proceedings, 13th International Symposium on Industrial Robots and Robots 7, April 17-21, 1983, Chicago, Ill. Robotics International of SME, Dearborn, Mich., 1983.

[6] J. F. Engelberger, Robotics in Practice, AMACOM, Division of the American Management Association, New York, 1980.

[7] Harmonic Drive Designer's Manual, Harmonic Drive Division, Emhart Machinery Group, Wakefield, Mass., 1983.

[8] J. Y. S. Luh, W. D. Fisher, and R. P. C. Paul, "Joint Torque Control by a Direct Feedback for Industrial Robots," IEEE Transactions on Automatic Control, vol. AC-28, no. 2, February 1983, pp. 153-161.

[9] H. P. Moravec, "The Stanford Cart and the CMU Rover," Proceedings of the IEEE, vol. 71, no. 7, July 1983, pp. 872-884.

[10] J. C. Quinlan, "Robot Wrists and Grippers," Tooling and Production, January 1983, pp. 85-88.

[11] M. E. Rosheim, "Robot Wrist Actuators," Robotics Age, Vol. 4, No. 6, November-December 1982, pp. 15-22.

[12] W. R. Tanner, ed.: Industrial Robots, 2nd ed., Vol. 1, Fundamentals, Vol. 2, Applications. Robotics International SME, Dearborn, Mich., 1981.

[13] M. Takano and G. Odawara, "Development of New Type of Mobile Robot TO-ROVER," University of Tokyo, Tokyo, Japan, Reported in [10], pp. 20-81 to 20-89.

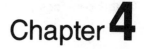
Drive Methods

An important distinguishing feature of robots is the drive method chosen. Drives are the source of motive power that drives the links to a desired position. Usually power is applied at the joints, either directly or through cables, gearing, belts, or other means. There are four major types of drive in use today: hydraulic, pneumatic, DC electric motors, and electric stepper motors. Each will be covered in detail in a separate section. Next, the criteria for selection of drive types are outlined, and the advantages and disadvantages of each type for different applications are discussed.

4.1 Hydraulic Drives

Hydraulic drives in robots use oil under high pressure as the working fluid. Drives may be open loop or closed loop and may be either linear or rotary. A simplified diagram of a hydraulic cylinder controlled by a servo valve is given in Figure 4.1.

Open loop drives can go from point to point accurately but cannot be controlled to stop at points in between. They come under the classification sometimes called <u>bang–bang</u> because they move from one position to another and "bang" up against a stop at the other end of their movement.

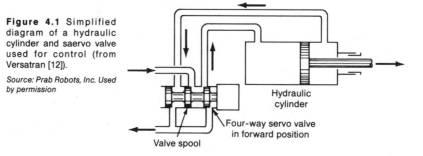

Figure 4.1 Simplified diagram of a hydraulic cylinder and saervo valve used for control (from Versatran [12]).

Source: Prab Robots, Inc. Used by permission

Hydraulic cylinder

Four-way servo valve in forward position

Valve spool

4.1.1 Linear Drive Cylinders

Linear hydraulic cylinders controlled by solenoid–operated valves are the simplest and least expensive open-loop hydraulic drives. It should be noted that a controlled stop is possible in linear cylinder operation by providing a controlled orifice to slow down the flow of oil when the piston approaches the end of its motion. Also, many pieces of equipment are controlled by manual valves so that controlled movement is possible. In this case, the human operator becomes part of the closed–loop system, and it is no longer an open–loop system. Lift masts on fork lift trucks and construction equipment such as back hoes are of this type.

Hydraulic pistons mounted in large–diameter cylinders are inexpensive to make and are capable of exerting large forces in a small space. Oil pressures of the order of 2,000 pounds per square inch are used. Therefore, a piston one inch in diameter can exert a force of 1,570 pounds.

Both linear and rotary hydraulic drives operate due to the pressure of oil—against a piston in the linear cylinder or against a rotary vane in the rotary hydraulic motor. Oil is admitted to one end of the piston by the action of the hydraulic valve, as illustrated in Figure 4.1. Valves in open-loop systems are opened and controlled by magnetic solenoids; closed-loop systems use electromagnetic servo valves or manually operated valves.

One of the most popular robots, the Unimate, used hydraulic drives exclusively for many years. Another early robot, the Versatran, illustrated in Figure 2.3, used hydraulic cylinder drives in a cylindrical coordinate configuration.

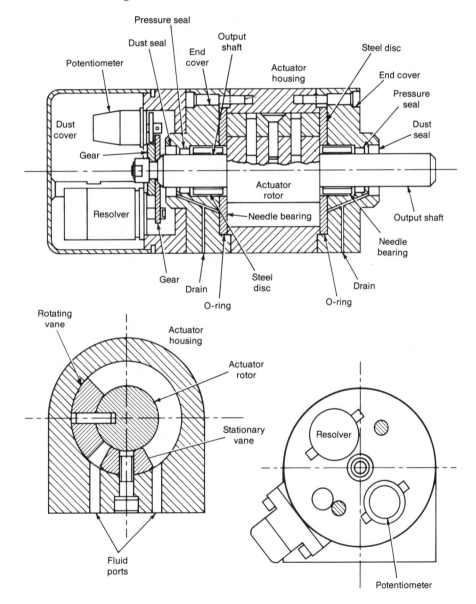

Figure 4.2 Rotary Actuator (from the TR-3500 Equipment Manual [15]).

Source: The DeVilbiss Company. Used by permission.

**4.1.2
Rotary
Actuators**

A well–designed rotary actuator is shown in Figure 4.2. The actuator housing is made from solid aluminum alloy, and the actuator rotor is machined from steel. Pressure seals and dust seals are used to contain the hydraulic fluid and to protect bearings, respectively. Oil is allowed to enter through the fluid ports, under control of the electrohydraulic valve, and impinges on the rotating vane fixed to the actuator rotor, causing the rotor assembly to move. A stationary vane blocks the direct oil path and forces oil to operate the rotating vane. Position information is obtained by both a potentiometer and a resolver operated by an antibacklash gear train. Coarse position is supplied by the potentiometer signal, while fine position is measured by the resolver. In this way, the high accuracy and limited range of the resolver are augmented by the lower accuracy and larger range of the potentiometer. Of course, the overall accuracy is no greater than that of the gear train that drives the potentiometer and resolver.

**4.1.3
Electrohydraulic
Control Valves**

Electromechanical or electrohydraulic servo valves (both terms are in common use) are actually quite complicated. There are two major types: the flapper valve shown in Figure 4.3 and the jetpipe valve shown in Figure 4.4. Most industrial robots today use the flapper valve but the jetpipe valve, which has been more expensive, is now being used. Its manufacturer states that it has higher reliability and greater efficiency than the flapper valve (Atchley [1]).

Figure 4.3 Flapper servo valve cross section (from the TR-3500 Equipment Manual [15]).

Source: The DeVilbiss Company. Used by permission.

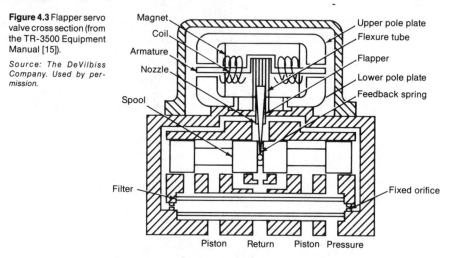

In either type of valve, it takes only a few milliseconds to reverse the flow of fluid in the system. Each valve has a torque motor, a hydraulic amplifier first stage, and a four-way spool valve second stage. The torque motor has an armature that moves either a flapper valve or a jetpipe assembly to control the flow of oil to the second stage, which controls the movement of the spool valve that actually controls the large oil flow to the driving cylinder or rotary motor. A relatively small current flow in the torque motor controls a flow of oil that causes the spool valve to move to control a large oil flow.

1. Flapper Servo Valve

In a flapper servo valve, the flapper is rigidly attached to the midpoint of the armature. The flapper passes between two nozzles, creating two variable orifices between the nozzle tips and the flapper.

An electrical signal generates a magnetic field that moves the armature and the flapper to open one orifice and close the other. This action causes an oil flow that sets up a differential pressure on the ends of the spool valve and causes the spool valve to move. As the spool moves, it bends the feedback spring that opposes its motion. When the force due to oil pressure differential equals the spring force, the spool valve stops moving. Movement of the spool valve opens the main piston oil paths and allows oil to flow, driving the piston in the desired direction.

Figure 4.4 Jet pipe servo valve (from Atchley [1]).

Source: R.D Atchley, "A More Reliable Electro-hydraulic Servovalve," Robots 6 Conference, March 2-4, 1982, page 164, Figure 1. ©*1982 Society of Manufacturing Engineers. Used by permission.*

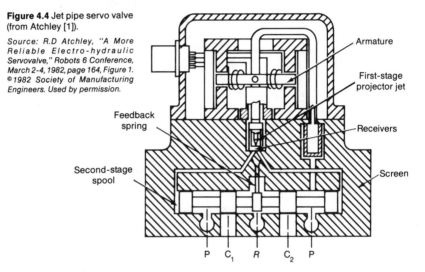

Armature

First-stage projector jet

Feedback spring

Receivers

Second-stage spool

Screen

P C_1 R C_2 P

2. Jetpipe Servo Valve

A jetpipe servo valve differs from the flapper servo valve in the way oil flow to the spool valve is controlled. When the torque motor is energized, it causes the armature and jetpipe assembly to rotate. More fluid is injected into one receiver than the other, causing the spool to move. Otherwise the action is essentially the same. An advantage of the jetpipe valve is that the openings through which oil passes are larger and less subject to blockage by tiny metal particles in the oil.

One of the major problems in hydraulic systems is the need to filter the oil in order to remove contaminants. Despite extreme care in manufacturing, it is possible for particles a few microns in diameter to break off from welded spots or rough spots in the cylinders, tubing, and pistons. Careful filtration and frequent cleaning of the filters are required to reduce the potential for blockage of the sensitive servo valves.

4.1.4 Closed-Loop Servo Control

Precise control of a hydraulic system is obtained by use of a complete closed-loop servo system, as shown in Figure 4.5. An electrical servo amplifier, usually an operational amplifier, provides an electrical signal to the electromechanical servo valve located in the hydraulic fluid system. The pilot valve which is controlled electrically in turn controls one or two stages of hydraulic amplification, generating enough power to drive the mechanical linkage.

A feedback potentiometer, resolver, or encoder mounted on the mechanical drive generates an output signal that is fed back to the comparator. In the "repeat" mode, the encoder output is compared to the output from the memory drum, and an error signal is generated that is sent to the servo amplifier to act as a control signal. When the encoder output signal is equal to the command signal from memory, the servo valve is allowed to close and the mechanical motion stops. In a computer–controlled system, the comparator and memory are part of the computer but perform the same types of functions.

Included in Figure 4.5 are the teach control, the memory drum, and the comparator used for teaching, as described in Section 4.1.5. Only one degree of freedom is shown for the robot.

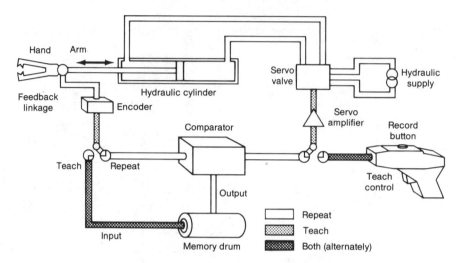

Figure 4.5 Simplified diagram of a hydraulic servo system (Unimate).

Source: UNIMATION, Inc. A Westinghouse Company. Used by permission.

**4.1.5
Training a
Hydraulic
Robot**

In training a hydraulic robot, the pressure in the hydraulic system is released by auxiliary valves or other means so that the robot operator is not working against the powerful actuators. In some systems, the "Train" switch operates all of the necessary auxiliary and bypass valves to allow the operator to guide the robot manually. Even though the hyraulic pressure has been released, a large robot arm would be difficult to move. Therefore, in some systems, springs are used to counterbalance the weight of the arm.

As the arm is moved through a cycle by the operator, he or she can record the coordinates of each joint at selected positions along the path or can have the control computer take sample coordinates periodically along the path. Coordinates are taken for all of the joints simultaneously and stored either in the computer memory or on the magnetic drum or disk used in the system. After training is completed, the operator can switch to the repeat mode and watch the robot go through the same sequence of steps just completed during the training mode (including all of the mistakes made.)

A further refinement in some of the newer robots is the use of a complete training arm that duplicates all of the position sensors of the actual robot arm and is much easier to move. Even when the robot arm is counterbalanced, it still has a considerable mass that is difficult to start and stop smoothly when training. The lightweight training arm avoids this difficulty and allows the operator to train the robot with greater speed and precision. During training, it is mounted beside the actual arm and at a measured distance from it. Compensation for the separation between arms is provided by a computer program that calculates the necessary correction for all trained movements.

**4.1.6
Hydraulic
Power
Source**

As illustrated in Figure 2.3, the initial power source for a hydraulic system is a powerful constant-speed electric motor driving a hydraulic pump. Oil under pressure is stored in an accumulator at pressures on the order of 1,500 to 2,000 psi. Cooling of the oil under pressure is provided by water from the local water supply, which is piped through the accumulator and back to the drain.

Filtering of the oil is necessary to prevent contamination in the oil from affecting the sensitive electrohydraulic valves, drive cylinders, and rotary actuators. Particles only a few microns in diameter break loose from the interior of the hydraulic system and cause jamming or scoring of valves and cylinders if not quickly removed.

**4.1.7
Advantages
and
Disadvantages
of Hydraulic
Drive**

Electrohydraulic servo valves used to control oil flow are quite expensive and require the use of filtered, high-purity oil to prevent jamming of the servo valve. In operation, the electrohydraulic valve is driven by a low-power electric servo drive that is relatively inexpensive. This low cost has offset the cost of the servo valve itself and the expensive and somewhat messy hydraulic tank, tubing, and filter system. Because of the high pressure, there is always a potential for oil leaks. Oil at 2,000 psi can quickly cover a large area with a film of oil, so this is a problem to be watched. As a result, the tubing and fittings used are expensive and require excellent maintenance for reliability.

Since the hydraulic cylinder provides precise linear motion, it is advantageous to use linear drives in the robot as often as possible. However, hydraulic rotary vane motors have been designed that provide good operation, although at somewhat higher cost. Rotary hydraulic motors can be made smaller for a given power output than electric motors. This is an advantage when the motor must be carried by a robot arm such as in a jointed configuration. The size and weight advantage is offset somewhat by the need for rotary joints to carry the high-pressure oil to other parts of the arm. Also, the newer designs of electric motors are beginning to be competitive in size and weight because of the use of new magnetic materials. Although expensive, the electric motors are more reliable and require less maintenance.

Perhaps the most significant advantage of hydraulic drives over electrical drives is the intrinsically safe operation. In explosive atmospheres such as in paint spray booths, there is a rigid requirement for safety. OSHA requires that electric equipment in certain defined areas carry no more than 9 volts because of the possibility of arcing and explosive action. Hydraulic systems do not have a problem with arcing and are now chosen almost exclusively for use in explosive atmospheres. An alternative would be the use of sealed electric motors, but at present the cost and weight of such motors are prohibitive for the power required.

4.2 Pneumatic Drives

Pneumatic drives, the simplest of all drives, are widely used in industry for many types of applications. Many of them are not robots by any of our definitions, but there are also many pneumatic systems that can be classed as robots. Both linear cylinders and rotary actuators are used to provide the motion required.

Several robot manufacturers are making very flexible robots using pneumatic drives. These are much like hydraulic systems in principle, but vary considerably in detail. The working fluid is compressed air, but the valve is simpler and less expensive and the pressures are much smaller.

Most pneumatic drives provide motion between stops. Air is so compressible that fine control of motion is difficult. Even when high-pressure air is applied to the opposite end of a piston, the inertia of the piston and load causes it to continue to move until it hits a mechanical stop or the air force finally overbalances the inertial force.

Mechanical stops can be used to provide high accuracy of positioning in pick-and-place application. Accuracies of 0.005 inch are achieved readily. Cushions in the air cylinder or associated with the mechanical stops slow the pneumatic cylinder at the end of its motion to prevent damage to the equipment or parts being handled. With this simple, easily understood, and easily programmed equipment, it is possible to do a large number of pick-and-place tasks. Pick and place is used to describe the operation of picking up a part at one location, moving to a new location and placing the part for a particular operation. This simplicity of operation is one of the great virtues of pneumatic systems.

High precision is difficult to achieve using pneumatic servo drives, but where the precision is adequate, pneumatic drives are lightest in weight and lowest in cost of all available robots.

One manufacturer, International Robomation/Intelligence, has developed a new approach to pneumatics using air servo motors. These are vane-type motors controlled closely by individual micro-processors. The manufacturers expect to obtain high positioning accuracy. The maximum load to be handled is 50 pounds. Low cost is the major advantage of this technique. The manufacturers are quoting a price of less than $10,000 per complete robot arm, including the computer control system. Repeatability of 0.040 inch is specified. Such a robot will be very competitive with the older hydraulic and electric motor-drive robots if it can be made to work with high accuracy and reliability.

Modularity is one of the biggest assets of pneumatic drives. Since the operating fluid is air, it is easy to run many air hoses to separate drives and build up an arbitrarily complex system from standard parts. This modularity is illustrated in the Auto-Place Series 50 robot in Figure 4.6.

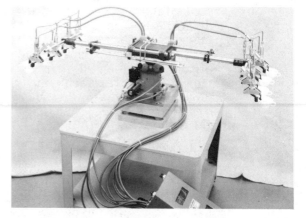

Figure 4.6 Auto-Place Series 50 robot with two parallel arms used to transfer eight hot glass parts.

Courtesy of Rimrock Corporation. Used by permission.

Power for the pneumatic system is supplied by compressed air from a good-quality air compressor. This air supply is shared by all the modules of the pneumatic system through a common air chamber called a <u>manifold</u>. Solenoid valves mounted on the manifold control the air flow to individual pneumatic actuators. In the simplest systems, the solenoid valves are controlled by stepping switches, limit switches, or part sensing switches. Several actuators can be assembled to provide three to six separate motions. Cylinders of different lengths can be used, for example, to build up a positioning mechanism.

Programmable controllers are frequently used for control with pneumatic systems. These controllers are usually microprocessors programmed to emulate relay systems. Relay systems are well understood by a generation of shop foremen and electricians. They find it easier to plan a relay sequence and enter it into the programmable controller than to learn to program a microprocessor. As a result, many simple control operations in factories use programmable controllers.

Pneumatic robots can be trained like other robots. Point-to-point operations can also be controlled by a pendant controller, which has control switches to move the robot through a sequence of tasks.

4.3 DC Electric Motors

We distinguish between electric direct current (DC) motors and stepper motors because of their inherently different methods of operation. DC motors run continuously in one direction, reverse perhaps, and run continuously in the opposite direction. Motion is smooth and continuous. There is no inherent control of position in a DC motor. Stepper motors are able to step precisely to a designated position but have other disadvantages, as described in Section 4.4.

Precise control of positioning in DC motors requires that a closed-loop servo be used with some type of positional feedback. Because of the closed-loop control, the smooth operation possible, and the ability to generate large torques, electric DC motors are used in those robots in which precise control and high power are required.

DC motors are driven and controlled either by relay switching or by the use of power amplifiers that electronically switch the direction of current flow through the motor armature to reverse the direction of operation. Either the field or armature current or both can be controlled.

DC motors today can achieve very high torque-to-volume ratios, much higher than those of stepper motors, and are competitive with hydraulic motors except for very high power drives. They are also capable of high precision, fast acceleration, and high reliability. The present DC motors have benefitted from the development of rare earth magnets, which have made possible large magnetic fields in compact motors. Also, the improvements made in the fabrication of brushes and commutators have made DC motors reliable. Another important factor is the improvement in the power-handling capacity of solid-state devices, which has made possible the control of large currents at a reasonable cost.

4.3.1 [*]
Types of
DC Electric
Motors

DC motors are torque transducers that convert electric energy to mechanical energy. The torque developed by the motor shaft is directly proportional to the magnetic field flux in the stator field and the current in the motor armature. In practice today, the stator field

---

*Information in this section closely follows Kuo [9] with additional considerations introduced by Koren and Ulsoy [7].

in DC motors used for robots is nearly always a rare earth permanent magnet (PM) that provides a strong, stable magnetic flux in which the armature shaft rotates. Controllable field motors with wound field coils are used so rarely that we will not discuss them in this text. Interested readers are referred to standard texts on control such as Kuo's [9].

Three types of PM motors are in use, each with specific advantages and disadvantages. These are discussed in the following sections.

1. Iron-Core PM DC Motors

Iron-core PM DC motors have a laminated iron rotor structure that has slots, as in conventional DC motors. Armature conductors or windings are placed in these slots. On the periphery of the motor is the permanent magnet material, which may be made of Alnico, ferrite, or a rare earth alloy. Motors of this type have high inductance, high inertia, high reliability, and low cost.

Due to the action of the slots, these motors have "cogging," or a tendency to start and stop as each slot passes through the edge of the magnetic field. Since the inertia is high, they cannot be controlled to start and stop as precisely as is sometimes required for robot use. Figure 4.7 shows this motor configuration.

Figure 4.7 Iron-core PM DC motor in cross section (from KUO [9]).

Source: Benjamin C. Kuo, Automatic Control Systems, 4th Edition, ©1982, pages 178, 179, and 180. Used by permission of Prentice-Hall Inc. Englewood Cliffs, N.J.

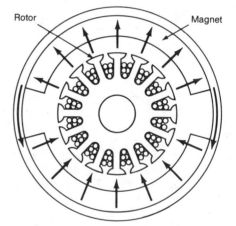

2. Surface-Wound PM DC Motors

In an attempt to develop more responsive motors for control tasks, engineers developed the surface-wound DC motors. Armature conductors are not placed in slots but are bonded to the surface of a cylindrical rotor. The rotor is made up of iron laminations to reduce the flow of eddy currents induced in the motor, as in the iron-core motor. However, since there are no slots, there is no cogging effect. These motors have higher inductance, higher cost, and larger outside diameters, and require a more powerful magnet than the iron-core motor. Figure 4.8 shows this motor configuration.

3. Moving-Coil PM DC Motors

In order to attain very low armature inductance and small moments of inertia, engineers eliminated all magnetic material in the armature. This step required the armature conductors to be supported

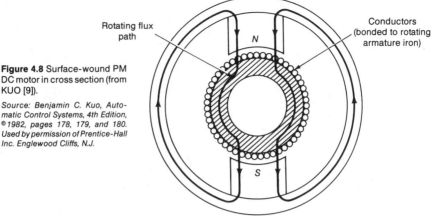

Rotating flux path

N

Conductors (bonded to rotating armature iron)

S

Figure 4.8 Surface-wound PM DC motor in cross section (from KUO [9]).

Source: Benjamin C. Kuo, Automatic Control Systems, 4th Edition, © 1982, pages 178, 179, and 180. Used by permission of Prentice-Hall Inc. Englewood Cliffs, N.J.

by nonmagnetic, nonconducting material such as epoxy resin or fiberglass. Since there was no magnetic material to provide a magnetic flux path, it was necessary to use larger, more powerful magnets to obtain the desired torque.

Armature inertia is extremely low due to the small mass. These motors are called moving-coil DC motors. They provide the most rapidly responding actuators for high-performance systems. Motor inductance less than 100 microhenries is commonly achieved with the configuration shown in Figure 4.9. Also available are thin, large-diameter motors in which the armature is a flat disc about 0.020 inch in thickness and less than 1 foot in diameter. These are called printed-circuit motors by their manufacturer and provide excellent response, high torque, and good reliability.

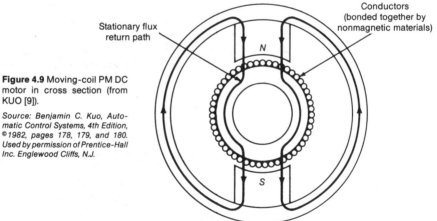

Stationary flux return path

N

Conductors (bonded together by nonmagnetic materials)

S

Figure 4.9 Moving-coil PM DC motor in cross section (from KUO [9]).

Source: Benjamin C. Kuo, Automatic Control Systems, 4th Edition, © 1982, pages 178, 179, and 180. Used by permission of Prentice-Hall Inc. Englewood Cliffs, N.J.

4.3.2
DC Motor
Analysis and
Modeling

As previously discussed, the output torque of a DC motor is proportional to the magnetic field of the PM and the current in the armature. Motion of the armature due to the generated torque is opposed by the inertia of the armature rotor, the voltage generated by the moving coil in the magnetic field, and the load on the motor shaft due to friction, damping, or work done.

Voltage generated by the moving coil in the magnetic field is called back electromotive force (back emf). This voltage rises with rotor velocity until the back emf plus the voltage lost in the armature due to current flow through the armature resistance becomes equal to the driving voltage of the motor. Initally, the back emf is zero, so that extremely high currents can flow through the low resistance of the armature. Some type of current limiter is required in the motor controller to protect the motor under these conditions.

An example calculation of torque, angular velocity, and other parameters in a PM motor follows. In this example, we have neglected the effects of motor inductance, viscous damping, and nonlinear effects. However, this example will serve to illustrate the relationships among the parameters in these motors. A mathematical model of the motor is given in Figure 4.10 using the nomenclature provided in the next paragraph.

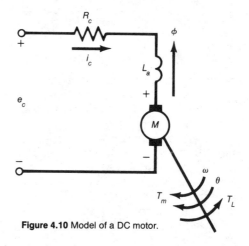

Figure 4.10 Model of a DC motor.

Terminology in motor and servo system analysis has been standardized to a considerable extent. Some standard terms are given below:

J_m = rotor inertia of the motor, fixed
J_r = inertia of the robot link being driven, time varying
R_a = electrical resistance of the motor armature, fixed
V = voltage applied to the motor
K_t = torque constant, fixed
K_b = back emf constant, fixed
T_m = torque developed by the motor, time varying
T_s = load torque due to friction or other causes, time varying
Φ = magnetic flux in the field, fixed by a permanent magnet
Θ = rotor angular displacement, time varying
i_a = armature current, in amperes, time varying
w = rotor angular velocity, radians per second, time varying
a_m = rotor angular acceleration, radians per square second, time varying
e_a = voltage across the armature, time varying
e_b = back emf, time varying
B_m = viscous frictional coefficient, fixed
L_a = armature inductance, fixed

Using the terms just defined, we can list the equations that govern the action of the motor and its load.

Torque, T_m , varies with time and is determined by multiplying the current in the armature at any specific time, i_a , by the torque constant, K_t .

$$T_m = K_t i_a$$

Back emf, the voltage generated by the armature rotating in the magnetic field, is $e_b = K_b w$

since it is a linear function of the rotor angular velocity.

Other equations allow us to calculate the current and angular acceleration from the applied voltage and the known constants.

$$V = i_a R_a + K_b \qquad T_m = K_t i_a - J_m a_m$$

Some tasks performed by robots require speed control. One example is spray painting, in which the thickness of the paint coat depends on both the spray rate and the speed of movement of the robot arm. Other tasks require control of torque or force, as in tightening a bolt or lifting an object. As can be seen from the preceding equations, rotor velocity, which determines the speed of the arm link, is itself determined by the armature current and the voltage across the armature. Koren and Ulsoy [7] point out that it is not possible to control both the speed and the torque of a robot arm at the same time. We can control one or the other, but not both at once, at least with the simple model we are considering.

4.3.3
Open-Loop
Torque
Control with
Current
Amplifier

Torque control can be achieved by controlling current in the DC motor with the use of a current amplifier. In this case, the voltage applied to the motor is forced to vary in such a way that it maintains the current through the armature at a constant amount. As can be seen in the preceding discussion, torque depends on both the current and the changing moment of inertia of the load. However, by measuring these values and compensating for them in the controller, it becomes possible to maintain a constant torque (Koren and Ulsoy [7]).

4.3.4
Open-Loop
Torque
Control with
Voltage
Amplifier

The speed of the robot arm link can be controlled by using a voltage amplifier to apply a controlled voltage to the drive motor. Variations in load inertia do not affect the final position error of such a system but do affect the time constant. However, torque is not controlled and the current can rise to high values if the motor is stopped or resisted so that the back emf is reduced. A current overload circuit can protect the motor from excessive current at the cost of stopping the operation (Koren and Ulsoy [7]).

4.3.5
Closed-Loop
Torque
Control

Either voltage amplifiers or current amplifiers can be used to drive robot arm links in closed-loop or servo control. Closed-loop servo systems are discussed in Section 6.7 after the necessary theoretical groundwork has been provided in Sections 6.1 through 6.4, and will not be covered further in this chapter.

4.3.6
DC Power
Sources
and Power
Amplifiers

Power sources and power amplifiers are an important and complex subject. In this section, we will review briefly the types of circuits used in robots and refer the reader to the references given for more detail.

Usually the power supply available in industrial plants is single-phase or three-phase, 60-cycle AC at nominal voltages of 110, 220, or 440 volts. Power converters are required to convert this AC to DC in order to operate electric drive motors on the robot joints. Then the DC must be regulated to a reasonably constant voltage level and applied to a power amplifier or switching circuit, which controls the current or voltage and direction of power applied to the drive motors.

Three types of circuits are used: rectifiers, power regulators, and power amplifiers or switching circuits. Input from an AC source is fed into a transformer that changes the input voltage to the desired level (usually lower) required to operate the power control circuits.

1. Rectification

The transformer output is still AC and must be converted to DC by being passed through a rectifier. The rectifier allows current to pass through in only one direction, so its output is a DC of varying voltage level. Either half-wave or full-wave rectification can be used. Half-wave rectification uses only one solid-state diode, so the voltage is off half of the time and much energy is lost. A full-wave rectifier circuit, as shown in Figure 4.11, uses two diodes in a bridge arrangement, so the current goes to a peak twice per cycle, or 120 times per second, for a 60-cycle input.

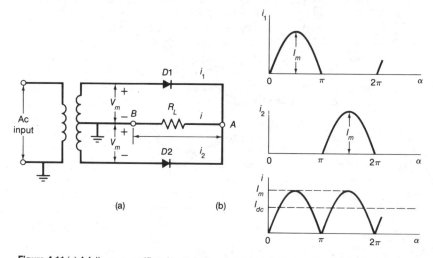

(a)   (b)

Figure 4.11 (a) A full-wave rectifier circuit. (b) The individual diode currents and the load current i. The output voltage is $v_o = iR_L$ (from Millman [10]).

Source: J. Millman, Microelectronics: Digital and Analog Circuits and System, 1979. © McGraw Hill Company. Used by permission.

2. Regulation

This varying voltage from the rectifier is termed the <u>unregulated</u> power. It is passed through another circuit that stabilizes the voltage level to a nearly constant value and applies it to the load. The voltage level can be stabilized either by filtering or by providing negative

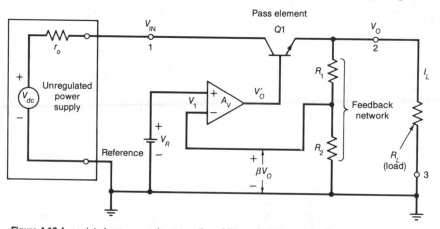

Figure 4.12 A regulated power supply system (from Millman [10] Figure 18-11).

Source: J. Millman, Microelectronics: Digital and Analog Circuits and System, 1979. © McGraw Hill Company. Used by permission.

feedback that opposes the high voltages and enhances the low voltages to generate a nearly constant output voltage level. Filtering is done by putting large inductances in series with the current flow and large capacitors in parallel with both the input and output. In the circuit of Figure 4.12, the negative feedback method is used to produce a regulated voltage across the load resistor, R_L .

An improved regulation method is the switching regulator shown in Figure 4.13. The conversion efficiency of this circuit can exceed 90%. This type of circuit can be modified to control the direction and amplitude of the drive current applied to the motor. In the diagram, the pulse-width modulator (PWM) generates a constant amplitude pulse of varying width so that the average current can be varied from zero to the maximum available. This output is filtered by the inductance, L, and the capacitor, C, to provide a smoothly varying output current. Since the power switch is switched between ground level and the V_{IN} level, the current can be reversed in direction and controlled in magnitude. In the figure, the outlined area is provided on a single integrated-circuit chip. Other components are discrete elements added externally.

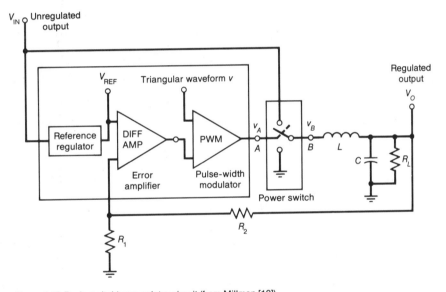

Figure 4.13 Basic switching regulator circuit (from Millman [10]).

Source: J. Millman, Microelectronics: Digital and Analog Circuits and System, 1979. © *McGraw Hill Company. Used by permission.*

3. Power Amplifiers and Switching Circuits

Early robots used power supplies with bipolar transistors to drive electric motors. The feedback from the position encoder was converted to an analog voltage and applied to the one input of a differential amplifier, a form of operational amplifier, while the control signal, also in analog form, was applied to the other input of the differential amplifier. This differential amplifier was called the error amplifier because its output was a measure of the difference between the desired position and the position actually existing at that instant. It operated in the same way as the error amplifier shown in Figure 4.13.

The error voltage could be used to control the direction of current flow through the motor by turning on either of two power amplifiers. Two power supplies, one negative and one positive, were used; they were connected to a common ground connection. One side of the motor was connected to the ground, and the other side was connected to the output of both amplifiers. Only one amplifier could be turned on at any one time; otherwise they would short each other out and cause great damage. It was therefore necessary to control carefully the switching of the power amplifiers so that only one was connected at any time.

A new type of switching converter developed by S. Cuk (pronounced "chook") and R. Middlebrook of Cal-Tech [3] can be used either as a switching regulator or as a combined regulator and power amplifier. An ingenious switching scheme is used to switch the AC input through large inductances so that the current through the load can be controlled in both magnitude and direction. Output voltages can be larger or smaller than the input, efficiency is high, and the electromagnetic interference (EMI) is reduced to a small value. Although the actual circuit is relatively simple, the theory of operation is complex; the reader is referred to reference, [3] for further details. The circuit of a bidirectional power amplifier is given in Figure 4.14. A drive motor can be used where the speaker is shown as the load. Switching can be done with power transistors or with the power MOSFETs described in the next section. A circuit similar to the full-wave rectifier circuit of Figure 4.11(a) could be used to provide the DC supply voltage identified as V_g.

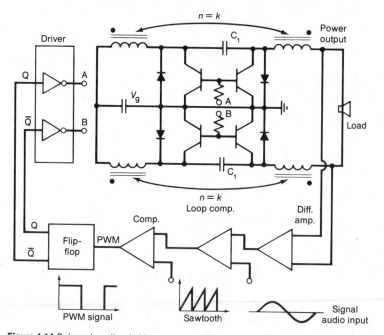

Figure 4.14 Cuk push-pull switching power amplifier.

Source: © 1983 The Hayden Book Company. and Robotics Age Magazine. From the book Robotics Age–In the Beginning. Use by permission.

**4.3.7
Solid-State
Switching
Devices**

Bipolar transistors were used in conventional power amplifier designs for several years. They were then used in pulse–width modulators and other improved circuits to improve switching performance.

Recently, a number of new semiconductor devices have become available, as listed in Table 4.1, which was prepared by B. J. Baliga of the General Electric Company [2]. The MOSFET devices have turned out to be ideal for power switching circuits because of their high switching speed, low drive power requirement, and low cost. Several companies now produce MOSFET devices at costs ranging from $5 per unit up to $30 for devices capable of handling 30 amperes at 100 volts.

Power dissipation in switching devices is an important factor. Since the MOSFET switches at a 2-megahertz rate, it has a dissipation about one-tenth that of a bipolar transistor. It is also considerably cheaper, since it can be made by simple (MOS) masking techniques. Although the (GaAS FET) devices appear to have better capability, they are not yet available at low cost. MOSFET's, bipolar transistors, and gate turn-off transistors are normally "off," while the other three devices may be "on." It is safer and easier to turn on a normally off device than to ensure that a device is off by other means.

As the costs of these devices decrease, we can expect improved drive and control circuits for robot motor drives.

4.4 Electric Stepper Motors

Stepper motors are used in robots when open-loop precise control is required and the torques required are small. These motors can step in precise increments of motion under the control of electrical pulses. Computer printers and disk drives often use steppers to provide positioning of the print head or the disk head. In small robots, stepper motors are sometimes used for the main drive motors. Encoders or potentiometers can be used to provide position feedback, so that stepper motors can also be used in closed-loop servos.

**4.4.1
Stepper Motor
Operation**

Stepper motors, like other electric motors, have a stator, or stationary magnetic element, and a rotor that rotates in the varying magnetic field produced by the stator and drives the load, as shown in Figure 4.15. (Sigma [14]). Usually the stator is on the periphery of the motor and is made up of multiple electromagnetic poles. Stator poles have a central magnetic core and are wound with many turns of wire to produce a strong magnetic field. In stepper motors, there may be a large number of poles. Each pole can be energized to be either a north or south pole by applying current in the specified direction.

The multiple magnetic poles, arranged like gear teeth around the periphery of the stepper motor armature, provide separate latching points, or steps, for the armature. Motion from step to step is caused by pulsing electric currents into the motor stator coils, which cause the polarity of the stator coils to change from north to south. It is this change of polarity that provides the motive power of the stepper motor. Since stepping is inherent in the operation, it is possible to move from step to step and go to a desired position merely by moving a specified number of steps. A price paid for this simplicity is the relatively large size of the stepper motor as compared to a DC motor for the same power output.

Rotors have two sets of poles machined in two separate sections along the length of the rotor. Each set of teeth looks like a gear on a separate hub along the rotor, with one set of teeth offset by the width of half of a tooth from the other set. One set of teeth comprise the north poles of the rotor, the other set the south poles. In reality, there is one magnet with multiple teeth or poles. There is an odd number of teeth in each set of poles on the rotor and an even number of wound poles on the stator, so that they cannot all line up at the same time.

Table 4.1 Gate turn-off devices.

Device characteristic	Bipolar transistor	Gate turn-off thyristor	MOSFET	JFET and static Induction transistor	Field-controlled thyristor (FCT)	GaAs FET
Normally on/off	Off	Off	Off	On	On	On/off
Reverse blocking capability, volts	< 50	500 to 2,500	0	0	500 to 2,500	0
Breakdown voltage range	50 to 500	500 to 2,500	50 to 500	50 to 500	500 to 2,500	50 to 500
Forward conduction* current density, A/cm²	40	100	10	10	100	100
Surge-current-handling capability	Poor	Good	Poor	Poor	Good	Poor
Maximum switching speed(approximate)	200 KHz	20 KHz	2 MHz	2 MHz	20 KHz	2 MHz
Gate-drive power	High (large base-drive current required during on-state and for turn-off)	Medium (large turn-off gate currents required)	Low (only small capacitive charging currents required)	Low	Medium	Low
Operating temperature, degrees centigrade	150	< 125	200	200	200	200

*The forward current densities are compared here for 500-volt devices operating at a forward voltage drop of 1.5 volts.
Source: Baliga [2].

Source: B.J. Baliga, "Switching Lots of Watts at High Speed," IEEE Spectrum, December, 1981, Volume 18, No. 10, page 44. ©1981 IEEE. Used by permission.

Each stepper motor is built to rotate through a specified number of steps in one revolution, although it can rotate in half-steps under proper drive conditions at reduced torque. Since there are many poles in both the stator and the rotor, there are many possible steps of rotation.

Figure 4.15 Stepper motor structure.

Courtesy of Sigma Instruments, Inc. Used by permission.

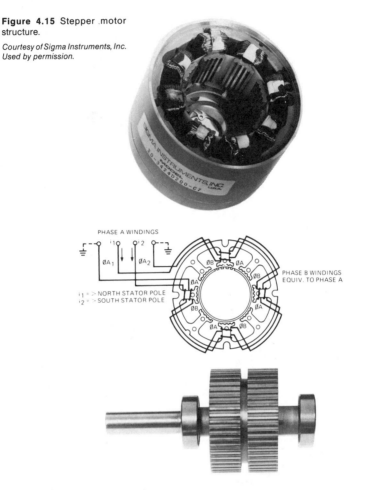

When a drive current is applied to a selected set of stator windings, the north poles of the rotor line up with the south poles of the stator. By changing the stator poles from north to south, the rotor is forced to step from one stable position to another. A typical energization sequence is shown in Figure 4.16. When the polarity of current applied to the stator windings is changed, the PMs on the rotor are made to step from one pole to another. Each step is a separate movement. Reversal of the polarity sequence causes the rotor to move in the reverse direction. Steps are precise and typically range from 1.8 to 30 degrees apart. When the electromagnets are changed rapidly, the stepper motor steps at an essentially continuous rate. However, there is always a cogging, or step effect, in even the smoothest stepper drives.

Stepper motors can be moved in either direction of rotation by properly phasing the currents applied to the stator windings. Most stepper motors are designed for two-, three-, or four-phase operation.

Two types of pulse sequences are commonly used. In <u>wave drive</u>, only one phase of the stator is pulsed each time. In <u>two-phase</u> drive, pulses are applied simultaneously to two stator phases. Two-phase drive produces greater torque from the motor but is somewhat more complex to provide. It is possible to mix the two pulsing modes to get half-step driving, which produces a smoother response.

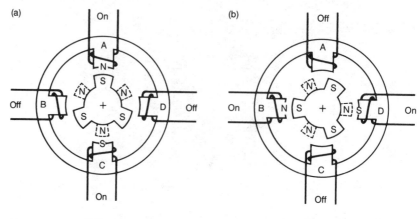

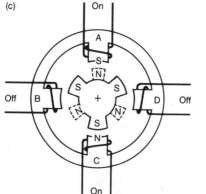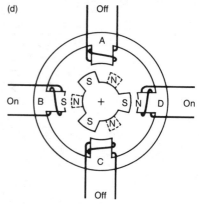

Figure 4.16 Energization sequence for a stepper motor.

Source: Paul Giacomo, "A Stepping Motor Primer," BYTE Magazine, February 1979. © 1979 McGraw-Hill, Inc. New York 10020. Used by permission.

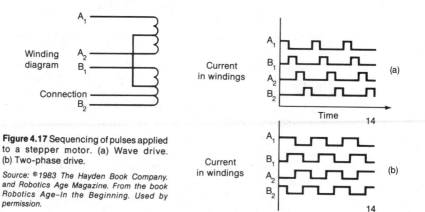

Figure 4.17 Sequencing of pulses applied to a stepper motor. (a) Wave drive. (b) Two-phase drive.

Source: © 1983 The Hayden Book Company. and Robotics Age Magazine. From the book Robotics Age–In the Beginning. Used by permission.

Resonance is an inherent tendency to oscillate at a particular frequency due to the mechanical and electrical characteristics of a design. It may occur in stepper motors. It can cause strange effects, including backward rotation, low torque, and vibration. Usually, resonance occurs at 50 to 150 steps per second. Half-step driving helps prevent resonance. Friction damping and special control circuitry can also be used. The best solution is to avoid prolonged operation at the resonance frequency by careful design. It is possible to accelerate or decelerate through resonance without causing problems.

Figure 4.17 shows the winding diagram and the current in the windings for a wave drive and a two-phase drive. Pulses are applied to the stator windings in a specific sequence, as shown, to control the operation of the stepper motor. Controllers are necessary, therefore, to provide the correct sequence of drive currents in the proper phase relationship.

Controllers contain a combination of electronic circuits to generate and apply the drive currents required. Several manufacturers produce stepping motor electronic circuit drive cards that accept two input signals—a drive pulse and a direction of rotation signal (Sigma [14]). Every time the drive pulse is applied to the drive card, the drive card generates the pulse sequence in the correct sequence to drive the motor in the direction specified by the direction of rotation signal. The pulse sequence is selectively applied to solid state power amplifiers, which supply current to the stator windings as required. Power amplifiers used may be single power transistors or paralleled transistor drivers controlled by operational amplifiers, depending on the power required.

Powerful MOS-FET power transistors now available can control more than 25 amperes at voltages up to 100 volts. Thus, the drive problem has become much easier to solve.

**4.4.2
Types of
Stepper
Motors**

PM-type motors have a PM rotor as previously described, and one set of windings on each pole. There may be only a few pole windings on the stator, but each may have multiple teeth machined into the pole pieces, so that there are, in effect, as many as 200 or 400 poles per revolution. This arrangement reduces the cogging effect and smoothes the operation.

Bi-filar motors are similar to permanent magnet (PM) motors but have two sets of windings on each pole. Each winding is of smaller wire than in the PM motor, and so has higher resistance. Therefore, the value of L(Inductance)/R(Resistance), which affects motor response, is smaller and the response time is improved. Only one power supply is required to drive the bi-filar motors, since the direction of rotation can be switched by switching power to different windings rather than by reversing the direction of current flow in one winding. Both a negative and a positive supply are required to drive PM motors.

A third type of stepper motor is the variable reluctance motor. It uses an unmagnetized rotor, so that the rotor position is independent of the polarity of the stator excitation. Only one power supply is needed. Direction of motion is controlled by the phasing sequence only. It has a slower response to pulsing than the PM motor but has a lower inertia, which is desirable in some cases. However, the variable reluctance motor has no holding torque when power is turned off, as the motors with PM rotors do.

An excellent discussion of stepper motor types and methods of operation is given by Giacomo [6]. The theory of operation and some advanced techniques are discussed by S. Motiwalla of the General Electric Corporation [11]. Kuo [8] provides a complete analysis of stepper motor design and application.

4.4.3
Open-Loop
Operation
The pulsing sequence shown in Figure 4.17 is usually generated with integrated circuits that have been designed to perform this function, as described in Section 4.4.1. Typically, the integrated circuit has two inputs and four outputs. Inputs specify the direction of motion and supply a timing pulse to control the timing of the motion. Outputs are the square wave pulses to be applied to the stator windings.

In the simplest type of stepper motor operation, it is only necessary to hook this circuit to the stepper motor and enter a series of pulses from an oscillator or a counter circuit. Pulsing may be done under computer control by setting up a digital number in the computer and counting it down to zero, emitting a pulse each time the number is decremented. It is this simplicity that has made the stepping motor valuable for open-loop operation. Although no direct position feedback is used in these circuits, it is possible to position the stepping motor to high accuracy. However, positioning accuracy depends on the stepping motor's moving one step for every pulse applied. If this does not happen—if the motor is stalled or blocked—there is no information regarding the position of the object being driven.

4.4.4
Closed-Loop
Control
Position information is required to provide assurance of high-precision positioning with a stepper motor. Several types of position feedback can be used, as described in Section 5.5.6. Potentiometers and incremental encoders are the least expensive. Either can be used.

Potentiometer output is passed through an analog-to-digital converter to generate continuously a digital count corresponding to the position of the potentiometer shaft, which is geared to the stepper motor shaft. The incremental encoder is counted into a register to keep a record of how many steps have occurred. In either case, a register or counter has a record of how many steps the stepper motor has made. Assume that an encoder is used to count the steps made and that this count is stored in an encoder register.

Positioning is initiated by setting a computer (or other position control device) to a specified digital number corresponding to the distance to be moved. A pulse is emitted from the computer and applied through the stepper motor controller to the stepper motor, causing it to rotate. The digital number from the encoder register is continuously compared to the computer register specifying the distance to be moved. When the two numbers agree, the comparator output goes to zero and signals that the movement has been completed as specified.

Since an arithmetic element is used as the comparator, there is automatic correction of overshoot. A negative number is generated that changes the direction of motion of the stepper motor to drive it back into the correct position.

4.4.5
Specialized
Control
Electronics
for Stepper
Motors
Maximum operating efficiency is obtained from stepping motors by <u>ramping</u>—controlled acceleration to a high speed, followed by controlled deceleration to a lower speed. Either linear or exponential acceleration is used. During acceleration, speed is increased rapidly so that the pulse rate is high enough to generate the minimum torque required to move the load. Then it is cut back so that a desired average speed is obtained. The reverse process is used during deceleration. Special ramping circuits are available to produce the proper pulse rate so that the total number of pulses, which control position, are obtained. In operation, the desired total number of pulses is loaded into a register; then the ramping circuit converts these

pulses to a series of fast and slow pulses so that the total number of pulses is obtained as desired. Figure 4.18 is the block diagram of a typical ramping circuit. The pulse source shown in the diagram could be the control register in the computer (Sigma [15]).

The basic reason special circuits are needed for optimal control is that the stepper motor acts like an oscillating pendulum. It applies torque to a load and then stops, so that the spring force of the load tends to backdrive the stepper motor. Therefore, it is necessary to maintain an average torque above a minimum value to obtain a reasonably smooth response.

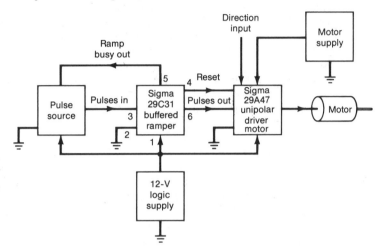

Figure 4.18 Ramping circuit for a stepper motor (from Sigma [15]).

Source: Sigma Instruments. Used by permission.

4.5 Selection of Drive Methods for Different Operations

For any application of robots, we must decide which of the available drive methods is most suitable. Positioning accuracy, reliability, speed of operation, cost, and other factors must be considered. This section summarizes and outlines some of these important factors and their importance in each type of operation.

**4.5.1
Hydraulic
versus
Electric Drives**

Electric motors are inherently clean and capable of high precision if operated properly. In contrast, hydraulic motors require the use of oil under pressure and pneumatic drives are not capable of high precision for continuous-path operation.

Hydraulic drives can generate greater power in a compact volume than can electric motor drives. Oil under pressure can be piped to simple motors capable of extremely high torque and rapid operation. Also, the power required to control an electrohydraulic valve is small. Essentially, the work is done in compressing the oil and delivering it to the robot arm drives. All the power can be supplied by one powerful, efficient electric motor driving the hydraulic pump at the base of the robot or located some distance away. Power is controlled in compact electrohydraulic valves. However, high-precision electrohydraulic valves are more expensive and less reliable than low-power electric amplifiers and controllers.

Electric motor drives must have individual controls capable of controlling the power for those drives. In large robots, this requires switching of 10 to 50 amperes at 20 to 100 volts. Current switching must be done rapidly; otherwise there is a large power dissipation in the switching circuit that will cause excessive heating of the switching system. Small electric motors use simple switching circuits and are easy to control with low-power circuits. Stepper motors are especially simple for open-loop operation.

The single biggest advantage of hydraulic drives is their intrinsically safe operation. They can be used in hazardous atmospheres, such as are found in paint booths and some chemical operations.

To summarize: Hydraulic drives are preferred where rapid movement at high torques is required, at power ranges of approximately over 5 horsepower unless the slight possibility of an oil leakage cannot be tolerated. Electric motor drives are preferred at power levels under about 2 horsepower unless there is danger due to possible ignition of explosive materials. At ranges between 1 and 5 horsepower, the availability of a robot in a particular coordinate system with specific characteristics or at a lower cost may determine the decision. Reliability of all types of robots made by reputable manufacturers is sufficiently good that this is not a major determining factor.

**4.5.2
Comparison
of Electric
DC Motors and
Electric
Stepper Motors**

Suppose we have made the tentative decision that electric motors are required. How do we choose between electric DC motors and electric stepper motors? Table 4.2 summarizes the previous discussions of these two types of motors to aid in making this evaluation. High positioning accuracy, on the order of 0.01 to 0.001 inch is assumed to be required.

**Table 4.2 Electric DC motors and electric stepper motors:
Advantages and disadvantages**

DC MOTORS FOR HIGH-ACCURACY POSITIONING	
Advantages	Disadvantages
Linear torque–to–power ratio.	Must operate in a closed loop for accurate positioning
Larger torque–to–volume ratio than stepper motor.	Reverse current switching required to reverse motor direction (or two windings).
Precise positioning possible.	Standard motors use commutators and brushes that wear. (Brushless operation is possible.)
Rapid response due to low inductance in armature.	Simple DC positioner is more expensive than for stepping motor drive.
High holding torque at a specified position.	

STEPPING MOTORS FOR PRECISE POSITIONING

Advantages	Disadvantages
Fully synchronous, so can be locked in speed to any reference with no long-term speed position error.	Larger for a given torque requirement than a DC motor.
Can operate to position precisely with open-loop control. Stable operation.	Stiffness, or position holding strength, depends on position.
Either unipolar or bipolar drives can be used. (Will operate with one battery, for example.)	Ramping and special circuits required for maximum efficiency. Not very efficient without them.
Excellent for light loads and precision open-loop positioning.	Not suitable for heavy loads, high torques, and high disturbance torques.
Less expensive motor and drive for simple positioning tasks.	Accelerates and decelerates at each step. Not a smooth drive.

**4.5.3
Pneumatic
Drives for
Lowest
Cost**

As mentioned before, if a pneumatic drive will meet the requirements for an application, it will almost certainly involve the lowest cost. Available point-to-point robots, such as the Auto-Place 10, which has six axes of motion, sell for about $12,000, and the basic International/Robomation robot sells for less than $10,000. These prices do not include the air compressor required, however.

**4.5.4
Drive
Selection
Calculations**

Some simple mathematical calculations are needed to determine the torque, velocity, and power characteristics of drive motors for different applications. These values are all related, of course.

Torque is defined in terms of a force times distance or moment. A force, F, at distance, L, from the center of rotation has a moment or torque, T. (Both T and M are used to designate the value of the torque, we will use T in this book.)

$$T = FL \tag{1}$$

Torque is in ft.-lbs. when L is in feet and F is in pounds.

In general terms, power P is transmitted in a drive shaft is determined by the torque, T, multiplied by the angular velocity, w, in radians per second. (Angular velocity in degrees per second may be converted to radians per second (by dividing by 57.3).

$$P = TW \tag{2}$$

The torque capability of a motor is determined by use of the equation

$$T = 63,000(hp/N) \tag{3}$$

where hp = number of horsepower, N = revolutions per minute (rpm) of the motor, and T = inch-pounds (in-lbs). Given the torque, T, and the revolutions per minute, N, we can calculate the horsepower by solving equation (3) for hp:

$$hp = TN/63,000 \tag{4}$$

where T is in in-lbs. and N is in revolutions per minute.

In International System (SI) units, this equation is written
$$T = 9.55(P/N) \qquad (5)$$
where T is in newton-meters, P is in watts, and N is in revolutions per minute. The conversion between units is:
$$1\ hp = 746\ \text{watts} \qquad (6)$$
and
$$\text{newton} = 0.225\ \text{pounds} \qquad (7)$$
The horsepower hp of any drive is defined in terms of ft-lb. per second or per minute (ft-lb/min):
$$1\ hp = 550\ \text{ft–lb/sec} = 33{,}000\ \text{ft–lb/min} \qquad (8)$$
For example, a calculation can tell us what horsepower is required in a motor used to drive a 6-foot robot arm lifting a 50-pound weight at 60 degrees per second. We will assume that the arm has zero mass.

Conversion of w degrees per second (deg./sec) to rpm or revolutions per minute (rev/min) is done by
$$N = w\ \text{deg/sec} \times (1\ \text{rev}/360\ \text{deg}) \times (60\ \text{sec/min}) = w/6 \qquad (9)$$
Using equation (4) and substituting N from equation (9) when $w = 60$, we obtain
$$hp = \frac{TN}{63{,}000} = \frac{50 \times (6 \times 12) \times (60/6)}{63{,}000} = 0.576\ hp \qquad (10)$$
The use of simple equations of this type is often sufficient to make a useful approximation of a needed value. More detailed calculations can take in all of the appropriate data, using the equations of statics and dynamics that apply. See, for example, Shelley [13].

Exercises

These exercises will aid in reviewing the topics discussed in this chapter and emphasize some of the more important concepts.

1. What is the typical oil pressure used in hydraulic systems for robots? What force will be generated on a hydraulic piston 0.5 inch in diameter as a result of applying this pressure?
2. Describe the two main types of electrohydraulic values. What are the advantages of each?
3. What is the most important single problem affecting the reliability of hydraulic drives, and how can it be eliminated or reduced in importance?
4. What are the functions of the "train" switch in a hydraulic robot?
5. What advantage of hydraulic drives would be difficult to duplicate in electrical drives at the present time?
6. What is the chief advantage of pneumatic systems? What is the chief disadvantage?
7. Why are programmable controllers still used in factory control systems when more flexible and more powerful computers are available?
8. Which type of electric motor provides the highest performance for robot use? Why is this true?
9. A DC electric motor is to be used to drive a robot joint that requires a torque of 50 in.-lbs. If the torque constant of the motor is 2.5 in. lbs per ampere, how much current is required from the drive amplifier at peak torque?
10. If a stepper motor has 200 slots cut in the stator poles and 199 slots in the rotor, what is the smallest step that can be made?

What positioning accuracy could you be sure of obtaining in stepping this motor through 10,000 steps lightly loaded so that the motor torque was always adequate to drive the load? Neglect effects of temperature and bending.

11. What is ramping in a stepper motor, and why is it used?
12. What is the chief advantage of the stepper motor? What is its chief disadvantage?
13. If you were specifying a robot joint drive that required a positioning accuracy of 0.002 inch and a peak torque of 175 ft. lbs at an angular velocity of 180 degrees/second, what type drive would you choose? Assume that it is to be used in a material-handling application in a normal factory environment and should be as compact as possible.
14. In Exercise 13, what type of drive would you use if size was not a problem?
15. Using equation (6) and the definition of horsepower in equation (8), verify that equation (5) is correct.
16. Why were so many early robots designed to use hydraulic power? Name some advantages of hydraulic power compared to DC Electric servos.
17. An industrial robot with a revolute coordinate system is assigned to do spot welding using a welding gun weighing 50 pounds. The DC servo motor driving the elbow joint (the last joint before the wrist) has a peak power capacity of 2 horsepower (hp). Assume that the 4-foot long arm and wrist assembly are equivalent to a point mass of 50 lbs. at 2 feet from the center of rotation and that the welding gun is equivalent to a point mass 4 feet from the center of rotation, as in the following diagram. What is the maximum vertical velocity with which the elbow motor can move the welding gun?

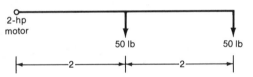

References

[1] R. D. Atchley, "A More Reliable Electrohydraulic Servovalve," Robots 6 Conference, March 2-4, 1982. Robotics International of SME, Dearborn, Mich., 1982, pp. 159-165.
[2] B. J. Baliga, "Switching Lots of Watts at High Speed," IEEE Spectrum, December 1981, vol. 18, no. 10, pp. 42-48.
[3] S. Cuk and R. D. Middlebrook, "Advances in Switched-Mode Power Conversion," Robotics Age, Winter 1979, pp. 6-19.
[4] J. F. Engelberger, "Robotics in Practice," AMACOM, Division of the American Management Association. New York, 1980.
[5] "Flexible Automation," GBA4925/370 AMF Versatran Industrial Robots (now a division of Prab Conveyors, Inc., Kalmazoo, Mich.), 1970.
[6] P. Giacomo, "A Stepping Motor Primer,": Part 1: "Theory of Operation," Byte, February 1979, Part 2: p. 90; "Interfacing and Other Considerations," Byte, March 1979, p. 1420.
[7] Y. Koren and A. G. Ulsoy, "Control of DC Servo-Motor Driven Robots," Robots 6 Conference Proceedings, March 2-4, 1982, Detroit, Mich. Robotics International of SME, Dearborn, Mich., 1982, pp. 590-602.

[8] B. J. Kuo, Theory and Application of Step Motors, West Publishing Company, 1974.

[9] B. J. Kuo, Automatic Control Systems, 4th ed. 1982, Prentice-Hall, Englewood Cliffs, N.J., 1982, section 4.6, pp. 176-182.

[10] J. Millman, Microelectronics—Digital and Analog Circuits and Systems. McGraw-Hill, New York, 1979.

[11] S. Motiwalla, "Continuous Control with Stepper Motors," Part 1: "Designing a Preview Controller," Robotics Age, July-August 1981, p. 28; "Part 2: Hardware and Software Implementation for Path Control," Robotics Age, September-October 1981, p. 16 (with Richard Tseng, Measurex).

[12] "Principles of Operation," Versatran Division of Prab Conveyors, Inc., Kalamazoo, Mich., 1972.

[13] J. F. Shelley, Engineering Mechanics: Statics and Dynamics, McGraw-Hill, New York, 1980.

[14] Sigma Stepping Motors Catalog, MC215, 10-80 Rev. Sigma Instruments, Inc., Braintree, Mass., 1980.

[15] Sigma 29C31 Ramping Card or Bulletin MC218, Sigma Instruments, Inc., Braintree, Mass., 1981.

Chapter 5

Sensors for Robots

One of the greatest needs in robots is improved and more versatile sensors of every kind. In this chapter, we will discuss a number of sensory applications and sensor types that can be expected to aid in meeting that challenge.

Operation and characteristics of common nonvisual sensors will be described. However, description of the operation of visual sensing systems and equipment will be deferred. That topic is so big and important that the main developments and requirements of visual sensing will be summarized separately in Chapter 9.

An early example of the use of sensors was the development of a touch sensor for a robot hand by Heinrich Ernst at MIT. For his doctoral thesis at MIT in 1961, Ernst demonstrated the use of touch sensing and computer control for picking up objects with a manipulator (Ernst [3]). He called this early manpiulator the *MH-1* for mechanical hand number 1. This is considered to be the first use of sensing on a robot hand. Figure 5.1 is a diagram of the sense organs on the hand. Note that some sensors are labeled *b* for binary output and others *c* for continuous output. Binary output refers to what is called simple touch. It is a signal that is either on or off, 1 or 0. Continuous output provides multiple levels of signal.

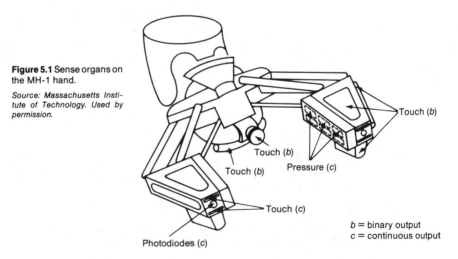

Figure 5.1 Sense organs on the MH-1 hand.

Source: Massachusetts Institute of Technology. Used by permission.

Touch (*b*)

Touch (*b*)

Touch (*b*)

Pressure (*c*)

Touch (*c*)

Photodiodes (*c*)

b = binary output
c = continuous output

Other pioneering work in sensors was done at Stanford University and at SRI, a separate organization in Menlo Park, California. Dr. Charles Rosen and his co-workers at SRI, now SRI International, were instrumental in developing complete sensory systems for industrial robots, as described in Rosen et al. [14] and subsequent papers. SRI developed the first industrial robot with computer vision and touch sensors.

Now, at many universities and industrial laboratories here in the United States and in many foreign countries, a considerable amount of research and development is going on to improve sensory capability for robots. Touch and tactile sensing, force sensing, and computer vision are the areas with the greatest current activity. Computer vision especially has become commercially feasible and is used for many industrial purposes, either alone or in association with robots.

Perhaps the most advanced work in tactile sensing is the High-resolution imaging touch sensor developed at MIT by Hillis [5] and the VLSI Tactile Sensing done at CMU and Cal-Tech by Marc Raibert and John Tanner [13] which are discussed further in Section 5.3.7.

5.1 Sensory Needs of Robots

Robots, like humans, must gather extensive information about their environment in order to function effectively. They must pick up an object and know that it has been picked up. As the robot arm moves through space, it must avoid obstacles and approach items to be handled at a controlled speed. Some objects are heavy, others are fragile, and others are too hot to handle. These characteristics of objects and the environment must be recognized and fed to the computer controlling robot movement.

The major capabilities required of robots have been classified:

1. Simple touch—the presence or absence of an object. 2.Taction or complex touch—the presence of an object plus some information on its size and shape. 3.Simple force—measured force along a single axis. 4.Complex force—measured force along two or more axes. 5.Proximity—noncontact detection of an object. 6.Simple vision—detection of edges, holes, corners, and so on. 7.Complex vision—recognition of shapes.

In addition to these capabilities, there is a need to measure temperature, humidity, slip, pressure, chemical activity, and other more esoteric characteristics. Note that humans can measure all of these characteristics and do so routinely in carrying out different tasks.

5.1.1
Robot Control
and
Self-Protection

Measurements of acceleration and velocity of the robot arm and its individual links are required to enable the control computer to predict the movements of the arm and the way it will respond to applied forces. As discussed in Chapter 3, the arm will flex due to the acceleration forces applied. This flexion will cause an error in position unless it is compensated for. Velocity information is needed to predict the time required for movement and to compute centrifugal forces.

In order for a robot to work safety, it should never exceed specified force limits in any part of the mechanism. Humans have the same kind of feedback and normally will not apply muscular force in excess of the force tolerated by the bone structure and tendons. Robot safety therefore requires forces in each link and structural part to be monitored continuously. Alternatively, each item must be so strong that it cannot fail under any reasonable circumstance. Most industrial robots today fall into the second category. As a result, they are much heavier than necessary. There is an opportunity for considerable improvement in capability and reduction in cost when force information is routinely available and is factored into the control operation.

5.1.2 Programmable Automation

Programmable automation, or robot operation in factories, is applied to tasks in material handling, inspection, assembly, painting, welding, grinding, and a host of other applications. Each application has somewhat different sensory requirements. Most of today's industrial robots have no external sensory capability and perform many important tasks quite well. However, it is clear that sensory information would be valuable in many of these tasks.

5.1.3 Manipulation of Workpieces

Material-handling and assembly tasks require many controlled operations. Grasping, holding, turning, inserting, aligning, fitting, screwing, and orienting are common operational requirements. When a robot has been trained in a structured environment, many or most of these functions can be done without sensory feedback. It is desirable, in future robots, to reduce the amount of training required and to provide more general direction to the robot through programming.

Sensory feedback is used by humans to do both simple and complex manipulation because humans do not have the ability to perform highly precise manipulation. Since robots do have this ability, the need for sensory feedback has not been essential. Today sensory feedback is becoming more important. Both contact and noncontact sensors are required for many tasks, especially when the robot is working in a poorly structured environment.

Sensing may be divided into coarse and fine sensing. Coarse sensing may be of the noncontact type, perhaps using vision or sonic sensing, while fine sensing may depend on taction. Workpieces, the objects being handled, may be randomly positioned in space and require orientation before being placed in a machine tool or assembly. Vision may be capable of providing coarse sensing to enable a robot end effector to grasp a part, but tactile sensing may be required to determine part location with sufficient accuracy. Vision may be very precise, of course, but may still not be suitable for all tasks because the field of view of the vision system is obscured by the robot gripper. It is therefore advantageous to use multiple sensory techniques in several types of applications.

5.1.4 Collision Avoidance

Nature provides whiskers on the head of a cat to allow the animal to sense obstructions in the dark. Robots have a need for this or some other way to sense obstructions so that they may avoid collisions with objects in their environment. Some robots have been supplied with wires or bumpers to detect obstructions in time, so that they can signal the controller to stop before damage occurs to the robot or surrounding objects.

5.1.5 Assembly Operations

An increasing number of robots are being used to insert pins, shafts, screws, and bolts into holes. Touch, force, and torque sensors are needed to control this type of operation. Vision is used in such systems to perform the coarse location of parts, but other sensors are needed to complete the task. Another useful capability is compliance, or the ability to give a little in response to the environment. Compliance may be provided by springs in the gripper or by the use of linkages of various kinds. Remote center compliance (RCC), as described in Section 5.3.7, provides a valuable means for precise insertion of small parts into assemblies.

5.1.6
Inspection

Visual inspection is one area of increasing robotic application. Parts or assemblies may be examined to detect burrs, cracks or voids; determine surface finish and cosmetic qualities; count the number of holes; assess the completeness of assembly; and determine the accuracy of assembly. One visual system used in the automotive industry measures parts location to an accuracy of 0.005 inch from a distance of 5 feet or more. This and other techniques and equipment for visual inspection are described in Chapter 9.

5.1.7
Search and
Recognition

In today's factory environment, operated mainly by humans, it is common to put parts into bins in order to group them together for movement from one machine to another. Humans have no difficulty in picking parts from bins and placing them in a machine or on a conveyor. Until recently, robots were not able to recognize parts visually and pick them from bins. It is still an expensive and difficult operation for a robot. Finding and orienting parts, then, is an important problem for robots.

Searching is defined as detecting a part by the use of sensors. Recognition is defined as determining the identity of the part and its position and orientation. Grasping is acquiring the part by using a gripper or fingers. Usually grasping requires sensors on the fingers or grippers. Placement of a part includes moving the part to a specified location and joining or inserting it into an assembly or machine tool. Force sensors are valuable and may be required for placement activity.

5.1.8
User Safety

Any automated system in an industrial environment must provide safety sensors to protect the operators and other humans nearby. This is mandated by OSHA rules as well as by good practice. Intrusion sensors detect the presence of a human and either shut down the equipment or send out a warning to the intruder. Area sensors are the most common sensors. Floor pads, with switches operated by pressure, also can be used. Photoelectric cells, the "electric eyes" of media reporting, can sense the interruption of a light beam. Another common method is the use of ultrasound fields to detect an intruder. Fences around robots are required during operation. Gates allowing ingress to the area must be equipped with switches that shut off power when the gates are open.

In some cases it may not be necessary to shut down the robot completely. Instead, its speed may be limited when someone is nearby. During the training operation, the operator must work directly with the robot. At that time, the safety switches must be bypassed or the operator will be unable to work. Even then, however, there must be as much protection as possible for the operator. Protection can be achieved in some cases by setting up trip wires that shut down the robot when it leaves the specified area. In all cases, it is necessary to provide sensors that detect failure of the control system and cause the robot to operate in an uncontrolled manner.

Safety is discussed at some length by R. D. Kilmer [9] of the National Bureau of Standards.

5.2 Sensor Evaluation and Selection

We can broadly categorize sensors by their characteristics, by types of physical phenomena, or by application. These categories will overlap but will still be useful in understanding the many types of sensors and the way in which they can be evaluated, selected, and applied.

Sensors can be broadly divided into two types: contact and noncontact. Contact sensors must touch an object directly in order to operate, such as switches, probes or feelers. Noncontact sensors can operate at a distance by sensing magnetic fields, sound waves, light, x-rays, infrared light, and so on.

Contact sensors may operate by closing an electrical switch, by moving a contact on a potentiometer, or by generating a voltage in a piezoelectric crystal, a change in resistivity in a piezoresistive material or some other physical phenomenon. Usually these sensors convert a mechanical movement to an electrical or electronic change. In fact, since all input to a computer must eventually be a digital signal, all of these other signals must be converted to digital codes before being used by a control computer.

Noncontact sensors measure electromagnetic or sonic phenomena. Magnetic fields, electrical fields, visible light, ultraviolet light, infrared light, and x-rays are all electromagnetic phenomena. Ultrasonic distance measurement depends on the emission of a high-frequency sound wave at a frequency above the human hearing range (hence ultrasonic) and a wait for the echo signal to return from an object. Optical sensors that use simple light beams are described in Section 5.3. Use of multiple light beams and complex viewing techniques is described in Chapter 9.

In using sensors, we must first decide what the sensor is to do and what results we expect. This section discusses some of the criteria that must be considered in selecting and using different kinds of sensors for the many applications in a robot system. The most important parameters to be considered are discussed in the following sections.

1. Sensitivity

Sensitivity is defined as the ratio of change of output to change in input. As an example, if a movement of 0.001 inch causes an output voltage to change by 0.02 volt then the sensitivity is 20 volts per inch. It is sometimes used to indicate the smallest change in input that will be observable as a change in output. Usually maximum sensitivity that provides a linear, accurate signal is desired.

2. Linearity

Perfect linearity would allow output versus input to be plotted as a straight line on graph paper. Linearity is a measure of the constancy of the ratio of output to input. In the equation:
$$Y = bX$$
the relationship is perfectly linear if b is a constant. If b is a variable term, then the relationship is not linear. For example, b may also be a function of X, such as $b = a + dX$ where the value of d would introduce a nonlinearity. A measure of the nonlinearity could be given as the value of d, to a first approximation.

3. Range

Range is a measure of the difference between the minimum and maximum values measured. A strain gauge might be able to measure values over the range from 0.1 to 10 newtons, for example.

4. Response Time

Response time is the time required for a change in input to be observable as a stable change in output. In some sensors, the output oscillates for a short time before it settles down to a stable value. We measure response time from the start of an input change to the time when the output has settled to a specified range.

5. Accuracy

Accuracy is a measure of the difference between the measured and actual values. An accuracy of $\pm$ 0.001 inch means that, under all circumstances considered, the measured value will be within 0.001 inch of the actual value.

In positioning a robot end effector, verification of this level of accuracy would require careful measurement of the position of the end effector in reference to the base reference location with an overall accuracy of 0.001 inch under all conditions of temperature, acceleration, velocity, and loading covered by the accuracy specification. Precision measuring equipment, carefully calibrated against secondary standards, would be necessary to verify this accuracy.

6. Repeatability

Repeatability is a measure of the difference in value between two successive measurements under the same conditions and is a far less stringent criterion than accuracy. As long as the forces, temperature, and other parameters have not changed, we would expect the successive values to be the same, however poor the accuracy is.

7. Resolution

Resolution is a measure of the number of measurements within a range from minimum to maximum. It is also used to indicate the value of the smallest increment of value that is observable.

8. Type of Output

Output can be in the form of a mechanical movement, an electrical current or voltage, a pressure or liquid level, a light level or another form. To be useful, it must be converted to another form, as discussed in Section 5.3.

5.2.3 Physical Characteristics

1. Size and Weight

Size and weight are usually important physical characteristics of sensors. If the sensor is to be mounted on the robot hand or arm, it becomes part of the mass that must be accelerated and decelerated by the drive motors of the wrist and arm, so that it directly affects the performance of the robot.

It is a challenge to sensor designers to reduce the size and weight to reasonable proportions, but it can be done. An early wrist force–torque sensor, for example, was 5 inches in diameter but was reduced to 3 inches in diameter through careful redesign.

2. Reliability

Reliability is of major importance in all robot applications. It can be measured in terms of Mean Time to Failure (MTTF) as the average number of hours between failures that cause some part of the sensor to become inoperative.

In industrial use, the total robot system is expected to be available as much as 98 or 99% of the working day. Since there are hundreds of components in a robot system, each one must have a very high reliability. Some otherwise good sensors cannot stand the daily environmental stress and therefore cannot be used with robots. Part of the requirement for reliability is ease of maintenance. A sensor that can be easily replaced does not have to be as reliable as one that is hidden in the depths of the robot. Maintainability is measured in terms of Mean Time to Repair (MTTR).

3. Interfacing

Interfacing considerations are often a determining factor in the usefulness of sensors. Nonstandard plugs or requirements for non-standard voltages and currents may make a sensor too complex and expensive to use. Also, the output and control signals from a sensor must be compatible with other equipment being used if the system is to work properly.

5.3 Available Sensory Techniques

5.3.1
Contact Sensing
Contact sensing can be done by direct mechanical contact, such as closure of a switch either mechanically or by a contact that induces an electrical or magnetic secondary operation.

1. Switches

Electrical contact switches are used routinely for such functions as determining whether contact has occurred or whether a part is in place, or to signal that a movement has reached a defined limit. These switches can be heavy-duty switches that switch large currents or smaller switches that handle small currents and voltages that are used for signaling. These switches are often mounted on auxiliary equipment that must interact with the robot. Switch closures are read into the robot system as required. Microswitches are small switches that operate on a minimum amount of movement and handle very small currents. When small signals are to be switched, microswitches are commonly used because they can be mounted compactly and cheaply.

Switches commonly have a resistance of a fraction of an ohm when closed and effective impedances of over a million ohms when open. They are rated by current-carrying capacity and their ability to withstand voltage across their contacts. Typical currents for signaling switches are 100 milliamperes; voltages ratings are 25 to 50 volts. Larger switches may handle several amperes and voltages of 200 to 500 volts.

2. Piezoelectric Transducers

The piezoelectric effect can be used to make dynamic pressure transducers and to measure surface roughness. One example is the use of piezoelectric crystals for phonograph pickups. Tactile sensors can use multiple crystals to sense a surface contour. The induced charge on a crystal is given by Holman [7] as

$$Q = dF$$

where Q is in coulombs, F is in newtons, and the constant d is the piezoelectric constant. Output voltage of the crystal is given by

$$E = gtp$$

where t is the crystal thickness in meters, p is the impressed pressure in newtons per square meter, and g is the voltage sensitivity. Voltage sensitivity for quartz (X cut, length along Y length) is 0.055 volt-m/N. For a thickness of 2 millimeters and a pressure of 200 psi, the generated voltage is about 152 volts.

**5.3.2
Position and
Displacement
Sensing**

Measurement of the position of robot joints and links can be done by several methods, depending on the accuracy, repeatability, and range of movement required.

1. Potentiometers

Potentiometers are mechanically variable resistors. Potentiometers are used primarily to control signal currents and small control currents. They can be made of many turns of wire wound around a form or can be deposited as thin films of carbon, metal, or other conductive materials. A slider element makes contact with the resistive material along one edge of the supporting form. Usually the resistive material is covered with insulation, except along the contact edge.

1. Linear Potentiometer. The resistance element is mounted so that the slider element moves in a straight line. The linear potentiometer is commonly used to measure the amount of linear movement of a mechanical device attached to the slider element. A fixed voltage is applied across the potentiometer terminals. As the slider moves, it picks off a voltage proportional to the amount of mechanical movement. Good potentiometers have a linearity error of less than 0.1%. Figure 5.2 shows a typical linear potentiometer which has good linearity and long life.

Figure 5.2 Linear potentiometer.

*Courtesy Waters Manufacturing
Company. Used by permission.*

2. Rotary Potentiometers. In these potentiometers, the resistance element is bent into a circle or helix. Single–turn potentiometers or <u>pots</u> (the short name commonly used) cannot turn a full 360 degrees because of the space occupied by the contacts. However, this is often not an important factor. If greater resolution is desired, the potentiometer is made in the form of a 5- or 10–turn helix to increase the effective length. Gearing may be used to convert a linear motion to a rotary motion so that a rotary potentiometer can be used (see Figure 5.3).

Potentiometers provide absolute position information that is maintained even when power is turned off and back on. Good-quality potentiometers can be purchased for $5 to $25; they are the least expensive way to measure position accurately. However, since there are moving and sliding parts, the life of potentiometers was less than that of noncontacting devices. Recent developments in potentiometers by Waters Manufacturing Company and others have made them competitive with optical encoders.

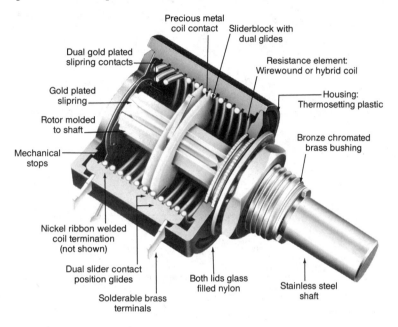

Precious metal coil contact   Sliderblock with dual glides

Dual gold plated slipring contacts

Resistance element: Wirewound or hybrid coil

Gold plated slipring

Housing: Thermosetting plastic

Rotor molded to shaft

Mechanical stops

Bronze chromated brass bushing

Nickel ribbon welded coil termination (not shown)

Dual slider contact position glides

Both lids glass filled nylon

Stainless steel shaft

Solderable brass terminals

Figure 5.3 A 10-turn rotary potentiometer.

Reprinted with permission from Beckman Industrial Corporation, Fullerton, California.

Figure 5.4 shows a representative potentiometer circuit. When a voltage V is applied across the terminals, the output signal $e(t)$ is proportional to the angular or linear displacement of the slider.

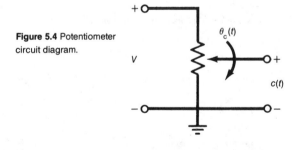

Figure 5.4 Potentiometer circuit diagram.

$\theta_c(t)$

V

$+$

$c(t)$

2. Linear Variable Differential Transformers (LVDTs)

LVDTs are made up of three coils, as shown in Figure 5.5. AC supplied to the input coil at a specific voltage E_i generates a total output voltage across the secondary coils that is a linear function of the displacement of a magnetic core inside the coils. The structure of the coils is shown in Figure 5.6, which also shows the relationship between the coils and the core for various output voltages. The more accurate plot in Figure 5.7 shows that the output is not completely linear near the center of the LVDT.

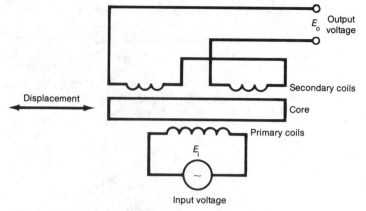

Figure 5.5 Schematic diagram of an LVDT.

Source: J. Holman, Experimental Methods for Engineers, © 1978 McGraw-Hill Book Company. Used by permission.

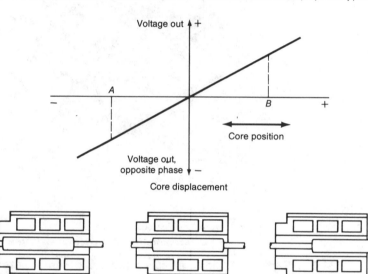

Figure 5.6 Structure and output characteristics of an LVDT as a function of displacement.

Source: J. Holman, Experimental Methods of Engineers, © 1978 McGraw-Hill Book Company. Used by permission.

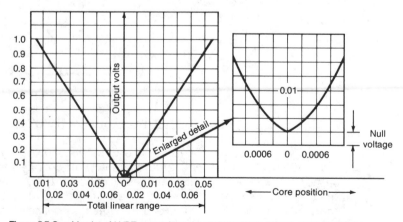

Figure 5.7 Graphic plot of LVDT output. Note the slight nonlinearity near the null region.

Source: J. Holman, Experimental Methods of Engineers, © 1978 McGraw-Hill Book Company. Used by permission.

The frequency response of LVDTs is limited by the inertia of the core but is satisfactory for many robot applications. The frequency of the applied voltage should be about 10 times the required frequency response. Note that LVDTs can be used as force or pressure transducers by applying a suitable mechanical conversion device.

3. Resolvers

Resolvers are rotational position measuring devices that work on electromagnetic principles. Stator and rotor windings are driven by clock-controlled two-phase sine waves, and the phase between the reference and rotor signals is measured and converted to digital output. A resolver and its associated electronics, as shown in Figure 5.8, can provide 12-bit resolution (1 part in 4,096) directly in binary or binary-coded decimal (BCD) code.

Resolvers are absolute encoders, since there is continuous reporting of position as long as power is on, and position information is not lost when power is turned off. Resolvers are also called electromagnetic encoders.

Multiturn resolvers are available up to 100,000 counts (17 bits). Accuracy of 0.1% is quoted by one manufacturer, Astrosystems, Inc.

Brushless resolvers can be used instead of conventional resolvers, which have brushes that can arc and wear. For robot use, where reliability is of prime importance, the use of brushless resolvers is common.

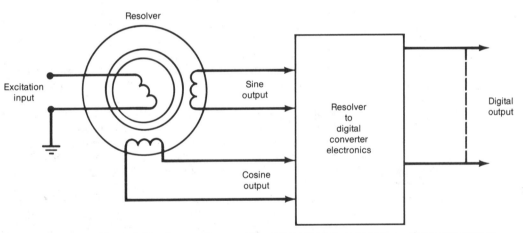

Figure 5.8 Resolver used to convert the shaft position into electrical form, which is then digitized.

4. Absolute Optical Encoders

Absolute encoders are usually high-precision rotary devices that are mounted on the shaft of a rotary drive to generate a digital word identifying the actual position of the shaft measured from a zero position. Multiple concentric rings of binary code are photographed or etched on the code wheel such that the code appearing along any radial line represents the absolute value of the position. It is necessary only to read one radial line of code to know the position of the code wheel. Therefore, the absolute encoder cannot lose its position reference and can be made very accurate. Linear absolute encoders are sometimes used but are less common.

Gray code, a code that allows only one of the binary bits in a code sequence to change between radial lines, is normally used in absolute encoders to prevent confusing changes in the binary output when the encoder oscillates between points. This code is not in strict binary form, and so must be recoded after readout by use of a table lookup in the readout circuitry or computer. An example of a 4-bit gray code is given in Table 5.1.

Table 5.1 A 4-bit gray code

Gray Code	Decimal Value	Gray Code	Decimal Value
0000	0	1100	8
0001	1	1101	9
0011	2	1111	10
0010	3	1110	11
0110	4	1010	12
0111	5	1011	13
0101	6	1001	14
0100	7	1000	15

Source: B.J. Kuo, Automatic Control Systems, 4th Edition, 1982. ©Prentice-Hall Inc. Used by permission.

Note, in Table 5.1 that the usual binary sequence is not followed. Instead, each binary code changes in only one binary position so that there can never be a position uncertainty of more than 1 bit between two successive codes. This arrangement provides greater accuracy and less chance for errors in positional logic. Some encoder disks may use natural binary codes instead of the gray code. The absolute encoder example in Figure 5.9, for example, uses natural binary code rather than gray code.

A large absolute encoder may have 10 to 20 concentric rings of binary code. Angular position can be measured to one part in 2^{10} or 2^{20} as from one part in 1,024 to one part in 1,048,576 depending on the design chosen. This presupposes that the bits can be laid down to that accuracy, of course. Photolithography of this accuracy becomes expensive. In addition, the mounts and bearings must be equally accurate. Thus, good absolute encoders start at a price of $300 each.

Obtaining position information from an optical encoder, either absolute or incremental, requires the basic components shown in Figure 5.10: a light source, the rotating code disk, multiple light detectors, and the signal-processing electronics. Light passes through the open areas of the code disk, is sensed by the detectors, and immediately generates a signal that can be digitized directly through the digital circuits to produce a digital output for transmission to the control section of the robot.

Figure 5.9 Absolute optical encoders. A 5-bit encoder. Each ring represents a binary bit in a digital code.

Source: B.J. Kuo, Automatic Control Systems, 4th Edition, 1982. ©Prentice-Hall Inc. Used by permission.

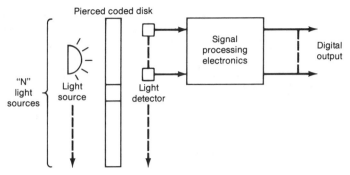

Figure 5.10 Basic components of optical encoders.

Source: B.J. Kuo, Automatic Control Systems, 4th Edition, 1982. © *Prentice-Hall Inc. Used by permission.*

5. Incremental Optical Encoders

Incremental encoders are simpler and less expensive than absolute encoders. Two photocells are required to sense both the individual steps and the direction of motion. In one design, two rings of slots or openings are used. Each ring has slots that are equal in length to two sensing positions of the light source and photocell being used. But the slots in the outer ring are displaced from the slots in the inner ring by one hole position, as shown in Figure 5.11.

When the encoder disk rotates in one direction, the outer photocell first sees light as the first half of the slot passes by; then both photocells see light at the same time on the next increment of motion. On the following increment of motion, the light to the outer photocell is cut off, but the inner photocell still sees light. Note that the slots are twice as long as the width of the photodetector-sensitive area, and the space between slots is equal to the length of the slots.

When the direction of motion is reversed, this sequence is reversed. As a result, the output of the photocells goes up and down, as shown in Figure 5.12. The output is up when the light reaches the photocell and down when light is cut off. Note that the same result could be obtained with two photocells and one row of slots, since the output is not dependent on the hole by which the direction sensing photocell is operated.

(a)   (b)

Figure 5.11 Incremental encoder disks. (a) Two-track disk. (b) One-track disk.

Source: B.J. Kuo, Automatic Control Systems, 4th Edition, 1982. © *Prentice-Hall Inc. Used by permission.*

Two photocells, A and B, are used for each disk. On the two-track disk, cell A is on the outer track and cell B is on the inner track on the same radial line. Output of the two photocells will be as just described and shown in Figure 5.12(a). Both cells are on one track on the one-track disk but displaced from each other by one and one-half slot widths or three photocell widths. As an exercise, the reader should determine why the same output is obtained.

Interpretation of these output waveforms can be done with simple logic circuits to obtain the direction and count of the number of holes that have passed by a photocell; hence, the position of the joint can be obtained. Usually a small gear ratio is maintained between the joint and the encoder, so that high positional accuracy for the driven member can be obtained with a reasonable encoder size and accuracy.

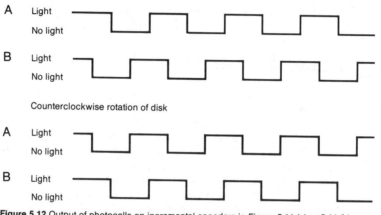

Figure 5.12 Output of photocells on incremental encoders in Figure 5.11 (a) or 5.11 (b).

Source: B.J. Kuo, Automatic Control Systems, 4th Edition, 1982. ©Prentice-Hall Inc. Used by permission.

**5.3.3
Force Sensing**

Control of the robot frequently requires force sensing. Measurement of the force on a link or the whole arm can be used to determine the angular deflection of the link, and allows for correction of applied force and angle to ensure that the end effector is moved to the correct position. It also allows detection of loads too heavy for the arm or collision of the arm with an obstacle. Another application of force sensing is in handling parts. It is then necessary to measure the force exerted and ensure that the part is not being gripped too tightly.

As robots become more complex and work in constrained environments under unpredictable conditions, more force and torque sensing will be required. Some methods for sensing force are discussed in this section. These are contact sensing methods, since they are internal to the robot arm itself.

1. Elastic Elements

Elastic elements are those whose dimensions change as a result of applied forces so that a strain, or change in some dimension, occurs. Force is measured directly by measurement of the strain. In strain gages, there is a change primarily in the resistance of the strain gage material so that an electrical output can be obtained. Of course, the underlying phenomenon is the same in both cases; only the magnitude of the strain and the method of determining it are different.

Simple Elastic Elements—Springs and Bars

Springs show considerable displacement when a force is applied. Displacement y due to the applied force F is related to the spring constant k by the equation

$$F = ky$$

For a simple bar or rod, the force is given by Figure 5.13:
$$F = (AE/L)y$$
where A = cross-sectional area of the bar, L = length of the bar, and E = Young's modulus for the bar material.

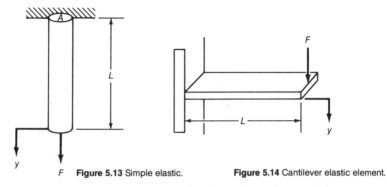

F   **Figure 5.13** Simple elastic.   **Figure 5.14** Cantilever elastic element.

Source: J. Holman, Experimental Methods For Engineers, 1978 © McGraw-Hill Cook Company. Used by permission.

Cantilever Beams.

Cantilever beams can also be used as elastic elements, as demonstrated in Figure 5.14. The force is then
$$F = (3EI/L^3)y$$
where I is the moment of inertia of the beam.

Micromechanical Accelerometers

Micromechanical accelerometers using cantilever beams have been developed at the International Business Machines (IBM) Research Laboratory in San Jose, California. As shown in Figure 5.15, these beams can be made very small. The beams shown have lengths of about 0.15 millimeter. Beams are formed from silicon dioxide by solid-state fabrication techniques similar to those used for integrated circuits. Bumps of gold have been formed on the ends of the beams to increase their inertia and therefore the sensitivity of the sensor device. All four beams were formed together, including capacitive sensors to measure the position of the beam under acceleration loads (Angell et al. [1]). Several other laboratories are working on related devices, as discussed in Angell et al. [1]).

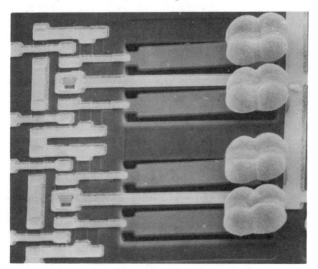

Figure 5.15 Micromechanical accelerometers fabricated by in-tegrated-circuit techniques.

Courtesy of Kurt Peterson, Trans-sensory Devices. Previously printed in Scientific American Magazine, April, 1983. Used by permission.

It will probably be necessary to build in a microprocessor element along with the micro-beams in order to calculate the amount of movement and include the necessary linearizing and correction terms because of the complexity of the mathematics needed to describe the sensory assembly. However, this is certainly a reasonable approach with today's technology and would provide valuable miniature force sensors.

Any of the three methods just described can be used as force sensors by converting the displacement to an electrical signal by use of any of the displacement sensors; potentiometers, LVDTs, or encoders. Force-versus-signal output can be calibrated and stored in the computer memory associated with the robot. Calibration methods are discussed in Section 5.4.

2. Strain Gages

Electrical resistance <u>strain gages</u> are widely used to measure strains due to force or torque. Gages are made of electrical conductors, usually wire or foil, bonded to the beam or other object whose strain is being measured. As the wire is deformed by the strain, its electrical resistance changes in a nearly linear fashion. Resistance of the conductor varies with deformation as

$$R = p(L/A)$$

where p = the resistivity of the material, and L and A are length and area, respectively.

Gage Factor

J. P. Holman [7] derives a gage factor for strain gages by differentiating the resistance equation and substituting the definition of the axial strain and Poisson's ratio as follows:

$$dR/R = dp/p + dL/L - dA/A$$

The area is related to the square of the transverse dimension, for example, the diameter of the wire. If the wire diameter is D, we have

$$dA/A = 2dD/D$$

Axial strain of a bar or wire is defined as

$$e_a = dL/L$$

A result of the axial strain is a related deformation in the cross-sectional area of the bar. The ratio of these two strains is defined as Poisson's ratio and must be determined experimentally.

$$u = -(dD/D)/(dL/L)$$

The value of Poisson's ratio for many materials is about 0.3.

Substituting for the differential values the values in terms of strain and Poisson's ratio, we obtain

$$dR/R = e_a(1+2u) + dp/p$$

Manufacturers of strain gages determine the gage factor for each gage and provide it with the gage as delivered. The <u>gage factor,</u> G, is defined as

$$G = (dR/R)/e_a$$

Substituting, we obtain

$$G = 1 + 2u + (1/e_a)(dp/p)$$

This allows us to express the observed strain as

$$e = (1/G)(-R/R)$$

in terms of the change in resistance, the original resistance, and the gage factor. Normally, the resistivity of the material does not change appreciably with strain, so we can simplify to

$$G = 1 + 2u$$

Gage factors vary from 2 for wire to 50 or more for semi-conductor gages. By measuring the change in resistance and applying the gage factor, the strain can be calculated.

Application of Strain Gages

Application of strain gages is a highly skilled art. Gages must be bonded to a clean surface with the proper type of cement. They must then be aligned properly and temperature compensated for the conditions under which they are to be used. Wire strain gages are most satisfactory when they are backed by a material such as Bakelite to maintain the wire configuration. Typical resistances for strain gages are 50 to 100 ohms. Sizes range from 1/8 x 1/4 inch to several inches. Foil gages are typically less than 0.001 inch thick; others are slightly thicker. Semiconductor gages use silicon as the material. It has the advantages of a large gage factor but the disadvantages of brittleness and a large temperature coefficient of resistance. Three types of gages are shown in Figure 5.16.

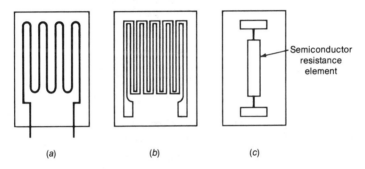

Figure 5.16 Three types of resistance strain gages. (a) Wire gage. (b) Foil gage. (c) Semiconductor gage.

Source: J. Holman, Experimental Methods For Engineers, 1978 © McGraw-Hill Book Company. Used by permission.

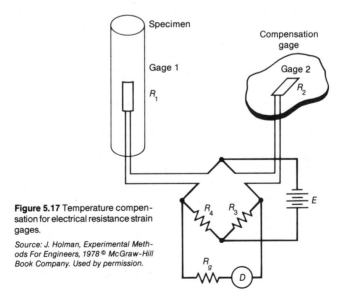

Figure 5.17 Temperature compensation for electrical resistance strain gages.

Source: J. Holman, Experimental Methods For Engineers, 1978 © McGraw-Hill Book Company. Used by permission.

Temperature Compensation

Temperature compensation is necessary whenever reasonable

accuracy is required. A bridge-type circuit, as shown in Figure 5.17, is suitable for most applications.

Multidirectional Strain Gages—Rosettes

All of the preceding discussions referred to linear strain gages. What do we do when strain is to be measured in other dimensions? Rosettes are the answer. Two types of rosettes are used. They can provide the amount of strain and the principal axis of strain along the surface to which they are applied. Rectangular strain gage rosettes are made up of three strain gages oriented as shown in Figure 5.18(a). Delta strain gage rosettes are made up of three strain gages formed into a triangle like the Greek letter delta, as in Figure 5.18(b), hence the name.

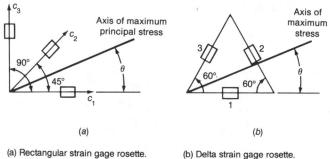

(a)   (b)

(a) Rectangular strain gage rosette.   (b) Delta strain gage rosette.

Figure 5.18 Strain-gage Rosettes.

Source: J. Holman, Experimental Methods For Engineers, 1978 © McGraw-Hill Book Company. Used by permission.

Strain gages may also operate on piezoelectric and piezo resistive principles. Quartz crystals are piezoelectric and can be mounted in much the same way as the resistive strain gages. Bending of the crystal generates an electrical signal that is a measure of the force applied. Such crystals are familiar to us due to their use in microphones and phonograph pickups. They will not stand large deformations, and thus are of limited use in robotics applications.

The piezoresistive effect is a change in electrical resistance resulting from mechanical stress. The magnitude of the effect depends on the crystalline orientation of the material, such as silicon, that is used as the sensor. Silicon pressure sensors have been built using the piezoresistive effect. At present, they are not suitable for large force measurements. Etched silicon sensors are described by Angell et al. [1]. A considerable advantage of these sensors is that the required amplifiers and digital logic can be fabricated in the silicon chip with the sensor.

5.3.4
Torque Sensing

Torque sensing may be done by use of strain gages, as described previously for force sensing. In fact, as discussed in the next section, combined force-torque sensors are extremely valuable in robotic applications.

An example of the use of a hollow cylinder for torque measurement is shown in Figure 5.19. Torque, or moment, may be calculated from

$$M = (3.1416\, G(r_o^4 - r_i^4)/(2L))\theta$$

where G = shear modulus of elasticity, r_i = inside radius, r_o = outside radius, L = length of the cylinder, and θ = angular deflection.

Since torque or moment is defined as

$$M = FL$$

where F = Force and L = moment arm, any of the methods for measuring force can also be used for measuring torque with the right geometric arrangement.

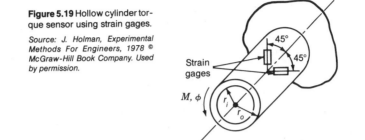

Figure 5.19 Hollow cylinder torque sensor using strain gages.

Source: J. Holman, Experimental Methods For Engineers, 1978 © McGraw-Hill Book Company. Used by permission.

5.3.5
Force-Torque
Sensing Systems

A robot arm is a rigid body that can be rotated about an axis or translated along a path. Since there are multiple links to a robot arm, various parts of it may be rotating or translating at the same time. This means that any link of the arm is subjected to many forces and torques as it moves as part of the robot arm. All of the torques and forces at any point can be summed and described by giving the forces in three Cartesian coordinates and the torques about three orthogonal (mutually perpendicular) axes. Force-torque sensing systems provide means for measuring all of these forces and torques at the same point in order to provide the necessary information to control the action of the robot.

Figure 5.20 Strain gage wrist sensor.

Source: C.A. Rosen and D. Nitzan, "Use of Sensors in Programmable Automation," Computer Magazine, December, 1977, IEEE, page 15. © 1977 IEEE.

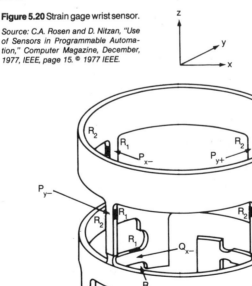

Manipulation of an object by a robot hand requires that these forces and torques be known for the wrist. Thus, several research laboratories and manufacturers have developed wrist sensors. As discussed previously, there are several different types of transducers used for sensing: strain gages, piezoelectric, magnetic, magnetostrictive, and several others. Since they are reliable, inexpensive, and rugged, strain gage sensors are most common.

Figure 5.20 shows the design of one type of sensor system developed at SRI for the Stanford arm. It is milled from an aluminum tube 3 inches in diameter. Eight narrow elastic beams, four running vertically and four horizontally, have each been necked down at one point, so that the strain in each beam is concentrated at the other end of the beam. Strain gages cemented to each point of high strain measure the force component at that point.

Foil gages, R_1 and R_2 are used to measure the strain. Two strain gages are mounted on each beam, so that the effects of temperature may be compensated for. This wrist sensor measures the three components of force and three components of force in a Cartesian coordinate system. An advantage of this design is the 7-bit resolution (1 part in 128) that was attained.

Pairs of strain gages are identified in Figure 5.20 as P_{x+}, P_{x-}, P_{y+}, P_{y-}, Q_{x+}, Q_{x-}, Q_{y+}, and Q_{y-}. Each pair consists of the two strain gages R_1 and R_2 connected as in Figure 5.21. Their outputs may be added as shown here to provide voltage outputs proportional to the force and torque components on the wrist (Rosen et al. [14]).

$$F_x = k_1(P_{y+} + P_{y-})$$
$$F_y = k_2(P_{x+} + P_{x-})$$
$$F_z = k_3(Q_{x+} + Q_{x-} + Q_{y+} + Q_{y-})$$
$$M_x = k_4(Q_{y+} - Q_{y-})$$
$$M_y = k_5(-Q_{x+} + Q_{x-})$$
$$M_x = k_6(P_{x+} - P_{x-} - P_{y+} + P_{y-})$$

The k values are the constants of proportionality for each equation.

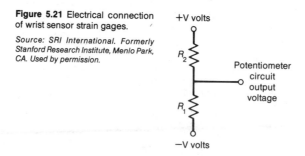

Figure 5.21 Electrical connection of wrist sensor strain gages.

Source: SRI International. Formerly Stanford Research Institute, Menlo Park, CA. Used by permission.

+V volts

R_2

Potentiometer circuit output voltage

R_1

−V volts

Other wrist sensors have been made that use extensional strain gages and strain gage bridges, so that three beams are sufficient to determine the forces and torques. Rosen and Nitzan [15] reference several designs.

Use of a wrist sensor allows constant monitoring of the actions of the robot gripper. When the gripper encounters an object, the forces and torques measured are fed to a computer, which determines the action to be taken. This procedure allows the robot to "feel" its way and, for example, to put a screw or bolt in a hole or pick up a part and be aware of the force required to do so.

Proximity sensors detect the presence of a nearby object before coming in contact with it. Some proximity sensors are sensitive over a very short range but can measure that range precisely, others have longer range but less accuracy.

1. Optical Proximity Sensor

A convenient way to detect objects in the path of a robot arm or manipulator is by use of a light source such as a light–emitting diode (LED) and a photosensitive cell. A lens is used to image the light source at a point in space where the light will be reflected from an object back through another lens to the photosensor. This arrangement does not act as a range finder but is a simple way to detect nearby objects. The geometry of the system is shown in Figure 5.22. Since the cost is low, it is feasible to use several sensors focused at different ranges.

2. Eddy Current Sensor

An eddy current is induced within a conductor whenever that conductor moves through a nonuniform magnetic field or is in a region where the magnetic field (magnetic flux) is changing. Eddy currents can be induced in any conductor but are most noticeable in solid conductors. Such currents cause losses in electric generators and motors but can also be useful.

An eddy current sensor projects a low-level field, usually at a low radio frequency (200 kilohertz is typical) that generates eddy currents in a nearby conductive target. These eddy currents reduce the impedance of the sensor in a nearly linear manner. This impedance can be converted to a voltage by suitable electronic circuits to provide a measure of the distance to the target object. Sensor calibration is different for different conductors, but linearity of $\pm 0.5\%$ is achieved using steel or aluminum targets. Output voltages of 15 to 400 millivolts per thousandth of an inch are obtained. These sensors are attractive because of their low cost, small size, and high reliability. Their effective range is short, typically less than 0.5 inch (12.7 millimeters). Distance information can be converted to torque, force, or pressure by suitable design. Sensors may be designed to be directional if desired. Differential motions of 0.001 inch are easily detected, and are therefore useful in many applications.

Simple, sensitive proximity detectors use solid–state oscillator circuits that oscillate until they reach a critical point and then cease oscillating, thereby providing a strong proximity signal. These units are commercially available in ranges of 1, 2, 5, 10, or 20 millimeters.

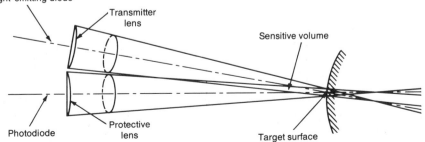

Figure 5.22 Optical proximity sensor.

Source:: C.A. Rosen and D. Nitzan, "Use of Sensors in Programmable Automation," Computer Magazine, December, 1977

3. Ultrasonic Echo Ranging

Use of pulsed acoustic beams has now become feasible and inexpensive. Perhaps the best-known sensor of this type is the Polaroid camera sonic sensor, which is used to measure the distance to objects being photographed and to set the lens opening and timing in response. This sensor is available commercially and is used in several robots. It produces pulses between seven and eight times per second. Actually, each pulse is a "chirp" with a duration of about 1.2 milliseconds; it is made up of 56 cycles at four distinct frequencies: 8 cycles at 60 kilohertz, 8 cycles at 57 kilohertz, 16 cycles at 53 kilohertz and 24 cycles at 50 kilohertz. This spread of frequencies prevents errors due to standing waves that might occur if the distance to the object was an exact multiple of a wavelength of sound. Ranges of 0.9 to 35 feet can be achieved with these sensors when appropriate time-varying gain and filter circuits are used.

Directional characteristics of sonic ranging sensors are quite good. As shown in Figure 5.23, the signal level is down 10 decibels (db) at an angle of 10 degrees. R. D. Kilmer at the National Bureau of Standards has used multiple ultrasonic sensors for safety systems (Kilmer [9]).

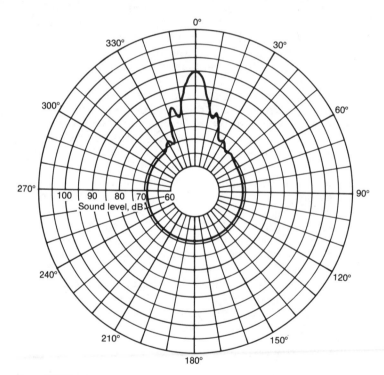

Figure 5.23 Directional characteristics of radiated ultrasonic waves from an electrostatic transducer.

Source: R.D. Kilmer, Robots 6 Conference, 1982. Courtesy of Society of Manufacturing Engineers and U.S. Bureau of Standards.

Other sensors designed for specific purposes are also available. Although the technology is relatively straightforward, several characteristics of the sound source and pattern must be taken into account.

Ranging accuracy is a function of the frequency of the pulsed sound wave and of the characteristics of the electronics used for control. One manufacturer, Massa Products Corporation, makes both the Model E-200 Ultrasonic Ranging Module, which operates at 26 kilohertz with an accuracy of $\pm$ 0.5 inch and the E-201 modules, which operate at 200 kilohertz with an accuracy of ± 0.003 inch.

Greater accuracy in the measurement of distance is obtainable at higher frequency because the wavelength of the sound wave in air is inversely related to its frequency. We can calculate the wavelength of a sound wave in air using the following information:

V_m = Velocity of sound in a medium—air, water, steel, and so on.
V_a = Velocity of sound in air (1,129 feet per second at 20°C and 1 atmosphere pressure.
w = wavelength of the sound wave in a medium
f = frequency of the sound wave in hertz (cycles per second)
w = V_m/f for air, $w_a = V_a/f$

For f = 26,000 hertz, w_a = 1,129/26,000 = 0.0434 ft. = 0.52 in.
For f = 200,000 hertz, w_a = 1,129 x 12/200,000 = 0.067 in.

Range information is obtained by pulsing the sound wave on for a very short time, usually only a few cycles, then turning it off and waiting for the echo to return. Timing circuits determine the time required for the pulse to go out and return. Distance is determined by multiplying half of the measured time by the known velocity of sound in the medium.

Sound can be used for thickness measurement when the acoustic transducer is tightly coupled to the object being measured. When the transducer is in contact with a solid, it is feasible to use higher frequencies to improve the resolution of measurement. Such sensors are not useful for direct proximity measurement because the attenuation of sound in air is very high for frequencies greater than a few hundred kilocycles. We could conceive of the higher-frequency sensors being very useful for underwater work, as are the sonar systems used by submarines. Sonic sensors could be valuable aids to location for robots working underwater. Positioning accuracies of a fraction of a wavelength are achievable at frequencies of several megahertz. This corresponds to an accuracy of a few microns, as limited by the sonic characteristics. Other limitations must be considered, of course.

4. Magnetic Sensors

Several types of magnetic sensors are in use as proximity devices. One of the simplest of them is reed relays. A fixed conductor and a reed made of a magnetic material are sealed in a glass tube. The magnetic reed is moved by the approach of a small permanent magnet and closes the electrical circuit. Essentially, these are relays in which the actuator magnet is separated from the contacts and armature. Their current-carrying capacity is low, but they are suitable for signaling or controlling low-power devices and are often used as limit switches on slow-moving mechanisms.

Another magnetic sensor is the Hall effect device, which depends on the generation of a potential in a semiconductor plate carrying a current in the presence of a magnetic field perpendicular to the plate. The potential difference generated between the two sides of the plate is orthogonal to the direction of both the current and the magnetic field. It is called the Hall voltage and is given by the equation

$$E_H = K_H(IB/t)$$

where I is the current in amperes, B is the magnetic field in gauss, and t is the thickness of the plate in the direction of the magnetic field. K_H is the Hall coefficient and is 4.1×10^{-8} volt-centimeters per ampere-gauss for silicon at a temperature of 23°C. K_H varies over a wide range of values for other materials.

5. Capacitive Sensors

Capacitive transducers have two parallel conducting plates. When the plates are displaced relative to each other, the capacitance changes so that the impedance to passage of an alternating current varies. They are not generally useful for robotic applications because of the variations in capacitance due to ambient conditions.

**5.3.7
Touch Sensing
and Taction**

Touch sensing is defined here as simple contact with an object. Taction is defined as the ability to determine the outline of an object by using multiple touch sensors. In the first item in this section, a multi-axis touch sensor is described. The last two items describe arrays of sensors that can determine outlines.

Touch sensing and taction are needed for several reasons even though a robot has good vision capability. Some good reasons are:

- When inserting small parts into assemblies, the hand may obscure the vision system, so that the robot loses its source of guidance.
- Accuracy of the vision system or other sensors may not be sufficient to detect small inaccuracies of movement.
- When inserting parts into a hole, there may be an obstruction that is detectable only by touch sensing.
- Training of robots during on-line teaching may require the operator to be in an unpleasant or hazardous area. Elimination of training may be possible if the robot is provided with good tactile capability.

1. Instrumented Remote Center Compliance Sensor (IRCC)

A new type of tactile sensor, the Instrumented Remote Center Compliance (IRCC) sensor, is based on the passive Remote Center compliance (RCC) device previously developed by the Draper Laboratory. As a passive compliance device, the RCC uses an ingenious mechanical linkage to allow a robot gripper to move either horizontally or at an angle as required to adapt itself to an insertion problem. The IRCC takes this design one step further and provides an active sensory feedback along with the passive mechanical structure.

When a rigid part contacts another rigid part, there is no give possible, and very large local forces can build up whenever there is an error in positioning the parts. When one of the parts can give or bend, we have a compliant relationship and the large forces do not occur. Humans typically handle objects in a compliant way. Muscles are tensed enough to handle the expected force and no more. Sometimes we are surprised by preparing to handle a heavy box and finding that it is empty, or vice versa. Robots can operate more accurately and effectively and can adapt to situations better if forces are known and prepared for.

As first developed in 1976, the RCC was the result of careful study and analysis of the forces developed and the motions required for assembling small objects. Examples of assembly operations are the insertion of a pin in a hole or a piston into a cylinder. It was found that both lateral and angular misalignment occurred. Gripper mechanisms with compound compliance, able to move separately laterally and in angle, were found to be surprisingly effective. In one example, it was possible to insert a bearing into a housing with a clearance ratio of .0004 in 0.2 second, starting from a lateral error of 1 millimeter and an angular error of 1.5 degrees. The RCC device and the research leading to it are described by J. L. Nevins and D. E. Whitney [11]. Drawings of the insertion problem and alternative solutions are provided in Figures 5.24 and 5.25.

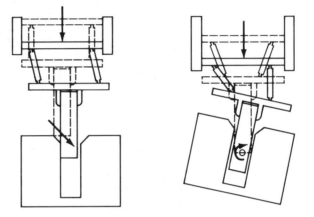

Figure 5.24 Grippers with linkage supports.

The linkage shown in Figure 5.24 can correct for errors in lateral position (left view) or in angular position (right view) but does not correct for both types of errors at the same time. Lateral errors are corrected by allowing the misaligned peg to move laterally without tilting. Placing a peg in a tilted hole is possible by rotation of the linkage without lateral motion.

Grippers with compound compliance, as shown in Figure 5.25, can correct for both lateral and angular misalignment. Because of the

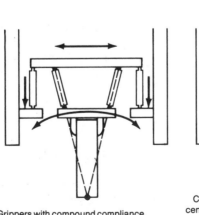

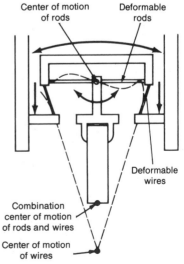

Figure 5.25 Grippers with compound compliance.

RCC linkage (left view), the effective center of rotation and lateral movement are at the tip of the peg being inserted, so the linkage can respond to either rotational or translational forces at the tip of the peg. In the right view, a set of deformable wires and rods is used instead of the linkage arrangement. The center of motion is not fixed by geometry but is controlled by the relative stiffness of the rods and wires.

An RCC used for instrumentation consists of a rigid mounting structure attached to the robot, a second rigid structure on which the end effector is mounted, and a compliant structure that joins the two.

Since the lateral deflections of the RCC represent lateral position error and the angular deflections represent angular error, it was natural to take the next step and provide sensors for the RCC. RCCs have five degrees of freedom: two lateral degrees, X and Y, perpendicular to the main axis, two angular degrees, θ_x, and θ_y, in the plane through the axis, and a rotation θ_z about the main axis. For many applications, it is not necessary to monitor the θ_z axis, so the number of transducers was reduced to four.

These transducers can be mounted in several ways, but the preferred arrangement is shown in Figure 5.26. Two of the transducers, X_1 and Y_1 are mounted in an XY plane, aligned with the X and Y axes. A second pair of transducers, X and Y_2 are aligned similarly in a plane parallel to the first plane and distance S away. As shown, the second plane is at distance L from the center of compliance.

These transducers measure the distances between the center shaft and the support case of the RCC.

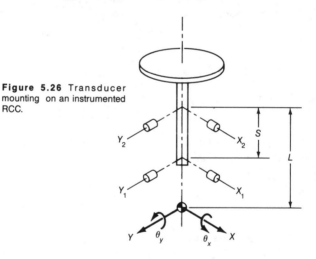

Figure 5.26 Transducer mounting on an instrumented RCC.

It can be shown, using linear approximations, that the relationship between the RCC parameters $(X, Y, \theta_x, \theta_y)$ is given in homogenous matrix form by:

$$\begin{bmatrix} X \\ Y \\ \theta_x \\ \theta_y \end{bmatrix} = \begin{bmatrix} L/S & 1-(L/S) & 0 & 0 \\ 0 & 0 & L/S & 1-(L/S) \\ 0 & 0 & -1/S & 1/S \\ 1/S & -1/S & 0 & 0 \end{bmatrix} \begin{bmatrix} X_1 \\ X_2 \\ Y_1 \\ Y_2 \end{bmatrix}$$

Either analog or digital transducers can be used. Analog units are proximity detectors made by Kaman Scientific that measure the inductive coupling to the metal surfaces of the reference frame. Digital measurements are made with optical transducers consisting of a LED light source, a thin metal shutter, and a Reticon linear diode array with 256 elements. Deflections of the compliant structure cause the shutter to move, allowing light to hit some or all of the photoelements. Scanning of the photoelements provides a count of the number of elements illuminated and therefore a measure of the distance moved. Scanning is done 50 times per second, but the photoarrays could be scanned 4,000 times per second if desired.

Resolution of the analog transducers is better than 0.5 mils over an operating range of 200 mils (5 millimeters) or approximately one part in 400. Digital resolution is 256 or, as sometimes stated, 8 bits over a 5.4 millimeter range.

Further information on the IRCC can be found in Seltzer [16,17].

2. Tactile Sensor Arrays

Humans have the ability to pick up an object, feel its shape, and identify it. Small objects—screws, pins, coins, cotter pins, and so on—can be identified and oriented entirely by the sense of touch. Robots require this type of ability if they are to become able to handle complex assembly tasks. The term <u>taction</u> will be used, as discussed by Harman [8], to mean the ability to identify objects by use of multiple touch sensors.

One approach to this capability is the tactile sensory array. W. D. Hillis [5], at the Artificial Intelligence Laboratory of MIT, has developed a monolithic array of 256 tactile sensors that fits on the end of a finger. This device was programmed to distinguish between several different fasteners—nuts, bolts, lock washers, flat washers, dowel pins, and set screws. The sensor array mounted on the end of a "finger" touches the object. Information obtained by the tactile sensors is analyzed by a computer to form the outline of the object and determine its characteristics. Note that the processing required is very similar, but not identical, to that of vision systems. Taction has the further problem of sensing and determining structure in three dimensions. For most applications, however, the three-dimensional data may be ignored. Figure 5.27 illustrates the array and the changing contact area on the conductors caused by pressure. The change in area reduces the resistance, so that a considerable range is available.

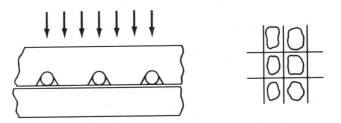

Figure 5.27 Contact resistance change with pressure. Resistance changes as pressure deforms conductive rubber.

Source: W.D. Hillis, "A High Resolution Imaging Touch Sensor," Robotics Research, Volume 1, Number 2, page 35, MIT Press © 1982. Used by permission.

Two conductive components are used in the touch array: a flexible printed circuit board and a sheet of anisotropically conductive silicone rubber (ACS). ACS has the property of being electrically conductive in only one direction along the plane of the sheet. The printed circuit board has fine parallel conductors that are placed at right angles to the ACS axis of conduction. This arrangement forms an array of pressure sensors, one at each intersection of the printed circuit board and the ACS. A resilient separator was needed to push the two layers apart when contact pressure was removed. A nylon stocking with a woven mesh was found to provide the best combination of sensitivity and range.

Figure 5.28 shows the mechanical structure of the touch sensor. ACS is the anisotropically conductive silicone rubber, and PC1 and PC2 are printed circuit boards. ACS is made up of multiple layers of silicone rubber impregnated with either graphite or silver. Note that PC2 is used to make contact with the ACS. Wire contacts were soldered to PC1 and PC2 to carry the sensor information to the computer; 32 wires were used, 16 on each coordinate direction, to provide a 256–sensor array.

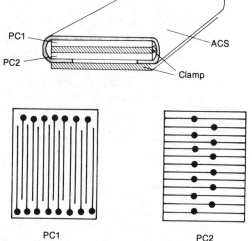

Figure 5.28 Mechanical drawing of a touch sensor.

Source: W.D. Hillis, "A High Resolution Imaging Touch Sensor," Robotics Research, Volume 1, Number 2, page 36, MIT Press ©1982. Used by permission.

PC1

PC2

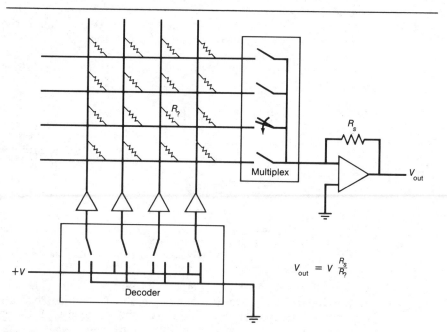

$$V_{out} = V \frac{R_s}{R_?}$$

Figure 5.29 Scanning the array.

Source: W.D. Hillis, "A High Resolution Imaging Touch Sensor," Robotics Research, Volume 1, Number 2, page 39, MIT Press ©1982. Used by permission.

Reading of the array is done by placing a fixed voltage on a selected column and grounding out all other colums to prevent false paths from being measured. Rows are all held to ground potential by injecting the necessary current to cancel the current injected by the active column. The current required for cancellation is a measure of the resistance in the sensor caused by pressure. Columns are scanned in sequence to measure the resistance of all the cross points. This is an ingenious way to avoid the use of diodes at each intersection, the normal way to prevent spurious paths from occurring when scanning an array. The scheme is illustrated in Figure 5.29.

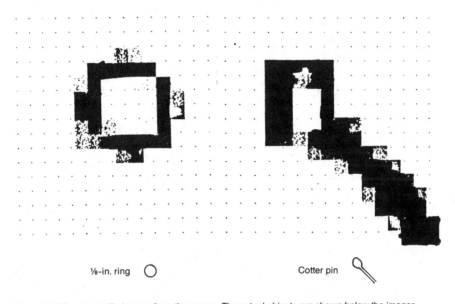

⅛-in. ring ○   Cotter pin

Figure 5.30 Sample tactile images from the sensor. The actual objects are shown below the images.

Source: W.D. Hillis, "A High Resolution Imaging Touch Sensor," Robotics Research, Volume 1, Number 2, page 39, MIT Press © 1982. Used by permission.

The results obtained were excellent, as illustrated in Figure 5.30. Even very small objects are clearly recognizable in the horizontal position. This device is still in the research stage but does demonstrate one approach to taction.

Figure 5.31 Architecture of a VLSI tactile sensor. A layer of pressure-sensitive rubber is in direct contact with the VLSI wafer.

Source: M.H. Raibert and John E. Tanner, "Design and Implementation of a VLSI Tactile Sensing Computer," Robotics Research, Volume 1, Number 3, page 4. © 1983 MIT Press. Used by permission.

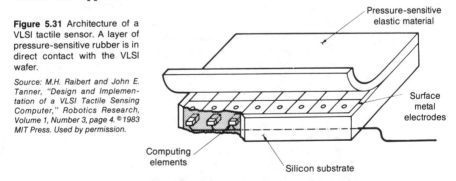

Pressure-sensitive elastic material

Surface metal electrodes

Computing elements

Silicon substrate

3. Very-Large-Scale Integration (VLSI) Computing Array

Marc Raibert and John Tanner [13] have designed and implemented a Very-Large-Scale Integration Tactile Sensing Computer that has some new properties. Instead of a passive substrate for the sensor array, each sensor has its own logical and control unit using VLSI techniques. Pressure sensing using a conductive plastic is the source of input, as illustrated in Figure 5.31. The additional functions of tactile image processing and communication are performed by the VLSI substrate.

Each sensor is connected to a computing element, which is essentially a simple microcomputer. As shown in Figure 5.32, there is an analog comparator, a data latch, an adder, an accumulator shifter, an instruction register, and a two-phase clock. Instructions are sent in from a control computer over a <u>global</u> bus that communicates with each sensor element. Instructions control all parts of the computing element, including communication to computing elements connected to neighboring sensors.

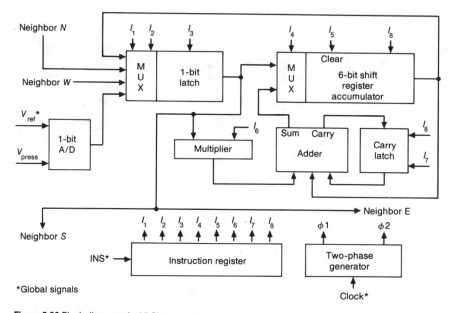

Figure 5.32 Block diagram of a VLSI computing element.

Source: M.H. Raibert and John E. Tanner, "Design and Implementation of a VLSI Tactile Sensing Computer," Robotics Research, Volume 1, Number 3, page 4. ©1983 MIT Press. Used by permission.

Capabilities of the VLSI computing element allow each sensor cell to sample the local pressure, store the transduced value, pass data to neighboring cells, and perform computations on the data. A valuable aid in analyzing images is the use of convolution, a mathematical procedure, between the image and a programmable mask. The computing elements in each cell can carry out this procedure in parallel so that the result can be available at a high rate. Image processing in the tactile image uses some of the same techniques used in visual images. Refer to Chapter 9 for a description of the image-processing procedures that are applicable.

This technique is still experimental but shows great promise. It illustrates the trend toward incorporation of computer capability at the lowest level of operation. Interestingly, this has been found to be the way in which the human vision system works. In the retina of the human eye, there is preprocessing of the information from large numbers of photosensitive cells, so that the number of nerves in the optical nerve bundle is much smaller than the number of photosensitive cells in the retina.

A satisfactory taction array should have about 25 elements in each dimension, with each element approximately 1 millimeter square (0.040 inch). This arrangement would approximate the capability of the human finger and make possible many tasks requiring location, identification, and handling of small parts. Speed is not critical. Manipulators operate in the 5- to 20-millisecond range. Solid-state circuits operate in the range of nanoseconds or microseconds. It is therefore desirable to trade speed for simplicity wherever possible.

5.4 Conditioning Sensor Output

Conditioning is a catch-all term referring to the ways in which sensor output can be made useful for measuring and controlling. Before sensor output can be applied usefully, it must be put in a form that is compatible with the rest of the system. There are several older techniques that can be used to improve the usefulness of sensors, plus the new techniques made possible by the availability of low-cost microprocessors.

Some of the techniques used are described and discussed in the following sections.

5.4.1
Conversion

Changing information from one form to another is called conversion. An example is changing a mechanical deflection to a voltage indication so that it can be handled electrically. This operation is done by various means: optical, magnetic, capacitive, and so on, as described previously. Conversion is also used to mean a change from analog to digital form by use of analog-to-digital converters, or the reverse conversion, from digital to analog, by use of digital-to-analog converters.

It is often necessary to convert from one digital code to another. Absolute optical encoders use gray code, which changes only one bit at a time when going from one position to another. Internally, in a computer a binary or BCD code is used, so a code converter is required to make this change.

Sometimes a change in voltage level is required because the voltage of the sensor output is different from that required. This is one of the classic cases of conditioning, in which operational amplifiers are used to set and compare voltage or current levels. Another use of operational amplifiers is to convert current levels to voltage levels or vice versa.

Inputs from photosensitive cells almost always require an amplifier to increase the current or voltage level and to adjust it to the

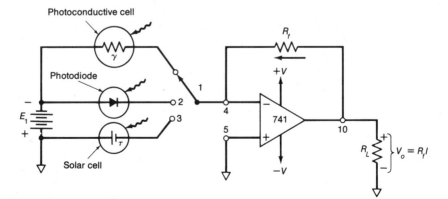

Figure 5.33 Using an operational amplifier to measure output current from photodetectors (from Coughlin and Driscoll [2]).

Source: R.F. Coughlin and F.F. Driscoll, Operational Amplifiers and Linear Integrated Circuits, 2nd edition, page 86. ©1982 Prentice-Hall, Inc., Englewood Cliffs, NJ. Used by permission.

desired input level for use in another circuit. This change is often made by operational amplifiers, as illustrated in Figure 5.33.

There are three types of photodetectors shown in this figure:

1. The solar cell directly converts light into electrical energy, which generates a voltage input into the 741 operational amplifier.
2. Photoconductive cells act like resistors whose resistance varies with the amount of light hitting them.
3. Photodiodes conduct current in only one direction, but current flow is nearly proportional to light intensity over a useful range.

Note that the source of energy for the photoconductors and the photodiodes is from the battery, while the solar cell generates its own current and voltage.

1. Operational Amplifiers

Operational amplifiers are balanced amplifiers (with two inputs, positive and negative), which can have a gain or multiplying factor of over a million times. That is, a very small change in input can cause a very large change in output. Output versus input could be nonlinear , but this situation can be prevented by feeding back some of the output (in this circuit, through resistor R1) to the input to counteract some of the input. By properly adjusting the circuit values, a stable, high-gain, very linear amplifier results.

This high linearity and stability were valuable in early analog computers for performing mathematical operations. Therefore, these amplifiers were called operational because they could be used reliably to perform mathematical operations. New operational amplifiers are capable of high-speed response, excellent linearity, and stable operation. They are widely used in robot control and sensory circuits, and in computer control circuits for various kinds of conversions and control.

5.4.2
Calibration

Calibration is the process of determining the relationship of sensor output to the actual value of the input. It is desirable to calibrate the complete sensor system if maximum accuracy is required. One way to calibrate is to move the input values to known points and record the outputs obtained. Then, if these input-output relationships are stored, we know that the sensor is correct at these points, at least.

Calibration may require the use of precision equipment such as primary standards, or secondary standards if maximum accuracy is desired. Primary standards are based on direct physical laws or on the basic standards of length, mass, and other values that are maintained by the National Bureau of Standards (NBS). Secondary standards have been certified by the NBS as accurate within a specified range.

Interpolation may be used between calibrated points, in most cases, to determine intermediate values between the calibrated points with suffiicent accuracy.

5.4.3
Linearizing

Some years ago, it was difficult to modify a circuit so that the sensor output bore a linear relationship to the input. Some sensors are inherently nonlinear. Logarithmic and square law relationships, for example, are common. Before computers were widely used, it was necessary to design circuits that could accept a nonlinear sensory output and produce a linear result as input to computations and control.

Microcomputers have simplified the linearizing of sensor outputs. Now it is only necessary to calibrate output versus input and store the values in the computer memory. Whenever an output is received by the computer, it can "lookup" the value in its memory and determine the actual value that caused that input. When the output value falls between the calibrated values, the computer program automatically calculates the interpolated value corresponding to the actual input value.

In principle, at least, it should be possible to linearize or correct for any environmental or loading condition by the use of a computer. By measuring the temperature, force, torque, or other disturbance and applying known mathematical relationships, we should be able to correct for all of the factors that cause inaccuracy. Developing this capability to a useful degree is one of the challenges of sensory and control development.

**5.4.4
Amplifying,
Comparing,
and Buffering**

Operational amplifiers, discussed briefly in Section 5.4.1, can be used to amplify weak sensory signals before they are sent to the computer or other control means. Signals may be compared as analog voltages by using operational amplifiers as comparators. Outputs from comparators may be the difference of two signals or merely a high or low value telling us which input was the larger. Details of circuits of this type are described in Coughlin and Driscoll [2].

Buffering is a means of temporarily storing sensory or other signals until they can be used for their intended purpose. Analog circuits (sample-and-hold circuits) that store a charge on a condenser may be used or the signal may be converted to digital form and stored in a digital storage register until it can be used. Registers are described in Chapter 7.

**5.4.5
Demodulation**

Particular types of sensors may generate frequency modulation (FM), in which the information is contained in a varying frequency of constant amplitude, or amplitude modulation (AM), in which the frequency of the signal is constant and its intensity or amplitude is changing to transmit information. In these cases, the signal must be passed through a demodulator to separate the information from the "carrier" frequency and generate a slowly varying signal that can be analyzed or digitized.

5.5 Analyzing Sensor Data

Converting sensor data from raw output data to useful information may require several steps. We have discussed the conditioning processes that provide the data in useful information form. However, it is often necessary to do further processing to determine the relationships between data from different sensors. Moving a robot end effector to a defined position and, at the same time, applying the required forces and torques requires the integration of information from several sensors with information about the environment and the physical capabilities of the robot itself, for example.

**5.5.1
Analysis
Programs**

Analysis programs are subprograms that integrate data for one or several sensors and pass it on to a higher-level program for further use. A good example is given in Section 5.3.7, which describes the VLSI computing array and the way in which local computing elements analyze part of the data before passing the summary results to the next higher level. This type of analysis can be done either in hardware or in software, as illustrated by some of the vision system hardware and software described in Chapter 9.

5.5.2   As yet, there has been little verification of sensor information
Verification for reasonableness, consistency, or comparative value.   This is a
necessary step if a robot is to carry out complex tasks reliably. An
example might be to check that a robot end effector is, in fact,
touching an object, as verified by the vision system, when the touch
sensors or the gripper indicate that an object is being touched or
gripped. In more complex systems it is being found that some kind of
verification is a necessity.

5.6 Sensors for Special Applications

Specialized sensors are required for certain applications in
addition to the more general tactile, force, torque, and other sensors
that have been described in this chapter. Of particular interest are
the specialized sensing techniques used for welding, described in
Chapter 10, and the vision sensor systems used for assembly and
inspection, described in Chapter 9.

5.7 Future Work

Sensors of all types will be used in increasing quantities as they
become more sophisticated and less expensive. Especially needed at
present are improved force, torque, and tactile sensors similar to
those being developed at SRI International, Cal-Tech, CMU, and MIT,
to name some of the many research organizations working in this field.

Another need is for integration of sensors and controls into much
more complex systems. This process will require microcomputers to
be built into the sensor itself and to do preprocessing of information at
the sensor level before transmitting information to the next higher
level.

Exercises

Do the following exercises to verify that you understand the concepts
discussed in this chapter.

1. What types of materials are used for potentiometers?
2. How does an incremental encoder differ from an absolute en-
    coder?
3. Which type of proximity sensing would you use for measurements
    over a range of 0.1 to 1.0 inch when the desired accuracy should
    be ± 0.001 inch?
4. Discuss the differences between accuracy and repeatability speci-
    fications for a robot end effector. Which is most influenced by
    changes in temperature?
5. How are newly developed sensor systems emulating the way in
    which sensor systems work in the human body?
6. Write an algebraic equation that describes the shape of a sensor
    response curve which has the linearity
$$Y = bX$$
where the value of $b = 8 + 0.1$.   Determine the equation for the
maximum deviation from perfect linearity in this sensor. Where does
this maximum deviation occur? If the range of X is from 2 to 50
inches, what is the numerical value of the maximum deviation from
perfect linearity?
7. If you are an advanced student who understands dynamics and the
    use of homogeneous matrix theory, show that the transducer in

Figure 5.26 has the relationships presented in the homogeneous matrix given in the text.

8. What sensory technique would you suggest trying as an alternative to the use of conductive rubber and similar compressible materials in touch sensors?

9. If you are familiar with logical circuits, design a digital logic circuit that will convert the 4-bit Gray code Part 4 of Section 5.3.2 to a 4-bit pure binary code. See Chapter 7 for logical circuit information and truth tables for logical circuits.

10. Describe the way in which two photocells or other optical sensors can be used on one incremental encoder track. Incremental encoders with two tracks are described in Section 5.3.2.

11. Several new terms have been introduced and defined in this chapter. A review of the text to find and understand these terms will be valuable to the reader. Some terms that should be reviewed are:

Potentiometer	Sensitivity
LVDT	Linearity
Resolvers	Range
Encoders	Response time
Strain gages	Accuracy
Force—Torque Sensing	Repeatability
Proximity sensing	Resolution
Ultrasonic echo ranging	RCC and IRCC
Taction	

References

[1] J. B. Angell, S. C. Terry, and P. W. Barth, "Silicon Micromechanical Devices," Scientific American, April 1983, p. 44.

[2] R. F. Coughlin and F. F. Driscoll, Operational Amplifiers and Linear Integrated Circuits, Prentice-Hall, Englewood Cliffs, N.J. 1982. This text has much valuable information on sensory data handling and robot control.

[3] H. A. Ernst, "MH-1, A Computer-Operated Mechanical Hand," Ph.D. thesis, Cambridge, Mass. Massachusetts Institute of Technology.

[4] C. C. Geschke, "A System for Programming and Controlling Sensor-Based Robot Manipulators," PAMI, Vol. 5, no. 1, January 1983, p. 1.

[5] W. D. Hillis, "A High Resolution Imaging Touch Sensor," Robotics Research, Vol. 1, No. 2, MIT Press, Cambridge, Mass., 1982.

[6] R. E. Hohn and J. G. Holmes, "Robotic Arc Welding—Adding Science to Art," Robots 6 Conference Proceedings, March 2-4, 1982, Detroit, Robotics International of SME, Dearborn, Mich., 1982, p. 303.

[7] J. Holman, Experimental Methods for Engineers, 3rd: McGraw-Hill, New York, 1978. This is an excellent source book for sensory theory and applications.

[8] L. D. Harmon, "Automated Tactile Sensing," Robotics Research, Vol. 1., No. 2, MIT Press, Cambridge, Mass., 1982.

[9] R. D. Kilmer, "Safety Sensor Systems for Industrial Robots," Robots 6 Conference Proceedings, March 2-4, 1982, Detroit, Mich. Robotics International of SME, Dearborn, Mich., 1982, p. 479.

[10] B. C. Kuo, Automatic Control Systems, 4th ed. Prentice-Hall, Englewood Cliffs, N.J., 1982.   This text covers servos and control systems thoroughly.

[11] J. L. Nevins and D. E. Whitney, "Computer Controlled Assembly," Scientific American, February 1978, p. 62.

[12] T. Okada, "Development of an Optical Distance Sensor for Robots," Robotics Research, Vol. 1, No. 4, MIT Press, Cambridge, Mass., 1982, p. 3.

[13] M. H. Raibert and J. E. Tanner, "Design and Implementation of a VLSI Tactile Sensing Computer," Robotics Research, Vol. 1, No. 3, MIT Press, Cambridge, Mass., 1982, p. 3-18.

[14] C. A. Rosen et al., "Exploratory Research in Advanced Automation," Second Report, Stanford Research Institute. Prepared for the National Science Foundation under Grant GI-38100X1, 1974.

[15] C. A. Rosen and D. Nitzan, "Use of Sensors in Programmable Automation," Computer, IEEE, December 1977, p. 12.

[16] D. S. Seltzer, "Use of Sensory Information for Improved Robot Learning," Robots IV Conference, October 1979. Reprinted in [19], Vol. 1, p. 424.

[17] D. S. Seltzer, "Tactile Sensory Feedback for Difficult Robot Tasks," Robots 6 Conference, Proceedings, March 2-4, 1982, Detroit, Mich. Robotics International of SME, Dearborn, Mich., 1982, p. 467.

[18] J. F. Shelley, Engineering Mechanics—Statics and Dynamics, McGraw-Hill, New York, 1980. Much good discussion of basic statics and dynamics, with many fine examples.

[19] W. R. Tanner, ed: Industrial Robots, Second Edition, Vol. 1, Fundamentals, Vol. 2, Applications, Robotics International of SME, Dearborn, Mich., 1981.

Controls and Control Methods

This chapter will classify and describe the most important methods and techniques for robot control. Also included is a discussion of alternative means for measuring and controlling the positioning of the individual links of a robot arm. After the description of alternative methods for control, some of the mathematical methods of analysis for servo mechanisms will be discussed.

There are several ways to classify control methods, depending on how the information is to be used.

6.1 Classification of Robots by Control Method

The two major control methods for robots are non-servo and servo control.

6.1.1 Nonservo Robots

Early robots were of the nonservo type. Powered by pneumatic cylinders, hydraulic cylinders, or electric motors, they moved from one position to another under the control of cams, mechanical stops, limit switches, valves, or relays.

1. Bang-Bang Controlled Robots

Bang-bang control robots were the simplest of the early robots and were capable of moving only between fixed stops at each end of their path. These robots were called <u>bang-bang robots</u> because they banged into stops at either end of their motion. An example of this type of robot is shown in Figure 4.6. Stops were adjustable mechanically but could not be changed during operation.

Some of the characteristics of these robots are:

- Movement from one fixed point to another (point-to-point operation).
- High accuracy possible; limited primarily by the precision of location of stops.
- Limited flexibility
- Low cost.

2. Sequence-Controlled Robots

Some limitations of bang-bang robots were overcome by providing sequence control of arm motion. Multiple stops were placed along the motion path, with the insertion of the stops controlled by auxiliary means such as pneumatic cylinders, solenoids, or other devices. Stepping switches could be used selectively to cause the stops to be inserted or removed as desired. More complex control was possible using programmable controllers. These controllers initially used relays

151

to control sequencing but graduated to the use of programmable controllers, which appear to operate like relay systems but incorporate a small microprocessor to provide flexible sequence control.

Some open-loop robots or nonservo robots could be positioned in many places by using stepping motors that could move in discrete steps. Stepping motors are quite accurate and do not require position feedback, such as is provided in closed-loop or servo-controlled robots. (Stepping motors were described in Section 4.4).

**6.1.2
Servo-
Controlled
Robots**

Improved control and the ability to stop at intermediate positions was made possible when fully servo-controlled or servo robots were developed. Servo-controlled robots use continuous feedback from position-measuring sensors to control the movement of joints. Each joint to be servoed has a separate position sensor. Usually the sensor output is electrical or is converted to an electrical signal that is amplified and used to control an electromagnetic valve or electric motor. Potentiometers, resolvers, optical encoders, or other devices are used to sense position, as described in Section 6.5.6.

A simple servo system is depicted in Figure 6.1. Note that this system also has a velocity feedback, which is often necessary to stabilize the servo operation so that it will stop accurately at the designated position. (See Section 6.4 for identification of terms.)

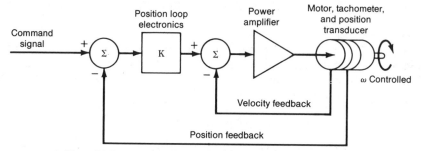

Figure 6.1 Servo system with position feedback and velocity feedback.

A separate servo control system is required on each joint to be controlled. It is possible to use servo control on some axes and nonservo control on other, less critical axes. In most high-accuracy robots, there are six separate servo systems controlling six joints to provide the necessary flexibility and accuracy.

1. Point-to-Point Operation

In an industrial environment, servo robots are further differentiated by the classifications point-to-point and continuous path.

End effectors of point-to-point robots move from one point in space to another without regard to the path taken between points, following a path similar to that of the nonservo robots. This type of operation is suitable for picking up a part at one fixed location and placing it in a machine or fixture at an arbitrary location several inches or feet away. These robots can go to a large number of different points, but the total number of points is limited by the control equipment.

2. Continuous-Path Operation

Continuous-path robots move the end effector from point to point but are controlled to follow a desired path that is made up of many points only fractions of an inch apart. Path control capability is discussed in Section 6.2.2.

Point-to-point and continuous-path methods are illustrated in Figure 6.2. Even if the point-to-point robot is servo controlled, there is no guarantee that it will follow a particular path between points P_1 and P_2. Drive motors and linkages may operate at different speeds on two sequential moves and cause different paths to be taken. In applications in which the robot must position an end effector through an opening or move around obstacles, the point-to-point robot may hit something. In continuous-path operation, the robot moves a short enough distance between points so that there are no accidents even though it may deviate slightly from the desired straight-line path. Figure 6.2 also shows why polar and cylindrical coordinate systems were favored for early robots: Straight-line motion could be obtained by driving one link at a time. Revolute or jointed robots had to move at least two links to generate a straight line.

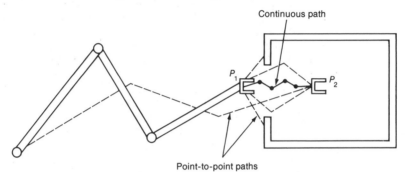

Figure 6.2 Point-to-point versus continuous path.

3. Categories of Servo System Operation

In servo control, each joint is controlled by an independent position servo, with joints moving from position to position independently. There are three possible categories of motion:

1. Sequential Joint Control. Joints are activated one at a time while the other axes are held fixed. This procedure simplifies control, but the operating time is longer than it would be if all joints operated at the same time. It provides better path control because there is no interaction between joint movements.
2. Uncoordinated Joint Control. All joints are allowed to move together, so that they follow a path determined by the relative speeds of each joint movement. There is no coordination between joints. Thus, prediction of the end-of-arm path is difficult or impossible, since varying arm loads or varying friction might change the joint velocities in a random way.
3. Terminally Coordinated Joint Control. This is the most useful type of point-to-point control but requires more control equipment. Individual joint motions are coordinated to reach their endpoints simultaneously.

6.2 Control of Servo Robots

6.2.1
Point-to-point
Control

Two methods were used for point-to-point control in the first industrial servo-controlled robots.

1. Potentiometer Control

Early point-to-point robots were controlled by potentiometer feedback. A potentiometer was assigned to each degree of freedom for each point to which the robot was to move. A system with five degrees of freedom and 20 potentiometers could go to four points. By adjustment of the individual potentiometers, the points could be set precisely.

Sequencing of movements from point to point was controlled by a stepping switch or counter that controlled relays to switch in the proper potentiometer at the correct time. An example of this type of circuit is shown in Figure 6.3. Note that the parallel line symbols represent relay contacts, not capacitors. For five degrees of freedom, this system required five servo amplifiers. With three potentiometers per link as shown, this servo amplifier could be controlled to move to three separate points.

Operation of this circuit was: Command potentiometers were set to the value needed for each position to which the arm link was required to go. This operation was relatively simple since the arm would move as the command potentiometer was adjusted until the feedback signal from the feedback potentiometer was equal to the signal from the command potentiometer. When the arm link was in the desired location, the command potentiometer setting was locked in place. Note that there was one feedback pentiometer for each link or actuator, but there were as many command potentiometers as there were points to be set. When all potentiometers had been set, the operation was ready to begin.

Each command potentiometer was connected to the input of the servo amplifier by a command relay contact that was actuated by a relay. Contacts, shown as parallel lines in Figure 6.3 were closed by the corresponding command relay (K1 controlled contact K1, etc.) Relays, in turn, were controlled by the electronic switching (counter) system.

Detailed sequencing of this system is:

1. At system turn-on, the counter is reset to Step 1. This causes relay K1 to close and connect command potentiometer (pot) R1 to the input of the servo amplifier shown in Figure 6.4.
2. When the specified arm link reaches the position set on command pot R1, the feedback signal is equal to the command pot signal and the net value out of the servo amplifier is zero. This zero error signal causes a step to occur from counter 1 to counter 2 so that relay K1 is opened, disconnecting pot R1. At the same time, Step 2,which operates relay K2 and connects pot R2 is set.
3. As each link completes the move to its next position, the switching system steps to the next relay to control the next step in the sequence.

A separate set of counters and relays was required for each link. The expense of this setup limited the number of steps used on point-to-point robots.

2. Point-to-Point Control Using Computer Memory

Improved controls store the required points in computer memory. Points must specify three positions and three directions to specifically determine the orientation of a link in space. However, each link is

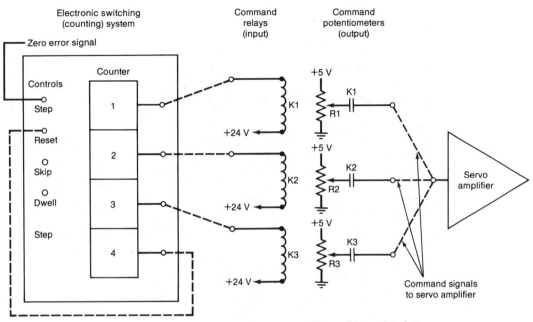

Figure 6.3 Potentiometer control for a point-to-point robot.

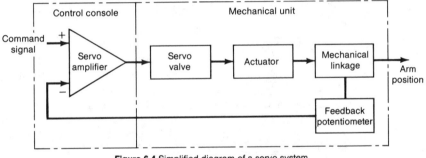

Figure 6.4 Simplified diagram of a servo system.

limited in position and direction by the associated links, so that it is only necessary to specify a movement relative to the preceding link, in order to specify the position of any link.

Training of the robot requires that the robot arm be moved to a succession of points and the coordinates of those points be stored in the digital memory of the computer. Coordinate information is obtained by reading the output of the feedback potentiometer on each joint into the memory.

Analog-to-Digital Conversion

Potentiometers generate an analog signal; the output from the slider is a voltage proportional to the value stored. This analog voltage must be converted to a digital number for storage in the computer's memory. Circuits known as <u>analog-to-digital converters</u> are used for this purpose. They accept a voltage input and generate a digital output, in the form of binary bits, that represents the same value. Digital input is then read into the computers memory and stored. A byte containing 8 bits of informaton can specify a position to 1 part in 256 in binary form or 1 part in 100 if stored in decimal form (see Appendix A.1 for a discussion of this point).

Memory-Size Required

Many manufacturers quote positioning accuracies of 0.04 inch for a robot with a 5-foot reach. This accuracy can be specified by using $60/0.04 = 1,500$ points, so that it would be necessary to use 11 binary bits (since $2^{11} = 2,048$ is the next binary number larger than 1,500). Binary numbers are discussed in Appendix A.1. For this reason, it is common to use 2 bytes per point to accomodate the various positioning accuracies required. Since each link position and direction requires at least 2 bytes of information, there is a need for storage of 12 bytes of information for each point for a robot with six links or degrees of freedom. Memory size in the computer limits the number of points that can be stored. If multiple programs are to be stored, the computer memory required may quickly exceed the memory available. Many older systems were limited to 16K bytes of memory (1 K equals 1,024 bytes). This memory would accomodate approximately 48 programs with 28 points each ($48 \times 12 \times 28 = 16,128$ bytes) for a six-degree-of-freedom arm or one program of 1,344 points for a continuous-path arm. Twenty-eight points are not enough for continuous-path operation, which may require 1,000 points or more, but are enough to allow a robot to service many machines in sequence. A part could be picked up from one machine, moved to another, and so forth. Each movement would be made up of several shorter movements. To pick up a part, for example, the robot must approach the part, move down to grasp it, raise up again, and then move to the next part location.

Digital-to-Analog Conversion

Conversion of the stored digital memory information to link movement is accomplished by digital-to-analog conversion in the simplest case. Since the potentiometers have been replaced by the stored memory, we must generate an analog voltage signal into the servo amplifier. This is done by use of a Digital-to-Analog Converter, a commercially available circuit that uses the digital code to control the current flowing through a network of precision resistors. Current through the resistors adds in a binary fashion to produce the required analog voltage equivalent to the digital number stored.

**6.2.2
Teaching
and Pendant
Control
(Teach Boxes)**

Teaching in point-to-point systems is done by moving the robot arm to each specific point manually and pressing a button to record the coordinates of that point. Position measurement is done by encoders, potentiometers, or other means, as described in Section 5.3.2. These devices, through suitable electronics, generate a set of digital numbers for each coordinate that are routed by electronic switching to the computer's memory. Large robot arms are often counterbalanced to make manual movement easier. Some large robots have an auxiliary lightweight arm that can be used for training or guidance.

Instead of moving the robot arm physically through the desired sequence of points, it is possible to use electronic control circuits to achieve the same result. This method is called pendant control because a control box (Teach Box) hanging from a cord, like a pendant, contains the push buttons and switches necessary to control the sequence of movements. Pressing the proper control button applies a signal to the servo amplifier that causes the arm link to move. After the arm has been moved to the desired point, a "record" button can be pressed to store the coordinates of the point in memory. A teach box with multiple control buttons and switches is shown in Figure 6.5. It is connected to the controller by a long cable and can be used for testing as well as training.

Pendant controls are especially useful for large robots that might otherwise be difficult to take through the training cycle. They are also useful in applications in dangerous environments. In addition, they are more convenient for many purposes other than manual movement, so the use of pendant control has increased in the last few years. In some cases, the pendant has been converted to a gun-shaped control box that has buttons placed for easy control of multiple links.

The teach pendant can control all axes of the robot arm in terms of position and movement velocity. In addition, it has controls for operating mode, tool type used, vacuum controls, and other functions. Different manufacturers provide capabilities to operate the robot for particular applications.

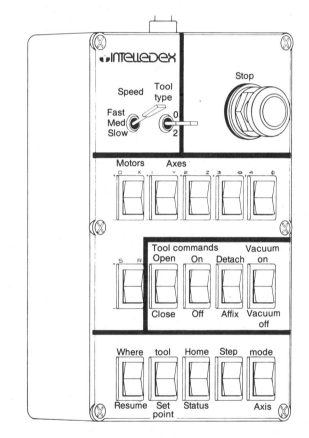

Figure 6.5 The Teach Pendant (courtesy of Intelledex Corp.)

**6.2.3
Continuous
Path**

1. Path Types

Three basic approaches to continuous-path servos are in use:

Standard Servo Control

Standard servo control is the simplest type of path control. A representation of the path is stored in memory, and the servo controller moves the arm through the path. All calculations made are based on past and present servo errors. Most industrial robots use this approach.

Feedforward or Look-ahead Control

A more complex approach is the feedforward control, which attempts to predict where the path will go as a result of the previous path. Such prediction can provide improved path control. It is similar to driving a car: The driver looks ahead and makes small corrections before reaching a desired point. There is then less need for large corrections in the path taken.

Controlled-Trajectory Control

The third approach to continuous-path control is so different that it has been given a new name by some researchers. It is variously called path planning, trajectory calculation (Paul [32], Edwal [12] or controlled path (Hohn [15]). In order to emphasize the important intrinsic difference, the term controlled trajectory would be better. Trajectories are often thought of as freeflight, with no control after release, as in the trajectory of an artillery shell. Controlled path is not sufficiently differentiated from other continuous-path approaches. In this book, therefore, we will use the term trajectory with the understanding that this is a precomputed and controlled, not a free-flight, trajectory. This topic is discussed in more detail in Section 6.3.1.

2. Teaching of a Continuous Path

The operator teaches a continuous path by moving the robot arm through the path just as he would do if actually carrying out the task. As the arm is moved, its coordinates are stored automatically at frequent intervals along the path. Storage frequency is dependent on the accuracy required and can be specified by the operator in some systems.

Large numbers of coordinate points are required for continuous-path control, so that a large memory is needed. One manufacturer of painting robots, which require continuous-path operation, specifies a memory size of over 400,000 bytes of semiconductor memory. This is Random Access Memory (RAM), which is capable of accepting information at high rates (on the order of a million bytes per second or more). In this system, there is enough memory to run the robot arm for 4 hours in a continuous-path spraying operation.

3. Programming

Programming of a continuous-path robot can be done with any of the robot-programming languages that have been developed. These languages simplify the programming procedure by providing English-like words to describe the operations being performed. Examples are words such as MOVE, OPEN, CLOSE, ROTATE, and similar ones. At present, there are at least 16 different programming languages used for robot control. They may be entirely new languages or modifications of existing ones. Various programming languages and methods are explored in depth in Chapter 8.

6.3 Advanced Control Methods

As robots became more sophisticated, it was found that even the continuous-path robots were unable to carry out tasks requiring smooth control of the path.   As a result, some researchers and manufactuers developed more powerful control techniques. One of the first of these was the Controlled-Path technique described by R. E. Hohn [15] of the Cincinnatti Milacron Corporation. Other names for similar approaches are path planning and trajectory calculation. We have chosen the more precise name controlled trajectory for the same approach. This technique is described in Section 6.3.1.

Robots must be able to avoid obstacles in their path. Either they must have knowledge of their surroundings (a map), or they must have sensors capable of measuring the distance to nearby objects and programming capability for detouring around objects in their path. Much work has been done in this area, as discussed in Section 6.3.2.

Improved sensory capability allowed the robot to respond to its environment in new ways, so that Adaptive control became feasible. Much pioneeering work was done in this area at SRI and reported by C.A. Rosen and D. Nitzan (Rosen [37]). This and other techniques are described in Section 6.3.3.

Intelligent robots are not yet available, but the current work on Artificial Intelligence (AI) at many locations suggests that in the near future there will be robots that can sense their surroundings, make measurements of important parameters, and decide for themselves the action to be taken in a particular situation.   In order to make these decisions, it will be necessary for them to have a model of their environment and of their own structure and capabilities.

Some applications for which these capabilities would be useful and possibly available fairly soon are discussed in Section 6.3.4. Also included in this section is a discussion of Task Planning (Lozano-Perez [21]), which is a major step in achieving intelligence in robots.

**6.3.1
Controlled
Trajectory**

A controlled trajectory for the robot's hand or end effector can be achieved only by controlling each joint of the robot so that the desired trajectory is achieved. Forces and torques must be applied to the robot joints to generate the necessary positions of the hand. In many cases, it is also necessary to control the accelerations and velocities of the individual joints in order to generate a desired velocity or acceleration in the end effector.

If no sensory means is used to determine the characteristics of the robot's environment, it can be controlled only by precomputing its trajectory, or path through space, and applying the computed values to the control means. In this case, we measure any necessary characteristics of the robot and the forces, torques, and so on to which the robot is exposed, but make no measurements of the robot's relationship to objects in its environment.

When the robot arm, or manipulator, is made up of sufficiently rigid links, the control problem is simplified; bending can be ignored. Industrial robots today are designed to be rigid enough to maintain specified positional accuracies under any load within a specified range. At present, when bending is permitted, there is no control method that will maintain high accuracy unless sensory feedback is available. Present computation methods are not capable of providing the necessary control information at a real-time rate (fast enough to control the robot) to compensate for the highly nonlinear effects due to bending.

Computation of the path of the robot's end effector requires that the inverse dynamic problem be solved. Robot arm dynamics describes the dynamic behavior of the manipulator. Dynamic equations are the equations of motion of the robot arm based on the forces, torques, inertias, and other characteristics of the system. They are used for design of control equations, simulation of arm movements, and evaluation of the kinematic design and structure of the arm.

As discussed by Paul [32], the dynamic equations for a manipulator with six degrees of freedom are made up of six coupled, nonlinear differential equations that are impossible to solve directly except in some trivial cases. Luh [23] illustrates the complexity of the equations with examples of torque equations that contain as many as 3,600 terms describing the interactions of each degree of freedom with inertias of each link. It is therefore necessary to use approximations, omit second-order terms, use linearization techniques, and even neglect such items as friction and Coriolis effects to obtain equations that can be used effectively.

Lee [19] discusses robot arm kinematics, dynamics, and control. He also evaluates various ways to formulate robot arm dynamics. Four general approaches have been investigated: Lagrange-Euler (L-E), Newton-Euler (N-E), recursive Lagrange-Euler, and a generalized d'Alembert (D-A) principle formulation. Lee [19] describes the L-E and N-E methods and evaluates the usefulness of the L-E, N-E, and D-A equations.

1. Lagrange-Euler Formulation

L-E uses Lagrangian mechanics (briefly discussed in section 6.7.4) to develop the equations of motion for the manipulator. The resulting equations are complex, nonlinear, and usually computationally inefficient. About 7.9 seconds are required to compute the move between two adjacent points on the trajectory for a six-joint arm when using the complete solution running in FORTRAN on a DEC PDP 11/45 computer. This time is too long for real-time or on-line operations of a robot arm. By neglecting second-order terms such as Coriolis and centrifugal terms, other investigators have been able to increase the computation speed, but at the cost of loss of accuracy at high joint velocities.

2. Newton-Euler Formulation

Using Newton's Laws and Euler's equations together results in the N-E formulation of the dynamics of a manipulator. Increased speed and accuracy can be obtained by the use of this formulation of the dynamic equations. These equations can be used sequentially to control the robot links. Only 4.5 milliseconds were required to perform the control computations of the PDP 11/45 computer using this technique and a program written in floating point assembly language (Luh [26]). This computation speed allows the control signals to be recomputed at least 60 times per second so that the required control accuracy is obtained.

Another approach using the N-E formulation excludes the dynamics of the control device and the gear friction. A set of forward and backward recursive equations results which can be applied to the robot links sequentially. Lee [19] reports that, with this algorithm, about 3 milliseconds are needed to compute the feedback joint torques per trajectory set using a PDP 11/45 computer.

Table 6.1 provides a comparison of robot arm dynamics formulations where n is the number of degrees of freedom in the manipulator.

Several alternative control methods have been developed that are applicable to trajectory control. Some of these methods are described in Section 6.6.3 in our discussion of alternative control methods. It should be noted that the Modular Reference Adaptive Control method described there is not an adaptive method by our definition since it does not require or necessarily use sensor information about its environment. It is a powerful method that is applicable to adaptive systems, however.

Table 6.1 Comparison of robot arm dynamic formulations.

Approach	Lagrange-Euler	Newton-Euler	Generalized d'Alembert
Multiplications	$\dfrac{128}{3}n^4 + \dfrac{512}{3}n^3$ $\dfrac{739}{3}n^2 + \dfrac{160}{3}n$	$132\,n$	$n\left[\dfrac{31}{2}n^2 + 44n + \dfrac{63}{2}\right]$
Additions	$\dfrac{98}{3}n^4 + \dfrac{781}{6}n^3$ $\dfrac{559}{3}n^2 + \dfrac{245}{6}n$	$111n - 4$	$n\left[9n^2 + \dfrac{69}{2}n + \dfrac{53}{2}\right]$
Kinematics representations	4×4 homogeneous matrices	Rotation matrices and position vectors	Rotation matrices and position vectors
Equations of motion	Closed-form differential equations	Recursive equations	Closed-form differential equations

Source: C.S.G. Lee, "Robot Arm Kinematics, Dynamics and Control," Volume 15, Number 12, IEEE Computer. © 1982 IEEE. Used by permission.

6.3.2 Obstacle Avoidance

Any robot with some flexibility of movement will have to deal with obstacles in or near its operating path or trajectory. In order to incorporate obstacle avoidance in its programming, it must either have a model of its environment or some type of sensory information that gives it information about its environment. Assuming that the necessary information is available, how can a robot deal with obstacles and avoid collisions? All parts of the robot must be considered; an elbow can cause as much damage as an end effector.

Lozano-Perez [21] discusses three classes of algorithms that have been developed to plan for obstacle avoidance:

1. Hypothesize and test
2. Penalty function
3. Explicit freespace.

Hypothesize and test is the oldest and most obvious method for obstacle avoidance. The steps are: (1) hypothesize a possible path between the initial and final position and configuration of the robot arm; (2) test a selected set of configurations along the path for possible collisions; (3) if a collision possibility is found, devise a way to avoid it by examining the shape and position of the obstacles; (4) if no collision-free path is found, it is necessary to hypothesize a new path until a clear path is found or no additional paths are possible.

Geometric modeling of the environment and the manipulator of the type available in CAD programs is required. Generating a possible path can be a difficult task, especially if the space includes many obstacles.

Use of a penalty function is the second class of algorithm considered. Penalty functions are suitable for applications that require only a small modification to a known path, but can lead to situations in which no further progress can be made. A penalty is assigned for approaching an object. The penalty becomes infinite if the object is encountered and drops off rapidly with distance from the object. The total penalty function for the path is found by adding the penalties from the individual obstacles. In some algorithms there is a penalty for deviating from the shortest path. Figure 6.6 illustrates the use of the penalty function. This is a drawing of the penalty function for a simple circular robot and for a two-link manipulator. Note that the small circular robot can move safely in the direction of the decreasing penalty function. If the two-link robot is positioned as shown and the end effector is moved along the direction of decreasing penalty function, one of the robot links will collide with the square obstacle. This is a case in which the entire robot configuration must be considered.

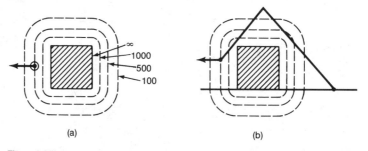

(a)   (b)

Figure 6.6 Illustration of penalty function for (a) a simple circular robot and (b) a two-link manipulator (from Lozano-Perez [21], p. 487).

Source: T. Lozano-Perez, "Task Planning," Chapter 6 of Robot Motion Planning and Control, Brady, et al (eds.), © *1982 MIT Press. All rights reserved. Used by permission.*

In avoiding collisions, it is usually unnecessary to consider details. If an object has an irregular outline, it is safer and easier to surround the volume it represents with a sphere, cube, or other regular figure to simplify computation. This procedure allows the penalty functions to be computed more easily.

A third approach to obstacle avoidance is the explicit free space method. Instead of listing obstacles, we can list available free space. One way to do this is to break up the volume into cubes or parallelepipeds (rectangular boxes) of varying sizes, and to store the coordinates of their corners. In this way it is possible to approximate the available free space to any desired degree. For simplicity of representation, all the parallelepipeds can be aligned with the coordinate axes of a Cartesian coordinate system. The robot links can be represented by simple cylinders or by parallelepipeds whose coordinates are kept in memory. It is then possible to consider whether or not the robot links are entirely surrounded by free space by comparing coordinates. Considerable computation is required to carry out this comparison unless an efficient algorithm is used.

Graphic and grid methods can also be used to list free space. Nodes of the graphs or grids can be entered in the free space list and then searched whenever the manipulator is within a specified range.

Free space methods can be used to guarantee that a path will be found if one exists. It is also feasible to search for the shortest path. These advantages are somewhat offset by the cost of computation. That problem is lessening as computers become faster and cheaper.

Lozano-Perez [21] points out that no efficient obstacle avoidance algorithm for general robots with rotary joints now exists. There is therefore a need for further development of algorithms for obstacle avoidance.

An interesting approach to obstacle avoidance is the configuration space (C-space) method developed by Tomas Lozanzo-Perez [20]. This is a well-organized mathematical approach. We will discuss only the basic ideas presented, as illustrated by Figure 6.7. Block A represents the robot traveling from point s to point g. We assume that it has a fixed orientation, as shown, and that the obstacles to be considered are represented by the inner outlines of each shaded area in the figure. Our goal is to find a path from s to g, the Findpath problem. If we shrink the robot to a point and enlarge each obstacle correspondingly, the same problem remains but the computational complexity is reduced. Only one point must be considered at each increment of the path. However, it is necessary to generate a suitable algorithm to enlarge the obstacles in the correct way. If the orientation of block A is allowed to change, there will be a <u>swept volume</u> that must be taken into account. These configurations are shown in two dimensions, but the third dimension must also be considered. This operation can be performed by taking slices of the volume at different levels and evaluating them separately. Algorithms for this analysis are given in Lozano-Perez [20].

Figure 6.7 Findpath problem and its formulation using C-space obstacles. The shortest path goes through vertices of C-space obstacles (from Lozano-Perez [20], Figure 2).

Source: T. Lozano-Perez, "Automatic Planning of Manipulator Transfer Movements," IEEE Transactions: Systems, Man and Cybernetics, SMC-11, Number 10, October 1981. © 1981 IEEE. Used by permission.

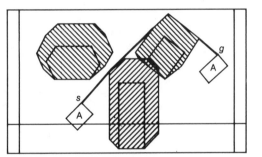

Grasping of objects with an end effector is related to the obstacle avoidance problem. Here we wish to grasp an object in the best location to enable it to be handled correctly. This subject is also considered by Lozano-Perez in the two papers previously mentioned.

Obstacle avoidance has been achieved by preprogramming as described by Myers and Agin [29]. Their program uses the <u>stepped move</u> approach. Prior to the actual move, the robot path is simulated on the computer in small increments or steps of motion. At every step, a collision check is made to see whether any part of the arm has contacted any of the obstacles that have been mapped into memory. If a contact has been made, another path is chosen until one is found that does not cause collision or no further paths are possible.

This method was chosen because of its greater generality than the swept volume approach, which is simpler to compute but not as powerful. In the swept volume approach, the volume to be swept out by the arm as it moves is inspected for obstacles and the path is altered accordingly. This method works satisfactorily only if the number of degrees of freedom is small.

Although the stepped move approach is capable of solving the obstacle avoidance problem, it requires considerable computational capacity. Myers and Agin found that computation of a solution using a powerful minicomputer, the Digital Equipment Corporation's VAX/11-

780, took longer than the actual move time of the arm. Therefore, this method is suitable only if the same path must be repeated many times.

Figure 6.8 shows the layout of a typical planning trajectory. Note that this approach allows the robot to pick items out of boxes, go over walls and other obstacles, and avoid collisions between any part of the arm and obstacles near the path.

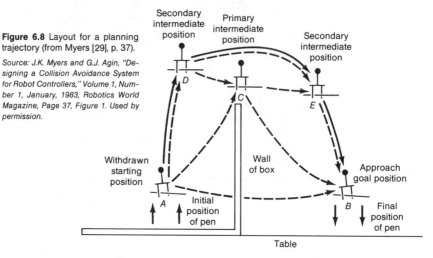

Figure 6.8 Layout for a planning trajectory (from Myers [29], p. 37).

Source: J.K. Myers and G.J. Agin, "Designing a Collision Avoidance System for Robot Controllers," Volume 1, Number 1, January, 1983, Robotics World Magazine, Page 37, Figure 1. Used by permission.

Future robots will require the ability to work in cluttered environments and avoid obstacles; more research and development in this area will be necessary to make this usage possible. Recent developments in multiprocessors can be expected to reduce computation costs and increase speed, so that even computationally intensive methods soon will be satisfactory.

6.3.3 Adaptive Operation

By <u>adaptive operation</u>, we mean that the robot modifies its operation in response to information obtained from its environment. Sensors on or associated with the robot detect or measure items in the environment and feed this information to a computer that is programmed to react to the sensed information in an appropriate way. Touch, vision, force, and proximity sensors have been used in adaptive operation. Touch, force, and proximity sensors are described in Chapter 5; computer vision techniques are discussed in Chapter 9. These sensory capabilities and well-planned computer programs makes possible the use of adaptive techniques.

In a factory, it may not always be possible to know in advance what path the robot should take or even what obstacles might exist; thus, adaptability is important in carrying out complex operations. Adaptability also reduces or even eliminates the need for and expense of training or precalculation of paths.

C. A. Rosen and D. Nitzan [37] at SRI used vision, force, and proximity sensors in 1975 to perform adaptive or corrective control of manipulation. Instead of relying entirely on the positioning capability of the robot, they used the sensory information to provide a new ability.

Although complete recognition of random parts in a factory is difficult, it is relatively easy, using computer vision, to locate previously identified parts with affordable computer capability and programming. When there are only a few possible parts to be located and picked off a conveyor, the task can be controlled with a small computer and relatively simple sensors. Rosen and Nitzan report that an automative casting randomly placed on a conveyor could be picked up by a Unimate Series 2000A, transported, and deposited in a specified position and orientation. In another experiment, a number of washing machine pumps were packed neatly in a tote box by the Unimate. Force sensors, plus a vision system, were used to control the motions of the end effector in this operation. A proximity sensor mounted on one finger of the end effector controlled the initial placement of the end effector in acquiring the next water pump.

Since the time of this pioneering work, computer vision systems have been developed at a number of laboratories, and adaptive operations are being carried on at a number of factories in the United States and Japan.

**6.3.4
Intelligent
Robots**

Intelligent robots are defined as those that have sensors, adaptive capability, and judgment capability. They are one step beyond adaptive robots in having expert knowledge capability as part of their programming. Also needed is a large data base of information about their environment.

These capabilities are characteristic of the many expert systems developed by Artificial Intelligence scientists for applications in other fields. Dana Nau [30] describes 14 expert computer systems that are in use today. These systems are being applied in areas as diverse as oil exploration, medical consulting, mineral exploration, analysis of electrical circuits, mathematical problem solving, and hypothesizing of molecular structure from mass spectrogram data.

In robots, expert systems can be expected to be able to identify an initial state, set up a goal and proceed to the goal by an appropriate means. Robots will have self-programming and learning abilities and will be able to respond to their environment appropriately.

The intelligent robot is not a reality today, but all of the component operations are being carried out either in a factory or in a laboratory environment.

6.4 Closed-Loop Servos

This section will describe the servo systems that control robots and manipulators. All of these systems use feedback— a signal from the output used to modify the input in a desired way. Servo systems are systems that use feedback to produce a controlled output. Because servo system analysis is a large subject, we will present only enough of the fundamentals of analysis to enable the reader to understand the examples given in this chapter. More complete treatment is available in several textbooks (Lathi [18], Kuo [17]).

Feedback control systems have been in use for over 50 years (Hazen [14], Brown [7]). Initially, they were used to control remotely steering devices such as the rudder of a ship. In those applications they were called servomechanisms because the remote device was the "slave" of the controlling steering device. (In Latin, the word servus means slave.) Over many years, feedback control systems were used for many purposes, including the control of electrical circuits, so that the mechanisms part of the name was not suitable. Therefore, the name was shortened to the more general servo to designate all types of closed-loop control systems.

Servos are systems that continuously compare their output with their input and attempt to change the output to agree with the input command. Early systems used potentiometers to change a control voltage in a desired way in order to control the systems. Disturbances of the output due to external conditions cause a change in the output signal that is fed back to the input, so that the system also tries to correct for external disturbances. <u>Closed-loop</u> and <u>servo</u> in the title of this section are therefore redundant; all servo systems have at least one closed loop in them.

Servo systems are used in aircraft, missiles, space vehicles, numerically controlled (NC) machines, disk storage units for computers, robots, and many other places where good control is required. They may be electrical, electromechanical, biological, or other systems as long as they operate by using a closed-loop feedback of information from the output to the input. In practice, there may be many closed loops in the system, all acting together or in opposition.

A standard set of nomenclature and symbols has been developed for servo systems. This standard has been approved by the American National Standards Institute (ANSI) under the sponsorship of the American Society of Mechanical Engineers (ASME). Figure 6.9 is a block diagram of a typical feedback control system illustrating the terminology used. This block diagram is made up of "cascaded" blocks, with the output of one block used as the input of another block. Use of (s) after each identifier indicates that the value is a function of the value s. As will be seen later, s may be a time value or a mathematical parameter.

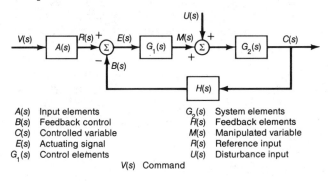

$A(s)$	Input elements	$G_2(s)$	System elements
$B(s)$	Feedback control	$H(s)$	Feedback elements
$C(s)$	Controlled variable	$M(s)$	Manipulated variable
$E(s)$	Actuating signal	$R(s)$	Reference input
$G_1(s)$	Control elements	$U(s)$	Disturbance input
	$V(s)$ Command		

Figure 6.9 Block diagram of a typical feedback control system.

If the output of a block is required to affect more than one other block, it is necessary to use a <u>summing point</u>, indicated by a circle. Usually the capital Greek letter sigma, Σ, is used inside the circle to indicate summation. See, for example, Figure 6.27. The input of block $G_1(s)$ is $E(s)$ where

$$E(s) = R(s) - B(s)$$

Note that the summing point acts to take $R(s)$ as a + value and subtract the value of $B(s)$ from it.

Many blocks may be used in a block diagram. Any set of blocks can be replaced by their equivalent mathematical value. Some simple systems may have only two or three blocks because some of the functions represented may be unnecessary in analyzing the servo system.

6.4.1
Nomenclature
and Symbols
(ANSI Standard)

Table 6.2 lists the ANSI standard nomenclature and symbols used extensively by control engineers and systems analysts. However, other symbols are used by some authors and analysts. Usually each user identifies the symbols used in the text of the material.

To illustrate the use of block diagrams, we will show how simple diagrams can be analyzed and interpreted.

1. Cascaded Blocks

$$F(s) \longrightarrow \boxed{G_1(s)} \xrightarrow{\ W(s)\ } \boxed{G_2(s)} \longrightarrow Y(s) \tag{1}$$

where $F(s)$ is an arbitrary input. $G_1(s)$ and $G_2(s)$ are called <u>transfer functions</u> because they change the input to produce a different output. Therefore,

$$W(s) = F(s)\, G_1(s)$$
$$Y(s) = W(s)\, G_2(s) = F(s)\, G_1(s)\, G_2(s)$$

Table 6.2 ANSI standard nomenclature for servo system elements

Symbol	Function	Interpretation
$A(s)$	Input elements	Input elements produce a signal proportional to the command. (This is normally an amplifier only.)
$B(s)$	Feedback signal	This is the primary feedback signal, which is a function of the controlled variable and is compared with the reference input in order to obtain the actuating signal.
$C(s)$	Controlled variable	This is the output of the servo system or feedback system that is being controlled.
$E(s)$	Actuating signal	This is the difference between the reference input and the primary feedback.
$G_1(s)$	Control elements	The control elements modify the actuating signal to generate the manipulated variable. Compensation circuits may be placed here.
$G_2(s)$	System elements	This is the robot arm drive or other device that is being controlled.
$H(s)$	Feedback elements	These elements produce the primary feedback from the controlled variable.
$M(s)$	Manipulated value	This is the quantity obtained from the control elements and applied to the controlled system.
$R(s)$	Reference input	This is a signal input proportional to the command. It is the driving signal for the servo system.
$U(s)$	Disturbance input	This is the sum of all of the undesired signals that attempt to affect the value of the controlled variable or output. May be made up of loading, heating, wind loads, bearing friction, etc.
$V(s)$	Command	The command is the external input that specifies the desired result. It is independent of the servo system.

Note that blocks may be consolidated into one block with the same total transfer function.

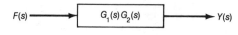

2. Parallel Blocks

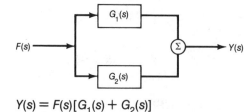

(2)

$$Y(s) = F(s)[G_1(s) + G_2(s)]$$

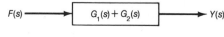

3. Feedback Loops

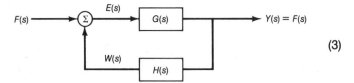

(3)

When feedback is used, the output, $Y(s)$, is forced to be equal to the input, $F(s)$, so the overall block diagram can be replaced by

$$F(s) \longrightarrow \boxed{\frac{G(s)}{1 + G(s)H(s)}} \longrightarrow Y(s) = F(s)$$

(4)

This result is obtained by noting that

$$E(s) = F(s) - W(s) \quad \text{but} \quad Y(s) = E(s)\,G(s) \quad \text{and} \quad W(s) = Y(s)\,H(s)$$

(5) (6) (7)

(5)   (6)   (7)

Substituting Equations (5) and (7) in Equation (6) and solving for $Y(s)$ results in

$$Y(s) = \frac{G(s)}{1 + G(s)\,H(s)}\ F(s)$$

which agrees with Equation (4).

**6.4.3
Laplace
Transform—
Frequency
Domain**

We will assume that the closed-loop control system is operating in a linear fashion. By <u>linear</u>, we mean that the change in output is directly proportional to the change in input that caused it. When the system is linear, we can add and subtract inputs and have the output respond in a proportional way.

A linear system can be analyzed by using the Laplace transform, a mathematical function, to convert the differential and integral functions describing its operation into algebraic equations. Manipulation of the resulting equations can be done algebraically, greatly simplifying the analysis. (This approach is analogous to the use of logarithms in numerical computations. Values may be converted to logarithmic form, added together, and converted back to numeric form in order to carry out multiplication, division, and other functions.) The term <u>integrodifferential equations</u> is used to designate equations containing both integral and differential terms.

Equations to be transformed are functions of time. Terms in the equations vary in value as time passes. In the case of robots, the position of the arm, for example, varies with time due to the forces applied by control motors and other devices.

The Laplace transform method will be presented without proof. This proof is available in many control theory texts (Lathi [18], Kuo [17]). However, a short discussion of the use of the Laplace transform tables is presented in this section.

For a given function $f(t)$ the Laplace transform $F(s)$ is defined as

$$F(s) = \int_0^\infty f(t)e^{-st}dt \tag{8}$$

Table 6.3 Selected properties of the Laplace transform

Operation	Time domain, $f(t)$	Frequency domain $F(s)$
Addition	$a_1 f_1(t) + a_2 f_2(t)$	$a_1 F_1(s) + a_2 F_2(s)$
Scaling ($a > 0$)	$f(at)$	$\dfrac{1}{a}F\left(\dfrac{s}{a}\right)$
Time shift ($t_0 > 0$)	$f(t - t_0)u(t - t_0)$	$F(s)e^{-st_0}$
Frequency shift	$f(t)e^{s_0 t}$	$F(s - s_0)$
Time differentiation	$\dfrac{df}{dt}$	$sF(s) - f(0^-)$
Time integration	$\displaystyle\int_{-\infty}^t f(t)\,dt$	$\dfrac{F(s)}{s} + \dfrac{\int_{-\infty}^0 f(t)\,dt}{s}$
Time convolution	$f_1(t) * f_2(t)$	$F_1(s)F_2(s)$
Frequency convolution	$f_1(t)f_2(t)$	$\dfrac{1}{2\pi j}[F_1(s) * F_2(s)]$

Table 6.4 Laplace transforms

Laplace transform, $F(s)$	Time function, $f(t)$
$\dfrac{1}{s}$	$u(t)$ (unit step function)
$\dfrac{1}{s^2}$	t
$\dfrac{n!}{s^{n+1}}$	t^n (n = positive integer)
$\dfrac{1}{s+a}$	e^{-at}
$\dfrac{1}{(s+a)(s+b)}$	$\dfrac{e^{-at} - e^{-bt}}{b - a}$
$\dfrac{\omega_n^2}{s^2 + 2\zeta\omega_n s + \omega_n^2}$	$\dfrac{\omega_n}{\sqrt{1-\zeta^2}}e^{-\zeta\omega_n t}\sin\omega_n\sqrt{1-\zeta^2}t$
$\dfrac{1}{(1+sT)^n}$	$\dfrac{1}{T^n(n-1)!}t^{n-1}e^{-t/T}$
$\dfrac{\omega_n^2}{(1+Ts)(s^2 + 2\zeta\omega_n s + \omega_n^2)}$	$\dfrac{T\omega_n^2 e^{-t/T}}{1 - 2\zeta T\omega_n + T^2\omega_n^2} + \dfrac{\omega_n e^{-\zeta\omega_n t}\sin\left(\omega_n\sqrt{1-\zeta^2}t - \phi\right)}{\sqrt{(1-\zeta^2)(1 - 2\zeta T\omega_n - T^2\omega_n^2)}}$

$$\text{where } \phi = \tan^{-1}\frac{T\omega_n\sqrt{1-\zeta^2}}{1 - T\zeta\omega_n}$$

Using this definition, all of the properties and transforms listed in Tables 6.3 and 6.4 can be proven by the use of calculus methods. Properties and transforms, respectively, are listed there for convenience of use. We will illustrate the derivation of one transform, that of the unit step function, as an example. The transformed value, initially a function of time, t ("in the time domain"), becomes a function of the complex frequency, s ("in the frequency domain"). Since s is a complex number, it has the form $a + jw$, where a and w are constants to be described later. Column headings in Table 6.4 identify the initial functions as being in the time domain and the transformed functions as being in the frequency domain.

1. Unit Step Function $U(s)$

By definition, a <u>unit step function</u> is an abrupt change from one value to another value that is one unit greater. It is usually written $u(t)$ and becomes $u(t) = 1$ in the Laplace transform equation. Then the unit step function $U(s)$ can be computed as:

$$U(s) = \int_0^\infty u(t)e^{-st}dt = \int_0^\infty 1e^{-st}dt = -(1/s)e^{-st}\Big|_0^\infty$$
$$= (1/s)(-e^{s\infty} + e^0) = 1/s \tag{9}$$

Therefore, the value of the Laplace transform is $L[u(t)] = 1/s$. Figure 6.10 shows the form of the unit step function.

The Laplace transform is obtained by inserting the value of $f(t)$ in Equation 8, and performing the integration between the limits of zero and infinity. The properties listed in Table 6.3 can be obtained by similar manipulation.

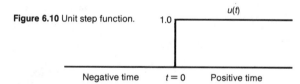

Figure 6.10 Unit step function.

2. Application of the Laplace Transform

Application of the Laplace transform to the solution of a mathematical equation can be done in three steps:

1. Transform the integrodifferential equation from the time domain, as a function of t, to the frequency domain, a function of s, by using the Laplace transform. This is done easily for simple terms by looking up Table 6.4 and applying the rules given in Table 6.3.
2. Manipulate the resulting algebraic equations until a solution is obtained for the desired output value.
3. Perform an inverse Laplace transform by looking up the s function in Table 6.4 and replacing it with the time function.

As an example, we will use the Laplace transform method to model and analyze a simple mechanical system.

**6.4.4
System
Modeling**

In order to study a system, it is necessary to make a model of it. Mathematical, electrical, or mechanical models can be used. Mathematical models are used extensively in the initial analysis of a system because they are easily modified and inexpensive. These models can be used to determine the general parameters and overall feasibility of the system. Simple models provide insight into system operation but may not be sufficiently complete to generate accurate values of the parameters involved.

A simple mechanical system can be modeled by considering only the mass, M, to be moved, the total friction opposing the motion, B, and the effective spring constant, K, represented by the structures involved, as seen in Figure 6.11. The graphic and alphanumeric symbols shown are standard in modeling mechanical systems. Some authors draw the damping symbol in a slightly different form, but the definition is usually clear from the context.

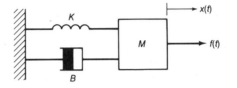

Figure 6.11 A simple mechanical system.

In the free-body diagram of Figure 6.12, the mass, M, is being acted on by three forces: f, a driving force; Kx, a spring force proportional to the spring constant, K; and a damping force or friction force, Bv, where v is the velocity of the moving mass. It is normally assumed, for analysis, that the friction force is proportional to velocity. All of these forces are functions of t, as indicated by the (t) values in the free-body diagram. The t values will be omitted hereafter for convenience.

Figure 6.12 Free-body diagram.

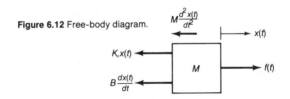

The overall force equation for this model can be written as

$$f = M \frac{d^2x}{dt^2} + B \frac{dx}{dt} + Kx \qquad (10)$$

Let's examine the effect of appying the force $f = Au(t)$ to this model, where A is a constant multiplier. Then

$$M \frac{d^2x}{dt^2} + B \frac{dx}{dt} + Kx = Au(t) \qquad (11)$$

To determine the motion of the mass, M, we substitute the Laplace transforms from Table 6.3. In the second term on the right, dx/dt is a time differentiation whose Laplace transform, as given in Table 6.3 is

$$B[sX(s) - f(0^-)] \qquad (12)$$

where $f(0^-)$ is the value of x at time, <u>zero minus</u>, the time just before the starting instant at zero. Initally, we will assume that $x = 0$, so that $f(0^-) = 0$. Repeating this process twice to handle the second derivative, we find that Md^2x/dt^2 transforms into

$$M\{s[sX(s) - f(0^-)] - f'(0)\} \qquad (13)$$

where $f'(0)$ is the rate of change of position, the initial velocity, at time zero. The prime indicates time differentiation. We will take the initial velocity as zero also.

The transform for Kx is $KX(s)$, and the unit step transform is A/s.

Substituting these values into Equation (11) and rearranging, the Laplace transform of this equation can be determined as follows:

$$Ms_2 + BsX(s) + KX(s) = \frac{A}{s} \tag{14}$$

Solving for $X(s)$, we obtain

$$X(s) = \frac{A}{s(Ms^2 + Bs + K)} \tag{15}$$

as the equation of motion in Laplace notation.

Now the solution for $x(t)$ can be obtained by doing the inverse Laplace transform. For this simple system, the inverse transform can be obtained by manipulating Equation (15) to a new form and substituting the inverse Laplace functions from Table 6.4. We need terms like those in the column headed $F(s)$ to obtain the inverse terms in the column headed $f(t)$ in Table 6.4 for substitution in Equation (15).

The simplest way to manipulate Equation (15) to meet our needs is to separate it into partial fractions of the form

$$X(s) = \frac{a}{s} + \frac{b}{(s + s_1)} + \frac{c}{(s + s_2)} \tag{16}$$

where s_1 and s_2 are the roots of $Ms^2 + Bs + K$ and a, b, and c are appropriate functions that maintain the equality of the equation.

For certain values of M, B, and K, the separation into partial fractions is simplified. The following values will demonstrate the method.

$$M = 1$$
$$B = 3$$
$$K = 2$$

Then

$$X(s) = \frac{A}{s(s^2 + 3s + 2)} \tag{17}$$

by factoring the denominator, we obtain

$$X(s) = \frac{a}{s} + \frac{b}{(s + 2)} + \frac{c}{(s + 1)} \tag{18}$$

To solve for the values of a, b, and c, we add the partial fractions by converting to a common denominator and obtain

$$X(s) = \frac{a(s + 2)(s + 1) + b(s)(s + 1) + c(s)(s + 2)}{s(s + 1)(s + 2)} \tag{19}$$

Multiplying the numerator out and setting it equal to the known numerator, A, we find

$$a(s^2 + 3s + 2) + b(s^2 + s) + c(s^2 + 2s) = A$$

and collecting coefficients of the terms in s^2, s and constants,

$$(a + b + c)s^2 + (3a + b + 2c)s + 2a = A$$

Because there are no s^2 and s terms in the numerator of Equation (20), but there is a constant term A,

$$a + b + c = 0 \tag{20}$$
$$3a + b + 2c = 0 \tag{21}$$
$$2a = A \quad \text{or} \quad a = A/2 \tag{22}$$

Substituting $a = A/2$ into Equations (20) and (21) and solving first for b and then c, we find

$$a = A/2 \quad b = A/2 \quad c = -A \tag{23}$$

These values can be checked by inserting them in Equations (20) and (21). The partial fraction expansion of Equation (18) becomes

$$X(s) = \frac{A}{2s} + \frac{A}{2(s+2)} + \frac{A}{(s+1)}$$

$$= \frac{A}{2}\left(\frac{1}{s} + \frac{1}{(s+2)} - \frac{2}{(s+1)}\right) \tag{24}$$

Now we can look up these values in Table 6.4 to obtain

$$x(t) = \frac{A}{2}(1 + e^{-2t} - 2e^{-t}) \tag{25}$$

This equation, representing the motion x of the mass as a function of time, is plotted in Figure 6.13.

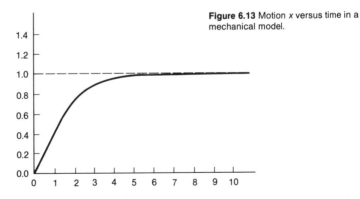

Figure 6.13 Motion x versus time in a mechanical model.

Returning to Equation (16), we can investigate another way to obtain information about the motion of an object. For convenience, we copy this Equation here with a new number.

$$X(s) = \frac{A}{s(Ms^2 + Bs + K)} \tag{26}$$

This equation can be rewritten in the form

$$X(s) = \frac{A}{K}\left(\frac{w_n^2}{s(s^2 + 2zw_n s + w_n^2)}\right) \tag{27}$$

by using two substitute variables, z and w_n, where

$$zw_n = \frac{B}{2M} \quad \text{and} \quad w_n{}^2 = \frac{K}{M} \tag{28}$$

then

$$w_n = \frac{B}{2Mz} \quad w_n{}^2 = \frac{B^2}{4M^2z^2} = \frac{K}{M}$$

and

$$z^2 = \frac{B^2}{4KM} \quad \text{so} \quad z = \frac{B}{2\sqrt{KM}} \tag{29}$$

and

$$w_n = \sqrt{\frac{K}{M}} \tag{30}$$

verifying that $zw_n = \frac{B}{2M}$ as defined.

It turns out that all of this manipulation has produced, as part of Equation (27) an equation that is called the characteristic equation of the second-order differential equation (27) when set equal to zero.

$$(s^2 + 2zw_n s + w_n{}^2) = 0 \tag{31}$$

This form of the equation is valuable because some direct insight can be obtained by inspection of equation (31). By factoring this equation into two terms, $s + s_1$ and $s + s_2$, the roots are found to be

$$\begin{aligned} s_1 &= -zw_n + jw_n\sqrt{1 - z^2} \\ s_2 &= -zw_n - jw_n\sqrt{1 - z^2} \end{aligned} \tag{32}$$

For future reference, we will identify new variables, α and w, which have the values

$$\begin{aligned} \alpha &= -zw_n \\ w &= w_n\sqrt{1 - z^2} \end{aligned} \tag{33}$$

There are three possible conditions in any closed-loop servo system:

1. Oscillatory and potentially unstable, when z is < 1. This is of the same form as (32).

$$\begin{aligned} s_1 &= -zw_n + jw_n\sqrt{1 - z^2} \\ s_2 &= -zw_n - jw_n\sqrt{1 - z^2} \end{aligned} \tag{34}$$

The system is underdamped and oscillates when a unit step is applied, as in Figure 6.14. If z becomes zero, the system is undamped. If z becomes negative, by the use of negative feedback, for example, the system oscillates continuously, which is useful for electronic frequency generators but not for robot servo systems.

2. Critically damped, when $z = 1$.

$$\begin{aligned} s_1 &= -zw_n \\ s_2 &= -zw_n \end{aligned} \tag{35}$$

The two roots of the equation, s_1 and s_2, are then of equal value.

When a unit step is applied, the mass moves directly to the unit level in the shortest time period without overshoot, as in Figure 6.14.

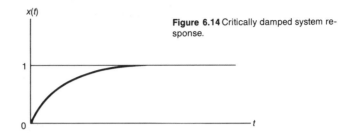

Figure 6.14 Critically damped system response.

3. Overdamped, when $z > 1$.

Substituting $z > 1$ in Equation (32), we see that when $z > 1$ the term on the right contains the square root of a negative number.   Since $j^2 = -1$· the second term becomes

$$w_n\sqrt{z^2 - 1} \tag{36}$$

and the values of s_2 and s_1 are

$$s_1 = -zw_n + w_n\sqrt{z^2 - 1}$$
$$s_2 = -zw_n - w_n\sqrt{z^2 - 1} \tag{37}$$

Figure 6.13 is an example of an overdamped system.   Note that in this case, using our model system as an example,

$$z^2 = \frac{B^2}{4KM} = \frac{9}{8} = 1.125 \tag{38}$$

so that $z > 1$ as expected.

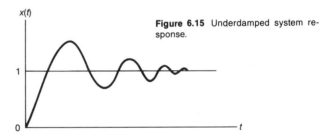

Figure 6.15 Underdamped system response.

By inspection of Equations (35) through (38), we see that α and w have the following meanings:

$$\alpha = zw_n$$

is a damping factor term because it determines how fast the output approaches the applied value.   It is an inverse measure of response time.

$$w = w_n\sqrt{1 - z^2}$$

is called the <u>damped natural frequency</u>.   When $z = 0$, it is the frequency of oscillation.

This analysis has outlined a general approach to servo analysis, using one method on a simple problem, to provide the reader with an overall awareness of the problem. Other methods of servo analysis are briefly discussed in the next section.

**6.4.5
Servo
Analysis
Methods**

Analysis of servo systems is a very complex subject. Robots and other equipment controlled by servo systems are becoming much more complicated, and their performance is critically dependent on the control techniques used. As a result, there has been much research on and development of various means of analysis. These topics are too complex to be covered here in detail. We therefore list them for reference and suggest that the interested reader refer to textbooks on control systems such as Kuo [17] and Lathi [18]. Some of the advanced work in robotic analysis is described in Luh [23-25] and in Paul [32]. There are also several advanced references in Brady [5].

The major types of analysis are:

1. Operator Methods. A simple application of the Laplace transform method was described in Section 6.4.4. Another operator method is the Z-transform method, which can be used for digital servo analysis.
2. Time Domain. This method, which deals with the time variable, t, is known as the time domain method. Operational notation is used to describe differential equations of motion of the system. The results can be expressed in matrix notation and solved by the use of transfer operators.
3. Frequency Domain. Analysis in the frequency domain is character- ized by the use of exponential functions. An advantage of this approach is that nonperiodic inputs can be represented by an exponential series of terms and analyzed by the use of Fourier transforms or other transform methods. A table of Fourier transforms is used much as the Laplace transforms were used in Section 6.4.4.
4. State Space. The state variable approach uses an internal descrip- tion of the system. State variables designate the condition of a system at any one time. If all state variables are known, then in principle, the condition of the system at any future time can be determined.

State equations of the system are set up in matrix form and can be solved by matrix methods. They are especially suitable for computer solution and/or simulation because of the matrix formula- tion. Frequency domain and Laplace transform methods can be used with the state space formulation, as can the time domain approach.

6.5 Servo System Components

This section will identify the major elements of servo systems and refer to the sections of this book where more detail on each element may be found.

**6.5.1
Inputs**

Inputs to servo systems are of two types: analog and digital. Analog inputs provide a varying voltage or current. Digital inputs provide either a serial sequence of fixed-height voltage or current pulses or a number of pulses in parallel. Information from nonelectri- cal inputs must be converted into electrical form before entry into the servo system.

Sensory input is of major importance to robots. Chapter 5 described some of the many types of sensors and their operating principles. These sensors provide both analog and digital inputs. A servo system using analog signals must have the inputs converted to analog form if they are not already in that form. This operation is performed by the use of analog-to-digital converters. Digital servo systems accept only digital inputs in a specific code, so that analog input must be converted to digital form by a suitable analog-to-digital converter. Vision systems are a particular form of sensor and operate under the same restrictions.

Commands are other inputs received by the servo system. These commands may come from a control computer, a teach box, a keyboard, or some other external device. These signals must also be converted to the proper type and level of signal to meet the servo system's requirements.

More detail on computers, conversion, and interfacing is provided in Chapter 7, which covers computer hardware operation. Because all of these inputs affect the computer system, it may be necessary to provide special software or programming capabilities. Programming languages and techniques are described in Chapter 8.

**6.5.2
Summing
Circuits**

Summing circuits perform addition, subtraction, and comparison functions on inputs. They are also used internally in the servo system to compare feedback signals with command signals or the output of a particular part of the system. Figure 6.27 is the diagram of a positional control system. Each circular symbol with the Greek letter sigma, Σ, inside it represents a summing circuit.

Summing can be done by analog circuits. Usually, these circuits use an operational amplifier to perform the summation. Addition, subtraction, and comparison may be done by the use of properly designed circuits.

Digital summing is done with arithmetic elements, usually solid-state devices designed for this purpose.

**6.5.3
Control
Elements**

Control elements are the internal modifiers of the signals in a servo system. They are signal amplifiers, compensating circuits, and function generators used to provide internal control. A typical application is an amplifier that multiples a signal by a fixed ratio. In the servo system analyzed in Section 6.7, there are several amplifiers that control the signal level in both the forward and feedback paths. See, for example, K_1, in Equation 6.7-25.

The compensating circuit modifies the frequency response of a servo system to improve its stability or to increase its response speed or other function. Luh [24] describes the use of a compensating circuit to reduce the effective friction and improve the response of a robot joint drive. The theory of compensatory circuits is described in Kuo [17], Lathi [18], and many other texts on control.

Adaptive systems use amplifiers and compensating circuits that vary with time and input conditions. Design and use of adaptive circuits are difficult but can provide greatly improved capability in a robot system. Adaptive systems are described in Section 6.3.3.

**6.5.4
Power
Amplifiers**

Electric motors are controlled in direction of rotation, angular acceleration, and angular velocity by power amplifiers. These amplifiers must provide high currents (10 to 100 amperes in a robot drive) typically in the range of 10 to 100 volts. For high-performance robots, the currents and voltages must be switched in direction within a few milliseconds at most. Therefore, the cost of amplifiers is a major part of the servo system cost. Because the performance of the robot depends directly on the power amplifier, the design of this part of the system is critically important. Section 4.3.6 describes the operation of DC power sources and power amplifiers and discusses some of the design parameters that are important.

6.5.5
Drive Motors

Hydraulic, pneumatic, and two types of electrical motors—DC electric and stepper motors— are used in robot systems. Chapter 4 provides a comprehensive discussion of alternative types of drive motors, drive methods, and the associated controls.

6.5.6
Position
Sensors

Measurement of the position of robot joints and links can be done by several methods depending on the accuracy, repeatability, and range of movement required. Detailed information on position sensors is presented in Section 5.3.2.

6.5.7
Conditioning
Circuits

Conditioning circuits are used to change voltage levels and other characteristics to match the servo system's inputs and outputs to external inputs and outputs. Sometimes a negative signal coming in must be inverted to have the same range of current or voltage, but to maintain a positive voltage at all times. Typically, this function is done with an operational amplifier with the proper characteristics. Nonlinear functions may be converted to linear functions by the use of conditioning circuits. Another use is to calibrate an input to match a particular desired characteristic.

6.5.8
Feedback
Circuits

Feedback circuits do not differ in type from other circuits, but are placed in the servo system to take an output from one part of the system and carry it back to an earlier point. Usually there is some amplification or compensation at the same time.

6.5.9
Velocity
Sensors

Measurement of the rotational velocity is valuable in stabilizing the servo system. Tachometers, which are essentially small voltage generators, are usually used for this purpose because they provide a voltage that is linearly proportional to the velocity. Incremental encoders could also be used to measure velocity, by counting the number of steps per second in each direction. In digital servos, the direct binary output from the counters is especially useful.

6.6 Kinematics Analysis and Control

Kinematics of a robot arm deals with the movements of the arm with respect to a fixed coordinate system as a function of time. In conventional analysis, the basis of the robot is taken as a reference, and all other movements are measured from the base reference point.

If the location of all of the joints and links of a robot arm is known, it is possible to compute the location in space of the end of the arm. This is defined as the direct kinematics problem. The inverse problem is to determine the necessary positions of the joints and links in order to move the end of the robot arm to a desired position and orientation in space. The inverse problem is much more difficult to solve, and there may be more than one solution.

Robots in commercial use today use a combination of rotary and sliding joints. These joints are the connections between the links of a robot arm or portions of the wrist assembly. Rotary joints allow two links to move only in an angular relation, while sliding joints permit only linear movement. In principle, other joint relationships are permissible, but in practice, only these two types of joints are being used.

Sequential combinations of joints and links are called chains. Chains may be open or closed. Chains composed of links joined by only one joint at each end, and with no connection to another link closer to the base link, are called open chains, while chains that have links connected back to a previous joint are called closed chains. Existing industrial robots primarily use open chain design.

Analyzing and controlling robot arms requires the development of analysis techniques and control algorithms. In simple robot arms, the need for analysis, as discussed in previous sections, was minimal. Arms with multiple joints subject to many forces and external influences require more complex analysis.

This section will summarize the developments in Denavit and Hartenberg [10] and Paul [32] as described in Rosen [36, second report], Paul [32], and Lee [19] on the use of homogenous matrices and coordinate transformations in kinematic analysis for robots. Control algorithms using these techniques are described in Section 6.3.

6.6.1 Homogeneous Transformation

J. Denavit and R. S. Hartenberg, in 1955, while at Northwestern University, wrote a major paper [10] that first proposed the use of homogenous matrices to describe the relationships between the links of mechanisms. Their work was based on and was an improvement of earlier work done in 1875 by F. Reuleaux in France. Since derivation of the homogeneous matrices is not essential in understanding their use, the reader is referred to the original paper for this information. Appendix B.2 describes conventional matrix notation and matrix operations.

The Denavit-Hartenberg (DH) Matrix is a 4×4 matrix that transforms a vector from one coordinate system to another. Each matrix can perform either or both of two actions: rotation or translation. The initial vectors must be expressed in homogeneous coordinates.

In computing the position of the end effector of a robot arm, each joint is taken as the center of a new coordinate system, starting with the base reference point of the robot. Base coordinates are expressed in a homogeneous coordinate matrix. Then a second matrix is written that describes the relationship between the center of the coordinates of the first and second joints. This process continues, so that a homogeneous matrix is written for the relationship between each subsequent pair of joints. When all of the matrices are written, it is only necessary to multiply them together, in the desired sequence, to transform a point on the end of the arm to a point in the base reference coordinates.

A homogeneous matrix has the general form given in Figure 6.16.

Figure 6.16 Homogeneous matrix.

$$
\begin{bmatrix}
A_{11} & A_{12} & A_{13} & P_1 \\
A_{21} & A_{22} & A_{23} & P_2 \\
A_{31} & A_{32} & A_{33} & P_3 \\
\hline
0 & 0 & 0 & 1
\end{bmatrix}
$$

In the upper-left-corner of the homogeneous matrix, there is a 3×3 matrix that describes the rotation between the two coordinate systems. In the upper-right-corner, a 3×1 matrix describes the translation, or vector distance, between the coordinate systems. P_1, P_2, and P_3 are the components of the vector joining the two coordinate systems. (The distance between coordinate system origins may be directly obtained by taking the square root of the sum of the squares of the vector components.) The zeros in the lower–left row of the matrix refer to perspective transformation along three axes, while the 1 in the lower–right-hand–corner performs a scaling function. This set of relations is shown in the matrix of Figure 6.17.

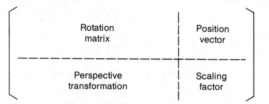

Figure 6.17 Submatrices and factors of a homogeneous matrix.

Perspective transformation and scaling factor are important parameters in CAD and in graphics, where they are used extensively. In robot manipulation, the perspective transformation values are always set to 0 and the scaling factor, w, is always 1.

By definition, a vector in vector notation is written as

$$p = ip_x + jp_y + kp_z$$

where the boldface characters p, i, j, and k indicate vectors and i, j, and k are the unit vectors in the x, y, and z directions, respectively. To simplify key entry, we will use these as vectors without making them boldface unless confusion could result by doing so.

In homogeneous coordinates, the p vector can be written as a column matrix in the form

$$p = \begin{bmatrix} x \\ y \\ z \\ w \end{bmatrix}$$

where

$$p_x = x/w = x$$
$$p_y = y/w = y$$
$$p_z = z/w = z$$

since the scale factor, w, is equal to 1.

For convenience, we will assign names to the elements of the homogeneous matrix as follows:

$$H = \begin{bmatrix} n_x & s_x & a_x & p_x \\ n_y & s_y & a_y & p_y \\ n_z & s_z & a_z & p_z \\ 0 & 0 & 0 & 1 \end{bmatrix} = \begin{bmatrix} n & s & a & p \\ 0 & 0 & 0 & 1 \end{bmatrix} \tag{38}$$

which is equivalent to identifying vectors n, s, a, and p with values obtained by reading down the first, second, third, and fourth columns, respectively, of Equation (38).

$$n = in_x + jn_y + kn_z$$

$$s = is_x + js_y + ks_z$$

$$a = ia_x + ja_y + ka_z$$

$$p = ip_x + jp_y + kp_z \tag{39}$$

A homogeneous matrix with no rotation and no translation is

$$N = \begin{bmatrix} 1 & 0 & 0 & 0 \\ 0 & 1 & 0 & 0 \\ 0 & 0 & 1 & 0 \\ 0 & 0 & 0 & 1 \end{bmatrix} \tag{40}$$

If we perform a matrix multiplication of N by any vector, there will be no change. The same vector will be obtained again.

$$\begin{vmatrix} a \\ b \\ c \\ d \end{vmatrix} = \begin{bmatrix} 1 & 0 & 0 & 0 \\ 0 & 1 & 0 & 0 \\ 0 & 0 & 1 & 0 \\ 0 & 0 & 0 & 1 \end{bmatrix} \begin{bmatrix} a \\ b \\ c \\ d \end{bmatrix}$$

Rotation Transformations

Rotation transformations are performed by rotation matrices in homogeneous coordinates. Matrices performing rotations about the X, y, or Z axes by an angle θ are

$$\text{Rot}(X,\theta) = \begin{bmatrix} 1 & 0 & 0 & 0 \\ 0 & \cos\theta & -\sin\theta & 0 \\ 0 & \sin\theta & \cos\theta & 0 \\ 0 & 0 & 0 & 0 \end{bmatrix} \tag{41}$$

$$\text{Rot}(Y,\theta) = \begin{bmatrix} \cos\theta & 0 & \sin\theta & 0 \\ 0 & 1 & 0 & 0 \\ -\sin\theta & 0 & \cos\theta & 0 \\ 0 & 0 & 0 & 0 \end{bmatrix} \tag{42}$$

$$\text{Rot}(Z,\theta) = \begin{bmatrix} \cos\theta & -\sin\theta & 0 & 0 \\ \sin\theta & \cos\theta & 0 & 0 \\ 0 & 0 & 1 & 0 \\ 0 & 0 & 0 & 0 \end{bmatrix} \tag{43}$$

Taking the simplest case initially, we will transform a vector u around the Z axis to a transformed vector v, as given in Figure 6.18.

For this example, the vector $u = 4i + 2j + 7k$. We wish to find the new vector that results from rotating u through 90 degrees around the Z axis to a new value, v . In this case, we use the matrix for Rot(Z, θ) with θ = 90 degrees, which is Equation (43) with sin θ = 1 and cos θ = 0.

The new column vector for v is found by multiplying the column vector for u by the Rot (Z, θ) matrix.

$$
\begin{bmatrix} -2 \\ 4 \\ 7 \\ 1 \end{bmatrix} = \begin{bmatrix} 0 & -1 & 0 & 0 \\ 1 & 0 & 0 & 0 \\ 0 & 0 & 1 & 0 \\ 0 & 0 & 0 & 1 \end{bmatrix} \begin{bmatrix} 4 \\ 2 \\ 7 \\ 1 \end{bmatrix} \tag{44}
$$

This result is diagrammed in Figure 6.18.

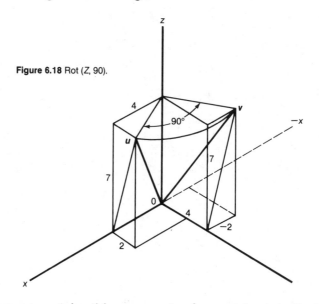

Figure 6.18 Rot (Z, 90).

We can now take this new vector for v and rotate it -90 degrees around the Y axis to a new value w by using the Rot (Y,θ) matrix of Equation (42), with sin θ = -1 and cos θ = 0. In this case, the rotation around the Y axis is in the negative direction because, by definition, rotations in X , Y , Z sequence are in the positive direction looking along the axis of rotation from the center of the coordinates. In the first rotation, around the Z axis, the rotation was in the positive X-Y direction.

$$
\begin{bmatrix} -7 \\ 4 \\ -2 \\ 1 \end{bmatrix} = \begin{bmatrix} 0 & 0 & -1 & 0 \\ 0 & 1 & 0 & 0 \\ 1 & 0 & 0 & 0 \\ 0 & 0 & 0 & 1 \end{bmatrix} \begin{bmatrix} -2 \\ 4 \\ 7 \\ 1 \end{bmatrix} \tag{45}
$$

The resulting transformation is shown in Figure 6.19.

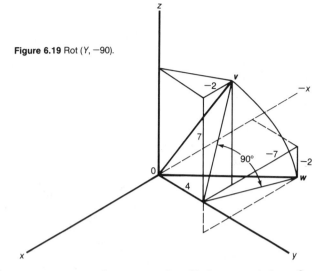

Figure 6.19 Rot $(Y, -90)$.

These two rotations were handled separately. One of the great advantages of the homogeneous matrix method is that single matrix operations can be used to compute multiple translations and rotations. In this case, we have carried out two rotations:

$$v = \text{Rot}(Z, 90)u$$
$$w = \text{Rot}(Y, -90)v$$

Solving for w in terms of u, we obtain

$$w = \text{Rot}(Y, -90)\text{Rot}(Z, 90)u$$

where the sequence of operations is important because matrix multiplication is noncommutative. We can multiply the two matrices together as indicated and obtain a new homogeneous matrix:

$$\text{Rot}(Y, -90)\,\text{Rot}(Z, 90) = \begin{bmatrix} 0 & 0 & -1 & 0 \\ 0 & 1 & 0 & 0 \\ 1 & 0 & 0 & 0 \\ 0 & 0 & 0 & 1 \end{bmatrix} \begin{bmatrix} 0 & -1 & 0 & 0 \\ 1 & 0 & 0 & 0 \\ 0 & 0 & 1 & 0 \\ 0 & 0 & 0 & 1 \end{bmatrix}$$

which results in a new matrix:

$$\text{Rot}(Y,-90)\text{Rot}(Z,90) = \begin{bmatrix} 0 & 0 & -1 & 0 \\ 1 & 0 & 0 & 0 \\ 0 & -1 & 0 & 0 \\ 0 & 0 & 0 & 1 \end{bmatrix} \qquad (46)$$

Multiplying this matrix by the original vector, u, results in the same value for w as obtained before. The combined result is plotted in figure 6.20.

$$
\mathbf{w} = \begin{bmatrix} -7 \\ 4 \\ -2 \\ 1 \end{bmatrix} \quad \begin{bmatrix} 0 & 0 & -1 & 0 \\ 1 & 0 & 0 & 0 \\ 0 & -1 & 0 & 0 \\ 0 & 0 & 0 & 1 \end{bmatrix} \quad \begin{bmatrix} 4 \\ 2 \\ 7 \\ 1 \end{bmatrix} \tag{47}
$$

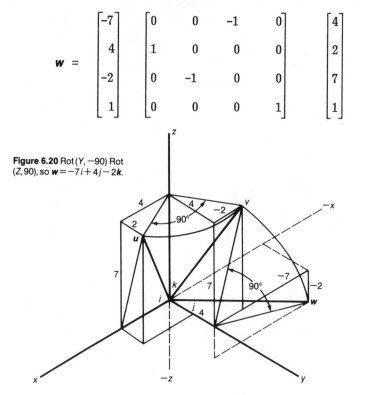

Figure 6.20 Rot $(Y, -90)$ Rot $(Z, 90)$, so $\mathbf{w} = -7i + 4j - 2k$.

The reader should perform the matrix multiplications in the reverse order to see that the result is quite different.

2. Translation

Translation of a vector from one coordinate center location to another can be accomplished by multiplying by a homogeneous matrix that has three 1s on the diagonal of the rotational matrix (to specify zero rotation about each axis) and contains in the 3×1 translational matrix area the vector value that identifies the center of the new coordinates as measured from the previous coordinate center.

$$
\text{Trans}\,(p_x, p_y, p_z) = \begin{bmatrix} 1 & 0 & 0 & p_x \\ 0 & 1 & 0 & p_y \\ 0 & 0 & 1 & p_z \\ 0 & 0 & 0 & 1 \end{bmatrix}
$$

Moving coordinates through the vector value $3i - 5j + 4k$ is done by the left matrix that follows. The complete new matrix value is obtained by postmultiplying this matrix by the former matrix incorporating the two rotations Rot$(Y, -90)$ and Rot$(Z, 90)$. This translation is shown as vector $\mathbf{wp}$ in Figure 6.21.

Trans (3,-5,4)
Rot(Y,-90)Rot(Z,90) =
$$
\begin{bmatrix}
1 & 0 & 0 & 3 \\
0 & 1 & 0 & -5 \\
0 & 0 & 1 & 4 \\
0 & 0 & 0 & 1
\end{bmatrix}
\begin{bmatrix}
0 & 0 & -1 & 0 \\
1 & 0 & 0 & 0 \\
0 & -1 & 0 & 0 \\
0 & 0 & 0 & 1
\end{bmatrix}
$$

$$
=
\begin{bmatrix}
0 & 0 & -1 & 3 \\
1 & 0 & 0 & -5 \\
0 & -1 & 0 & 4 \\
0 & 0 & 0 & 1
\end{bmatrix}
\tag{48}
$$

Now the original vector point $u = 4i + 2j + 7k$ transforms into the new value

$$
\begin{bmatrix}
-4 \\
-1 \\
2 \\
1
\end{bmatrix}
=
\begin{bmatrix}
0 & 0 & -1 & 3 \\
1 & 0 & 0 & -5 \\
0 & -1 & 0 & 4 \\
0 & 0 & 0 & 1
\end{bmatrix}
\begin{bmatrix}
4 \\
2 \\
7 \\
1
\end{bmatrix}
\tag{49}
$$

This result is shown in Figure 6.21. The new vector from the origin is $p = -4i - j + 2k$. The vector values are repeated here to illustrate the effects of two rotations and one translation on the final location of the vector p.

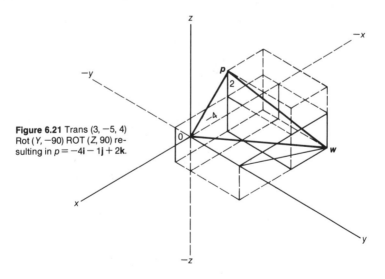

Figure 6.21 Trans (3, −5, 4) Rot (Y, −90) ROT (Z, 90) resulting in $p = -4i - 1j + 2k$.

$u = 4i + 2j + 7k$ (the original vector)

$v = -2i + 4j + 7k$ (rotated 90 degrees around the Z axis)

$$w = -7i + 4j - 2k \quad \text{(rotated -90 degrees around the Y axis)}$$

$$t = 3i - 5j + 4k \qquad \text{(translation vector to the new coordinate system measured in the original coordinate system)}$$

$$p = -4i - 1j + 2k \qquad \text{(the new point measured in the original coordinate system}$$

$$(50)$$

3. Interpreting the Homogeneous Matrix

Much valuable information can be obtained directly from a homogeneous matrix by interpreting the row and column values properly. For convenience, we repeat Equation (38) here, with a new name and an equivalent matrix form.

$$T = \begin{bmatrix} n_x & s_x & a_x & p_x \\ n_y & s_y & a_y & p_y \\ n_z & s_z & a_z & p_z \\ 0 & 0 & 0 & 1 \end{bmatrix} = \begin{bmatrix} n & s & a & p \\ 0 & 0 & 0 & 1 \end{bmatrix} \tag{51}$$

T is used as the name of the matrix that results from multiplying several matrices together in anticipation of further use of homogeneous matrices where this is the standard name for the matrix representing the transformation of the end of the arm to the base coordinates. Paul [32] and Lee [19], for example, use T in this way.

As will be seen later(Section 6.6.3, Part 2), the value of T can be developed by multiplying together the series of matrices that represent the coordinate transformations of a robot arm, with one matrix for each joint. Then T provides a direct specification of the location of the end-of-arm point in the robot base reference coordinates.

Whether T represents one coordinate matrix or a series of matrices multiplied together, it provides the total rotational and translational vectors directly. We will demonstrate how the homogeneous matrix identifies both the location and orientation of the end-of-arm point when properly interpreted.

As previously, the second matrix in Equation (51) represents the same values as the first matrix. It is merely a different notation for the same information. For convenience, we repeat the four vectors from this matrix as given in Equation (39):

$$n = in_x + jn_y + kn_z$$

$$s = is_x + js_y + ks_z$$

$$a = ia_x + ja_y + ka_z \tag{52}$$

$$p = ip_x + jp_y + kp_z$$

Now we can compare these vectors and the values in the matrix of Equation (50) with the results obtained in Equation (48) by equating each column of the matrix of Equation 51 with the corresponding column of the matrix of Equation (48). Doing this, we find that

$$n' = 0i + 1j + 0k$$

$$s' = 0i + 0j - 1k$$

$$a' = -1i + 0j + 0k \tag{53}$$

$$p = 3i - 5j + 4k$$

Equation (53) shows that the new coordinate system $(X'Y'Z')$ has an orientation in the old coordinate system (X, Y, Z) described by

$$n' = j \quad (X' \text{ parallel to } Y)$$

$$s' = -k \quad (Y' \text{ measured in the negative } Z \text{ direction}) \tag{54}$$

$$a' = -i \quad (Z' \text{ measured in the negative } X \text{ direction})$$

This orientation is shown by the primed coordinate system (X, Y, Z) in Figure 6.22.

We can interpret equation (54) as the vector directions for the new coordinate system, as noted, where primes indicate the new coordinate system. In order to interpret the matrix, we note that the original vector

$$u = u_x i + u_y j + u_z k = 4i + 2j + 7k \tag{55}$$

has not changed in length. However, its coordinate system has been rotated 90 degrees in a positive direction around the Z axis, then rotated 90 degrees negatively around the Y axis, and finally translated along a vector distance $t = 3i - 5j + 4k$. Therefore, in the new coordinate system it has the value

$$u' = u'_x i + u'_y j + u'_z k = 4i + 2j + 7k \tag{56}$$

where the primed values signify that they are measured in the $X'Y'Z'$ system.

We now determine a new vector,

$$q = q_x i + q_y j + q_z k \tag{57}$$

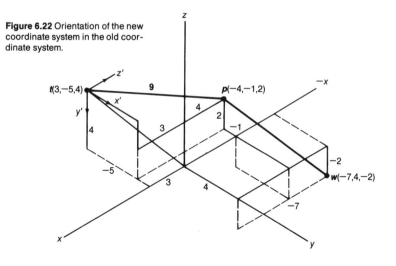

Figure 6.22 Orientation of the new coordinate system in the old coordinate system.

that describes u' in the X, Y, Z coordinate system. We wish to find where the tip of vector q will lie in the X, Y, Z system. To do this, we use the relations of Equation (54) to find the values of q_x, q_y, and q_z corresponding to u'_x, u'_y, and u'_z, respectively. These values are found to be

$$q_x = -u'_z$$

$$q_y = u'_x \quad\quad\quad (58)$$

$$q_z = -u'_y$$

These relationships can be seen in Figure 6.22.

Substituting these values in Equation (57), we find that

$$q = -7i + 4j - 2k \quad\quad\quad (59)$$

as measured from the X', Y', Z' center of coordinates located at the vector distance

$$t = 3i - 5j + 4k$$

from the X, Y, Z center of coordinates due to the translation performed.

Therefore, by performing the addition, the location of the tip of q in the X, Y, Z system is found to be

$$p = q + t = -4i - j + 2k$$

which agrees with the value in Equation (50) computed using matrix transformations and verifies that the n, s, and a vectors provide the orientation of the new coordinate system in the original coordinate system. By inspection of the final homogeneous matrix, we can find both the orientation and the center of the coordinates of the new coordinate system.

Further analysis using homogeneous matrices including matrix inversion is given in Paul [33] and Lee [19]. The introduction provided here should be of value in understanding those references.

6.6.2
Coordinate
Transformation

A robot is made up of links connected by joints. In kinematic analysis, it is considered to be a chain of links and joints. At one end of the chain is the supporting base; at the other end is the end effector or hand.

Controlling a robot requires that the end effector or hand be moved to a specific point in space to carry out a task. In performing the task, the robot's end effector must move through a particular designated path. This section discusses a simplified mathematical method that is useful in describing the relationship of the end effector to the robot base coordinates. Both the position in space and the orientation or direction of the robot's end effector must be described and controlled.

Determining the position and orientation of any joint in a robot relative to its base coordinate set requires the transformation of coordinates through all other joints between the base reference and the joint whose coordinates are being determined. If the robot has six joints, or degrees of freedom, it is necessary to set up six coordinate transfers, one for each joint. Each transformation relates the coordinates of one joint to the coordinates of the previous joint in the chain of links and joints.

1. Joint-to-World Transformations

World coordinates are defined as the base reference coordinates of the robot. These coordinates are taken through the base joint of the robot or at a known distance from it. Base coordinates, by convention, are defined as x_0, y_0, and z_0 in the 0 coordinate frame.

Joint coordinates are defined as the set of coordinates centered on a particular joint. In a sliding or prismatic joint, one coordinate of the coordinate set is along the direction of motion. In a rotary or revolute joint, one coordinate is parallel to the axis of the joint.

Figure 6.23 shows the relationship between the successive joints in a robot. Joint i -1 could be taken as the base reference joint because it could be any joint. The next joint toward the end effector (along the chain of links) has the number 1 assigned; each successive joint has an assigned number identification that is one greater than the preceding joint.

Links between joints are assigned in a similar way. These assignments are shown in Figure 6.23. In this particular case, a Cartesian (x, y, z) system is used. If we choose to set i equal to 1, the first link is link 1, and the first joint has the coordinates of x_0, y_0, z_0 and can be considered fixed in the reference base.

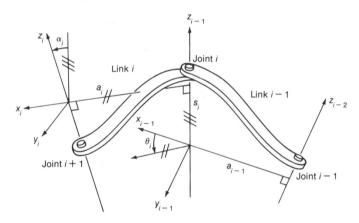

Figure 6.23 Definition of joint parameters (from Rosen [39], Second Report).

2. Coordinate Frame Parameters

Each coordinate set is also called a frame. Every frame is determined by four parameters that describe how it relates to a previous frame. This mathematical approach was first developed by Denavit and Hartenberg in 1955. It provides the necessary parameters in homogeneous matrices to perform the transformations between coordinate systems in a remarkably simple way (Denavit and Hartenberg [10], as described in Section 6.6.2.

There are two distances and two angles in each set of four parameters. These parameters are called theta sub-i (θ_i) , s sub-i (s_i) , a sub-i (a_i) , and alpha sub-i (α_i) , as shown in Figure 6.23.

Frames are oriented and determined by two rules:

1. The z_{i-1} axis lies along the axis of motion of the ith joint.
2. The x_i axis is normal (perpendicular) to the z_{i-1} axis and is pointing away from it.

In a Cartesian coordinate system, the x axis is always normal (perpendicular) to the z axis. Thus, rule 2 above means that the x_i axis is normal to both the z_i axis and the z_{i-1} axis and goes from the z_{i-1} axis to the z_1 axis. Trace out this relationship in figure 6.23 to make sure that this point is clear.

Note that the coordinate system is attached to the corresponding link even though its z axis direction is determined by the direction of the joint. The ith frame moves with the ith link; the nth frame (assigned to the hand or end effector) moves with the hand. There are $n+1$ frames on an n-joint arm.

The four parameters for each ith joint are defined as follows (check each of these in Figure 6.23):

1. The parameter θ_i is the angle from the x_{i-1} axis to the x_i axis measured around the z_{i-1} axis going in the positive direction specified by the right-hand rule. The right-hand rule specifies that moving from the x axis to the y axis around z in the clockwise direction (the direction of advance of a right-handed screw) will be positive. Moving from y to z clockwise around x is also positive, as is the movement from z to x clockwise around y. Going in the counterclockwise direction gives a negative value for the angle.
2. The parameter s_i is the distance from the origin of the $(i-1)$th frame to the intersection of the z_{i-1} axis with the x_i axis. This distance is measured along the z_{i-1} axis.
3. The parameter a_i is the shortest distance from the z_{i-1} axis to the z_i axis. Note that this distance is measured along a line which makes an angle θ_i with the x_{i-1} axis. The two slashes (//) through these two lines in figure 6.23 indicate that they are parallel. If the center of the coordinates for the z_i axis lies on the z_{i-1} axis, which is often true, the value of a_i is zero. In the Stanford arm, for example, illustrated in Figure 6.24 all the a_i values are zero.
4. The parameter α_i is the <u>offset angle</u> from the z_{i-1} axis to the z_i axis about the x_i axis, again using the right-hand rule. Three back slashes (\\\\) on each line are used to indicate that the reference line for α_i is parallel to the z_{i-1} axis.

In the simple robot configurations that we will consider, the distances between axes may become zero. In most cases also, the angles will be either zero or multiples of 90 degrees.

Any point in the ith coordinate frame (x_i, y_i, z_i) can be transformed to the $(i-1)$th system by using the homogeneous matrix defined by the parameters previously determined in the matrix equation

$$\begin{bmatrix} x_{i-1} \\ y_{i-1} \\ z_{i-1} \\ 1 \end{bmatrix} = A_{(i-1)i} \begin{bmatrix} x_i \\ y_i \\ y_i \\ 1 \end{bmatrix} \tag{60}$$

where $A_{(i-1)}$ is the 4×4 homogeneous matrix derived from the values of the four joint parameters. Multiplying the coordinate vector of the $(i-1)$th frame by the $A_{(i-1)}$ matrix generates the coordinate vector of the ith frame. The value of the generalized A matrix is given by Paul [32] as

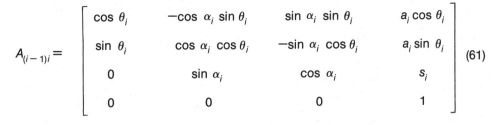

$$A_{(i-1)i} = \begin{bmatrix} \cos\theta_i & -\cos\alpha_i\sin\theta_i & \sin\alpha_i\sin\theta_i & a_i\cos\theta_i \\ \sin\theta_i & \cos\alpha_i\cos\theta_i & -\sin\alpha_i\cos\theta_i & a_i\sin\theta_i \\ 0 & \sin\alpha_i & \cos\alpha_i & s_i \\ 0 & 0 & 0 & 1 \end{bmatrix} \quad (61)$$

Transformation of hand coordinates to world coordinates can be done by successively multiplying together the individual homogeneous matrices, the A matrices, representing the transformations between coordinate frame. These A matrices are obtained by substituting in the preceding matrix the values found for the four parameters in Table 6.5.

The T matrix is the total transformation matrix and is defined by

$$T = (A_{01})(A_{12})(A_{23})\cdots(A_{(n-1)n}) \quad (62)$$

where the subscripts for the A matrices indicate the initial and final coordinate systems, respectively.

Using the T matrix, we can convert the hand coordinate vector to the world coordinate vector:

$$\begin{bmatrix} x \\ y \\ z \\ 1 \end{bmatrix} = T \begin{bmatrix} x_n \\ y_n \\ z_n \\ 1 \end{bmatrix} \quad (63)$$

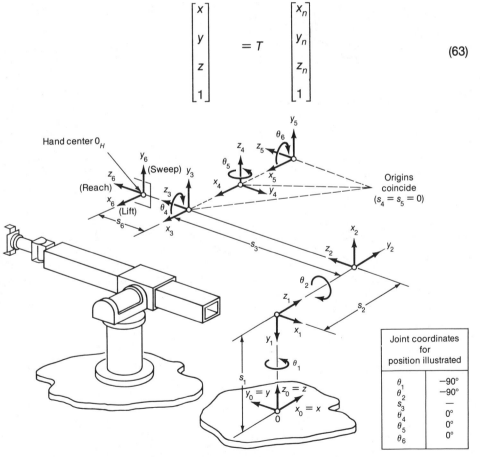

Figure 6.24 Joint coordinate system for the Stanford arm (from Rosen [39], Second Report.

This matrix was first applied to the Stanford Arm, designed by V. Sheinman in 1969. The Stanford Arm is shown in Figure 6.24. Nodes in the schematic diagram represent the six joints of the arm. Five of the joints are rotary (hinge joints) with angles of rotation θ_1 , θ_2 , θ_4 , θ_5, and θ_6 . The remaining joint, at position 3, has a prismatic (sliding) joint with a length, s_3 , that is variable. Note that three of the rotary joints coincide, so that s_4 and s_5 are zero. These three joints act as a ball-and-socket joint to support the hand. There are four links in the arm: the vertical column of length s_1 , the horizontal cylinder with length s_2 , the sliding boom with variable length s_3 , and the wrist-to-hand length s_6 .

Cartesian coordinates are used to describe the locations of each joint. These coordinates are defined as described in the four para-meter rules illustrated in Figure 6.23. Note that all the α_i angles are fixed at either 0, –90, or +90 degrees, so that their sines and cosines are constants of values 0, 1, or –1.

By examining Figure 6.24 carefully and using the four parameter rules of Section 6.6.2, Part 2, we can tabulate the values of each of the four parameters at each joint as listed in Table 6.5.

Table 6.5 Parameter values for Stanford arm

Joints	Θ_i	Parameters		
		s_i	a_i	α_i
1	-90°	s_1	0	-90°
2	-90°	s_2	0	90°
3	0°	s_3	0	0°
4	90°	0	0	-90°
5	90°	0	0	90°
6	0°	s_6	0	0°

By inserting the parameter values from Table 6.5 into the $A_{(i-1)i}$ matrix of Equation (61) for each joint we obtain the D-H matrices for the Stanford Arm. It has become customary to substitute C_i for $\cos\theta_i$ and S_i for $\sin\theta_i$ in these matrices in order to reduce the space required and improve the readability of the matrix equations. Those substitutions have been made in the following matrices for the Stanford Arm.

$$A_{01} = \begin{bmatrix} C_1 & 0 & -S_1 & 0 \\ S_1 & 0 & C_1 & 0 \\ 0 & -1 & 0 & s_1 \\ 0 & 0 & 0 & 1 \end{bmatrix} \tag{64}$$

$$A_{12} = \begin{vmatrix} C_2 & 0 & S_2 & 0 \\ S_2 & 0 & -C_2 & 0 \\ 0 & 1 & 0 & S_2 \\ 0 & 0 & 0 & 1 \end{vmatrix} \qquad (65)$$

$$A_{23} = \begin{vmatrix} 1 & 0 & 0 & 0 \\ 0 & 1 & 0 & 0 \\ 0 & 0 & 1 & S_3 \\ 0 & 0 & 0 & 1 \end{vmatrix} \qquad (66)$$

$$A_{34} = \begin{vmatrix} C_4 & 0 & -S_4 & 0 \\ S_4 & 0 & C_4 & 0 \\ 0 & -1 & 0 & 0 \\ 0 & 0 & 0 & 1 \end{vmatrix} \qquad (67)$$

$$A_{45} = \begin{vmatrix} C_5 & 0 & S_5 & 0 \\ S_5 & 0 & -C_5 & 0 \\ 0 & 1 & 0 & 0 \\ 0 & 0 & 0 & 1 \end{vmatrix} \qquad (68)$$

$$A_{56} = \begin{vmatrix} C_6 & -S_6 & 0 & 0 \\ S_6 & C_6 & 0 & 0 \\ 0 & 0 & 1 & S_6 \\ 0 & 0 & 0 & 1 \end{vmatrix} \qquad (69)$$

Multiplying these matrices together in the reverse sequence gives us the T matrix described by Equation (62). We start by multiplying A_{45} by A_{56} to obtain A_{46}, multiply A_{34} by A_{46}, and continue this process to finally obtain the T matrix. The first result is:

$$T_{46} = \begin{vmatrix} C_5C_6 & -C_5S_6 & S_5 & 0 \\ S_5C_6 & -S_5S_6 & -C_5 & 0 \\ S_6 & C_6 & 0 & 0 \\ 0 & 0 & 0 & 1 \end{vmatrix} \qquad (70)$$

Repeating this process, we obtain the T_{06} matrix, which relates the position of the end of the arm to the base as

$$T_{06} = \begin{bmatrix} n_x & s_x & a_x & p_x \\ n_y & s_y & a_y & p_y \\ n_z & s_z & a_z & p_z \\ 0 & 0 & 0 & 1 \end{bmatrix}$$

where

$$n_x = C_1[C_2(C_4C_5C_6 - S_4S_6) - S_2S_5C_6] - S_1(S_4C_5C_6 + C_4S_6)$$

$$n_y = S_1[C_2(C_4C_5C_6 - S_4S_6) - S_2S_5C_6] + C_1(S_4C_5C_6 + C_4S_6)$$

$$n_z = -S_2(C_4C_5C_6 - S_4S_6) - S_4S_6) - C_2S_5C_6$$

$$s_x = C_1[-C_2(C_4C_5S_6 + S_4C_6) + S_2S_5S_6] - S_1(-S_4C_5S_6 + C_4C_6)$$

$$s_y = S_1[-C_2(C_4C_5S_6 + S_4C_6) + S_2S_5S_6] - C_1(-S_4C_5S_6 + C_4C_6)$$

$$s_z = S_2(C_4C_5S_6 + S_2C_5) + C_2S_5S_6$$

$$a_x = C_1(C_2C_4S_5 + S_2C_5) - S_1S_4S_5$$

$$a_y = S_1(C_2C_4S_5 + S_2C_5) + C_1S_4S_5$$

$$a_z = -S_2C_4S_5 + C_2C_5$$

$$p_x = C_1S_2s_3 - S_1s_2$$

$$p_y = S_1S_2s_3 + C_1s_2$$

$$p_z = C_2s_3$$

In the preceding equations, the values for s_1 and s_6 have been set to zero, so the effective base is at the center of the cylinder and the end of the arm is taken as the end of the wrist. This arrangement simplifies and clarifies the equations.

The reverse problem, determining the joint angles required to position the arm to a particular location and orientation in space, is complex and will not be covered in detail here. In the next section, some useful control methods are described. More detail is available in the references cited there.

6.6.3
Control
Methods

Controlling an industrial robot arm along a desired trajectory is done by applying feedback control to the servo motors that drive the individual links so that a desired path will be followed despite external forces and torques.   Ordinary servo equations are unsatisfactory models of arm operations, so that improved methods of control are essential.

Several robot arm control methods have been worked out by Paul [31], Bejczy [4], Whitney [44], Albus [1], and many others, as described in Lee [19] and Brady [5,6]. these control methods are discussed briefly in the next sections.

1. Resolved Motion Rate Control

Resolved motion rate control (RMRC) was developed to provide a way to coordinate the simultaneous motion of several joints at once. Given a desired path to be followed, RMRC is used to control the individual joints in order to achieve the desired path. This method was developed by Whitney and his associates at Draper Laboratory [44]. Several alternative previous methods are reviewed in this reference.

RMRC allows commands to be applied in a wide variety of coordinate systems, and the axes being controlled can be placed wherever desired, even outside the manipulator. Several varieties of a command may be provided, so that the user may choose among them for a specific application. The example given is the alignment of the hand for grasping a tool such as a wrench, which has a preferred axis.

Computation is done using the D-H matrices (described in Section 6.6.1) to formulate the required velocity and rotation vectors for the hand. These vectors are resolved along the coordinate axes, and the Jacobian matrix for the relationship between the command rates along the axes and the rotation rates about the joints is determined. The Jacobian is a 6 × 6 matrix made up of the partial derivatives of the six individual positional or angular coordinates with respect to the six individual joint angles. The Jacobian matrix can be calculated by using vector cross-products and then inverted to obtain the desired joint angular velocities. Inversion of the matrix is difficult to do and may not be possible. An alternative is to use numerical interpolation methods to determine the joint velocities.   These methods were used successfully for some purposes but were not suitable for other applications, so that further work was required.

2. Cerebellar Model Articulation Control

James Albus, at the NBS, proposed the Cerebellar Model Articulation Control (CMAC) [1]. This was a table look-up method based on current understanding of the neurophysiological theory of operation of the human cerebellum. Control functions were computed on the basis of tables stored in computer memory. Essentially, a training function was performed that stored the joint angular velocities that resulted from moving the arm through a set of paths. If a large number of discrete paths were stored, it was possible to generate interpolated paths with a small amount of computation.   A problem with this approach was the need to control a large memory and to do the necessary interpolations with the required accuracy.

3. Computed Torque or Inverse Problem

Paul [31] described the inverse problem technique by which the required input torque is computed as a function of desired compared to actual accelerations, velocities and positions. Bejczy [4] called this the <u>computed torque technique</u>. It was necessary to satisfy several equations of motion simultaneously to ensure the validity of the solution. Under the right conditions, the solutions would converge and become valid. An improvement of this technique is described by Lee [19].

4. Resolved Acceleration Control

Luh et al. [26] describe an approach that combines the resolved motion rate control with the inverse problem approach (also called the computed torque technique). By preplanning the accelerations of the joints in addition to specifying the velocities, the resolved acceleration can be determined, so that certain mathematical problems are avoided. Measurement of the errors between the preplanned and actual accelerations is used to provide the control information needed in the servo system. The advantage of resolved acceleration control is that some computation is avoided and convergence of the solutions is not required.

A simulation of this method for the Stanford arm on a PDP 11/45 computer, using floating point assembly language, provided solutions at a rate sufficient to support a sampling frequency of 87 hertz.

5. Modular Reference Adaptive Control

Adaptive control can be used to obtain good performance over a wide range of motions and payloads. Modular reference adaptive control uses second–order, linear differential equations to describe each degree of freedom (Dubowsky and DesForges [11]). Control is obtained by varying position and velocity feedback gains to follow the model. The important modification here is the adjustment of gains in the servo loop to attain a desired result. In principle, varying loads and inertias can be compensated for in the control equations by suitable variation of the individual gains in the system. However, the actual design of gain control is difficult, and stability control is critical. Good results have been obtained by careful planning of the system for specific applications. An 8–bit microprocessor, the Mostek 6502, can perform a control update in 18 milliseconds, for example.

6. Compliant Motion Control

Many robot tasks require the robot arm to react continuously to contact forces or tactile stimuli in the environment. It is not sufficient, in these cases, to define a trajectory and expect the robot arm to follow it. Instead, the control function is defined in terms of a force or torque exerted or on the maintenance of specified stimuli. Turning a crank, opening a door, driving a screw, moving along a surface, and handling eggs are examples of activities requiring compliant motion control.

Two types of compliance exist: passive and active. Passive compliance can be provided by springs or other elastic members that bend to follow a path. Active compliance requires a control motion generated in response to stimuli on a sensor. The Remote Center Compliance (RCC) described in Part 1 of Section 6.3.7 is an example of passive compliance, while the Instrumented Remote Center Compliance (IRCC) described in the same section is an example of active compliance.

Compliant control of a robot arm in response to sensory stimuli allows joints to be force-servoed. An example would be placing an object on a table and sliding it along. Other examples would arise from the assembly of parts by fitting them together until they are properly aligned. Mason [27] surveys some applications of compliant control; Brady [5] has a section written by M. Mason with papers by R. Paul and others on compliant control.

6.7 Controller Design Example

This section contains a modified and summarized form of the first few sections of a tutorial on robot control by J. Y. S. Luh [25]. A brief summary of the remaining sections of the tutorial are included to provide the reader with additional background information. Organization of this section follows the tutorial to take advantage of the excellent organization of the subject provided by Luh. Additional material has been injected as needed to explain terms or concepts that might be difficult to understand and to provide references for further study on the topics discussed.

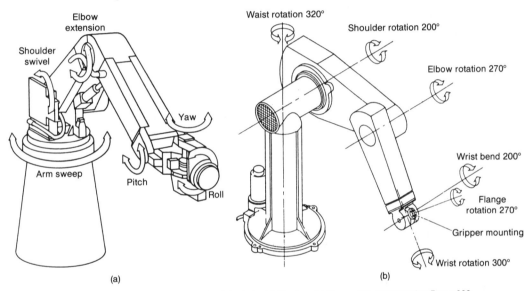

Figure 6.25 Examples of industrial robots. (a) Cincinnati Milacron T3. (b) Unimation Puma 600.

Source: (a) Cincinnati Milacron. Used by permission. (b) UNIMATION, Inc. A Westinghouse Company. Used by permission.

Industrial robots commercially available today typically have six joints, giving six degrees of freedom, with a gripper that is referred to as a <u>hand</u> or <u>end effector</u>. Examples are the Cincinnati Milacron T3 and the Unimation PUMA 600 shown in Figure 6.25. Each joint of these robots is driven with a feedback control loop similar to the loop shown in Figure 6.27, which is a block diagram of a joint control for the Stanford manipulator. The Stanford manipulator shown in Figure 6.26 is driven by a permanent magnet motor drive (Luh [23]), and has an optical encoder to provide positional feedback and a tachometer feedback for damping. An industrial robot is a positional device with a positional control servo for each joint, as these examples show.

6.7.1 Cartesian and Joint Coordinates

A robot task can be specified in terms of the movement of the hand in Cartesian coordinates. Position is identified by a position vector, p, referenced to the robot base reference and orientation vectors in three directions, as shown in Figure 6.28 and described in Section 6.6. The position of the hand with reference to the base coordinates may also be specified by a six-dimensional vector. If all of the joint positions are known, the position of the hand is deter-

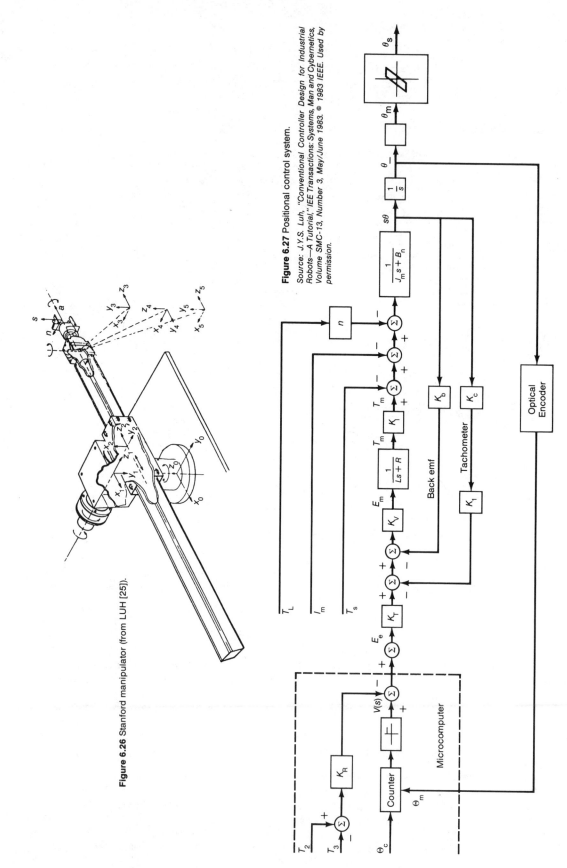

Figure 6.26 Stanford manipulator (from LUH [25]).

Figure 6.27 Positional control system.

Source: J.Y.S. Luh, "Conventional Controller Design for Industrial Robots—A Tutorial," IEEE Transactions: Systems, Man and Cybernetics, Volume SMC-13, Number 3, May/June 1983. © 1983 IEEE. Used by permission.

mined. If a specific position and orientation of the hand are required, it is necessary to specify the positions of each of the joints in the robot arm and wrist. This specification requires that the inverse vector function be solved in the *n* dimensions corresponding to the number of joints. Even if this solution is found, it might not be unique.

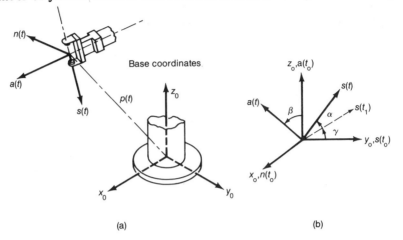

(a)   (b)

Figure 6.28 (a) Position and orientation vectors of the robot hand. (b) Euler angles of orientation.

Source: J.Y.S. Luh, "Conventional Controller Design for Industrial Robots—A Tutorial," IEEE Transactions: Systems, Man and Cybernetics, Volume SMC-13, Number 3, May/June 1983. © 1983 IEEE. Used by permission.

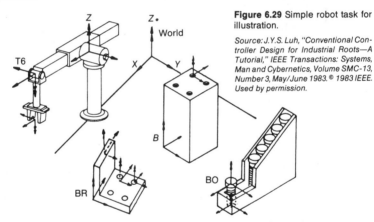

Figure 6.29 Simple robot task for illustration.

Source: J.Y.S. Luh, "Conventional Controller Design for Industrial Roots—A Tutorial," IEEE Transactions: Systems, Man and Cybernetics, Volume SMC-13, Number 3, May/June 1983. © 1983 IEEE. Used by permission.

A sample task to be accomplished is shown in Figure 6.29. The robot hand is to go in a straight line to the bolt feeder, pick up a bolt, and place it in a designated hole in the vertical block.

Means for controlling one joint are first studied in this section, and then a method is described to extend position control to all of the joints. Although a general solution may be infeasible, it is shown that solutions exist that are suitable for use with specific existing industrial robots using available microprocessor control systems.

6.7.2 Positional Control of a Single Joint

When there are no constraints on the path to be followed by the hand, it is only necessary to ensure that the hand passes through all of the corner points of the prescribed path. Corner points are those specified by the user as critical to the operation to be performed. In a teach-by-doing program, these are the points recorded by the operator. Between points, the robot is allowed to deviate from a completely controlled path.

If a path is specified to be a straight line, a specific curved or spline line, or a computed path, it is necessary to perform sufficient computation to ensure that the desired path is followed. In practice, this may mean that the controller must calculate the path coordinates 20 or 30 times per second.

1. Single–Joint Controller

In the following analysis, a single joint and its associated links will be considered. The links are assumed to be rigid bodies, with all movement due to the rotation (or translation) of the joint being studied. Referrring to Figure 6.30(a), a schematic representation of the actuator, gear, and load assembly, the following terms may be identified:

J_a	actuator inertia of one joint (oz.in.sec^2/rad.),
J_m	manipulator (robot) inertia of the joint fixtures on the actuator side of the gear assembly
J_l	inertia of the manipulator link
B_m	damping coefficient at the actuator side (oz.in.s/rad)
B_l	damping coefficient at load side
f_m	average friction torque (oz.in)
t_g	gravitational torque
t_m	generated torque at the actuator shaft
t_1	internal load torque
θ_m	angular displacement at the actuator shaft (rad)
θ_s	angular displacement at the load side

If the following are defined:

N_m	number of teeth on the actuator shaft gear
N_m	number of teeth on the load shaft gear
r_m	pitch radii of the actuator shaft gear
r_s	pitch radii of the load shaft gear

Then

$$n = r_m/r_s = N_m/N_s <= 1 \tag{71}$$

*Equation numbering follows Luh [24]. Some equations have been omitted as unnecessary in the limited development described here.

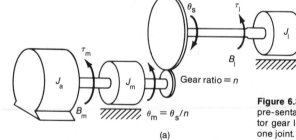

Gear ratio $= n$

$\theta_m = \theta_s/n$

(a)

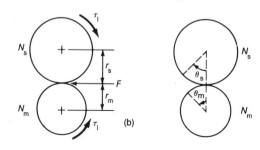

(b)   (c)

Figure 6.30 Schematic re-pre-sentation of an actua-tor gear load assembly for one joint.

Source: J.Y.S. Luh, "Convention-al Controller Design for Indus-trial Robots—A tutorial," IEEE Transactions: Systems, Man and Cybernetics, Volume SMC-13, Number 3, May/June 1983. © 1983 IEEE. Used by permis-sion.

The force F, as indicated in Figure 6.30(b), is transmitted from the actuator to the load through the contact point of the mating gears. Define

$$t'_1 = \text{equivalent internal load torque at actuator shaft}$$
$$= Fr_m$$

and
$$t_1 = Fr_s \tag{72}$$

so that
$$t'_1/t_1 = r_m/r_s = n \tag{73}$$

or
$$t'_1 = nt_1 \tag{74}$$

From Figure 6.30(c) the pitch angle of one tooth is

$$\theta_m = 2pi/N_m \tag{75}$$
$$\theta_s = 2pi/N_s \tag{76}$$

where $pi = 3.1416$, so that

$$\theta_s = \theta_m N_m/N_s = n\theta_m \tag{77}$$

Taking time derivatives on both sides of Equation (77), the angular velocities w_s and w_m, respectively, and the angular accelerations a_s and a_m, respectively, are found, with the following values:

$$w_s = nw_m \tag{78}$$
$$a_s = na_m \tag{79}$$

By D'Alembert's principle, the net torque must equal the product of inertia and angular acceleration in Figure 6.30(a) so the internal load torque t_1 minus the damping effect $B_1 w_s$ must be equal to the link inertia effect $J_1 a_s$, as follows:

$$t_1 - B_1 w_s = J_1 a_s \tag{80}$$

At the actuator shaft, the same principle is applied, resulting in

$$t_m - t'_1 - B_m w_m = (J_a + J_m)a_m \tag{81}$$

The torque relation at the actuator shaft is obtained by substituting Equations (78) and (79) into Equation (80) and then combining the result with Equation (74) to obtain

$$t'_1 = n^2(J_1 a_m + B_1 w_m) \tag{82}$$

Combining Equations (81) and (82) yields
$$t_m = (J_a + J_m + n^2 J_1)a_m + (B_m + n^2 B_1)w_m \tag{83}$$

Where $J_{eff} = (J_a + J_m + n^2 J_1)$ is the effective inertia

and
$$B_{eff} = (B_m + n^2 B_1)$$

is the effective damping coefficient at the actuator shaft.

To express the torque relation at the load shaft, eliminate a_m and w_m in Equations (78), (79), and (83) to obtain

$$t_m/n = [(J_a + J_m)/n^2 + J_1]a_s + (B_m/n^2 + B_1)w_s \tag{84}$$

where $[(J_a + J_m)/n^2 + J_1]$ is the effective inertia

and $(B_m/n^2 + B_1)$

is the effective damping coefficient at the load shaft. The parameter t_m/n is the equivalent generated torque at the load shaft.

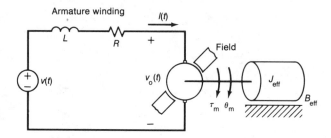

Figure 6.31 Schematic diagram of an electrical drive system.

Source: J.Y.S. Luh, Conventional Controller Design for Industrial Robots—A Tutorial," IEEE Transactions: Systems, Man and Cybernetics, Volume SMC-13, Number 3, May/June 1983. © 1983 IEEE. Used by permission.

Permanent magnet DC motors are used in the Stanford manipulator to drive the arm joints. Their speed is controlled by controlling armature current using the system shown in Figure 6.31. In the figure, the back electromotive force (EMF) generated in the armature winding due to rotation in the magnetic field is

$$v_b(t) = K_b w_m(t) \tag{85}$$

where K_b is the back EMF constant in Volt–sec/rad., L is the inductance in henries, and R is the resistance in ohms of the armature winding. Then, by Kirchoff's voltage law, one obtains

$$v(t) - v_b(t) = Ldi(t)/dt + Ri(t) \tag{86}$$

where $i(t)$ is the current through the armature winding.

This equation may be transformed into the frequency domain by using a Laplace transformation to obtain $\Theta_m(s)$, the transformed value of θ_m

$$V(s) - K_b s\Theta_m(s) = (Ls + R)I(s) \tag{87}$$

where s is the complex frequency in radians/second. The DC motor is operated in its linear range, so that the generated torque is proportional to the armature current. The relation in the frequency domain (using Laplace transforms) is

$$T_m(s) = K_1 I(s) \tag{88}$$

where K_1 is the torque constant in oz.in/(A = ampere of current).

The motor shaft is mechanically connected to an actuator gear load assembly, as in Figure 6.31, with an effective inertia J_{eff} and an effective damping coefficient B_{eff} at the actuator shaft. The relation among the mechanical components is described by Equation (83), which has a Laplace transform equivalence:

$$T_m(s) = (J_{eff}s^2 + B_{eff}s)\Theta_m(s) \tag{89}$$

with J_{eff} and B_{eff} as defined previously.

Eliminating T_m and $I(s)$ in Equations (87) to (89) yields

$$\frac{\Theta(s)}{V(s)} = \frac{K_I}{s[LJ_{eff}s^2 + (RJ_{eff})s + (RB_{eff} + K_IK_b)]} \tag{90}$$

This is the transfer function, or the feedforward gain, from the applied voltage to the DC motor (input) to the angular displacement of the motor shaft (output). From this relation, we can obtain the angular displacement of the output shaft as a function of the voltage applied to the motor.

To construct a positional controller for the angular displacement of the load shaft, we must convert the displacement into an electrical control voltage to drive the DC motor. For a closed-loop servo, the actuating signal is the error at time t between the desired and the actual displacements:

$$e(t) = \theta_d(t) - \theta_s(t) \tag{91}$$

Various position encoders, as described in Section 6.4.6, could be used to obtain the position of the output shaft. Using a potentiometer (or optical encoder and counter assembly), the displacement error is converted into an error voltage:

$$v(t) = K_\theta[\theta_d(t) - \theta_s(u)] \tag{92}$$

The transform of $v(t)$ becomes $(\Theta_d(s)$, and $\Theta_s(s)$ are the transformed values of θ_d and θ_s respectively)

$$V(s) = K_\theta E(s) \tag{93}$$
$$= K_\theta[\Theta_d(s) - \Theta_s(s)]$$

where K_θ is the conversion constant in V/rad.

Now all parts of the physical apparatus may be put together to construct a positional controller with the block diagram shown in Figure 6.32(a). The feedforward gain, or the open-loop transfer function is

$$\frac{\Theta_s(s)}{E(s)} = \frac{nK_\theta K_I}{s[LJ_{eff}s^2 + (RJ_{eff}) + (RB_{eff} + K_IK_b)]} \tag{94}$$

which is obtained using Equations (90) and (93) with the relation

$$\Theta_s(s) = n\Theta_m(s)$$

Improved performance of the drive motor in this system has been obtained by designing it for low armature inductance. As a result, the inductance of the armature winding is in tenths of millihenries and its resistance is approximately 1 ohm. Therefore, we can neglect the terms containing the inductance, L, in Equation (94), with the result that

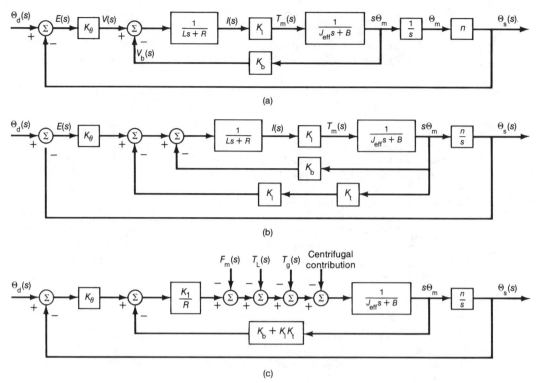

Figure 6.32 Positional controller.

Source: J.Y.S. Luh, "Conventional Controller Design for Industrial Robots—A Tutorial," IEEE Transactions: Systems, Man and Cybernetics, Volume SMC-13, Number 3, May/June 1983. © 1983 IEEE. Used by permission.

$$\frac{\Theta_s(s)}{E(s)} = \frac{nK_\theta K_l}{s[RJ_{eff} + (RB_{eff} + K_l K_b)]} \tag{95}$$

Using Equations (93) and (95) together, we find that the closed–loop transfer function of the unity feedback positional controller of Figure 6.32(a) is

$$\frac{\Theta_s(s)}{\Theta_d(s)} = \frac{\Theta_s/E}{1 + \Theta_s/E}$$

$$= \frac{nK_\theta K_l}{RJ_{eff}} \times \frac{1}{s^2 + (RB_{eff} + K_l K_b)s/(RJ_{eff}) + K_\theta K_l/(RJ_{eff})} \tag{96}$$

This transfer function results in a second–order system that is theoretically always stable. To improve the response time, the system gain can be increased, causing a faster response. Some damping of the system can be added to reinforce the effect of back EMF by adding a negative feedback of the motor shaft velocity from a tachometer or by computing the velocity by measuring the shaft angular displacement for a fixed interval.

Figure 6.32(b) is a block diagram of the resulting controller, where K_t is the tachometer constant in volts per volt (V/V). the feedback voltage is now $K_b w_m(t) + K_l K_t w_m(t)$ instead of $K_b w_m(t)$ alone. The proof is left to the reader. This and further developments are worked out in detail in Luh [24].

In figure 6.32(c), the effects of friction torque, external load torque, gravitational torque, and centrifugal contribution are considered, as discussed in Luh [24].

Parameter values for the joint 1 and 2 motors used to drive the Stanford manipulator are given in Table 6.6. From these values, the response of the system may be calculated by substituting in Equation (6.7-31). Joint 1 used a U9M4T, and joint 2 used a U12M4T DC motor, both with integral tachometers 030/106. These motors were made by the Photocircuits Corporation and have a flat armature with very low inductance to attain maximum response.

Table 6.6 Motor parameters*

Model	U9M4T	U12M4T
K_I(oz–in/A)	6.1	14.4
J_a(oz–in–sec^2/rad.)	0.008	0.033
B_m(oz–in–sec/rad.)	0.01146	0.04297
K_b(Volt–sec/rad.)	0.04297	0.10123
L (microHenries)	100.	100.
R(ohms)	1.025	0.91
K_t(Volt–sec/rad.)	0.02149	0.05062
f_m(oz–in)	6.0	6.0
n	0.01	0.01

*Motor-tachometer parameters supplied by the Photocircuits Corporation were taken from Luh [24].

The effective inertias of each joint are also required for the calculation of the robot response. Luh [24] gives the values shown in Table 6.7 which, of course, vary with the load.

Table 6.7 Joint parameters*

Joint No.	Minimum value (No Load)	Maximum value (No Load)	Maximum Value (Full Load)
1	1.417	6.176	9.570
2	3.590	6.590	10.300
3	7.257	7.257	9.057
4	0.108	0.123	0.234
5	0.114	0.114	0.225
6	0.040	0.040	0.040

*All load values are in Kilograms-meter^2.

The large changes in inertia with load complicate the control problem and must be considered in determining whether or not the system will be stable during all operations. This is discussed further in Luh [24]. It is important to be aware of the differences between the metric system and the English system in computing the inertia. Values of inertia are given in kg-m^2 in the table above which is equivalent to 141.61 oz–in–s^2/rad.

2. Determination of the Gain Constants

We saw in Equation (96) that the ratio of the output angular displacement, θ_s, to the angle commanded at the input, θ_d, was proportional to two constants, a conversion constant K_θ defined by Equation (93) and a gain K_1. K_θ is the ratio of voltage output from the position transducer to the angular difference between the input and output shafts. It is usually implemented as the gain of an electronic amplifier. In the block diagram of Figure 6.32 this appears as a separate block. The other value, K_1, is the gain of the driver amplifier that drives the motor, also in a separate block. These two values are important to the control function and must be determined with care.

After the tachometer feedback is incorporated into the servo system equation, as indicated in Figure 6.32(b), the transfer function for output versus input as a result of introducing the additional values into Equation (96) is

$$\frac{\Theta_s(s)}{\Theta_d(s)} = \frac{\Theta_s/E}{1 + \Theta_2/E}$$

$$= \frac{nK_\theta K_1}{RJ_{eff} + K_1(K_b + K_1 K_t)]s + nK_\theta K_1} \tag{97}$$

The denominator of Equation (97) is the <u>characteristic equation</u> of the transfer function when set equal to zero, because it determines the damping ratio and the undamped natural frequency of the system.

$$RJ_{eff}s^2 + [(RB_{eff} + K_1(K_b + K_1 K_t)]s + nK_\theta K_1 = 0$$

[Equations (98), (99), and (102) are omitted. They should be developed by the reader as an exercise; See Exercise 24.]

This equation may be rewritten as (see Section 6.4.4)

$$s^2 + 2\xi\omega_n s + \omega_n^2 = 0 \tag{100}$$

The value ξ is the damping ratio and ω_n is the undamped natural frequency. These values can be shown to be (Exercise 24)

$$\omega_n = (nK_\theta K_1/(RJ_{eff})^{.5} > 0 \tag{101}$$

$$\xi = [RB_{eff} + K_1(K_b + K_1 K_t)]/[2(nK_\theta K_1 RJ_{eff})^{.5}] \tag{103}$$

By using a value of 5 kg-m^2 for the inertia of joints 1 and 2, Luh [24] determines structural frequencies of 25 and 37 rad/s respectively. Paul [33] suggests that for a conservative design with a safety factor of 2, the undamped natural frqeuency should be set to no more than half the structural frequency by adjustment of parameters. This is further discussed in Luh [24].

3. Steady-State Error for Joint Controller

A continuation of the above analysis leads to the block diagram of Figure 6.32(c), which includes the effects of gravitational, load, and other torques. Luh [24] provides a detailed analysis of these effects.

6.7.3
Conveyor
Following
with a
Single-Joint
Controller

When a robot is required to pick up an object from a moving conveyor operating at a constant speed, the input signal, Θ_d, must be updated frequently depending on the allowable error of tracking. Because the conveyor velocity is known, an input can be generated to supply the necessary information. Updating at 60 times per second is often sufficient in industrial applications. The following sections summarize, somewhat inaccurately, the detailed analysis provided by Luh [24].

1. Velocity Error and Compensation

Compensation for the velocity error requires that the steady-state velocity error be obtained and then countered by a control signal. By use of feedforward signals, the steady-state error can be reduced to nearly zero while in the tracking mode. Compensation becomes more difficult in the manipulation mode in some cases. At slow speeds, the error due to centrifugal effects may be ignored, although it still exists.

2. Compensation for the Centrifugal Term

The centrifugal contribution can be computed from $D[w_s(t)]^2$, where D is a proportional constant and w_s is the rotational velocity of the robot link that can be measured with a tachometer. The value of D depends on the geometrical configuration of the robot.

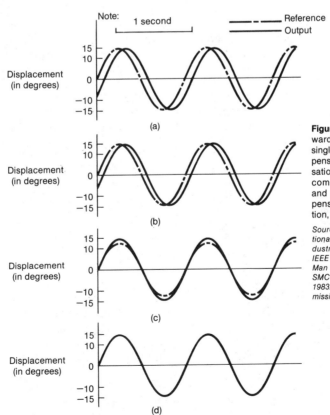

Figure 6.33 Effect of feedforward compensation with a single joint. (a) Without compensation. (b) With compensation for gravity. (c) With compensation for gravity and frction. (d) With compensation for gravity, friction, and inertia.

Source: J.Y.S. Luh, "Conventional Controller Design for Industrial Robots—A Tutorial," IEEE Transactions: Systems, Man and Cybernetics, Volume SMC-13, Number 3, May/June 1983. © 1983 IEEE. Used by permission.

3. Acceleration Error and Compensation

Feedforward compensation can be used to reduce the steady-state acceleration error. This compensation can be calculated at the cost of increased complexity in the servo system. Because it is affected by inertia, it is necessary to incorporate ways to determine the effective inertia as needed, as well as the centrifugal and other effects.

4. Effect of Feedforward Compensation

Figure 6.33 shows the effect of feedforward compensation as determined by a group at CMU. This figure is taken from the CMU report (Asada [2]) on the investigation.

**6.7.4
Controller for a
Robot with
Multiple
Joints**

The preceding analysis was based on the use of a single-joint robot. Complications occur when we use a robot with multiple joints, the only kind that is really useful. As discussed in Part 3 of Section 6.1.2, the simplest control is gained by moving one joint at a time to avoid coupling effects between joints. As more joints are controlled, the computation required increases rapidly due to the interactions of forces and moments between joints. To determine the necessary compensation, it is necessary to analyze the dynamic behavior of the robot.

1. Lagrangian Formulation of Dynamic Equations

A number of authors (Hollerbach [16], Luh [24], Lee [19], Freund [13], Paul [33]) have discussed the use of the Lagrangian approach to the formulation of dynamic equations for robots. This approach is mathematically complex and computationally difficult but gives results that provide insight into both the linear and nonlinear operation of robot arms.

The Lagrangian L is defined as the difference between the kinetic energy K and the potential energy P of the system. It may be expressed in any coordinate system (Paul [33]). Dynamic equations describing the system are obtained in the form of partial differential equations that are functions of generalized coordinates. Equations can be obtained describing either force, for a linear coordinate, or torque, for an angular coordinate.

Using the Lagrangian method makes possible the computation of the interaction of gravitational, loading, and centripetal torques and forces dependent on the external forces, the inertias of the various components of the robot arm, and the structure of the system. Coriolis acceleration is derived as part of the overall analysis.

Because the mathematical development of the Lagrangian approach is beyond the scope of this book, the reader is referred to Paul [33] for basic development of the approach and to Luh [24], Freund [13], and Lee [19] for a survey of advanced techniques used in outlining and solving nonlinear problems in robot control. Hollerbach [16] describes a recursive Lagrangian formulation that is computationally efficient. Some subjects discussed in detail by Luh in his tutorial are:

1. Coupling Between Joints and Compensation.
2. Computation of Compensation for Coupling Inertia.
3. Computation of Compensation for Gravity, Centrifugal, and Coriolois Terms

Path tracking is required to follow a conveyor or other external mechanism. Using a positional controller, path tracking can be accomplished by breaking down the path into segments and moving between the inputs of the segments under positional control. When the segments are very short, the differential transformation becomes nearly linear, and it is possible to use the Jacobian matrix as a sufficiently accurate approximation to the desired path. There are possible singularities in the matrix calculation when the robot configuration becomes degenerate. (<u>Degenerate</u> here means that there is a point to which two different joints could move to generate the same result, so there is no mathematical way to determine which joint to move.)

Specifying a point-to-point path, either by teaching (a better term may be <u>guiding</u>) or by specifying numerical coordinates, will cause the robot arm to stop at each point unless explicit instructions to do otherwise are provided. Frequent stopping is undesirable because it slows down the robot operation. By specifying nonzero velocities at each point (sometimes called a <u>corner</u> of the path), the Jacobian matrix can be used to generate a smooth path through the corner points. This motion is referred to as <u>resolved rate</u> motion by Whitney [44], and requires that a velocity control loop be added to the controller. As an alternative, a digital computer can be used to compute the joint torques to yield appropriate velocities at fixed acceleration. This is a time-consuming calculation because the couplings between joints are nonlinear.

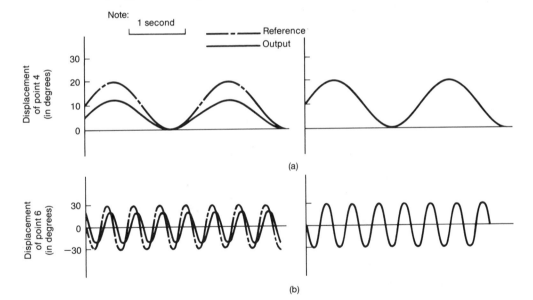

Figure 6.34 Effect of feedforward compensation for multiple joints with interaction. (a) Without compensation. (b) With compensation (from Asada [3]).

Source: J.Y.S. Luh, "Conventional Controller Design for Industrial Robots—A Tutorial," IEEE Transactions: Systems, Man and Cybernetics, Volume SMC-13, Number 3, May/June 1983. © 1983 IEEE. Used by permission.

A comparison of the joint trajectories with and without feedforward compensation is shown in Figure 6.34. This result is based on a study at Carnegie Mellon University (Asada [2]). Dynamic feedforward compensation of gravity force, friction torque, and centrifugal and Coriolis effects for a direct-drive arm with two joints was generated by extensive computation. It results in close agreement between the input command and the output when the full compensation is implemented.

Positional control of single joints can be obtained by the use of conventional methods using the damping factor and the structural resonant frequency. It is possible to reduce the steady-state position error and to eliminate the velocity and acceleration errors by feed-forward compensation. For multiple-joint controllers, further feed-forward compensation can be used to overcome the effects of force interactions between joints. Because of great computational difficulties, it is necessary to use approximation, simplification, and even parallel computation to perform the necessary computations rapidly enough to control an industrial robot.

More detail on many of the above subjects are given in the references listed at the end of this chapter. Brady et al. [5] contains many of these references and several others. In addition, there is a well-organized discussion of the many facets of robot motion by M. Brady and the other contributing editors.

Exercises

The following exercises provide a means for review of the subjects in Chapter 6. Specific topics are emphasized by more detailed questions or by mathematical problems.

1. What are the two major types of control for robots?
2. What are the advantages of simple bang-bang control?
3. Identify the parts of the servo system in Figure 6.1 by referring to the interpretation column of Table 6.2.
4. What are the chief advantages of continuous-path versus point-to-point robots? What applications require continuous-path operation?
5. Is sequential joint control advantageous under any conditions?
6. What memory size would be required to store 100 programs with 30 points each, using a six-degrees-of-freedom, point-to-point robot?
7. When is the use of a teach box advantageous?
8. What is the difference between standard servo control and feedforward control?
9. What is the major difference between controlled-trajectory robots and adaptive robots as defined in Section 6.3?
10. Describe the requirements for solving the inverse dynamic problem. Is it always possible to solve it? You may wish to refer to some of the references listed, which discuss this problem in detail.
11. What are the four methods for obstacle avoidance discussed in Section 6.3.3? Which of them would be best in a factory environment?
12. How does the availability of vision and other sensors simplify the control problem?
13. What is the significance of the Greek letter sigma, Σ, in a servo control loop?
14. Perform the matrix multiplication of Equation (46) in the reverse order and discuss the meaning of the resulting matrix.
15. Determine the resulting matrix if the matrix of Equation (48) was multiplied on the right by the following vector:

$$m = 8i - 6j + 5k$$

16. If the result of Exercise 15 was rotated by Rot(X, 90), what orientation would the new coordinates have relative to the initial coordinate center?

17. In the T_{06} matrix for the Stanford manipulator, we set s_1, and s_6 to zero. What would the n, s, a and p vectors be if these values were used as s_1 and s_6 instead? (This answer requires the multiplication of sequence of A matrices together to get a new T_{06} matrix and is therefore a lengthy problem unless done on a computer.)

18. Represent the block diagram of Figure 6.32(a) as an algebraic function of the same form as Equation (4). Use the name for the block term, $1/(Ls + R)$, and $J(s)$ for the following block term:

$$J(s) = \frac{1}{J_{eff}s + B}$$

Your answer should be in terms of the ratio of $\theta_s(s)$ to $\theta_d(s)$.

19. Substitute the values $M = 3$, $B = 9$, and $K = 6$ in Equation (4). Find the Laplace transform of the resulting equation when a unit step function is applied. Is this a stable or an oscillatory system?

20. What type of position feedback device would be most reliable in a servo system under factory conditions?

21. Prove that the addition of the tachometer as shown in Figure 6.32(b) results in the substituted term

$$K_b + K_1 K_t$$

instead of K_b alone, as in Equation (96).

22. Show that the values in Equations (101) and (103) can be obtained by the use of Equations (97) and (100).

23. Substitute the parameter values for the U9M4T motor, as listed in Table 6.6, into equation (97) and determine the characteristic equation terms, as in Equations (101) and (103). What is the value of the gain, K_1, to make the system critically damped for Joint 6 under no-load conditions? Use the joint parameters from Table 6.7.

24. Equations (98), (99), and (102) were omitted from the derivation in Part 2 of Section 6.7.2. Rewrite Equation (97) to provide a denominator of the form of Equation (100) and solve for the value of $2zw_n$ determined by the characteristic equation. Note that the answer must be consistent with the value of z found in Equation (103).

References

[1] J.S. Albus, "A New Approach to Manipulator Control: The Cerebellar Model Articulation Controller (CMAC)," J. Dynamic Systems, Measurement and Control, 97, (1975), 220-227. Also, "Data Storage in the Cerebellar Model Articulation Controller (CMAC)," J. Dyn. Sys. Meas. and Control, 97, (1975), 228-233.

[2] H. Asada, T. Kanade, I. Takeyama, "Control of a Direct-Drive Arm," The Robotics Institute, Carnegie-Mellon University, CMU-RI-TR-82-4, Pittsburgh, Pennsylvania 15213, March 9, 1982, pp. 1-15 plus figures.

[3] A. Balestrino, G. DeMaria, and L. Sciavico, "Adaptive Control of Manipulators in the Task Oriented Space," [9], 1983, pp. 3-13 to 13-28.

[4] A. K. Bejczy, "Robot Arm Dynamics and Control," Tech. Memo. 33-669, Jet Propulsion Laboratory, February 1974.

[5] M. Brady, J. M. Hollerbach, T. L. Johnson, T. Lorano-Perez, M. T. Mason, eds.: Robot Motion: Planning and Control, MIT Press, Cambridge, Mass. 1982.

[6] M. Brady, "Trajectory Planning," in [5], pp. 222-243.

[7[G. S. Brown and D. P. Campbell, Principles of Servomechanisms—Dynamics and Synthesis of Closed-Loop Control Systems. Wiley, New York, 1948.

[8] L. Clark, N. Webber, and T. Sutton, "Adaptive Control with The T^3 Robot," [9], 1983, pp. 13-1 to 13-12.

[9] Conference Proceedings, 13th International Symposium on Industrial Robots and Robots 7, April 17-21, 1983. Robotics International of SME, Dearborn, Michigan. Many papers are referenced from these proceedings.

[10] J. Denavit and R. S. Hartenberg, "A Kinematic Notation for Lower-Pair Mechanisms Based on Matrices." ASME Journal of Applied Mechanics, June 1955, pp. 215-221.

[11] S. Dubowsky and D. T. DesForges, "The Applications of Model Referenced Adaptive Control to Robotic Manipulators," ASME Journal of Dynamic Systems Measurement and Control, vol. 101, September 1979, pp. 193-200.

[12] C. W. Edwall, C. Y. Ho, and H. J. Pottinger, "Trajectory Generation and Control of a Robot Arm Using Spline Functions,".

[13] E. Freund, "Fast Nonlinear Control with Arbitrary Pole-Placement for Industrial Robots and Manipulators," Robotics Research Vol. 1, No. 1 (1982), pp. 65-78. Reprinted in [5].

[14] H. L. Hazen, "Theory of Servomechanisms," Journal of the Franklin Institute, Vol. 218, No. 3, September 1934, pp. 279-330.

[15] R. E. Hohn, "Application Flexibility of a Computer-Controlled Industrial Robot," Presented at First North American Industrial Robot Conference, October 1976. Reprinted in [40].

[16] J. M. Hollerbach, "A Recursive Lagrangian Formulation of Manipulator Dynamics and a Comparative Study of Dynamics Formulation Complexity," IEEE Trans. Sys. Man and Cyber., SMC-10, no. 11, Nov. 1980, pp. 730-736. Reprinted in [5].

[17] B. C. Kuo, Automatic Control Systems, Prentice-Hall, Englewood Cliffs, N.J. 1982. Fourth Edition.

[18] B. P. Lathi, Signals, Systems and Controls. Harper and Row, New York, 1974.

[19] C. S. G. Lee, "Robot Arm Kinematics, Dynamics, and Control." Computer, IEEE, December 1982, vol. 15, no. 12, ppo. 62-80.

[20] T. Lozano-Perez, "Automatic Planning of Manipulator Transfer Movements," IEEE, Vol. SMC-11, No. 10, Oct. 1981, pp. 681-689. Reprinted in [5].

[21] T. Lozano-Perez, "Task Planning." Published in [5], pp. 474-498.

[22] T. Lozano-Perez, "Robot Programming," Procedings of the IEEE, vol. 71, no. 7, July 1983, pp. 821-841.

[23] J. Y. S. Luh, W. D. Fisher, and R. P. C. Paul, "Joint Torque Control by a Direct Feedback for Industrial Robots," IEEE Transactions on Automatic Control, vol. 28, February 1983, pp. 183-161.

[24] J. Y. S. Luh, "Conventional Controller Design for Industrial Robots--A Tutorial, " IEEE Transactions on Systems, Management and Cybernetics, vol. SMC-13, No. 3, May/June 1983, pp. 298-316.

[25] J. Y. S. Luh, "Anatomy of Industrial Robots and Their Controls," IEEE Trans. Automat.Contr., Vol. 28, February 1983, pp. 133-153.

[26] J. Y. S. Luh, M. W. Walker, and R. P. C. Paul, "On-line

Computational Scheme for Mechanical Manipulators," 1980, ASME Journal of Dynamic Systems, Measurement and Control, vol. 25, no. 3, 1980, pp. 468-474. Reprinted in [5].

[27] M. T. Mason, "Compliance and Force Control for Computer-Controlled Manipulators," MIT Artificial Intelligence Lab. AIM-515, April 1979.

[28] V. Milenkovic, "Coordinates Suitable for Angular Motion Synthesis in Robots." In [35].

[29] J. K. Myers and G. J. Agin, "Designing a Collision Avoidance System for Robot Controllers," Robotics World, vol. 1, no. 1, January 1983, pp. 36-39.

[30] D. S. Nau, "Expert Computer Systems," Computer Magazine, IEEE, vol. 16, no. 2, February 1983, pp. 63-85.

[31] R. P. Paul, "Modeling Trajectory Calculation and Servoing of a Computer Controlled Arm" Stanford Artificial Intelligence Laboratory, AIM 177, 1972, Stanford University, Stanford, CA.

[32] R. P. Paul, Robot Manipulators: Mathematics, Programming and Control," MIT Press, Cambridge, Mass., 1981, pp. 119-155.

[33] R. P. Paul, B. Shimano, and G. E. Mayer, "Kinematic Equations for Simple Manipulators," IEEE Transactions on Systems, Man. and Cybernetics, vol. SMC-11, No. 6, June 1981, pp. 449-455.

[34] R. P. Paul, B. Shimano, and G. E. Mayer, "Differential Kinematic Control Equations for Simple Manipulators," IEEE Transactions on Systems, Management and Cybernetics, vol. SMC-11, No. 6, June 1981, pp. 456-460.

[35] Robots 6 Conference Proceedings, March 2-4, 1982, Detroit, Mich. Robotics International of SME, Dearborn, Mich. Many papers are referenced from these proceedings.

[36] C. A. Rosen, D. Nitzan, G. Agin, A. Bavarsky, G. Gleason, J. Hill, D. McGhie, and W. Park, "Machine Intelligence Research Applied to Industrial Automation," SRI International, Second through Eighth Reports. Reports not issued every year.

[37] C. A. Rosen and D. Nitzan, "Developments in Programmable Automation," Manufacturing Engineering, September 1975. Reprinted in [40], pp. 254-258.

[38] W. M. Silver, "On the Equivalence of Lagrangian and Newton-Euler Dynamics for Manipulators," Robotics Research, vol. 1, no. 2, Summer, 1982, pp. 60-70.

[39] W. R. Tanner, Editor, "Industrial Robots," Second Edition, Vol. 1/Fundamentals, Robotics International of SME, Society of Manufacturing Engineers, One SME Drive, Dearborn, Mich.

[40] W. R. Tanner, Editor, "Industrial Robots," 2nd Edition, vol. 2, Application. Robotics International of SME, Dearborn, Mich. 1981.

[41] D. E. Whitney, "The Mathematics of Coordinated Control of Prosthetic Arms and Manipulators," 1972, Journal of Dynamic Systems, Measurement and Control, pp. 303-309. Reprinted in [5].

[42] D. E. Whitney, "Use of Resolved Rate to Generate Torque Histories for Arm Control," memo MAT-54, Draper Laboratory, Cambridge, Mass., July 1972.

[43] D. E. Whitney, "The Mathematics of Coordinated Control of Prosthetic Arms and Manipulators," Journal of Dynamic Systems, Measurement and Control, ASME, December 1972, pp. 303-309. Reprinted in [ ].

[44] D. E. Whitney, "Use of Resolved Rate to Generate Torque Histories for Arm Control," C. S. Draper Lab, MAT-54, July 1972.

Chapter 7

Computer Hardware for Robot Systems

This chapter discusses the computers and computer components used to operate and control robotic systems. Digital computers are discussed that use the binary number system or the related octal and hexadecimal systems as described in Appendix A.1. Those not familiar with these number systems should review Appendix A.1 before proceeding with this chapter.

Some of the functions to be supplied by computer hardware in robotic systems are discussed first. The essential functions are supervisory control, trajectory calculation and control, vision and sensory information processing, several types of monitoring functions, communication with related equipment through the input/output interfaces, and peripheral equipment functions to provide for data storage, printing, and system support.

Fundamental concepts of logic circuits and their use for control and calculation in digital computers are covered briefly in this chapter. Then the overall organization of computer central processing units is described. Peripheral equipment used in computers—disks, tapes, printers, and similar data processing equipment—is described briefly. Then the more esoteric and advanced equipment used as part of robotic systems is discussed: vision systems, sensors, and robot servos. The important area of input/output equipment is described as it relates to robotic systems, along with consideration of the problems of interfacing, interrupts, and interrupt handling. Then overall system organization of computers, robots, and associated equipment is described. Examples of present and proposed robot systems are described to complete the overview of computer application to robotics.

7.1 Hardware Needs for Robot Systems

Some of the hardware needed by robotic systems is essentially the same as that of the usual data processing systems. Many of the functions required can be performed by microprocessors and minicomputers used for other commercial and industrial applications. Some vision and some of the more complex control calculations require high-speed computation capability, which is only available in parallel processors and large mainframe computers. These advanced systems are discussed in Section 7.8.

7.1.1 System Supervision

Digital computers provide a unique capability to supervise the many complex and interacting operations being performed, often simultaneously, in a robotic system. Operation of the robot manipulator arms must be controlled and synchronized with the operation of associated vision systems, sensors, and auxiliary equipment. At the same time, the supervisory function must gather operational data to record the status and activity of the system. In addition, it must communicate with other associated systems and perhaps with a higher-level control system.

**7.1.2
Trajectory Calcu-
lation and
Control**

Movement of the robot arm through a controlled trajectory as described in Section 6.3.1 may be relatively simple or may challenge the total computing capability of the computer system. In many robot systems, one or more separate computers are used for each degree of freedom of the manipulator. Calculations performed include matrix multiplication, matrix inversion, and other complex mathematical functions, as described in Section 6.3. Advanced systems must incorporate obstacle avoidance in the trajectory calculation.

**7.1.3
Vision**

Typically, the vision function is provided by a separate computer system that communicates its results to the supervisory control computer upon request. Some robotic languages have commands to specify to the vision system the location to be examined. The vision system then does the necessary scanning and analysis, and reports the result of this work to the supervisory or control computer in the form of tables placed in a specified location in memory accessible to both computers. This communication requires that both computers use an agreed-upon sequence of signals in a specified code.

**7.1.4
Sensor
Monitoring**

As robotic systems become adaptive and intelligent, more and more sensors will be used. Some of these sensors, as described in section 5.3.7, will have their own associated microcomputers or logic functions. Other sensors will require periodic scanning by the control or supervisory computer to gather the necessary information. Sensory information may cause high-priority interrupts of the main processing and control function. These interrupts must be handled quickly, in most cases, to ensure that the robotic system does not damage itself or its environment. Sensors monitoring force, torque, and position will provide a constant stream of input to be acted upon by the control computer, and perhaps analyzed and recorded by a computer program assigned to this task.

**7.1.5
Safety
Monitoring**

Safety monitoring depends on specialized sensors in addition to the many sensors used for motion control and position monitoring. Some systems now have a separate microcomputer assigned to this task. It has the ability to override the control computer and stop or reverse motion and other functions if necessary, much like the reflex actions in a human arm, hand, or finger. Light beams and acoustic sensors are especially useful for surveillance of the volume surrounding the robot, while temperature, pressure, and other sensors may warn of other types of danger. It is necessary to plan for safety and design in the required safety control systems as part of the basic system design in order to obtain a reliable, safe system.

**7.1.6
Input/Output**

Communication of data and control between the components of the robotic system requires that suitable interfaces and communication standards be established between these parts of the system. Data storage is commonly done on magnetic disks. Output data is sent out to printers, display consoles, or remote monitors. Status information is received from auxiliary equipment. Commands are sent to this auxiliary equipment over the input/output lines.

There may be multiple levels of control in the system. Disk files usually have their own controller, which controls several disk drives. This controller accepts data from the central processor or control computer, converts it appropriately, and stores it on the disk file. On request, it locates stored information and sends it back to the central processor. Other equipment has similar input/output needs. These topics are discussed in more detail in Sections 7.4 and 7.5.

7.2 Logic Circuits and Computer Elements

Variations of the same logic circuits and computer elements are used to perform the myriad tasks in a computer system that depend on digital logic and control. Some of these circuits and elements are described in this section. Logic circuits are the building blocks of all computer elements and are used in the registers, decoders, multiplexers, counters, arithmetic and logic units (ALUs), and even in the high-speed semiconductor memory elements now being used.

**7.2.1
Logic Circuits
(Logic Gates)**

It is interesting to note that all digital functions in a computer system can be built using only three types of logic circuits: the AND, OR, and NOT, or inverter, circuits. Even more surprising, perhaps, is the idea that either of two types of circuits, NANDs and NORs, can be used to provide all of the digital functions of a complete computer system. A description of these logical circuits will make this idea clearer. These circuits are often called logic gates because they act to open and close logic paths, as a gate does.

In describing logical circuits, it is convenient to compare them to conventional switching circuits and then to provide a truth table, the logic equation, and the standard symbol that is used to identify that function in a logic diagram. Each of the common logic circuits will be described in this way. Boolean algebra is used to write the logic equation. This is a simple and easily understood variation of ordinary algebra, but the symbols used have somewhat different meanings. Simple Boolean algebra is described in Section 7.2.2. Three-state switches, used for control of data switching in data buses, are described in the last part of this section.

1. AND Circuit

The meaning of an AND circuit is the same as the ordinary meaning of the word and. In Figure 7.1 there are two switches, A and B, connected to a battery or other electrical power source and a light, C. If both switches, A and B, are closed, the power reaches the light, C, and the light turns on. If either A or B is open, no power reaches the light and the light, C, remains off.

This same information is expressed by the truth table, where 0 is used to indicate that the switch is open and 1 is used to indicate that the switch is closed. Binary logic is used in computers because it can be represented by two conditions that are easily achievable by electronic means.

The Boolean equation is C = A and B, which is written C = A • B where the meaning of the dot is the AND operation now described. Often the dot is omitted if the meaning is clear from context, so that C = AB.

The AND function operation is summarized:

1. When both A and B are open (A = 0, B = 0), the light is off (C = 0).
2. With A open and B closed (A = 0, B = 1), the light is still off (C = 0).
3. With A closed and B open (A = 1, B = 0), the light is still off (C = 0).
4. When A is closed and B is closed (A = 1, B = 1), the light is on (C = 1).

These results are tabulated in the truth table for the AND circuit.

The standard symbol for the AND circuit and the truth table are given in Figure 7.1. Multiple inputs, corresponding to multiple switches in series, can be used.

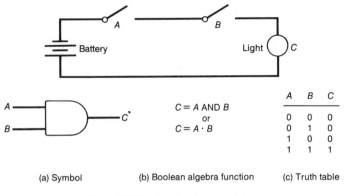

A	B	C
0	0	0
0	1	0
1	0	0
1	1	1

$C = A \text{ AND } B$
or
$C = A \cdot B$

(a) Symbol   (b) Boolean algebra function   (c) Truth table

Figure 7.1 AND circuit and representation.

2. OR Circuit

OR circuits represent the statement "if A or B is true, then C is true." In an OR circuit made up of switches A and B, as in Figure 7.2, the light C will be on if either A or B is on. The on state represents the 1 or true condition. The off state represents the 0 or false condition, just as it did in the AND circuit. We have introduced the ideas of true and false as another way to look at logic circuit parameters. A truth table can be constructed for the OR circuit in a way similar to that used for the AND circuit. This truth table is shown in Figure 7.2 with the logic symbol and the Boolean Algebra function for the OR circuit. Note that the light is on (C = 1) when either switch A or switch B is closed. We use the + sign to signify OR in the Boolean Algebra statement.

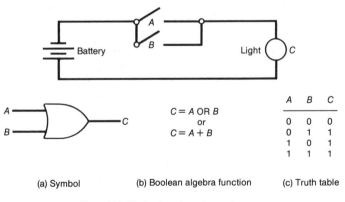

A	B	C
0	0	0
0	1	1
1	0	1
1	1	1

$C = A \text{ OR } B$
or
$C = A + B$

(a) Symbol   (b) Boolean algebra function   (c) Truth table

Figure 7.2 OR circuit and representation.

3. NOT Circuit (Inverter)

A necessary logic function is the NOT or inversion function. This circuit changes an incoming 1 to a 0 on the output and a input 0 to an output of 1. A circuit equivalent could be made up using a relay controlled by the A input to open the light circuit C when A was closed, so that the output C would always be the inverse of the input

A. This circuit will be left as an exercise. The symbol, algebra, and truth table for the NOT circuit are shown in Figure 7.3. Note the use of the prime (') to signify the inversion. Some writers use a bar over the top of the character:

$$C = \overline{A}$$

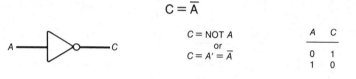

(a) Symbol	(b) Boolean algebra function	(c) Truth table

Figure 7.3 NOT or inversion representation.

4. NAND Circuit

If an AND circuit and a NOT circuit are put together, it is called a NOT-AND or NAND circuit. The AND symbol is used with a small circle at the output end to signify the inversion. Its truth table has the value of C inverted from that of the AND truth table. As discussed in Section 7.2.2, the NAND is equivalent to a combination of an inverter and two OR gates. With enough of these NAND gates, a complete computer could be made.

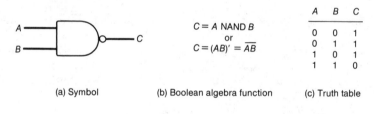

(a) Symbol	(b) Boolean algebra function	(c) Truth table

Figure 7.4 NAND circuit and representation.

5. NOR Circuit

NOR circuits are a combination of an OR and a NOT circuit to become a NOT-OR or NOR circuit. This circuit is equivalent to some AND circuits plus a NOT circuit. These circuits are also sufficient to build all possible logic circuits. The C values of the NOR truth table are the inverse of those in the OR gate. As in the NAND gate, the inversion is shown with a small circle at the output end.

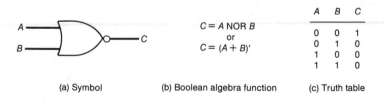

(a) Symbol	(b) Boolean algebra function	(c) Truth table

Figure 7.5 NOR circuit and representation.

6. Exclusive-OR (XOR) Circuit

One important use of the XOR circuit is in arithmetic adders for computers because it provides the logical function

$$C = A'B + AB' = A \oplus B$$

as shown in Figure 7.6. The XOR requires that the two inputs A and B be different to get C = 1 as the output. A double curve is used in the symbol to differentiate it from that of the OR circuit. A circle with a plus sign inside it is used to represent the XOR function in Boolean algebra.

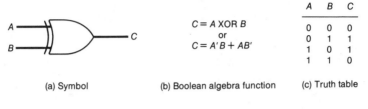

						A	B	C
			$C = A$ XOR B			0	0	0
			or			0	1	1
			$C = A'B + AB'$			1	0	1
						1	1	0

(a) Symbol   (b) Boolean algebra function   (c) Truth table

Figure 7.6 XOR circuit and representation.

7. Three-State Switches (Tri-State Operation)

Three-state switches are used extensively to switch data signals from buses to memory or other devices. They have the unusual property of being able to be 1 or 0 or off. In the off state, they have a very high impedance and look like an open or disconnected circuit. This third or <u>off</u> state allows many outputs to be connected to one bus or device without loading the circuit. Only the output whose control signal is on is allowed to send pulses into the input circuit. The symbol for the three-state switch is given in Figure 7.7. (The term <u>tri-state</u>, also used for this circuit, is trademarked by National Semiconductor Corporation.) NAND and NOR gates can also be operated in this way. Circuits are available with either positive or negative signals used for control.

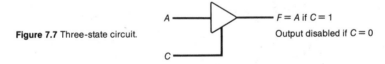

Figure 7.7 Three-state circuit.

7.2.2
Boolean Algebra

Boolean algebra deals with binary variables and logical operations. It was invented by George Boole, an English mathematician, about 1850. As illustrated in Section 7.2.1, variables are designated by letters and the relations are designated by plus signs, dots, and primes.

We introduce Boolean algebra only to illustrate some of the basic logical relationships and to enable understanding of some of the logical circuits encountered in the literature of computer architecture and design. There are 17 basic relations of Boolean algebra as listed in Table 7.1. The left column shows the OR-type relations, while the right column shows the AND-type relations. Equations (15) and (16) are especially useful because they are <u>DeMorgan's laws</u> and provide a way to convert AND circuits to OR circuits and vice versa, an important and valuable capability.

These relations are easy to derive by remembering that they represent switch functions. Truth tables can be used to prove the correctness of the functions. Some relations given are not true in ordinary algebra, although others are. Substitutions of variables are allowed; y could be replaced by y = p + q, for example.

DeMorgan's theorems can be used to convert a NOR gate to an inverted AND gate. A three-input NOR gate could be written as C = (x + y + z)' and drawn as shown in Figure 7.8.

Table 7.1 Functions of Boolean algebra.

OR Functions		AND Functions	
(1)	$x + 0 = x$	(2)	$x \bullet 0 = 0$
(3)	$x + 1 = 1$	(4)	$x \bullet 1 = 1$
(5)	$x + x = x$	(6)	$x \bullet x = x$
(7)	$x + x' = 1$	(8)	$x \bullet x' = 0$
(9)	$x + y = y + x$	(10)	$xy = yx$
(11)	$x + (y + z) = xy + xz$	(12)	$x(yz) = (xy)z$
(13)	$x(y + z) = xy + xz$	(14)	$x + yz = (x + y)(x + z)$
(15)	$(x + y)' = x'y'$	(16)	$(xy)' = x' + y'$
(17)	$(x')' = x$		

Figure 7.8 NOR circuit with three inputs.

$$C = (x + y + z)'$$

by application of DeMorgan's laws, the equation $C = (x + y + z)'$ can be written as $C = x'y'z'$ and drawn as shown in Figure 7.9, using the convention that "o's" represent inversion or complementation of the signal. In Figure 7.9 inputs x, y, and z, are converted to x', y', and z' by the inversions and then are ANDed together by the circuit presented by the AND part of the symbol.

Figure 7.9 Inverted-AND circuit with three inputs.

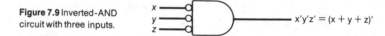

$$x'y'z' = (x + y + z)'$$

Proof of the conversion just discussed is simple.

1. Substituting $p = y + z$ and using Equation (15) from Table 7.1 we get

$$(x + y + z)' = (x + p)' = (x'p')$$

2. Replacing p with $y + z$, and using Equation (15) again, we obtain

$$x'(y + z)' = x'(y'z')$$

3. Using Equation (12) to remove parentheses, this becomes $(x'y'z')$, so that

$$(x + y + z)' = (x'y'z)$$

and can be drawn as shown in Figure 7.9. This type of conversion is frequently done and is shown in logic diagrams, so it is important that the reader be familiar with DeMorgan's laws. Other uses of Boolean algebra are primarily in design and will not be considered further here. There are many good books on the subjects of logic design and computer architecture that provide further detail. Mano [21] contains a good description of this and other related subjects at an introductory level.

7.2.3 Flipflops

      <u>Flipflops</u> are active storage elements for binary signals. They are "active" because they depend on a power supply in order to hold their signals. Four main types of flipflops are used to hold binary information temporarily for use in computer registers and computa-

tion. Most flipflops are <u>synchronous</u>; they only operate when signaled to do so by a <u>clock</u> signal provided by the main control circuit of the computer.   Clock signals are provided to all active elements in a synchronous computer.

Each flipflop has two inputs and two outputs in addition to the clock signal that controls the time at which it changes.   A basic flipflop may be built by the use of two NAND gates arranged as shown in Figure 7.10.   Two NOR gates could also be used.   This simple flipflop is nonsynchronous or <u>asynchronous</u>. However, it is seldom used this way in modern computers.   One input is the S or set input; it always sets the output to a value of 1.   The other input is the R or reset input; it resets the output to a value of 0.   There are two outputs, Q and Q',which are always the opposite of each other.   We consider the Q output as the main output and the Q' output as the complementary output.

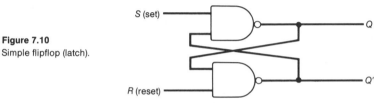

Figure 7.10
Simple flipflop (latch).

Flipflops are capable of storing 1 binary bit and are sometimes called <u>latches</u> because they latch into a particular position when actuated. The control signal is called a <u>latch signal</u> in some cases.

If additional gates are added to the simple flipflop, it can be made responsive to clock signals and different types of inputs.

Four main types of synchronous flipflops are in common use: RS (reset–set) flipflops, D (data) flipflops, which are modified RS flip-flops, JK flipflops, and T flipflops. Each has particlar advantages and disadvantages.   RS flipflops have an indeterminate state; when both inputs are 0, the output may be either a 1 or a 0 depending on chance factors.   JK flipflops have added gates that specify the outputs, so that they have four stable outputs for the four possible combinations of input signals. T flipflops are JK flipflops where the J and K inputs are tied together so that the output of the flipflop "toggles" between states 1 and 0.   T flipflops are commonly used in counters.   JK flipflops are used in shift registers and arithmetic registers.   D flipflops are used to store data temporarily while they are being held for further operations. The logic symbol and the truth table for each of these four types of flipflop are shown in Figures 7.11 to 7.14. Symbols used are the ANSI standards.

In Figures 7.11 to 7.14, Q (t + 1) indicates the state of the Q output after completion of the clock cycle with the inputs as noted. It is the next stable condition or stable state of the flipflop.   Outputs

Figure 7.11 RS flipflop.

Source: M. Morris Mano, Com-
puter System Architecture, page
56. ©1976 Prentice-Hall Inc.,
Englewood Cliffs, N.J. Used by
permission.

S	R	Q(t + 1)	Comments
0	0	Q(t)	No change
0	1	0	Clear
1	0	1	Set
1	1	?	Not allowed

Q'   Q

R   S

CP

(a) Graphic symbol   (b) Characteristic table from the Q and Q' terminals become inputs to subsequent stages of the logic operation, as shown in Sections 7.2.4 to 7.2.9, which illustrate applications of the flipflops to computer operational elements.

The D flipflop acts only as a storage unit. At each clock time, it samples its input and holds it until the next clock cycle. It is often used in holding input/output signals temporarily.

Figure 7.12 D flipflop.

Source: M. Morris Mano, Computer System Architecture, page 56. ©1976 Prentice-Hall Inc., Englewood Cliffs, N.J. Used by permission.

D	Q(t + 1)	Comments
0	0	Clear
1	1	Set

(a) Graphic symbol   (b) Characteristic table

Figure 7.13 JK flipflop.

Source: M. Morris Mano, Computer System Architecture, page 53. ©1976 Prentice-Hall Inc., Englewood Cliffs, N.J. Used by permission.

J	K	Q(t + 1)	Comments
0	0	Q(t)	No change
0	1	0	Clear
1	0	1	Set
1	1	Q'(t)	Complement

(a) Graphic symbol   (b) Characteristic table

Figure 7.14 T flipflop.

Source: M. Morris Mano, Computer System Architecture, page 26. ©1976 Prentice-Hall Inc., Englewood Cliffs, N.J. Used by permission.

T	Q(t + 1)	Comments
0	Q(t)	No change
1	Q'(t)	Complement

(a) Graphic symbol   (b) Characteristic table

**7.2.4
Decoders**

A typical application of decoders is the selection of one out of n lines by using $\log_2 n$ inputs. For example, if four possible output lines

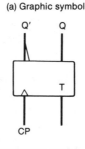

x	y	D_0	D_1	D_2	D_3
0	0	1	0	0	0
0	1	0	1	0	0
1	0	0	0	1	0
1	1	0	0	0	1

(a) Logic diagram   (b) Truth table

Figure 7.15 Simple 2 × 4 decoder.

Source: M. Morris Mano, Computer System Architecture, page 25. ©1976 Prentice-Hall Inc., Englewood Cliffs, N.J. Used by permission.

are available, they can be controlled with two control lines in the decoder. This circuit is called a 2 × 4 decoder. A simple logic circuit to perform this decoding is shown in Figure 7.15, which uses two inverters or NOT circuits and four two-way AND circuits. These circuits are typically available as prepackaged integrated circuits.

Multiple decoders are normally supplied in an integrated circuit package. Larger decoders, 3 × 8 or 4 × 16, may be supplied only one to a package. In complete microprocessors on a chip, the necessary decoders are designed into the system. There is a frequent need for decoders to control switching of input and output data as well. Decoders are used to select lines. They are not used to control data transfer directly, as multiplexers are.

7.2.5
Multiplexers

Multiplexers can be constructed from decoders by adding a signal path, as shown in Figure 7.16. Inputs I_0, I_1, I_2, and I_3 are signal lines whose input is selectively switched to the single output line by use of control lines S_0 and S_1. Two paths from each control line are used, with an inverter in one path to generate four control values. These inputs combine as shown to select one of four AND circuits and allow the signal on the related input line to pass to the output. With $S_0 = 1$ and $S_1 = 0$, input I_1 would be selected. One four-way OR circuit gathers up the four AND outputs to generate one output. Since the control signals are unique (only one of four is generated), only one signal is passed to the output. If multiple outputs are desired, it is necessary to use multiple sets of AND circuits, but one set of control lines may be sufficient depending on the electric circuit loading conditions. Many other designs are discussed in Mano [21] and other textbooks.

Multiplexers are widely used in control circuits for input and output, and are usually shown as blocks, as in Figure 7.16(a). Integrated circuits usually contain one or two multiplexers. Multiple packages can be used to form larger multiplexers if an additional control line called an enable line is provided. One way to do this would be to use a four-way AND in Figure 7.16 and control it with the enable line.

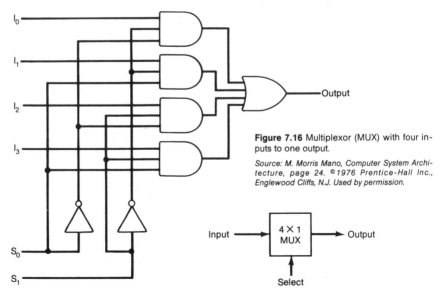

Figure 7.16 Multiplexor (MUX) with four inputs to one output.

Source: M. Morris Mano, Computer System Architecture, page 24. ©1976 Prentice-Hall Inc., Englewood Cliffs, N.J. Used by permission.

(a) Logic diagram   (b) Block diagram

It is often necessary, in using computers, to keep a count of events taking place. Usually the input is in the form of digital pulses and can always be converted to digital pulses if desired. Several flipflops may be connected together to form a counter that accepts a series of digital pulses and puts out a control pulse after a specified number of input pulses have occurred. Counters are useful for generating timing signals to control operation sequences in computers.

Counters can be designed to count forward or backward. By the use of suitable switching, the direction of counting can be controlled by an external signal. T flipflops are often used for counters because they toggle from one output state to another for each input pulse. The first T flipflop in a sequence will switch every time.

Its output will go to 1 with every other input pulse and can be used to trigger the next T flipflop in line, which will go to 1 with every fourth input pulse into the first flipflop. This binary progression continues, so that a sequence of pulses can be converted to a binary number expressed as 1 states and 0 states on the individual outputs of the T flipflops.

In the counter of Figure 7.17, a logic 1 pulse on the input of the first flipflop causes it always to be ready to count. Then every clock pulse into the clock input causes the first flipflop to toggle from 0 to 1 and back, as described. Outputs A_0 to A_3 provide the binary number equivalent of the number of clock pulses that have occurred. After 12 clock pulses, for example, the output values reading from A_3 to A_0 will be 1100, the binary equivalent of the number 12.

It is possible to load a counter with a binary number by suitable switching, so it can start at any value. Frequently, a counter will be loaded with a large positive number and counted down to 0. When the counter becomes 0, this condition is detected and used to initiate some other operation. It is simpler to detect just one condition than it is to wire up the counter to detect and trigger on a large number of different conditions. Counters may be set to all 0s or all 1s, if desired.

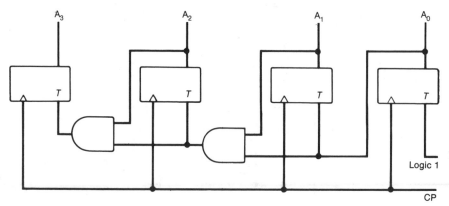

Figure 7.17 A 4-bit binary counter.

Source: M. Morris Mano, Computer System Architecture, page 24. © *1976 Prentice-Hall Inc., Englewood Cliffs, N.J.*

Registers are the working elements of the computer. They are made up of flipflops to provide different types of operations. Multiple inputs may be supplied to each binary digit of the register where control circuits operate gates or logic circuits to select the inputs provided to the register. Registers commonly have 8, 16, or 32 flipflops to handle word sizes of 8, 16, or 32 bits. Each flipflop stores 1 bit of information at a time.

1. Storage Registers

Storage registers are used to hold information temporarily while it is awaiting transfer or being used in an operation or a control function. Information from memory is transferred into a Memory Data Register (MDR), held there for a clock cycle, and transferred to another register, perhaps the accumulator register, which holds information for input to the Arithmetic and Logic Unit or ALU. These registers must have the ability to read in information from one source and read it out to another register, memory, or the ALU. They are therefore provided with input and output switching circuits of many types depending on the use to be made of the register.

2. Shift Registers

Many computer registers are shift registers; they are capable of accepting information at one end of the register and shifting it toward the other end, 1 bit at a time. Suppose an 8-bit register is set to 00000000 initially. Then if we apply a pattern such as the binary number 00110101 to the left end and shift it 1 bit at a time, we will see the following pattern of bits as the new number is shifted in. These steps assume that the right end of the new number is entered into the left end of the register.

Initial condition	0	0	0	0	0	0	0	0
Step 1	1	0	0	0	0	0	0	0
Step 2	0	1	0	0	0	0	0	0
Step 3	1	0	1	0	0	0	0	0
Step 4	0	1	0	1	0	0	0	0
Step 5	1	0	1	0	1	0	0	0
Step 6	1	1	0	1	0	1	0	0
Step 7	0	1	1	0	1	0	1	0
Step 8	0	0	1	1	0	1	0	1

After eight pulse times, the complete new number has been shifted into the register. It could have come from another register, or it could have been a serial input set of pulses from an external device. Shift registers usually have the ability to read in or out from either end. In addition, they can read in or out all binary digits at once, in parallel.

Some shift registers can do end-around shifting. The condition of the right flipflop of the register can be moved to the flipflop at the left end and all others shifted one step to the right. End-around shifting can be done in either direction. To provide for the carry operation in arithmetic, an additional separate flipflop is used to hold the high-order (left end) bit temporarily. This bit can then be tested to see if a carry operation occurred.

Instruction operations operate directly on registers for many of their functions, as described in Section 7.3.2. Control of shifting direction and amount is provided by the control registers of the computer, as discussed in Section 7.3.3.

7.3 Computer System Organization

Computer systems usually consist of five functional units, as shown by the block diagram of Figure 7.18.

In this diagram, the Central Processing Unit (CPU) is made up of two parts: the Control Unit and the ALU. The CPU is typically one integrated circuit chip in microprocessors and provides the central control, registers, buses, and the arithmetic and logic functions for the computer system. Peripheral equipment, the input and output equipment, is connected to the CPU directly or through auxiliary controllers. Memory is separate from but directly connected to the CPU.

Figure 7.18 Block diagram of a computer system.

Source: From Chips to Systems by Rodnay Zaks © 1977 SYBEX Computer Books, Berkeley, CA. Used by permission.

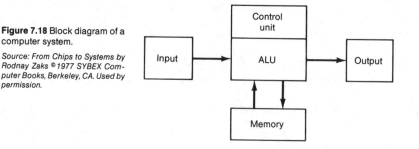

7.3.1
Organization of the
Central Processing
Unit (CPU)

Most robot control systems today use microprocessors to perform the control and operation functions required. Others use minicomputers that are similar to microprocessors in operation. Microprocessors and microcomputers differ only in that a microcomputer includes a microprocessor and the necessary memory, while the microprocessor alone has no memory except that necessary for certain control functions.

CPUs found in microprocessors are quite simple in principle. Their apparent complexity is caused by the large number of interacting components. We will use the common Intel 8080 8-bit microprocessor as an example because it is easier to understand than the advanced microprocessors such as the Intel 8086, Zilog 8000, Motorola 68000, National NS16000, and other 16- to 32-bit microprocessors. Another simple computer, the Motorola 6800, is very similar to the 8080 in principle, but the details vary. Both the 8080 and the 6800 microprocessors are still in extensive use, although they were developed 8 to 10 years ago.

The Intel Corporation, based in "Silicon Valley" (as the Santa Clara county and adjacent areas of California have been nicknamed), was the first to develop a microprocessor. Its Intel 4004, introduced in 1971, was a 4-bit word microprocessor, followed by the 8-bit Intel 8008 in 1972. The successor to the 8008, the Intel 8080, was introduced in 1973. Several other microprocessors were introduced in the next 2 years by other manufacturers. Motorola's 6800 family was one of the most successful. Zilog's Z-80, similar to the 8080, has also been used in large quantities. All of these microprocessors come in families of related devices that have similar characteristics and use several functional and input-output devices in common.

1. Architecture of the Intel 8080

The major functional units of a microprocessor CPU are registers, the ALU, buses, decoders, multiplexers, and buffers. Figure 7.19 shows the architecture of the Intel 8080 microprocessor. (Architecture is a term used to describe the organization and capabilities of a computer, just as it is used to describe a building. A computer architect is one who plans the overall features of a computer or computer system. Some system planners use architecture to mean the structure and capabilities of the computer system as seen by the programmers or other users of the system.)

Registers, decoders, and multiplexers have been discussed previously (Section 7.2). Buffers are registers used to store information temporarily, pending further action. Buses carry information from one place to another in the computer. They are usually 8, 16, or 32 electrical signal wires or lines in parallel. We speak of buses that are "8 bits wide," "16 bits wide," and so on to convey the idea that several bits are moved at the same time from one place to another.

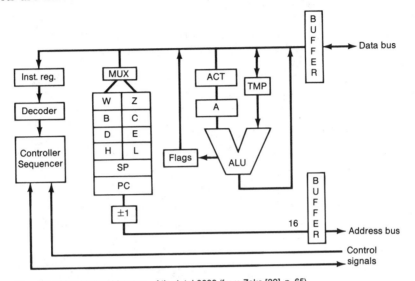

Figure 7.19 Internal architecture of the Intel 8080 (from Zaks [30], p. 65).

Source: From Chips to Systems by Rodnay Zaks © 1977 SYBEX Computer Books, Berkeley, CA. Used by permission.

There are three types of buses in Figure 7.19: a data bus, an address bus, and a control bus. These internal buses are implemented directly on the silicon chip as part of the complex circuitry formed on the chip. The data bus is used to carry data signals, the actual information being acted upon, between parts of the system. Address buses carry address information to control selection of data from memory or other storage locations. Control buses carry the command signals that tell each element in the microprocessor what function to perform. The clock signal, not shown but considered part of the control function, provides timing information to synchronize the activities of the individual registers and control elements. Each of the internal buses connects with an external bus through a buffer, or sometimes directly, to communicate information between individual chips or components of the computer. This activity will be described in a later section.

In Figure 7.19 the data bus is at the top of the diagram and can transmit information to or from the instruction register, the multiplexer (MUX) for the internal registers, the ACT register, and the TMP register. It accepts output from the ALU and the FLAGS register. The direction of information flow is indicated by the arrows. The data bus is 8 bits wide. Information from the external data lines, transmitted in parallel, is stored for a short time in the buffer register on the data bus. This buffer can store 8 bits of data when directed to do so by a control line (not shown).

An instruction is brought in from the memory over the data bus, stored temporarily in the data bus buffer, and then transmitted to the Instruction Register (IR). Control lines open up a path to the IR to allow an instruction to be placed in the IR. The instruction in the IR is decoded by the block of circuits labeled <u>decoder</u> and passed to the <u>controller-sequencer (CS)</u>, which sends out the required control signals at the correct time to carry out the instruction stored in the IR.

Data to be acted upon is brought in over the 8-bit-wide data bus and placed in one of the data registers connected to the MUX or multiplexer. Control signals from the CS cause the MUX to open up a path to the desired register at the correct time. There are eight registers capable of storing 8 bits each: W, Z, B, C, D, E, H, and L, and two registers, SP and PC, capable of storing 16 bits each. These are all connected to the multiplexer and can receive data from and transmit data to the data bus since the multiplexer can handle two-way communication.

An additional register, the FLAGS register, stores the status of the microprocessor as the result of the most recent instruction. It contains bits that signal whether the last instruction resulted in positive, negative, 0, overflow and other conditions. It can be tested by certain instructions that control branching of the program based on conditions in the FLAGS register. The FLAGS register is also called the <u>Program Status Word (PSW)</u> and can be stored on the stack by using a command called PUSH PSW. The stack will be described.

Data can be transferred between registers 8 bits at a time, or transferred to the ACT temporary register, or transferred to and from the TMP temporary register. The ACT register is a temporary storage register that holds incoming data destined for the A register, commonly called the <u>accumulator</u>. In the 8080, all arithmetic and logic instructions must use the A register for one of the words of input data. (In this computer, a word is 8 bits.) The TMP register is a temporary register used to hold information being transferred to and from the ALU. Registers ACT and TMP are required to handle certain timing and buffering problems encountered in a few of the many instructions carried out by the 8080.

Note that the ALU has two input paths and one output path, each with the ability to transfer 8 bits in parallel. (The FLAGS register stores status information only.) It can add, subtract, compare, and perform other operations on the two inputs and then place the result on the output and send it back to the data bus. On the data bus, the result is then transferred back to one of the data registers, into the TMP or ACT register, or to the data bus buffer to memory or to external devices attached to the data bus. Operation of the ALU is described in more detail in Section 7.3.2.

In the 8080, the W and Z registers are not addressable by the program. They are invisible registers used only in carrying out certain instructions as sequenced by the controller-sequencer. Essentially they act as temporary storage registers during instruction operation. The other six registers work together in pairs or independently. Registers B and C are one pair, D and E are another pair, and H and L are the third pair. These registers work together to handle some 16-bit functions and independently on 8-bit operations. Registers H and L derive their names from High (H) and Low (L), respectively, because they are used to store the high-order 8 bits and the low-order 8 bits of an address when external memory is being addressed.

Special registers, the Stack Pointer (SP) register and the Program Counter (PC) register, perform two essential functions. The SP stores the current address in memory of the top of the stack. The PC stores the address in memory of the <u>next</u> instruction to be carried out. These are both 16-bit registers and can therefore address 2^{16} or 65,536 bytes of memory or the whole range of the 64K memory of the

8080. (Remember that K in computerese has the value 1,024, not 1,000.)

A <u>stack</u> is a location in memory where data is stored temporarily for use in a program. (It derives its name from the spring–supported stack of plates found in some cafeterias and restaurants. Plates are pushed down on top of the stack and pop up when the top plate is removed.) There are instructions in the 8080 that are used to PUSH a byte (8 bits) on to a stack in memory, thereby adding 1 to the SP register, and to POP a byte off the top of the stack, automatically decrementing the SP register by 1. In this way, the number in the SP always "points" to the memory address of the top of the stack. (Actually, the address may be decremented rather than incremented if it is desired that the stack work down in memory addressing instead of up. The principle of operation is unchanged, however.) Stacks are very useful in performing certain instructions and in addressing subroutines.

Modification of the PC is an integral part of each instruction that affects where the next instruction is to come from. The small block in Figure 7.19 labeled "±1" (under the PC register) is used to increment or decrement the SP and PC as needed. When branch instructions or other instructions require a large change in the location of the next instruction, the PC is updated by an addition or subtraction as appropriate using the ALU. In this way, the PC is always ready to address the memory location of the next instruction to be executed. Output of the PC or SP is 16 bits, placed on the address bus at the proper time in the instruction cycle.

2. Fetch and Execution Instructions

Instructions in the 8080 are made up of one to five machine cycles taking place at the basic clock rate of the microprocessor depending on the instruction being executed. Some simple instructions, moving data from one register to another, for example, can be done in one machine cycle. Others require multiple accesses to memory. Each instruction requires at least one access to memory in order to fetch the instruction itself.

Instructions are separated into two parts: Fetch and Execution. Small microprocessors, such as the 8080, perform Fetch and Execution as separate operations. In some computers, the Fetch of the new instruction overlaps the Execution of the previous instruction to save time and increase processing speed.

The Fetch sequence is (refer again to Figure 7.19):

1. The controller–sequencer (CS) causes the 16 bits from the Program Counter (PC) to be placed on the 16–bit–wide address bus and stored in the buffer register on the address bus. This operation is initiated either by the start–up routine or by the completion of the previous instruction.
2. Address lines on the address bus cause the selected output in RAM to be addressed and read out of memory. This output is an 8–bit word or byte, and is placed on the data bus and stored in the 8–bit data bus buffer register.
3. Control lines from the CS open the gates to the Instruction Register (IR) and allow the 8–bit byte to move to the IR. Sometimes the instruction is more than 8 bits long. It could be 16 or 24 bits long and be stored in memory as 2 or 3 successive bytes requiring one or two additional accesses to memory. The first byte

is the instruction code and goes to the IR; the second byte may go to the TMP register temporarily if there are only 2 bytes. In 3-byte instructions, the second and third bytes are addresses and stored temporarily in the W and Z registers. Sequence control for these operations is provided by the CS.

4. The fetched instruction in the IR is decoded by the decoder logic and sets up the CS to execute the instruction that was fetched.

5. If a 2- or 3-byte instruction is specified by the instruction, the decoder logic recognizes this and causes the additional byte or bytes to be fetched and stored as required. In this case, additional memory cycles are generated by the control logic. In 3-byte instructions, the next address is read from the W and Z registers rather than from the PC. However, the PC is incremented by 1 for each memory access to keep track of where the instruction bytes are stored in memory. This operation is possible because the bytes of an instruction are stored in successive address locations in memory. Therefore, at the end of any type of instruction, the PC points to the byte to be addressed for the next instruction.

Instruction Execution

Instructions may be one, two, or three words in the 8080: One word instructions are the simplest. An example is the

MOV r1, r2

instruction, which is a register-to-register transfer. Data in the r2 register is transferred to the r1 register. (It is a quirk of the 8088's design that this backward notation is used. However, it has become standard and is not easily changed.) Transfers are possible between any two of the B, C, D, E, L, H, and A registers and can be done in one machine cycle time. The contents of the r2 register will not be changed; this is really a copy instruction.

MOV r1, r2 is the assembly language form of the instruction. It is a <u>mnemonic</u> or code for the actual binary form of the instruction. It is possible to write instructions in the assembly language of a computer and then use a program called an <u>assembler</u> to convert the assembly language program to the internal <u>machine language</u> of the computer. For the 8080, the machine language for the MOV r1, r2 command is a binary code represented in the form

01DDDSSS

where 01 is the machine code for the MOV command, DDD are 3 bits representing the Destination (D) register, and SSS are 3 bits identifying the Source (S) register. Since there are 3 bits, we can identify eight different locations: the seven available registers plus memory.

As an example, the instruction MOV A,B, which causes data to be moved from register B to register A, is transformed by the assembler to the binary number 01111000.

These same codes are used in all instructions as required. We have also identified a code for register pairs because it is possible to address registers in pairs, except for the accumulator register, A, which can only be addressed alone. Transfer to and from a memory location requires a two-address instruction and will be discused later.

Coding for the registers and memory, in the 8080, is

Code	Register	Code	Register Pairs
000	B	00	BC
001	C	01	DE
010	D	10	HL
011	E	11	SP
100	H		
101	L		
110	– (MEMORY)		
111	A		

Binary register codes are supplied only as information to those who may see machine language coding and wonder what it means. Instruction references in future will be to mnemonic codes in assembly language form as they are usually supplied by the manufacturer. Higher-Level Languages (HLLs) such as BASIC, FORTRAN, and PASCAL are one step higher than assembly languages and are discussed in Chapter 10.

Another example of a 1-word instruction is

ADD r

This instruction causes the contents of the r register to be added to the accumulator, (A). The result is stored in the accumulator. Note that the value of one of the operands is always in the accumulator. Instructions of this type are called implicit instructions because one value is implied by the instruction name itself. In using this instruction, it is necessary to load one value into the accumulator either directly or as the result of a previous operation, before using the ADD instruction.

Two-word instructions also occur. Some instructions, known as immediate instructions, carry numerical information in the instruction itself. These are examples of two-word instructions in which the first word is the instruction code and the second word is data.

An instruction used to add a specific value to the accumulator is the ADI instruction. ADI 06, for example, will add the value 06 to the accumulator. In the mnemonic ADI, AD specifies ADD and I specifies that this is an immediate instruction. The assembler interprets this as a command to add the next word of the two-word instruction to the accumulator directly. When a code or character is used directly as data, it is called a literal because it is treated literally as what it says it is, rather than being a symbol for some other quantity. A letter could be used, but it must be the name of a constant that has already been identified to the assembler program. That is, if X = 9 has already been identified, the instruction could be written ADI X.

There are other two-word instructions in which the second word in the instruction is a piece of data.

Instructions that must supply an address such as

LDA addr

will be made up of three words. The LDA instruction says "load the accumulator from the address specified by the next two words of this instruction." As the instruction is read in from memory, the first word is put in the IR and the next two words are put in the W and Z registers. Then the instruction causes the contents of the memory location addressed by the W and Z registers to be copied into the accumulator. The memory location is not altered.

The reader is referred to other references such as Zaks [30] or the Intel 8080 manuals for detailed operation of the many available instructions. Our purpose has been to provide an overview of 8080 architecture and to familiarize the reader with some nomenclature and basic operating methods.

**7.3.2
Arithmetic and
Logic Unit (ALU)**

The ALU, as the name implies, provides several arithmetic and logic functions. We will designate the two inputs as A and B, use S to indicate the sum of A and B, and C to indicate the carry result in arithmetic operations. A and B are binary numbers containing n digits each. ALUs commonly handle 4-, 8-, 16-, or 32-bit numbers depending on the computer type and the accuracy desired.

1. Binary Addition

This section will identify some terms and provide an example of a simple binary half-adder. For more complex circuits, the reader is referred to logic design textbooks.

Addition of two binary numbers is similar to addition of two decimal numbers. The only real difference is that there are only two values, 0 and 1, so a carry occurs often. If we consider only two inputs, x and y, the possible outputs are shown in the truth table of Table 7.2 along with simple examples of the addition of two values.

Table 7.2 Simple binary addition of two binary values

x	y	C	S					
0	0	0	0	x	0	0	1	1
0	1	0	1	y	0	1	0	1
1	0	0	1					
1	1	1	0		0	1	1	10

Table 7.3 Truth table and some examples of full binary addition

x	y	z	C	S
0	0	0	0	0
0	0	1	0	1
0	1	0	0	1
0	1	1	1	0
1	0	0	0	1
1	0	1	1	0
1	1	0	1	0
1	1	1	1	1

```
(a) A = 0 1 0 0          A = 0 0 0 1
    B = 0 0 0 1     (b)  B = 0 1 0 1
       ---------            ---------
        0 1 0 1              0 1 1 0

(c) A = 0 1 0 1          A = 1 0 1 1
    B = 1 0 0 1     (d)  B = 1 1 1 1
       ---------            ---------
        1 1 1 0             1 1 0 1 0
```

Truth Table

Examples of binary addition. Note that there is a carry in example (d) from the fourth to the fifth position, counting from the right, so C = 1 in the truth table.

Table 7.3 shows the results of adding single binary input values x and y and the carry from a previous stage, z, to obtain the sum, S, and the carry, C. These results are in the form of a truth table for single inputs, with some examples of addition applied to 4–bit binary numbers using the same rules. A and B, in this example, are 4–bit binary numbers.

A circuit to implement the arithmetic function shown in the truth table of Table 7.2 is called a half–adder because it handles only one binary digit. A full–adder can handle the sums and carries for several binary digits such as the examples given in Table 7.3.

The logic functions corresponding to a half–adder are:

$S = x'y + xy' = x \oplus y$ (This corresponds to column S in Table 7.2.)

$C = xy$ (This corresponds to column C in Table 7.2.)

They can be implemented with the circuit shown in Figure 7.20.

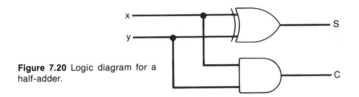

Figure 7.20 Logic diagram for a half-adder.

2. Arithmetic and Logic Operations

Both arithmetic and logic operations can be generated by the same circuit by using mode controls to select either the arithmetic or the logic mode. In the logic mode, the carry is suppressed. In the arithmetic mode, the carry propagates from stage to stage and causes the following stage to increment if a carry comes in from the preceding binary position.

A frequently used term in describing ALU operations is the complement of a number. The complement of a binary number is the NOT operation applied to the number. All 1s are changed to 0s and all 0s are changed to 1s. The simple complement is called the 1s complement of the number because another useful function is the 2s complement, obtained by adding 1 to the 1s complement of the number. Complements are used to provide subtraction capability and for some logical functions in ALUs.

As an example, the complement of 11001010 is 00110101, also called the 1s complement. By adding 1 to 00110101, we obtain 00110110, the 2s complement of 11001010. The rule for 2s complements is: Invert the binary digits to obtain the 1s complement and then add 1.

Incorporated in a complete ALU are some or all of the following operations:

Arithmetic instruction	Operation performed
Increment B	Add 1 to the binary value of B
2s complement B	Change all 0s to 1s and all 1s to 0s, and then add 1

Increment A	Add 1 to the binary value of A
Add A and B	Find the sum and carry of A and B
A plus 1s complement of B	Find the sum and carry of A plus the complement of B as defined
A plus 2s complement of B	Find the sum and carry of A plus the 2s complement of B
Decrement A	Subtract 1 from binary number A

These operations are carried out by a full-adder, capable of receiving the 2 binary digits being added plus the carry, if any, from the preceding stage. The full adder is made up of two half-adders and is implemented by a set of logical circuits. There are as many full adders in an arithmetic unit as there are binary digits in the numbers to be added.

Logical operation	Operation performed
Clear all bits	Sets all binary bits to 0
Transfer B	Place bits in B on the output
Complement B	Perform 1s complement of B
Set all bits	Sets all binary bits to 1s
Transfer A	Place bits in A on the output
XOR	Do XOR operation on individual bits of A and B and output
XOR	Do XOR and then invert all bits (perform the complement)

Logic operations are performed by the ALU when it is controlled by a signal from the control unit to be in the logic mode.

Individual arithmetic and logic operations performed by the ALU are called micro-operations and are controlled by microprogramming to perform the instructions such as ADD, SUBTRACT, AND, OR, NOT, and various kinds of compare functions, as discussed in Section 7.3.3.

7.3.3
Control Unit

Visible elements of the Control Unit in Figure 7.19 are the IR, the Decoder, and the CS. For convenience of reference, we will consider the storage registers and buses as part of the ALU and the remainder of the CPU as the Control Unit. This section summarizes the operation of the Control Unit. For more detail, the reader is referred to textbooks such as Mano [21].

Sequencing and control functions are the most complex operations in the CPU of a microprocessor. Sequence control starts with the clock circuit that supplies precise timing pulses to each part of the microprocessor. The clock circuit uses a thin quartz crystal oscillating at precise frequency to provide the basis for accurate time pulses. This circuit is like that found in most digital watches today and works on the same principles.

A typical microprocessor, such as the Intel 8080 previously described, carries out specific operational steps during each clock cycle. For control purposes, each clock cycle is divided into three to five micro-operations or microsteps. Individual steps are called internal states. During a microstep, certain paths between registers and buses are opened and information is allowed to move between locations. Logic circuits controlled by the CS apply control signals to selected gates in order to open and close the pathways between locations. To move information from the data bus to the TMP register, for example, it is only necessary for the control to supply one control line, which opens an 8-bit parallel path from the data bus to the TMP register during a specific microstep. During the next microstep, another control line is actuated to move data from the TMP register to the ALU.

Control of microstep sequencing is done in several ways. One way is to set up complex logic circuits controlled by counter circuits to generate the right pulses at the right time. A more efficient and flexible way is to use microprogramming, in which the control sequence is stored in a ROM memory and selected for each instruction cycle by reading information from the ROM into a control register. Instruction decoding consists of converting the 8 bits of the instruction into a series of control impulses by using one of these techniques. Since multiple lines are required to control all the separate functions of the computer, the control register itself may be as large as 32 bits or more, with each bit controlling the opening and closing of a particular path. Alternative control methods and analyzed and evaluated in Mano [21]. The interested reader is referred to this and other references for detailed discussion of this complex topic.

7.3.4
Memories

There are three main memory elements used in modern computers:

1. Random Access Memory (RAM). RAM can be rapidly written to and read from by electronic means. Most computers today use semiconductor RAMs because they are simpler and less expensive than the older core memories. Semiconductor memories are volatile; they lose their memory when power is turned off. Other types of RAM, bubble memories and core memories, store information magnetically and do not lose their memory when power is turned off. Batteries are used with many semiconductor RAMs for robotic systems to provide backup power in case of main power loss. Typical operation cycles for RAMs are 200 to 500 nanoseconds to store or retrieve a binary word. (A nanosecond is 10^{-9} seconds.)

2. Read-Only Memory (ROM). ROMs are nonvolatile semiconductor memories. They are built at the factory to the user's specifications and cannot be altered by the user. They are faster and less expensive in general than RAMs. Also, they provide a secure memory of some of the operational software and control programs that are essential to computer operation. Their fixed nature is thus a valuable feature in some instances. ROMs are supplied by the manufacturer in pluggable integrated circuit units.

3. Alterable Memories. There are several semiconductor memory packages or integrated circuits that can be read rapidly but are difficult to change. They are used when relatively permanent memory is desired but when changes must be made from time to time. During the development of a computer especially, these memories are valuable. They are also useful for storing software that is changed infrequently.

Three main types of alterable memories are used:

1. Programmable ROM (PROM). These memories can be slowly written into by electrical control units usually called <u>PROM-burners</u> because they use electrical currents to burn out <u>fusible links</u> in the memory units. PROMs can be written into only once; they then hold the memory information indefinitely like ROMs.

2. Electronically Programmable ROM (EPROM). These memories can be written into electrically and can be erased by exposing them to ultraviolet light through a quartz window on top of the chip. Writing can be done rapidly, but erasing may take several hours. They are also used like ROMs.

3. Electrically Erasable PROM (EEPROM). These memories can be written into and erased fairly rapidly but have the nonvolatile characteristic of ROMs. They are somewhat more expensive than the other types.

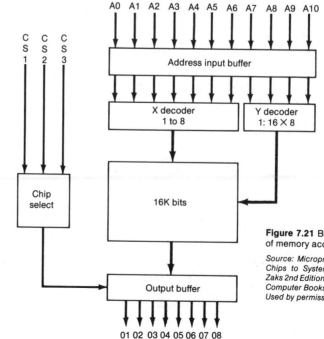

Figure 7.21 Block diagram of memory access.

Source: Microprocessors from Chips to Systems by Rodnay Zaks 2nd Edition, © 1977 SYBEX Computer Books, Berkeley, CA. Used by permission.

All memories are accessed by the use of some type of Memory Address Register (MAR) that applies signals simultaneously to selected X and Y lines of the memory matrix. Information is read out from the memory word at the intersection of the selected lines into an output buffer register sometimes called the Memory Data Register (MDR). Output of the complete word is in parallel. A block diagram of a memory with its associated registers is provided in Figure 7.21.

7.3.5
Bus Standards

All processors require some sort of standardization of their external buses. Internal buses are formed on the individual semiconductor chips, but the external buses are usually made up of multiple-wire cables in a particular configuration. Typically, microcomputers are housed in aluminum boxes or card cages that have printed wiring (the bus) along the bottom of the box and several equally spaced female connectors spaced along the wiring, as shown in Figure 7.22. This wired arrangement is called the motherboard. Individual flat cards carrying the semiconductor chips are plugged into the connectors of the motherboard, usually vertically. There may be 50 to 300 connecting lines in the bus built into the motherboard. Buses are more than wiring; they also perform a number of important control functions.

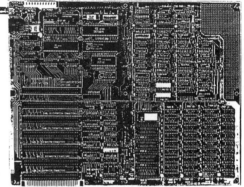

Figure 7.22 Motherboard.

Courtesy of Display Telecommunications Corporation© 1984. Used by permission.

External buses must carry data lines, address lines, control signals, interrupt signals, power lines for several levels of voltage, ground lines, clock signals, and communication signals. Data buses may be 8, 16, or 32 bits wide. Addressing requires at least 16 bits even in the small microcomputers. Larger microcomputers can address up to 4 megabytes and use as many as 32 address lines. Control lines, interrupt lines, clock signals, and communication lines are variable in number, but the total number of these lines can easily exceed 50 or more. Power may be required at $\pm$ 8 volts, $\pm$ 16 volts, and possibly at other voltage levels for special purposes. All of these needs must be served by whatever bus standards are selected.

As an indication of the number of buses available, a recent handbook on buses for microcomputers (Poe [22]) provides characteristics of 21 different buses and still does not include all of the buses in use. Characteristics of some of the most widely used buses are now given as examples of some items to be considered in selecting a bus for a computer system. It is possible to use a computer board on a bus not designed to handle it, but it is frequently difficult to achieve compatibility of signals and physical characteristics.

1. S-100 Bus (IEEE-696 Bus)

Many small microcomputers use the S-100 bus, which supplies

100 lines in a specified standard configuration. It became a de facto standard bus several years ago when several manufacturers started offering equipment for the 8080-based Altair kit made by MITS, Inc. As a result of the availability of equipment with the S-100 bus, other microcomputer manufacturers offered computer systems using this bus, so that its usage grew. As a result, the S-100 bus has become the most popular bus for small computers for hobbyist use and is widely used for other small computers. Thus, it is a good example of a standard bus.

This bus has been standardized by the Institute of Electrical and Electronic Engineers (IEEE), the professional society for electrical, electronic, and computer engineers. IEEE's standard is called the IEEE 696 Bus Standard, and specifies in detail the purpose of each of the 100 lines in the bus and the electrical and timing characteristics of each line. In order to prepare for future uses, the IEEE 696 standard (S-100) increases the number of standard address lines to 24 so that 16 megabytes of memory can be addressed. With this change the S-100 bus can be used for both 8- and 16-bit data transfers, so that both 8- and 16-bit microprocessors can be used on the same bus.

The physical characteristics of the circuit boards that plug into the S-100 bus are:

1. Dimensions—5.3 × 10 inches (13.4 × 25.4 centimeters).
2. Edge connections—50 pins on each side of the bottom edge of the board. Pins are spaced 0.125 inch (0.317 centimer) apart. Pins are flat and will fit into an edge connector with spring clips on each side of the board.
3. Pins are numbered from 1 to 50 on the component side of the board and from 51-100 on the opposite side.

S-100 systems have 4 to 22 card slots, so that many different types of compatible cards may be used. A permanent bus master is required in each system, usually the processor. The bus master decides which of the 22 cards can use the bus, normally only 2 at time. There is an "arbitration" process, performed digitally, which accepts requests from each temporary master and decides which master will get control of the bus. There are 4 bits assigned for bus priority, so that up to 16 temporary masters can request the use of the bus. Some devices, such as memories and some input/output devices, are not capable of being masters and are called slaves. Data flow may be unidirectional or bidirectional on data lines, as specified by two new control lines added by the IEEE-696 specification. Design of the standard bus has provided, in an ingenious way, for the use of most of the previous boards using 8-bit data transfer and the efficient handling of 16-bit data transfers. This was not an easy task.

Some electrical lines provided on the IEEE-696 (S-100) bus are (all buses must provide essentially the same or additional functions)

1. A voltage of +8 volts unregulated to supply +5 volts through an on-board regulator. Two lines supply this voltage for convenience in use.
2. Unregulated voltage lines of +16 volts and –16 volts.
3. An XRDY line to synchronize bus masters, which control the bus operation, to bus slaves that follow the master synchronizing signal.
4. Eight vectored interrupt lines. These lines range from V10, highest priority, to V17, lowest priority. All are open collector, so that

external resistors and power sources may be used to provide interrupt signals. Interrupts are used to signal external events such as switch closures or sensor conditions.

5. An interrupt signal line and an interrupt acknowledge line to request a software interrupt and to acknowledge an interrupt request. There is also an NMI line used for Non-maskable Interrupts. This line can always interrupt the processor. Other interrupt lines may be masked or controlled by software to prevent interrupts at the wrong time.

6. Twenty-four address lines to send address information to memory and to external devices.

7. Eight extended address lines for additional address control and bank selection.

8. Several lines to indicate clock cycles and phasing of clock signals.

9. Memory control lines to initiate reading and writing and to signal whether 8-bit or 16-bit words are to be written or read. Signals are also provided to indicate that address information is valid and can be used to address memory.

10. Sixteen data lines that can be used for either 8-bit or 16-bit data transfers. These lines may be bidirectional, data can be sent either into or out of the computer.

11. Four Direct Memory Access (DMA) lines that are used to assign priorities to bus access for up to 16 bus masters. Bus masters are those devices capable of independent operation and controlling of data transfer. The microprocessor is one device; others are disk controllers and communication controllers.

12. Other control lines are used to signal impending power failures, and errors, reset, disable and enable certain functions, power-on, clear, and miscellaneous other functions.

13. Three ground lines, connected together on most boards, used to make grounding more reliable and convenient.

14. There are three not defined NDEF lines and four reserved for future use (RFU) lines. Manufacturers are allowed to use the NDEF lines for approved functions. The RFU lines are not to be employed until the standard is revised to allow their use.

An example of some of the equipment that can be attached to an S-100 bus is given in Figure 7.23. Note that one of the devices that can be attached is the Gen-Purpose-Interface (GPIB) also called the IEEE 488, discussed subsequently.

Several other buses have been designed and used by various manufacturers. Some of them are discussed in the following paragraphs.

2. Multibus (IEEE 796)

The Intel Corporation developed the Multibus, which has 86 pins, for their single-board computers (SBC) line of printed circuit boards. These boards contain CPUs, memory, input/output functions, disk controllers, and other necessary boards for various uses. A power supply mounted in a chassis and provided with a card cage to hold the individual cards was made available to Original Equipment Manufacturers (OEMs) to enable them to use the Intel boards without a great amount of development work. The Multibus lines run across the bottom of the card cage to provide several connectors into which the individual boards may be plugged for use.

Typical Multibus boards are 12 inches (30.4 centimeters) long, 6.75 inches (17.1 centimeters) high, and 0.062 inch (0.157 centimeter)

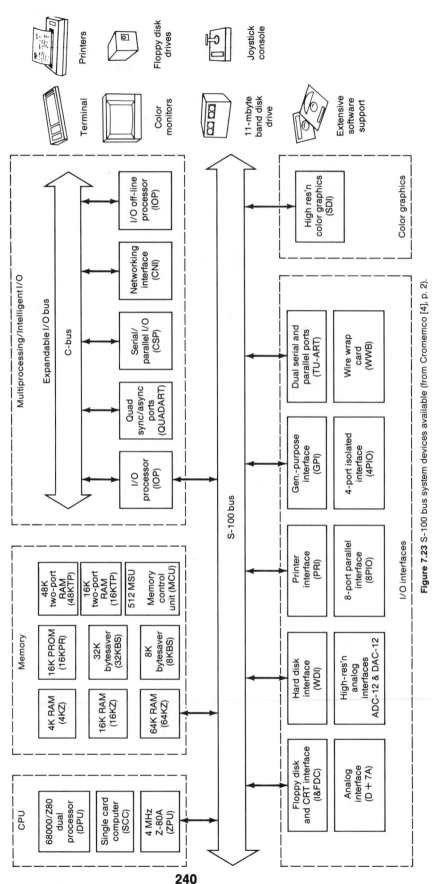

Figure 7.23 S-100 bus system devices available (from Cromemco [4], p. 2).

Courtesy of Crommemco, Inc. Used by permission.

thick. They have two edge connectors along the bottom rather than one long single line of connectors. They may also use up to three edge connectors along the top of the board.

The Multibus has 16 data lines and 16 address lines. In addition, it has all the necessary control functions provided by the S-100 bus, but without redundancy, so that 86 lines are sufficient for all functions performed.

Many board manufacturers are producing Multibus or IEEE 795 boards. Most of the well-known microprocessors are supplied on Multibus boards, and the necessary memories and peripheral controllers are also available.

3. Versabus

With 260 pins, the Versabus is the most versatile of the buses supplied for microcomputer use. It was developed by Motorola to work with their line of microprocessors, starting with the 8-bit 6800 and including the 16-bit 68000 and the 32-bit 68020. Each of these microprocessors has available a large number of other devices that are used with them. There are also configurations in which multiple microprocessors are used, as discussed in Section 7.10.

The Versabus has 32 data lines and 32 address lines. The 32 adddress lines allow addressing over 4 gigabytes of information directly, thus providing an important feature of the Motorola 68000 system, the ability to give the programmer a virtual memory of 4 gigabytes to greatly simplify programming and data-handling functions. True 32-bit data transfers can also be performed, instead of transferring 8 or 16 bits at a time using multiple memory cycles. This capability considerably reduces the running time of programs.

A variety of additional control signals are available on the Versabus. The complexity of these controls and their utilization are important only to the designers of the system. However, it should be noted that the flexibility provided can be extremely valuable in robotic systems.

4. IEEE-488 Bus (GPIB Bus)

The IEEE-488 bus is also known as the General-Purpose Interface Bus (GPIB). It was developed primarily for systems using extensive instrumentation, along with the computer for control and data gathering. The computer is used as a bus controller. It determines which instrument will put signals on the bus and where the signals will be routed. The Hewlett-Packard Company developed the Hewlett-Packard Interface Bus (HPIB) for use in instrumentation systems. The HPIB was modified and standardized as the IEEE 488 bus in 1975 and became known as the GPIB.

Devices operate in three basic modes on the IEEE-488 bus. A listener-only mode accepts data from the bus, a talker sends data over the bus, and a controller performs supervisory control of the bus. A controller can address other devices or grant permission for talkers to use the bus. Several listeners can be assigned, but only one talker, of course.

There are eight bidirectional data lines that carry a byte of address, data, or command; therefore, the data rate is slower than that of some other buses. Three data byte transfer lines and five general interface management lines perform control functions.

The IEEE–488 bus is valuable for connecting instruments into a system and could be used as an auxiliary bus, as shown in Figure 7.23. However, it would not be suitable for use as the main bus of a robotics system.

5. Other Buses

Other buses that are becoming more popular are the VME bus, designed for use with large systems, and the IBM bus used on the IBM Personal Computer (PC). Since it is likely that both of these buses will be used in robotics, they will become more important. Although the IBM PC has been used for robotic control, the bus has not yet been standardized and approved by the IEEE.

7.4 Peripheral Equipment

Around the edges or <u>periphery</u> of the computer for the robot system are the machines and devices that provide input to the computer and receive output from it. Note that input and output are referenced to the computer.

Computer input is stored information, vision input, sensory input, or status input from other devices. Output from the computer is data for storage, printing, or communication, robot servo control signals, and control signals for auxiliary equipment. All of the input/output devices are classified as peripheral equipment and make possible the useful operation of the robot system.

Figure 7.24 is an example of a complete system controlled by a main computer called the <u>host computer</u>. The LSI–11 Controller is used to control the robot arm itself. This system is described in Lechtman [12].

Figure 7.24 An integrated robot system with peripherals (from Lechtmen [12], p. 464).

Source: Society of Manufacturing Engineers, from the Robots 6 Conference Proceedings. Copyright, 1982, SME Technical Paper "Connecting the PUMA Robot with the MIC Vision System and Other Sensors," by H. Lechtman, R.N. Nagel, L.E. Piomann, T. Sutton, and N. Webber. Used by permission.

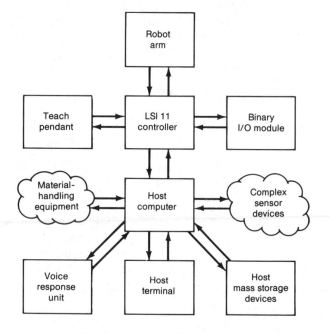

**7.4.1
Data Storage
Devices**

Magnetic disks and magnetic tapes are the two chief storage media for robot systems. There are two general types of disks: hard disks and flexible or floppy disks. Tapes are available in large reels 24 inches in diameter and in two or three sizes of tape cassettes similar to the common tape cassettes used for audio recording. The characteristics of these disks and tapes will be discussed.

Older systems used magnetic tapes to store control programs, but magnetic disks are being used extensively in newer systems. Tapes are still useful for backup storage and for small systems in which the low cost of tape cassette recorders is important. As prices decrease, the use of magnetic disks for control programs is expected to increase. Tape cassette storage now costs less than $80 per tape unit, but floppy magnetic disk systems are available for as little as $300, so that the previous cost advantage of tape cassette storage is rapidly being eroded. Either method is capable of storing several hundred thousand bytes of information, and thus is more than adequate for many simpler applications.

Disks have the advantage of direct access; it is possible to locate data or a control program by mechanically moving a magnetic read head directly to the disk track where it is stored. Access time is in the range of 25 to 150 milliseconds depending on the type of disk and drive mechanism used. Access to information on tapes requires reading the tape serially until the desired information is found. Serial scanning of the tape may require several seconds or even minutes depending on the length of the tape and the location of the information along the tape.

Disks are more reliable than tapes and usually easier to handle. Another advantage of disks is higher data transfer rates. Floppy disks transfer data at 180,000 to 360,000 bits per second, hard disks at more than 1 million bits per second. The reason is that the rotational speed of a disk can be higher than the linear speed of tape. When very high transfer rates are required, multiple heads can be used on disks, so that all tracks on a disk can be read at once.

1. Magnetic Disks

Large computers such as mainframes use 14-inch disks capable of storing as many as 600,000 bytes per disk. These disks have transfer rates of several megabytes per second. They are too expensive for most robot applications.

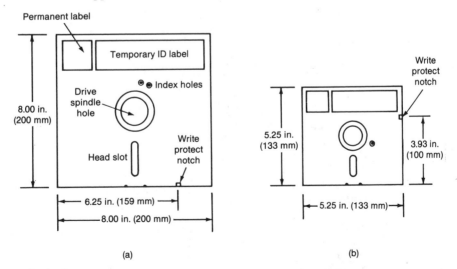

(a)   (b)

Figure 7.25 Floppy disks in a protective envelope. (a) Standard 8-inch disk. (b) Standard 5.25-inch disk.

Source: Microprocessor Interfacing Techniques, (Second Edition), by Austin Lesea and Rodnay Zaks, Figure 4-84. ©1978 SYBEX Computer Books, Berkeley, CA. Used by permission.

Robot systems typically use floppy disks. Floppy disks are made of a circular piece of magnetic tape with a center hole to accomodate the drive mechanism. They are flattened out by the centrifugal force due to the rotational speed of the drive mechanism. These disks are enclosed in a thin, square envelope of stiff paper lined on the inside with a soft material to protect the surface and keep it clean. In use, the floppy disk, in its envelope, is inserted into a drive mechanism that clamps the disk through a concentric drive hole in the envelope and rotates the disk inside the envelope. There are access holes in the envelope to allow disk heads to contact the magnetic surfaces and read information from the rotating disk. The standard form of the 8 inch and 5.25 inch diameter disks is shown in Figure 7.25. A smaller disk, 3.50 inches in diameter, is also available.

A standard enclosure for a 5.25-inch floppy disk is shown in Figure 7.26. Newer disk drives are half as high as these to save space.

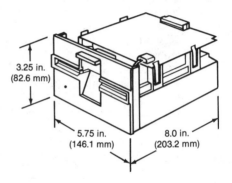

Figure 7.26 Mini-floppy disk drive (5.25 inches).

Source: Microprocessor Interfacing Techniques, (Second Edition), by Austin Lesea and Rodnay Zaks, Figure 4-81. ©1978 SYBEX Computer Books, Berkeley, CA. Used by permission.

3.25 in. (82.6 mm)

5.75 in. (146.1 mm)

8.0 in. (203.2 mm)

Floppy disks are identified as single–sided, single–density (SSSD), single–sided, double–density (SSDD), double–sided, single–density (DSSD), and double–sided, doubled–density (DSDD). Single–sided disks are recorded on only one side, double–sided disks on both sides. Similarly, single–density disks are recorded at 3,200 bits per lineal inch of track, double–sided disks at 6,400 bits per lineal inch. There are formatted and unformatted specifications because some of the recording area is required to put in control information. Also, there are hard–sectored and soft–sectored disks.

Hard–sectored disks have holes punched in a track around the disk to identify the beginning and end of a sector. Soft–sectored disks use a special binary code to identify the start of a sector. It is important, in purchasing disks, to buy the hard–sectored or soft–sectored disks if your system requires them. One type of disk cannot be run on a drive of the other type. Most blank disks can be run on double–sided drives, but there may be a reliability problem. There is a four–to–one difference in storage capacity between the SSSD and the DSDD disk formats, but the single–density disks are more reliable and allow more flexibility.

A disk is divided into sectors like a piece of pie. Each sector contains an identification mark and several tracks of information. Its characteristics are as follows:

1. Size: 8-inch diameter   Track density: 48 tracks per
                                             inch (tpi)
    Total tracks: 77   26 sectors per track

Format:	SSSD	SSDD	DSSD	DSDD
Capacity: (in bytes)	250K	500K	500K	1000K

Reliability: less than one soft error in 10^9 bits
less than one hard error in 10^{12} bits

Soft errors are detectable by software; hard errors are undetectable.

2. Size: 5.25-inch diameter   Track density: 48 tpi
    Total tracks: 23   18 sectors per track

Format:	SSSD	SSDD	DSDD	DSDD
Capacity:	89K	180K	180K	360K

Hard disks are also available for use with microprocessors in 8-, 5.25, and 3.50-inch diameters. These disks are rigid, usually made of aluminum, and mounted on a rigid shaft. As a result, the accuracy of bit placement is greatly enhanced. Many more tracks per inch (tpi) of disk radius are possible (192 tpi, for example), and bit densities up to 10,000 bits per lineal inch of track are used. They are modeled after the IBM design of hard disks called the Winchester disk after the San Jose, California, location of the original development laboratory. Standard capacities of 10, 15, and 25 megabytes of storage are available on these disk systems. Costs are in the $1000 to $3,000 range for complete disk drives, or roughly $100 per megabyte of storage. Access times can be as low as 25 milliseconds.

2. Magnetic Tapes

Magnetic tapes used in robotics are primarily one-eighth inch wide in standard tape cassettes. These tapes are identical to those used for video recording except that, in some cases, the magnetic material has been optimized for storing digital pulse information. Standard video tapes will work satisfactorily in most digital recording applications, however.

Data is recorded serially in magnetic tape cassettes. There are two tracks per tape, and each is normally recorded separately. Data rates are 185 bytes per seocnd, so that storing a large program may take several seconds. Some recorders are capable of storing data at higher speeds.

Tape data is read by serially searching the tape until the specified program is found, as identified by an identification code written on the tape when the data was recorded.

An early example of tape cassette use, in 1974, was the Unimation Type 501 Unimate Cassette Program Storage Unit. The Type 501 was a rugged device weighing 22 pounds and encased in a heavy-gauge aluminum case. This device could store 400 program steps at the rate of 1 program step per second. It was used to copy programs from a magnetic drum in the Unimate® robot for storage and later use. More recently, Unimate® has converted to the use of floppy disks for program storage.

7.4.2 Printers

Printers are used to provide information output from computers. Typical output is system status information, data processing reports, the results of computations, or a copy of stored information for archiving and backup use.

Major categories of printers are serial (or character) printers and line printers. These may each be subdivided into printers that form characters by mechanical pressure (impact printers) and nonimpact printers that form characters by the use of heat, light, sprayed ink droplets, or other non-mechanical means. Any of these printer categories can be further divided into dot matrix and typeface printers.

As an example, the familiar typewriter uses a ball or type bar with a typeface engraved on it to form images. Typewriters have a ribbon that is pressed against the paper by the type bar to form an image in the shape of the character on the type bar or ball. It prints one character at a time, serially, one after another. We would then classify the typewriter as a serial, impact printer using a typeface to form images.

1. Dot Matrix Serial Printers

Dot matrix printers form characters from an array or matrix of small dots. Typically, the dot matrix array is seven dots wide by nine dots high. In high quality dot matrix printers, the array may be 14 dots wide by 21 dots high to allow for maximum flexibility in forming characters such as *M* or *W* that are wider than the average character. Some dot matrix characters are shown in Figure 7.27.

In Figure 7.27(a), the dot matrix example has *x*s where the matrix is to print and *O*s to indicate a wire or hammer that will not print. In Figure 7.27(b), the appearance of the resulting four letters is shown, for the word *Ship*, in greatly enlarged form. Note the descender in the letter *p* that forms the portion of the character that descends below the print line.

```
OXXXXXO        OXOOOOO        OOOOOOO        OOOOOOO
XOOOOOX        OXOOOOO        OOOOOOO        OOOOOOO
XOOOOOO        OXOXXOO        OOOXOOO        OOXOXXO
OXXXXXO        OXXOOXO        OOOOOOO        OOXXOOX
OOOOOOX        OXOOOOX        OOOXOOO        OOXOOOX
XOOOOOX        OXOOOOX        OOOXOOO        OOXXOOX
OXXXXXO        OXOOOOX        OOOXOOO        OOXOXXO
OOOOOOO        OOOOOOO        OOOOOOO        OOXOOOO
OOOOOOO        OOOOOOO        OOOOOOO        OOXOOOO
```

(a)

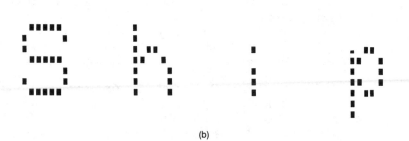

(b)

Figure 7.27 (a) Dot matrix layout. (b) Enlarged view of the resulting printed characters.

Dot matrix printers may have only one column set of hammers or print wires. This one set is used repeatedly to form the complete character. At each step, the dots to be printed in that column are selected. To print a matrix that is seven dots wide by nine dots high, the set of hammers would be seven high and be used nine times per character. Since these printers can operate at rates of 50 to 150 characters per second, each set of hammers must step across the paper and print at 450 to 1,350 times per second, less than a millisecond per step in some cases.

2. Daisy Wheel Serial Printers

The most common type of serial printer to provide high-quality typeface quality is the daisy wheel printer. There is a radial element, or "petal", for each character to be printed. These petals are arranged in a circle about 3 inches in diameter on a central rotating shaft. Each petal has a character molded on it and is flexible in a direction perpendicular to the plane of the wheel. Daisy wheels are rotated at constant speed, and printing is done "on the fly." A single hammer is controlled to hit the selected petal at the exact instant when it is in position in front of the paper. Figure 7.28 shows a typical print wheel.

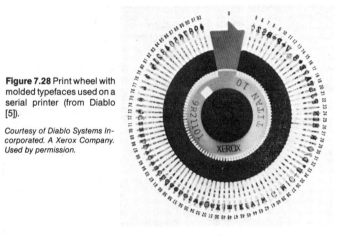

Figure 7.28 Print wheel with molded typefaces used on a serial printer (from Diablo [5]).

Courtesy of Diablo Systems Incorporated. A Xerox Company. Used by permission.

Control of printers is necessarily complex. Figure 7.29 is a block diagram of the Diablo HYTYPE II printer showing the logic control blocks, the servos for the print wheel and carriage, the power

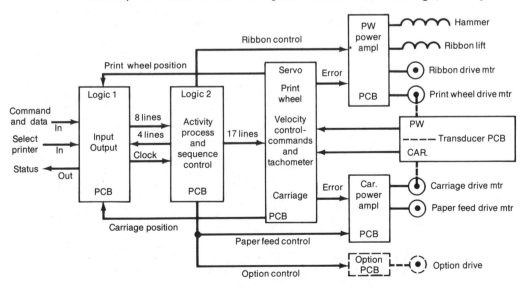

Figure 7.29 HYTYPE II printer block diagram (from Diablo [5]).

Source: Diablo Systems Incorporated. A Xerox Company. Used by permission.

amplifiers for the drive motors and controls, and the transducer for monitoring print wheel position precisely. An Intel 8080 microprocessor is used to monitor and control the printer's activities. It is of interest to note that the control capability provided is sufficient to control one joint of a robot arm.

This printer can move 48 increments per inch vertically and 120 increments per inch horizontally to provide for subscript and superscript printing and to print in the high-resolution graphics mode. It is also possible to control spacing between characters and lines in a very flexible way. Proportional spacing between characters can be obtained with this type of printer under the control of a word processing program such as Wordstar.

3. Nonimpact Serial Printers

Instead of requiring mechanical pressure and a ribbon to form a character image, the nonimpact printer may use heat, ink jets, or laser beams to form the image. Some fast, quiet printers use a dot matrix of wires that are selectively heated to form a character on heat-sensitive paper, typically a coated white paper with a black background. Ink jet printers form the character by spraying ink through an orifice and between plates that electrostatically deflect the ink jet vertically to form the character. Horizontal motion of the print head allows the complete character to be formed. This technique is illustrated in Figure 7.30. Ink not needed to form a particular character is recycled back to the ink reservoir.

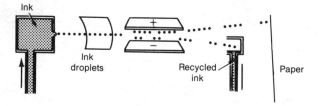

Figure 7.30 Ink jet print head (from Sanders [25]).

Source: Donald H. Sanders, Computers in Society, Third Edition, 1981. ©1981 McGraw-Hill Book Company. Used by permission.

Lasers are used on nonimpact line printers and will be discussed in Section 7.4.2, Part 5.

4. Line Printers--Impact Type

High-speed printed output from mainframe computers and minicomputers is provided by line printers. Impact printers are reasonable in cost and provide high-quality output. In addition, they can make up to six carbon copies of each page of output due to the force of the impact. Impact line printers are made in two general configurations: rotating drum and moving horizontal chain.

Drum Printers

Rotating drum printers have a set of raised characters around the periphery of the drum for every character position; there are usually 132 positions. The drum is behind the paper. In front of the paper is a horizontal ribbon and a set of print hammers. As the selected character rotates into position behind the paper, the hammer hits the ribbon and presses it against the paper and the raised character on the drum. If a line of As is to be printed, for example, all 132 print hammers would strike the paper at once as the As in each character position came into position. In some printers the hammers are movable to one of six or more horizontal positions, thus requiring fewer hammers and control circuits but slowing the print speed by a factor of six. Timing errors in operating the hammers cause the characters to be displaced vertically, producing an uneven line. Drum printers print at 500 to 2,000 lines per minute, one line per complete rotation of the drum.

Chain Printers

Horizontally moving chains carry several sets of raised characters on the individual chain links, typically five sets of 48 characters. The continuously moving chain is in front of the paper. Behind the paper is a set of hammers, one for each character position. The horizontal ribbon moves between the chain and the paper. When the selected character for each position is aligned properly, the hammer presses the paper against the ribbon and the chain to form the character. In this printer, an error in hammer timing causes a horizontal error that is less observable than the vertical position error found in the drum printer. For this reason and others, the chain printer has become the dominant impact line printer.

Line Printers—Nonimpact Type

Nonimpact line printers primarily use ink jet printing or laser printing. Ink jet printing was described in Section 7.4.2, Part 3. Laser printing is done by a high-intensity laser beam that is deflected by mirrors or other optical means to draw the character on the paper. Printing can be done on light-sensitive material or by directly burning the paper. Computer output microfilm (COM) is often created by the use of a laser beam because of the high printing speed possible. This microfilm can then be used for storage purposes, or the image can be enlarged and duplicated by an electrostatic printing process such as that used in copiers made by Xerox and others. Printing rates of 20,000 lines per minute are obtained. These high-speed laser printers are expensive, in the range of $100,000 or more each.

Another nonimpact printing process forms the image by depositing an electric charge on the paper from a high-density line of electrodes across the paper. Several thousand electrodes are formed in a horizontal line across the paper and selectively supplied with a high voltage to charge the paper at a particular point. As the paper moves upward, the characters are formed line by line by controlling the charge pattern on the paper. Development of the charge pattern is done by applying a fine resinous powder and heating, as in other electrostatic printing processes. This method produces high-speed character printing and is also excellent for forming diagrams, pictures, and graphic output. It is also quite expensive, but is especially useful for duplicating drawings and diagrams for engineering and scientific use due to the high resolution of the resulting output.

7.4.3
Visual Display
Terminals

Visual display consists of forming characters and figures on a screen to present information from the computer to human users. Such displays are also called <u>monitors</u> and are used wherever a human operator is involved in giving <u>direction</u> to or monitoring the operation.

Two types of visual display terminals are in general use: round and flat screens. Round screen Cathode Ray Tubes (CRT's) have been in use for many years. They are the same type of tube as the television tubes that are ubiquitous in our society. Other technologies are now being used to form a flat screen display that is more compact than the CRT displays. These flat screen devices use liquid crystals, electroluminescence, or gas plasmas to generate or control the image formed on the display. All of these displays depend on the control of electrical patterns on a grid or mosaic of individual elements. They operate like the dot matrix printers, except that the arrays are larger and require separate control lines for multiple X and Y positions, just as memories do.

Liquid crystal displays are similar to those used on many watches. An electric field is used to rotate tiny crystals so that they will pass or reflect light. This phenomenon is used to form flat screen images that are suitable for displaying characters and slowly varying phenomena. Typical screens are 60 to 80 characters wide and show up to 40 lines at a time. They are more expensive than CRTs but are compact and require only a small amount of power to operate.

Electroluminescent displays form a lighted image, usually in red or green, by selectively applying a relatively high voltage to a grid of cells and exciting them to produce light in a particular pattern. They are useful in making small displays but are more expensive than CRTs for large displays.

Gas plasma displays require a carefully controlled mixture of gases to be enclosed in a sealed glass container and excited by high voltages (100 to 500 volts) to form an image. They are capable of producing high-intensity light at rapidly varying rates in a flat, compact screen display unit. Control is expensive, but the resulting product is of sufficiently high quality to be useful in some display applications.

7.4.4
Vision

Vision is an increasingly valuable sensory input for robots. It enables the robot to orient itself in space, locate and identify objects, and make measurements on parts in its field of view.

Software in the robot control program can be used to send requests to a vision system that is a separate entity and receive back from the vision system the desired information on location, orientation, and the identity of objects. This input can then be analyzed and manipulated so that the robot control system can perform its assigned task.

Hardware and software for vision are described in detail in Chapter 9 and will not be discussed here. Interfacing of the video system with the robot control system has many of the same requirements as other input-output systems and will be discussed in Section 7.5.

7.4.5
Sensors

Position, temperature, pressure, magnetic field, proximity, force, velocity, torque, stress, and many other types of sensory elements are required for robot system applications. These elements may be part of the positioning control of the robot itself or may be associated with auxiliary equipment used with the robot system.

Sensors and their operation are discussed in considerable detail in Chapter 5. Interfacing is discussed in Sections 7.5.2 and 7.5.3. Sensory elements are especially likely to produce computer system interrupts to signal a particular situation. Therefore, their control systems must be carefully designed to handle the interrupt conditions that may occur.

**7.4.6
Robot Servos**

Servo motors may be controlled by either digital or analog servos. Each servo receives a command from a main control computer for the robot arm and carries out the command independently. Simple servos may perform their task with fixed or hard-wired electronic controls. However, the advanced robot systems use a microprocessor on each robot servo. These controllers are peripherals of the main robot control computer.

A good example of a well-designed robot controller is given in Figure 7.31. Separate digital servo boards are used, as shown, in the PUMA 550/560 controller chassis. Each board is plugged into a connector on the motherboard or back plane of the controller and connects to the bus supplying power and signals to all of the control boards. A microprocessor is used on each digital servo board to perform the necessary calculations and comparisons to control one joint of the robot in response to commands from the main control computer, the Digital Equipment Corporation (DEC) LSI-11/2 seen at the left side of Figure 7.31. The CPU, RAM memory, EPROM memory, and other control boards are plugged into the same bus and communicate as required with each other and with the digital control boards.

Position commands from the robot controller, the LSI-11/2, are sent to the digital control boards, where they are compared with the digital information from the encoders on each joint axis. The difference between the commanded position and the measured position is a digital number that is converted to an analog voltage (error signal) and sent to the analog servo amplifiers that directly control the drive motors. Motors rotate in such a way that the digital output of the encoders becomes the same as the digital input command, the error signal goes to 0, and the joint stops at the commanded position. Current feedback from the motors is used to stabilize the analog servos since it is a measure of motor velocity. This input is part of the Proportional, Integral, Derivative (PID) control in the motor control servo loop.

Output from the controller is sent to the CRT, floppy disks, manual control panel, and the auxiliary vision system through the Quad Serial Board, which converts the digital signals from the LSI-11/2 to the correct serial or parallel form, at the correct voltage levels, to control these peripheral devices.

Hydraulic servos work in much the same way, with either analog or digital signals to control the electrohydraulic valves used to control oil flow in the hydraulic system.

Clock pulses are generated by the Clock/Terminator board at a 1 megahertz (mhz) rate for input to the servo interface board. This device also supplies an event timing pulse to the CPU to synchronize the sampling of events by the CPU. This pulse is fast enough to allow the CPU to monitor other equipment properly and have time to perform the computations required to prepare commands for the next cycle of events.

**7.4.7
Auxiliary
Equipment**

Control of auxiliary equipment may be provided by either the host computer or the robot controller. As depicted in Figure 7.31, auxiliary analog inputs from other devices are converted, in the Analog-to-Digital Converters, to digital signals for input to the controllers. Auxiliary devices may be conveyors, numerically controlled machine tools, furnaces, forging hammers, or other equipment working with the robot. In this way, the robot controller can insure that all devices operate in a controlled, synchronous manner to attain

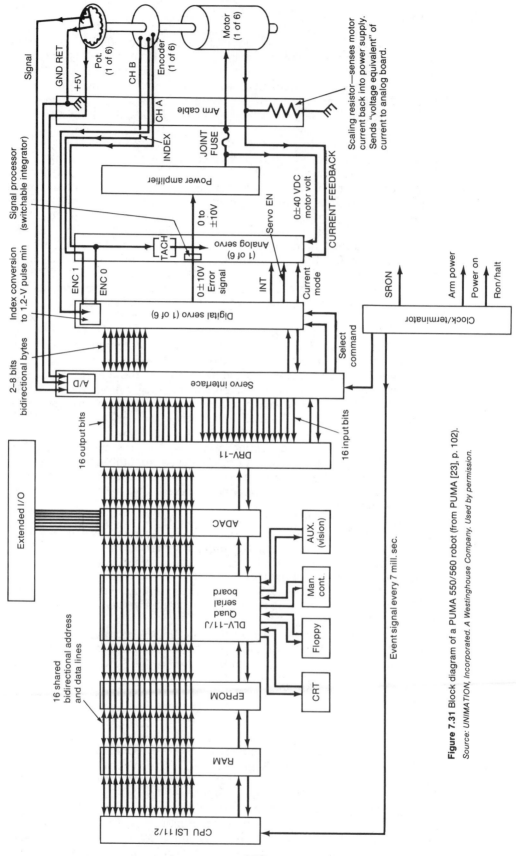

Figure 7.31 Block diagram of a PUMA 550/560 robot (from PUMA [23], p. 102).

Source: UNIMATION, Incorporated. A Westinghouse Company. Used by permission.

optimum efficiency. Vision systems may be controlled by a digital command through the AUX (vision) output from the Quad Serial Board.

7.5 Input/Output (I/O) Operations and Control

Input/output is a general term applied to the functions that provide communication of information between the computer and the various devices attached to it. Since there may be more than one computer or controller in a system, I/O is also used to refer to communication between other devices on the computer bus in addition to direct communication to the computer. External devices of many kinds may communicate with the computer system and attach to one of the system buses. Some of these peripheral devices were described in Section 7.4. In addition, the reader should be aware of the functions performed by standard computer buses, as described in Section 7.3.5.

A major consideration in I/O communciation is the provision of the correct interfaces, both hardware and software, to ensure compatibility of signals and control information to and from the devices attached to the system. Signals may be serial or parallel, and may appear at several different voltage levels and at varying data rates, so that the interface equipment must provide means for handling these differing signals to and from the computer. Both analog and digital signals may appear at the inputs or be required at the outputs of the system.

In this section, some of the problems of device addressing, interfacing, interrupts, and communication control will be discussed.

7.5.1
Device Addressing
How does a computer communication with a specific device? Of course, the answer is by use of a data bus. Many devices are connected to the common data bus and addressed by a common set of address lines. Usually, both the data bus and the address bus are sets of parallel lines. Address lines carry an address in the form of several binary digits or bits in a specific code recognized by logic circuits on the device. Data buses carry the actual data. When the device recognizes its address on the address bus, it prepares to read the information from the data bus or write information to it. Control lines from the computer determine whether the data is to be written to the device or read from the device to the computer, to memory, or to another device.

Three general methods are used. One method is a continuation of memory addressing, so that the device uses one of the memory locations as an address. This method is called memory-mapped addressing. The computer then has only one addressing method that is used for both memory and device locations. This method has the advantage of simplicity. It also allows the programmer to address devices in an easy way. A drawback is that some parts of the possible memory locations are used for device addressing and are not available for memory use. This may be a small price to pay. Out of 64K possible memory locations, it is unlikely that more than a few hundred will ever be used for control purposes. Motorola's 6800 and 68000 microprocessors make good use of memory-mapped addressing. Memory-mapped addressing is especially useful in graphic addressing of displays, since the memory in the display is part of the total memory of the computer and can be used just like any other memory if you don't care what is shown on the display. However, in this case, there is a large amount of memory used, perhaps 2K bytes, so that there is a loss of memory space for other uses.

Another method of addressing, which still uses the memory data bus, performs addressing with separate control lines so that none of the memory capacity is used. This operation is more difficult and may be somewhat awkward, but it does increase the total memory that can be used. Intel's 8080 and the Zilog Z-80 use this method.

A third method uses a completely separate bus, called the input/output bus, that is similar to the data bus. Addressing on this bus is controlled by a peripheral-address bus to read information to and from the data lines of the I/O bus. Control signals are carried by a separate peripheral-control bus. A new set of instructions, called I/O instructions, are used to operate this bus. Considerable additional hardware is required. Memory space is saved and the speed of I/O operations is increased.

Control lines determine that a device is ready to receive or transmit a signal, control access to the bus, and provide timing information to control the rate of information transmission on the bus.

**7.5.2
Interfacing and
Interface
Standards**

An interface is the boundary between two devices or programs. Mechanically, it is the set of connectors that must be plugged together to carry signals, power, and so on; in electrical terms, the voltages must be at the same level; logically the representation of a binary 1 or 0 on one side of the interface must be equal to the representation on the other side. Also, the codes used must be compatible and the data rates must be within an acceptable range. In programming software, the forms of data representation, instruction sets, coding, subroutine linkages, and the operating system must all be compatible. This section will discuss some of these considerations using simple examples. Since interfacing is a major problem, we can only outline the nature of the solutions being used.

1. RS-232 Interface

A standard interface widely used in computers, terminals, and peripheral equipment is the RS-232 interface. This standard was approved by the Electronic Industries Association (EIA) to define the electrical characteristics of an interface between a unit of data terminal equipment (DTE) and data communications equipment (DCE).

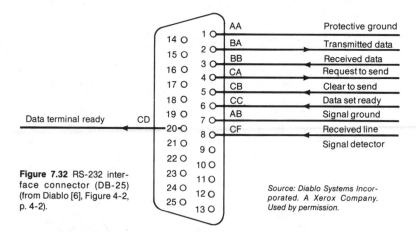

Figure 7.32 RS-232 interface connector (DB-25) (from Diablo [6], Figure 4-2, p. 4-2).

Source: Diablo Systems Incorporated. A Xerox Company. Used by permission.

Mechanically, the RS-232 interface is implemented by a 25-pin subminiature connector known as the DB-25 connector, available at most computer supply houses and normally supplied with equipment sold as RS-232 compatible. This connector has the appearance shown in Figure 7.32. A female DB-25 connector is to be used on the DTE data terminal equipment and a male connector on the DCE. Originally, the terminal was a teletypewriter and the communication equipment was a modem. Now the terminal may be a robot control system or a piece of auxiliary equipment.

Implementing the RS-232 interface requires the use of several logical and electrical level-setting circuits, as shown in Figure 7.33. In this example, the universal synchronous/asynchronous receiver/ transmitter (USART) is located in the terminal or in a computer, and the modem is connected to a telephone line. In between are the input and output level converters. Voltage levels on the RS-232 were set in the days of teletypewriters; they were high enough to provide plenty of power to operates relays and drive communication equipment. On the modem side,the voltages are +12 and -12 volts. On the terminal side, the voltages are TTL compatible and range from 0 to 5 volts. We will now examine these features in more detail.

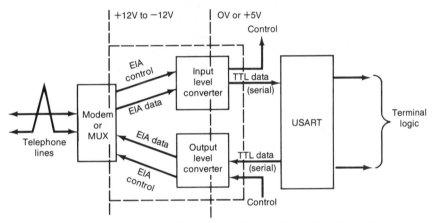

Figure 7.33 Interface functional block diagram (from Diablo [6], Figure 4-1, p. 4-1).
Source: Diablo Systems Incorporated. A Xerox Company. Used by permission.

2. Parallel-to-Serial and Serial-to-Parallel Conversion

The USART provides for conversion of parallel data from the computer to serial data for transmission to the modem and conversion of serial data from the modem to parallel data into the computer. Parity, stop bits, and other characteristics of the signal are generated or controlled by the USART.

Several steps are required in the computer software to prepare the USART for operation and to control communications. Internal registers in the USART must be set to control the direction of transmission on each computer port and to prescribe the bit rate, number of stop bits, parity bits, and other characteristics. These programs are usually provided by the monitor or control program that runs the computer system. Many commercially available programs have these controls built into the program. However, the provision of the proper interface coding must be checked with each new piece of equipment. These programs are often a source of problems if they are not correct.

Data format at the interface is given for three different bit rates in Figure 7.34. By convention, the positive-level pulses are called <u>space</u> and the negative level is called <u>mark</u>, a hangover from telegraph nomenclature. The first bit in a character is always a start bit, a positive pulse. In the diagram, the first bit of the character is a

0 value. The 1 values are represented by a negative level. Reading from left to right, we see the start bit and then the binary sequence 1 1 0 1 0 0 1 0 in the upper drawing (A. 110 baud, 2 stop bits). The first 7 bits after the start bit are the ASCII code for the letter; the eighth bit is labeled *P* for parity bit.

Note that bit 7 is the high-order bit of the character. Thus, the code sequence 1 1 0 1 0 0 1 0 must be read in the reverse order with the parity bit first to obtain 0 1 0 0 1 0 1 1, which is the ASCII code for the letter *K*. (See the ASCII code chart, Appendix A.2.)

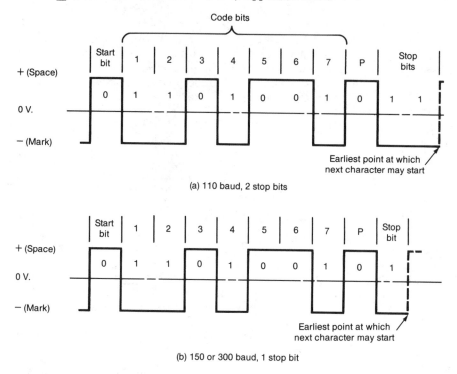

(a) 110 baud, 2 stop bits

(b) 150 or 300 baud, 1 stop bit

Figure 7.34 Data format on interface (from Diablo [7], Figure 4-4, p. 4-4).

Source: Diablo Systems Incorporated. A Xerox Company. Used by permission.

3. Parity

Even parity is being used here. The 1 bits are counted, and if the number is even (as it is here), the parity bit is set to 0 so that the total number of bits in the character is even. If the number of bits in the character had been odd, a 1 bit would be entered in the parity position to make the total number of bits even. Odd parity is also used where the total count would be forced to be an odd number by adding a parity bit if necessary. No parity is another option. In that case, the circuitry does not check the parity bit at all. Stop bits are added at the end of the character to separate the characters and to give the receiver time to assemble the bits of a character before another character comes in.

Standard bit rates are 50, 75, 110, 134.5, 150, 300, 600, 1,200, 2,400, 3,600, 4,800, and 9,600 bits per second on the RS-232 interface. Printers run at 300 to 1,200 bits per second; displays usually run at 1,200 to 9,600 bits per second. Baud rate is the number of actual bits per second when the start and stop bits are counted; thus, it is slightly faster than the corresponding bit rate.

Table 7.4 lists the RS-232 interface signals with the number of the connector pin used and the EIA coding as shown in Figure 7.32. Note that the names of signals are from the viewpoint of the DTE or data terminal. Sometimes it is difficult to decide which of two devices is the data terminal and which is the data communications equipment (DCE). In this case, it is essential that pin 2 of one device be connected to pin 3 of the other device. If necessary, they should be switched around. The standard is not clear for all applications, and some manufacturers use their own version.

Table 7.4 Interface signals

Connector Pin No.	EIA Circuit	EIA Signal Name	Function
1	AA	Protective Ground	Connected to pin 7, signal ground, or may be disconnected in some cases and left "floating."
2	BA	Transmitted Data	The serial ASCII data being transmitted from the terminal. In the mark state between characters.
3	BB	Received Data	The serial ASCII data being received by the terminal. Held in the mark state between characters.
4	CA	Request to Send	Goes high when a request to send signal is received. Often wired high to the power source.
5	CB	Clear to Send	Must be on (high) during data transmission. Can be wired high or controlled by computer.
6	CC	Data Set Ready	Must be on (high) for terminal operation. Often wired high to the terminal power source.
7	AB	Signal Ground	Ground reference for all other interface signals.
8	CF	Received Line Signal Detector	Carrier Detect. Signals that a carrier signal has been detected on the line.
20	CD	Data Terminal	On (high) when power is on.
23	CH	Data Signal Rate	Not often used.

4. Handshaking Signals

Handshaking and other signals are used in <u>handshaking</u> operations, in which the transmitter and receiver signal status to each other and perform sequence control operations. One important use of handshaking is to allow a slow-speed device such as a printer to communicate with a high-speed device such as a computer. If the computer transmitted data at its full speed, the slow-speed device could not keep up and some data would be lost. The solution to this problem is a signaling system such as handshaking.

Printers and other devices usually have a buffer memory that can hold a certain amount of information; most printer buffers can hold a complete line of text to be printed. When the computer wants to operate the printer, it sends a request code, saying, "are you ready?". If the printer is turned on and operating properly, it sends back a code, "ready, go ahead and transmit." Then the computer transmits data until the printer sends a code "Stop, my buffer is full," so that the computer will wait until the printer sends another code, "go ahead." In this way, the operation of a slow-speed device can be synchronized with that of the higher-speed device.

Although this discussion gives some idea of handshaking, it can become a somewhat complex routine; the reader is referred to the full RS-232 specification for details. Leibson [17] warns of some of the problems that can be encountered in utilizing handshaking between equipment provided by different manufacturers.

**7.5.3
Interrupts and
Interrupt Handling**

Input/output devices require servicing at irregular intervals depending on the operation being performed. Some servicing can be performed by the communications protocols or handshaking, as discussed in Section 7.5.2. Other devices must be serviced at a particular time or data will be lost. In robot systems, sensors guide and protect the robot itself, so that sensor input must be accepted and acted upon promptly. Efficient servicing of peripherals is necessary just to perform the large number of tasks required.

Two methods of servicing are in general use. In the <u>polled</u> method, as illustrated in Figure 7.35, the computer must test each device in sequence and ask "do you have anything to report?". Polling systems must get around to every device frequently enough so that all important input is collected and none of it is lost. This is an inefficient method; much computer time is spent switching to each device separately and asking it for input. Some devices may have data

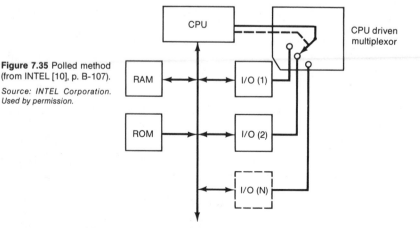

Figure 7.35 Polled method (from INTEL [10], p. B-107).

Source: INTEL Corporation. Used by permission.

frequently; others may not have input for hours. In polling, all devices must be scanned just in case they may have something.

A more desirable method of servicing is to communicate with a device only when it has something to say. In a classroom, a student raises a hand to request attention. A telephone bell rings to tell the user that someone is calling. In a computer system, a signal line called an interrupt is used to get the computer's attention. This method is illustrated in Figure 7.36. A circuit called the Programmable Interrupt Controller (PIC) is commonly used to accept the interrupt signals from the individual input/output devices and present them, in an orderly way, to the CPU. When an interrupt request comes from the PIC, the computer is also provided with necessary information to handle it. This capability is part of the circuitry in the PIC chip.

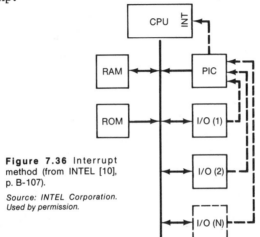

Figure 7.36 Interrupt method (from INTEL [10], p. B-107).

Source: INTEL Corporation. Used by permission.

Figure 7.37 is a block diagram of a typical PIC. This diagram is for the INTEL 8259A, used with the INTEL 8086/8088 family of microprocessors such as those used in the IBM PC and several other microcomputer systems.

Using interrupts to get the computer's attention means that devices can get immediate attention when necessary but do not otherwise take up computer time providing service. It is a more efficient method and provides better response and control at the cost of a small amount of hardware, the PIC, and the associated lines.

The PIC accepts requests from the peripheral devices, determines which of the devices has the most important information (highest priority), ascertains whether the device requesting service has higher priority than the device currently being serviced (if any), and issues an interrupt to the CPU based on this analysis. Usually, there is a special routine or program stored in memory and associated with each peripheral device. When that device is to be serviced, this routine is called into the CPU and controls the action taken by the computer system. The INTEL 8259A is designed to operate in real time, interrupt-driven microcomputer systems. It can manage eight levels of requests and can expand to 64 levels by adding additional PICs to work with it since the necesary addressing and control is part of the PIC.

Individual elements of the 8259A and their functions are: (see Figure 7.37):

1. Interrupt Request Register (IRR). Stores all of the interrupt levels that are requesting service.

Pin configuration

Pin names

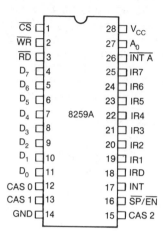

$D_7 \cdot D_0$	Data bus (directional)
$\overline{RD}$	Read Input
$\overline{WR}$	Write Input
A_0	Command Select Address
$\overline{CS}$	Chip Select
CAS2 CAS0	Cascade Lines
$\overline{SP/EN}$	Slave Program/Enable Buffer
INT	Interrupt Output
$\overline{INTA}$	Interrupt Acknowledge Input
IR0–IR7	Interrupt Request Inputs

Block diagram

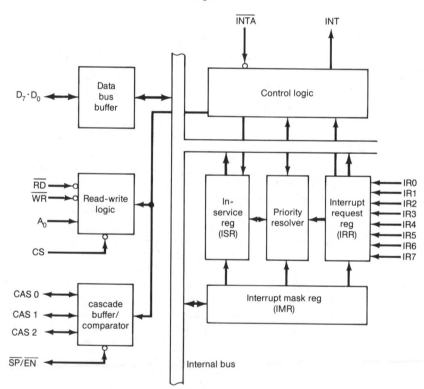

Figure 7.37 PIC block diagram (from INTEL [10], p. B-106).

2. **In-Service Register (ISR).**   Stores all of the interrupt levels that are being serviced.
3. **Priority Resolver.** A logic block that determines the priorities of the bits set in the IRR. The highest priority is selected and sent to the control logic block for transfer to the CPU.
4. **Interrupt Mask Register (IMR).**   Stores the bits that control the interrupt lines to be masked. Masking effectively turns off the particular interrupt line. It is used, under program control, to

prevent interrupts from certain lines at the wrong time.

5. INT (Interrupt). The signal to the CPU interrupt input that controls the initiation of the interrupt sequence.

6. INTA (Interrupt Acknowledge). The signal back from the CPU that tells the PIC to put vectoring information on the Data Bus to tell the computer what action to take.

7. Data Bus Buffer. A typical three–state, bidirectional, 8–bit buffer used to interface the PIC to the system Data Bus. (See Section 7.2.1, part 7 for a description of tri–state operation.)

8. Read/Write Logic. The control block that accepts output commands from the CPU and controls formats for various operations.

9. Cascade Buffer/Comparator. Stores and compares the identification codes of all of the PICs used for an interrupt system. It controls which PIC, if more than one, is allowed to send the interrupt to the CPU.

1. Vectored Priority Interrupt

Actual operation of the PIC is somewhat complex. However, its basic task is to accept interrupt requests from the various devices and service them in a particular priority order that has been assigned by a computer program. It also provides the capability for vectored priority interrupts. In this case, vectoring means that the interrupt signal provides the address of the routine to be used to service the interrupt. The subroutine address and the routine itself must have been preprogrammed and assigned to the particular interrupt line. Subroutines are stored in memory at equally spaced intervals, so that a simple identification code associated with the interrupt line is sufficient to cause the CPU to jump to a particular location and start executing the subroutine. When an interrupt routine is completed, the PIC is notified by the CPU and the next interrupt, in priority sequence, is accepted. Priorities and interrupt masking can be changed by the CPU as required.

If the CPU is running a program when an interrupt occurs, the interrupt signal forces the CPU to save the current program information in specified memory locations and branch to the interrupt routine in the memory location specified by the interrupt vector. It then executes the interrupt routine, which may be as complicated as necessary, and returns to the current program unless another interrupt has occurred meanwhile. It is quite possible for the CPU to spend all of its time handling interrupts and never complete its main program. This, of course, means that the system design is inadequate and must be changed.

**7.5.4
Direct Memory
Access (DMA)**

We have discussed handling devices that operate at a slower rate than the CPU does. What about high–speed devices: how do we handle them? A good example of a device with a high data rate is the disk file; others are vision systems and some sensor systems. The disk file is a good example of a device where Direct Memory Access (DMA) is desirable.

When devices require input or output at rates approaching the speed of the memory, it is wasteful to use the CPU registers to handle each byte individually. In these cases, it is more efficient to allow the high–speed device to access the memory directly, just as though it was a CPU. To provide this capability, disk controllers have memory addressing capability, memory control commands, and data buses that can connect directly to the memory data bus. High–speed transfer of data is accomplished by sending it directly from the registers in the disk controller to the memory at an address specified by the CPU. In this operation, the CPU tells the disk controller to transfer a certain block of information from a particular disk file to a specified address in memory. The disk controller then takes control of the memory bus and transfers the data as required. When it has completed the operation, it notifies the CPU that it has completed the command given and that the memory is now available for other use.

A DMA machine can be a disk controller with direct memory access capability or another device. Some mainframe computers use channels that are essentially DMA machines and can handle many different types of input or output. In these systems, the disk controller can be one channel, a tape controller can be another, and a high-speed communication channel may also be used. DMA machines can operate in burst mode, where they take control of the memory bus and transfer all their information at high speed, or can operate in cycle stealing mode, where they have maximum priority for access to the system memory but the CPU may be able to use a few memory cycles whenever they stop.

Figure 7.38 shows how a DMA machine operates in parallel with the CPU and under the CPU's control to transfer information from an I/O bus to memory. After the CPU has directed the DMA machine to transfer data, the DMA machine acknowledges this command and requests control of the memory bus with a Bus Request. If the CPU is ready, it releases control of the bus to the DMA machine by issuing a Bus Grant command, and the DMA machine performs the operation requested. In issuing the information transfer command, the CPU transfers the required address information and control information to registers in the DMA machine. If the CPU was not ready at the time of a bus request (perhaps because it was handling an interrupt), it would not issue the bus grant command until it was ready.

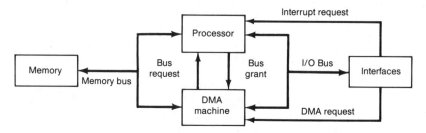

Figure 7.38 Direct Memory Access (DMA) machine (from Leibson [15], Figure 5, p. 140).

Source: "Direct Memory Access Machine," Part 2: "Interrupts and Direct Memory Access," by Leibson in the March, 1982 issue of BYTE magazine. Copyright © 1982 by McGraw-Hill, Inc. New York NY 10020. Used by permission.

7.6 Advanced Computer Systems

Earlier in this chapter, we discussed the simple microprocessors and their architecture. In this section, we will describe some of the new concepts in microprocessors and the organization or architecture of advanced computers. Microprocessors are now growing in capability. Many can be considered to be 16 or 32-bit devices where word size, register length, or bus capability are measures of size, as discussed in Section 7.6.1.

New types of system organization are becoming common. Pipelined processors run programs through a serial sequence of processors, each performing a different operation. Parallel processors allow many processors to work cooperatively on parts of the same problem at the same time. And new organizations more complex than these are being used in increasing quantities. New architectures are discussed in Section 7.6.2.

Computers can also be organized into communication networks to take advantage of common data or programs, to provide higher reliability and efficiency, or to provide communication between different functions or operations in different areas of the country or world. Networks are described briefly in Section 7.6.3.

**7.6.1
Improved
Microprocessors**

More complex computer applications tend to require the use of more main memory, more powerful instructions, and higher computing speed. Also, the control capabilities must be improved to handle more input/output operations and more complex external devices. As a result, microprocessor manufacturers have worked to provide these capabilities. Cost reductions in semiconductor chips have been rapid and improved designs have been made, so that many of these goals have been realized. Gupta and Toong [8] survey existing microcomputers and discuss the advantages and disadvantages of the features provided.

An 8-bit microcomputer, using two words for addressing, can address 2^{16} or 65,536 (64K) bytes of memory. This can be a severe limitation when larger programs requiring more data are to be run. By increasing the word size to 16 bits, the computer can use two words and address 2^{32} bytes of memory, or 4,294,967,296 bytes. Since this is more than enough, it is feasible to use 20 bits and address 2^{20} or 1000K, or to use 24 bits and address 16 megabytes of memory. Both 20-bit and 24-bit are used in 16-bit computers.

Other measures of microprocessor capability are the widths of internal instruction paths, registers, and buses. Larger registers can handle more complex instructions, provide more addressing modes, and transfer information faster. Wider buses can transfer data and addresses faster, 32-bit-wide buses are twice as fast as 16-bit-wide buses.

A true 32-bit microcomputer would have all of its internal buses and registers, and its external buses, capable of handling 32 bits at a time. However, there is a limit to the number of connections that can be made to a microprocessor chip, so that it is difficult to bring out 32 bits on the data bus and enough address bits at the same time. As a consequence, the Motorola 68000 and the National NS 16032 microcomputers have 32-bit internal paths and 16-bit external paths. Other manufacturers have made other compromises.

Different addressing modes can be used in microprocessors to increase the speed and efficiency of program coding and execution. Modes commonly provided are indirect, indexed, and direct addressing. Indirect addressing allows the actual memory address to be stored in another memory location. Indexing allows a register to be used as a reference address. Both of these addressing types are useful in speeding up the search of memory for information or for sequencing through a large input file to carry out the same instructions on many pieces of data. Some computers provide autodecrementing and autoincrementing, so that the next address is automatically specified when these commands are used.

One of the features of the Motorola 68000 is its ability to address directly up to 16 megabytes of memory in a linear way. This action permits the use of a memory management unit, an auxiliary chip or circuit, to provide the virtual memory capability. When virtual

Table 7.5 Specifications of 16-bit microprocessors

	TI 9900	Intel 8086	Zilog–Z8000	Motorola 68000	NS 16032	Intel 80286
Year of Commercial Introduction	1976	1978	1979	1980	1982	1982
No. of Basic Instructions	69	95	110	61	82	121
No. of General-Purpose Registers	16	14	16	16	8	14
Pin Count	40	40	48/40	64	48	68
Direct Address Range (Bytes)	64K	1M	48M*	16M/64M	16M	16M
Number of Addressing Modes	8	24	6	14	9	
Basic Clock Frequency	3MHz	5MHz (4–8MHz)	2.5–3.9MHz	5–8MHz	10MHz	8MHz
System Structures						
Uniform Addressability	x	x	x	/	/	x
Module Map and Modules	x	x	x	x	/	x
Virtual	x	x	x	x	/	/
Primitive Data Types						
Bits	/	x	/	/	/	x
Integer Byte or Word	/	/	/	/	/	/
Integer Doubleword	x	x	/	/	/	x
Logical Byte or Word	/	/	x	/	/	x
Logical Doubleword	x	x	/	/	/	x
Character Strings (Byte, Word)	/	/	/	/	/	/
Character Strings (Doubleword)	x	x	x	x	/	/
BCD Byte	x	x	x	x	/	x
BCD Word	x	/	/	/	/	/
BCD Doubleword	x	x	x	x	/	x
Floating Point	x	x	x	x	x	x

Feature						
Data Structures						
Stacks	✓	✓	✓	✓	✓	✓
Arrays	x	✓	x	x	x	✓
Packed Arrays	x	✓	x	x	x	✓
Records	✓	✓	✓	✓	✓	✓
Packed Records	x	✓	x	x	x	x
Strings	✓	x	✓	✓	✓	x
Primitive Control Operations						
Condition Code Primitives	x	✓	✓	✓	✓	✓
Jump	✓	✓	✓	✓	✓	✓
Conditional Branch	✓	✓	✓	✓	✓	✓
Simple Iterative Loop Control	✓	✓	✓	✓	✓	✓
Subroutine Call	✓	✓	✓	✓	✓	✓
Multiway Branch	x	✓	x	x	x	x
Control Structure						
External Procedure Call	✓	✓	x	x	x	x
Semaphores	✓	✓	x	x	x	x
Traps	✓	✓	✓	✓	✓	✓
Interrupts	✓	✓	✓	✓	✓	✓
Supervisor Call	✓	✓	✓	✓	✓	x
Compatibility with other microprocessors	x	✓	x	x	x	x

*Six segments of 8M each.

Source: Gupta and Toong [8], Table 1, p. 1239.

memory is available, the programmer does not have to consider the actual memory size. Instead, the memory management unit, (MMU) controls the memory and disk files together, so that the apparent memory is the full size addressable by the internal registers, which may be in the gigabyte range (a gigabyte is 1,000 megabytes). This very large effective memory capability greatly simplifies many programming tasks. Several other microprocessors also have this capability.

Another characteristic of microprocessors that must be considered is the capability of handling various data types. Data types are defined as follows:

1. INTEGER.   The numbers to which we are accustomed.   Whole numbers without fractions or decimals. Examples: 123, 64, 89000.
2. REAL. Numbers that have decimal points in them such as: 456.25, 64.95, 12345.6789.
3. FLOATING POINT. A way to express numbers in exponential form and to handle them inside the computer as a number with an exponent and a mantissa.   Exponential notation is written in the form 12.345E3, for example. This is a real number equivalent to 12345 because the E3 means that the number is multiplied by 10 three times.   Most microprocessors do not handle floating point arithmetic internally because it requires considerable computing capability. Instead, it is either simulated by software or a separate floating point circuit is used to provide the capability. Floating point is a valuable capability in robotics because of the complex mathematical analysis required for manipulator arm control.
4. CHARACTER.   Letters of the alphabet, both uppercase and lowercase, and some special characters such as dashes, question marks, and so on.
5. STRING.   Groups of characters, which may include numbers. Examples are "This is a string,"   "The price is $56.25," and "123456789abcde."
6. BIT. The ability to address and test bits individually. This is a valuable feature for robotic systems. It allows efficient reading and testing of sensory information, which may be expressed in individual bits.
7. LOGICAL. A data type that can have either of two values, TRUE or FALSE (may be represented by a 1 or a 0 in some cases). This data type is useful in doing any logical analysis and is especially important in Artificial Intelligence applications.
8. BCD. Binary coded decimal representation of numbers. Used more often in commercial operations than in scientific and control operations.

Any of the preceding data types (except bit) may be handled as bytes, words, or double words.   Many computers provide all three capabilities. Bytes are nominally 8 bits, words are 8, 16, or 32 bits, depending on the computer; and doublewords are twice the length of single words. Doublewords are useful in handling very large numbers or complex numbers when multiple-digit accuracy is desired.

Table 7.5 lists some of the important characteristics of existing microprocessors. Of these, control structures are especially important in robotics applications. Semaphores are means for synchronizing operations working together or concurrently (operating at the same time). When a robot arm is to hand something to another robot arm, for example, a fairly complex sequence of signals must be transmitted between the robot control systems to ensure that they cooperate and do not collide with each other. Semaphores are protected signals used for this type of signaling, and provide increased security and speed of concurrent operation. Although they can be simulated by software, system efficiency is decreased. Interrupts, described in Section 7.5.3, and traps are similar. They are necessary for efficient handling of external operations and sensory information. Procedure calls are useful in efficient handling of subroutines because they provide automatic handling of the addressing required. Supervisor calls allow the program to address the operating system efficiently and are a valuable feature in robotic systems.

7.6.2
Advanced System
Organization

Humanity never seems to be satisfied. As we get faster computers and more machines to serve us, the need for more and better machines seems to expand. This is also the situation with robots. The present robots are simple compared to the complex machines we can expect to see in a few years. In this section, we will look briefly at some of the needs for improved systems and possible ways to approach them. It is clearly beyond the scope of this book to describe how these machines might work, but it is of interest to discuss existing needs and the developments in process to fill them.

As we saw in Chapter 6 and as illustrated in Table 6.1, the computation required to control one robot arm can quickly exceed the capacity of a powerful microcomputer, a minicomputer, or even a mainframe computer. When we consider the computation required for the interaction of several robots, each with a vision system and tactile sensors, we see that it may become necessary to have more computing capability than is readily available in one computer. Even more computing power becomes necessary if we consider artificial intelligence, speech recognition, or other new areas of robot application.

Two basic approaches are used to provide more computing power. We can make faster individual computers or use more computers together. Faster individual computers are certainly feasible. New semiconductor circuits have been made that operate in 5 picoseconds, which is 1,000 times faster than circuits in common use today. However, it appears that using several computers together may be a quicker and less expensive way to solve some problems.

Multiple computers can be used in parallel, each doing part of a computation; this is called <u>multiprocessing</u>. Alternatively, they can work in serial and process information in assembly-line fashion, this is called <u>pipelining</u>.

Some computations can be broken down into pieces and done all at once. Matrix multiplication is an example. We can multiply a row of one matrix by a column of another matrix without interference between the operations. In principle, the matrix multiplication of Equation 46 in Section 6.6.1 can be done by four microprocessors working together if we figure out how to coordinate their operation, a problem we will not discuss here. Figure 7.39 illustrates one way of doing matrix multiplication with four processors and two data streams. A systolic array is made up of processors connected to do a certain class of problems as efficiently as possible.

A systolic array can be designed to do pipelining efficiently. In Figure 7.40, one processor element (PE) can do 5 million operations per second (MOPS). With six PEs in series, it may be possible to do 30 MOPS with the same type of processor.

In general, there is no linear increase in speed with multiple processors on the same problem, that is, six processors may not speed

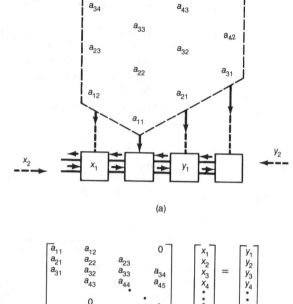

Figure 7.39 Linearly connected systolic array for computing the product of matrix *a* by vector *x*. (a) Linear systolic hardware; (b) Computation (from Haynes et al. [9], Figure 1, p. 10).

Source: L.S. Haynes, R.L. Lau, D.D. Siewiorek, and D.W. Mizell, "A Survey of Highly Parallel Computing," IEEE Computer, Volume 15, Number 1, January, 1982, Figure 1, page 10. ©1982 *IEEE. Used by permission.*

$$\begin{bmatrix} a_{11} & a_{12} & & & 0 \\ a_{21} & a_{22} & a_{23} & & \\ a_{31} & a_{32} & a_{33} & a_{34} & \\ & a_{43} & a_{44} & a_{45} & \\ & & & & \cdot \\ 0 & & & & \cdot \end{bmatrix} \begin{bmatrix} x_1 \\ x_2 \\ x_3 \\ x_4 \\ \vdots \end{bmatrix} = \begin{bmatrix} y_1 \\ y_2 \\ y_3 \\ y_4 \\ \vdots \end{bmatrix}$$

(b)

up the operation by six times. Problems of synchronization, overhead processing required for control, development of a suitable program (or algorithm), contention between processors for access to memory, or input/output problems may reduce the speedup to half as much as would be expected. If overhead is a big problem, there may be a limit on the total number of processors that can be used.

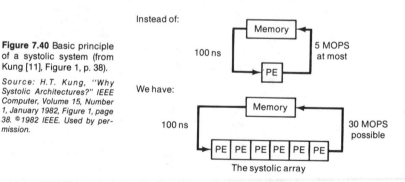

Figure 7.40 Basic principle of a systolic system (from Kung [11], Figure 1, p. 38).

Source: H.T. Kung, "Why Systolic Architectures?" IEEE Computer, Volume 15, Number 1, January 1982, Figure 1, page 38. ©1982 *IEEE. Used by permission.*

Figure 7.41 is a graph of the speedup obtained for some actual multiple processor systems. Harpy is a speech recognition system developed at Carnegie Mellon University. Quicksort is an algorithm or procedure widely used for sorting large files. Both of thse flatten out at a speedup of about four, even though six to eight processors are used together.

It is possible, for some types of problems, to get speedups of thousands of times. The Massively Parallel Processor (MPP), described in Chapter 9, has 16,384 processors working in parallel. It has a speed of about 6 billon operations per second, a speedup on the order of 5,000 times. This high speed makes certain types of vision processing feasible.

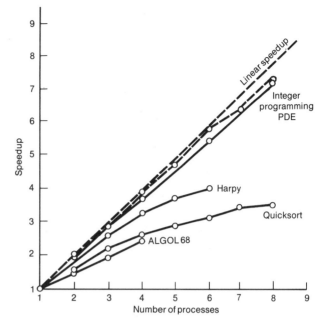

Figure 7.41 Performance of some sample programs using multiple processors (from Haynes et al. [9], Figure 8, p. 14).

Source: L.S. Haynes, R.L. Lau, D.D. Siewiorek, and D. W. Mizell, "A Survey of Highly Parallel Computing," IEEE Computer, Volume 15, Number 1, January 1982, Figure 8, page 14. ©1982, IEEE. Used by permission.

7.6.3
Networks

There are two basic types of networks: local area and long-distance networks. A local area network (LAN) is a data communications system that provides communication capability between a number of independent local devices. It might be used in one building, in several buildings on a site, or within a local area usually limited to a few thousand feet. LANs are usually owned by one organization. A LAN for a robotic system might allow communication between all of the robots, machines, and control units on a factory floor, for example. Data rates in a LAN are in the moderate to high range up to as much a 10 million bits per second.

One popular LAN is the Ethernet. It is a passive system of stations located at specific devices or computers. Each station has a transmitter and receiver (transceiver) hooked to a common coaxial cable. They use a system called Carrier–Sense Multiple Access/ Collision Detection (CSMA/CD). In operation, all the receivers listen on the coaxial line for a specified period of time. If they do not detect a carrier signal from some other device's transmitter, they start to send. Since there is a finite transmission time on the cable, it is possible for two stations to transmit at the same time. In that case, there is a "collision". Each detects the collision and waits a random length of time, and then they try again. This system has been found to be simple to implement, relatively inexpensive, and reliable in operation.

There are other LANs using various operating rules. Another method is a master–slave arrangement where one processor controls the network and determines which stations will transmit.

Long–distance networks may be provided by a common carrier such as Telenet or by one of the telephone companies such as American Bell. In this case, they operate like any other long–distance network, and customers pay tariffs for their use. Switching, message routing, flow control, and other functions are provided by the long-distance network.

Local networks may be connected to long–distance networks, as shown in Figure 7.42. Interfaces are required between the local

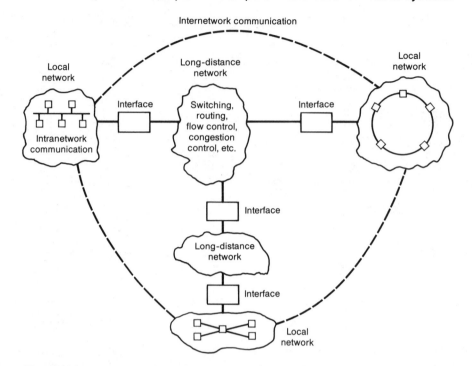

Figure 7.42 Connection of local networks and long-distance computer networks (from Schneidewind [27], Figure 2, p. 16).

Source: N.F. Schneidewind, "Interconnecting Local Networks to Long Distance Networks," IEEE Computer, Volume 16, Number 9, September 1983, Figure 2, page 16. © 1983 IEEE. Used by permission.

networks and the long distance networks to provide a way to match capabilities, data rates, signals, and so on.

A comprehensive discussion of local networks and long–distance networks is given in Schneidewind [27]. As robotic systems become more complex, they can be expected to use both local networks and long–distance networks to provide data processing and control functions.

Exercises

These exercises will test your knowledge of the information and concepts covered in this chapter.

1. Name the major robot system functions performed by computer hardware.
2. Without looking at the book, write out the truth tables for the AND, OR, and NOT logic circuits with inputs A and B and output C. Do all of them require two inputs?
3. Could a complete computer, including memory and control, be made from a sufficient quantity of AND, OR, and NOT logic circuits? Assume that there are printed circuit boards to connect the circuits properly and, that suitable power sources are available.
4. Write out DeMorgan's laws and show that an AND gate can be implemented by use of the proper configuration of OR and NOT gates. What logical circuit would you suggest to provide the equivalent of the AND circuit?
5. Make a truth table for the digital logic circuit shown in Figure 7.43. Use A and B as inputs and X as the output.

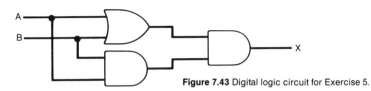

Figure 7.43 Digital logic circuit for Exercise 5.

6. What is the primary use of flipflops in a computer system?

7. Name four types of flipflops.

8. What are the functions of multiplexers and decoders? In what important characeristics do they differ?

9. Shift registers can shift bits in at one end and out the other, with bits moved in either direction. Although the subject received little discussion, what arithmetic instructions might make use of this capability?

10. What are the four main types of memory? Which types of memory are volatile and which are nonvolatile?

11. Describe the instruction Fetch sequence in the Intel 8080 microprocessor. Would it be possible to overlap the Fetch and Execute cycles in a microprocessor? How might this be done? (This is not discussed in the book and requires some original thinking.)

12. Add the following binary numbers:

```
A = 1 0 1 0 1 0 0 0 1
B = 0 1 0 1 0 0 1 1
```

```
C =
```
Convert the value of C obtained into an octal number using the values in Appendix A.1. What is the hexadecimal representation of C?

13. Using binary value A given in Exercise 12, derive the 1s complement.

14. Determine the 2s complement of binary value B in Exercise 12.

15. The S–100 bus is one of the most popular buses used for small microcomputers. What functions can be performed by the eight vectored interrupt lines on this bus?

16. What are the advantages and disadvantages of magnetic disks and tapes in a computer system? Which type of storage is most useful in robotic control systems?

17. Matrix printers are generally much faster than daisy wheel printers, and also less expensive. Why are daisy wheel printers used, and where are matrix printers best?

18. The RS–232 interface is widely used in computers and in robotic systems. What are its characteristics?

19. What is parity, and what types of parity can be used in the RS–232 interface?

20. What are the advantages of using interrupts compared to polling to determine which external devices require service?

21. Read the description of the Westinghouse APAS system. (Chapter 10 and Ref. [28].) What significant factor was overlooked in the system's planning and became very important in the final system?

22. What is meant by a data type? Which data types, in your opinion, would be most often used in a robot control system?

23. Several kinds of multiprocessors are being used in robotics for vision and control. Why are multiprocessors needed?

24. How could a local network be used to improve the efficiency of a robotic system?

References

[1]   W. Barden, Jr., The Z-80 Microcomputer Handbook, Howard W. Sams, Indianapolis, Ind., 1978.

[2]   A. B. Bortz, "Robots in Batch Manufacturing - The Westinghouse APAS Project," Robotics Age, January 1984, pp. 12-18.

[3]   Conference Proceedings, 13th International Symposium on Industrial Robots and Robots 7, April 17-21, 1983, Chicago, Ill. Robotics International of SME, Dearborn, Mich. 1983.

[4]   Cromemco Inc., Excellence in Microcomputers, 1984, Mountain View, California.

[5]   Series 1300 HyType II Printer, Models 1345A (etc.) Maintenance Manual, 82403-02, July 1977.   Diablo Systems Inc., A Xerox Company, 24500 Industrial Blvd., Hayward, California, 94545.

[6]   Diablo   Systems,   Inc.,   HyTerm   Communications   Terminal, Model 1610/1620, Product Description, 82332 Rev. G. December, 1978. Hayward, California.

[7]   Diablo Systems Inc., Maintenance Manual 82403-03, 1982.

[8]   A. Gupta and H. D. Toong, "Microprocessors - The First Twelve Years," IEEE Proceedings, Vol. 71, No. 11, November 1983, pp. 1236-1256.

[9]   L. S. Haynes, R. L. Lau, D. D. Siewiorek, and D. W. Mizell, "A Survey of Highly Parallel Computer," IEEE Computer, vol. 15, no. 1, January 1982, pp. 9-24.

[10]   The 8086 Family User's Manual, October, 1979, 9800722-03, Literature Department Intel Corporation, Santa Clara, CA 95051.

[11]   H. T. Kung, "Why Systolic Architectures?" IEEE Computer, vol. 15, no. 1, January 1982, pp. 37-46.

[12]   H. Lechtman, R. N. Nagel, L. E. Piomann, T. Sutton, and N. Webber, "Connecting the PUMA Robot with the MIC Vision System and Other Sensors." In [23], 1982, pp. 447-466.

[13]   S. Leibson, "The Input/Output Primer," Byte. The six parts are referenced separately in [13] to [18].

[14]   Part 1, "What is I/O?" February 1982, pp. 122-145.

[15]   Part 2, "Interrupts and Direct Memory Access," March 1982, pp. 126-138.

[16]   Part 3, "The Parallel and HPIB (IEEE-488) Interfaces," April 1982, pp. 186-208.

[17]   Part 4, "The BCD and Serial Interfaces," May 1982, pp. 202-220.

[18]   Part 5, "Character Codes," June 1982, pp. 242-258.

[19]   Part 6, "Interrupts Buffers, Grounds, and Signal Degradation," July 1982, pp. 35-46.

[20]   A. Lesea and R. Zaks, "Microprocessor Interfacing Techniques," 2nd ed., Sybex, Inc., Berkeley, California 1978.

[21]   M. M. Mano, Computer System Architecture. Prentice-Hall Inc., Englewood Cliffs, N.J., 1976.

[22]   E. C. Poe and J. C. Goodman, The S-100 and Other Micro Buses. Howard W. Sams, Indianapolis, Ind., 1981.

[23]   PUMA (1981) UNIMATE® PUMA™   Robot, Vol. 1 Technical Manual 398H1A, October, 1981 Unimation, Inc., Danbury, Conn. 06810.

[24]   Robots 6 Conference Proceedings, March 2-4, 1982, Detroit, Michigan, Robotics International of SME, Dearborn, Mich. 1982.

[25]   D. H. Sanders, Computers in Society, 3rd ed., McGraw-Hill, New York, 1981.

[26] A. C. Sanderson and G. Perry, "Sensor-Based Robotic Assembly Systems," IEEE Proceedings, vol. 71, no. 2, July 1983, pp. 856-871.

[27] N. F. Schneidewind, "Interconnecting Local Networks to Long Distance Networks," IEEE Computer, vol. 16, no. 9, September 1983, pp. 15-24.

[28] R. N. Stauffer, "Westinghouse Advances in the Art of Assembly," Robotics Today, February 1983, vol. 5, no. 1,pp. 33-36.

[29] P. Villers, "Recent Proliferation of Industrial Artificial Vision Applications." In [3], 1983, pp. 3-1 to 3-20.

[30] Zaks, Microprocessors, From Chips to Systems, 2nd ed., Sybex, Inc., Berkeley, California 1977.

Robot Software

In this chapter, we will discuss the software required for control and programming of robotic systems. Chapter 7 described some of the hardware used and the ways in which the hardware system could be organized. Now we will categorize the various classes of control and programming software that have been found to be most useful and discuss some problems encountered and potential solutions.

8.1 Robot Software Requirements

Software is defined as all of the procedures, programs, and control methods used in a system that are not hardware. Therefore, the operating manuals and procedures would also be part of software. Other important software includes the overall control systems, diagnostics, debugging aids, and software development support programs.

8.1.1 Need for Programming and Control

Earlier chapters described the control of nonservo robots by setting points, and the teach-by-guiding and teach-by-teach-box methods used for servo-controlled robots. All of these methods require the operator to spend considerable time and effort preparing the robot for a task.

Sometimes it is not practical to teach a robot how to perform a task by any of the preceding methods. Examples are robots that are performing tasks at high speed where speed is critical to the success of the operation, robots working in hazardous areas, and large robots that have reaching abilities greater than a human's.

Spray painting, for example, must be done at a controlled speed to ensure that the amount of paint is not excessive, causing runs, or insufficient, so that the surface is not properly covered. Robots are taught by guiding, but this can be done only at human speeds. High-performance operations must be taught in some other way. When teach boxes are used, it is difficult to get the fine control that is required. Another example is robot welding. In both of these cases, the optimum speed may be beyond the capability of human operators to teach, and the necessary scaling of speed is not easy to predict or control.

In hazardous operations, it is necessary to keep the operator safe while training the robot. This may be time-consuming, expensive, or impossible. Also, it is obviously impossible for a small human to lead a larger than human-sized robot through its tasks. Again, we must seek a better way to control the robot.

Teaching by teach boxes has been used in all of these areas. This method cannot provide the close control needed in many cases. When sensory information is required to assist in robot control, there is a further problem of transferring the sensory information to the robot.

For these and other reasons to be discussed, it is desirable and often essential that robots be controlled by computer programming done in advance. Some authors have called this off-line programming or preprogramming to differentiate it from the programming that occurs during teaching. However, in most computer applications, we are accustomed to programming the operations in advance. Thus, in this book, we will use programming to mean the usual function of providing computer instructions in the proper sequence to carry out a predefined task. In cases where confusion could result, we will use preprogramming for programming done in advance using a program language.

8.1.2 Monitors and Operating Systems

Control of the computer system itself is provided by monitors and operating systems. Monitors are used in smaller systems where there are only a few control functions to be performed. Operating systems are used where the functions to be performed are complex and involve large amounts of data manipulation. Most microcomputer systems use monitors. Most minicomputers and large mainframe computers use operating systems.

Both monitors and operating systems are provided by the computer system manufacturer and are designed to optimize the characteristics of a particular system.

1. Monitors

Some commands used by monitors to perform useful functions are:

- Memory Commands. Cause specific blocks of memory to be displayed, modified, moved, filled with specific characters, compared to other blocks, tested, or located by identified contents.
- Input/Output Commands. Read data from selected input ports, send data to selected output ports, or examine the contents of storage registers in a selected port. Ports are the points of entry and egress between the external world and the computer. They have specific characteristics for receiving, storing, and transmitting data, as described in Chapter 7.
- Program Control Commands. Examine a designated computer register, set a breakpoint in a program, go to a specific location in a program, single-step (one program step at a time) through a program, and do hexadecimal airthmetic on data in locations to aid in program control.
- Disk Utility Commands. Startup of the monitor system by reading it in from a disk storage location (this is usually called booting up the system or booting) is a major function. Other important functions are reading from specific disk locations, writing to specific locations, and formatting a disk to a specific layout of sectors and tracks.

Booting up the system is required every time it is turned on. Usually a part of the monitor program is stored in ROM. This part of the monitor [called the Initial Program Load (IPL) or boot program] has enough capability to control reading of the complete monitor from a specified location in the disk file to a specified location in the high-speed memory. In the complete monitor are all of the instruction steps required to carry out the functions just listed.

In some systems, the monitor is started with a specific command that causes the computer to go to the ROM location, read in the initiating part of the monitor, and then transfer control to the boot program. The boot program calls in the remainder of the monitor from disk files as needed and sets all the values required to ready the computer to carry out its tasks. It is possible to make the boot-up function automatic so that turning on the computer causes it to prepare for operation automatically.

2. Operating Systems

Operating systems provide the same basic functions as monitors but provide many others in addition. Task scheduling, control of program execution, loading of programs, and program call-up are some basic capabilities of an operating system. Larger operating systems, such as the IBM Operating System, will perform all input/output functions on command from the programmer, manage memory space and disk file space, and interleave operation of multiple programs in memory at the same time, all without further intervention from the programmer. Other functions provided are library lookout of programs, compiling of programs written in higher-level languages (HLLs) to provide object code at the level of machine code, control of multiple terminals, and allocation of memory space and disk space as required. Larger computers running multiple programs at the same time and serving relatively inexperienced programmers need powerful operating systems to make them run efficiently. Languages such as ADA, to be discussed later, need support from a large operating system in order to perform their programming tasks.

8.1.3 Assembly Languages and Higher-Level Languages (HLLs)

All computers operate at the level of machine code and logic levels. As described in Chapter 7, the commands in a computer are made possible by logical circuits that carry out a sequence of operations. Machine-level commands do simple tasks and are controlled by coded binary (or hexadecimal) numbers that identify a particular command.

The next step above machine code is Assembly Language, which has a one-to-one correspondence with machine code. For example, the assembly language command ADD may be converted to the hexadecimal machine code AB, which is further converted to the binary number 1010 1011 and used to control the switching of logic circuits in the computer. Early computers worked directly in machine code, and it was necessary to code all commands in 1s and 0s, a very tedious task. Then someone decided to give the binary codes a mnemonic name such as ADD and write an Assembler, which would accept words such as ADD and SUB and convert them to the proper machine code. As a by-product, it was found possible to let the computer keep track of the location of commands in memory and look up various locations when another location referred to them in its program. In assembly language, there is an alphanumeric code for every binary code, so that the assembler program can be kept simple. This means, however, that a separate line of assembly language code must be written for every machine instruction. Also, the programmer is required to know many details of machine operation in order to write assembly language code which will work properly.

Higher-level languages (HLLs) provide much stronger programming capability and simplify the task of programming. FORTRAN was the first HLL to be written, and several others have been developed since. Well-known commercial data processing HLLs are COBOL, PL-1, APL, and PASCAL. These languages all have the capability to express in 1 line of a program the same information that would require

2, 5, or even 25 assembly language steps. For example, the FORTRAN statement

$$M = N + K$$

will generate the following assembly language commands:

LOAD N,A   (Go to memory location N, get the number there, and put it in register A in the computer)

LOAD K,B   (Go to memory location K, get the number there, and put it in register B in the computer)

ADD A,B   (Add the contents of register A to the contents of register B by using the Adder logic unit)

STORE B,M   (Store the contents of register B in memory location M)

Here we have one line of FORTRAN generating four lines of assembly language code. In addition, the FORTRAN compiler has kept track of all of the variable names and has a neat little table of their actual locations in memory. Also, the compiler checks for many errors that the programmer might make and flags them. A great saving of time occurs, and the programs are much easier to write and much more likely to run without error.

A price is paid for this service; a compiler must be written for each language and adjusted to work with a particular computer design. Also, some computer time is taken to do the compiling. In addition, the assembled language generated by the compiler is usually less efficient (takes longer to run and uses more memory) than the same program written directly in assembly language. When HLLs were first invented, computers were expensive, memory was expensive, and programmer time was comparatively inexpensive, so that many computer owners were reluctant to use HLLs. Now computers and memory are much cheaper and programming time is expensive, so that direct assembly language coding is done only when speed and memory space are critical. Even then, only part of the program is hand-coded; the rest is still written in an HLL.

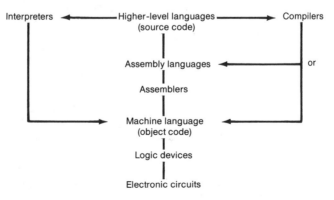

Figure 8.1 Levels of languages.

Some of the important relationships between the levels of language are illustrated in Figure 8.1. At the top are the HLLs in which programs can be written most efficiently. These can be converted to assembly language and to machine code by the use of compilers, interpreters, and assemblers. Then the machine code is seen to be controlled by logic devices that are made possible by the circuitry of the computer and related equipment.

8.1.4
Compilers and
Interpreters

Compilers accept HLL input, called source code, such as FORTRAN, COBOL, or ADA, and generate assembly language code or machine language code (object code) directly. They must work on a whole program at a time because they typically make several passes through the program to do the compiling. On the first pass through a program, the compiler makes up a list of all of the variable names and address names used in the program. On the second pass, it checks for references between names and addresses, and makes up a cross reference table. On the third pass, it assigns memory addresses and sets up the detailed addresses in memory for every instruction used and every constant value. With this approach fairly efficient object code is generated, but the process may be long and involved for a large program. Also, every time a change is made in the program, the program must be recompiled. In some language compilers, it is possible to recompile only part of the program. However, it is still a separate operation, so that a change cannot be made in a program while it is operating.

Interpreters are simpler programs to write than compilers and easier to use. An interpreter takes one line of code at a time, generates the equivalent assembly language or machine language, and then causes it to be executed by the computer. If a mistake is found, the interpreter will not carry out a particular instruction but will instead signal the user that there is a mistake. It is possible for a programmer to sit down at a terminal and work interactively on a program. This means that a few instructions can be written to describe a task and then run on the computer to see if they will work. If not, it is a simple matter to change them and try again. The programmer is in control of the machine.

But interpreters work slowly and must do some tasks repeatedly. If they have interpreted a program segment once and find another just like it a few times further on, they must reinterpret the lines of code. A compiler would discover that two parts of the program are the same and just recopy those parts rather than repeat the same steps. As a result of this and other factors, interpreted code is much slower than compiled code. However, the good news is that, for many robot tasks, high speed is not important and the flexibility of interpreters is. In those cases in which speed is important, it is possible to compile portions of the program separately, and keep them on file. Usually the interactive parts of a program, where the programmer wants to go step by step, are the parts where slow computer operation is acceptable.

In the sections describing robot languages, we will discuss the use of compilers versus interpreters for individual languages. Some languages have both available; some have one or the other. There are advantages for each type.

8.1.5
Types of
Users

Robot software must be designed for use by many types of users. Some people who may use robots are: factory workers who know how to do a task but have little or no knowledge of computers; factory foremen experienced in factory operations; manufacturing and industrial engineers with some technical background; experienced part programmers for numerically controlled (NC) machines; computer-aided designers who are familiar with graphics and design programming; and computer science graduates and Ph.Ds with formal training in computer science but little experience in applications. All of these people will approach the robot control task differently. There is a need for language capability to allow all of them to do their jobs.

There have been some advocates of very simple languages for use by the factory worker. RPL, designed at SRI International, is a simple, easy-to-use language, but it does not have the capability to do complex tasks requiring concurrent operations and system integration. At the other extreme is AUTOPASS, a language partially developed at IBM, which would allow complex tasks to be specified in simple language. AUTOPASS has not yet been made workable but has developed some useful ideas.

As we review the available languages, it is important to keep in mind who the users might be. Powerful languages are required for some applications, but it is not feasible to have all programming done by computer science Ph.Ds; there won't be enough of them. Probably the best approach is to provide two or three levels of language that are compatible so that the right level is available for each application.

If the languages have the proper capabilities, it should be possible for sophisticated programmers to work with knowledgeable users to provide special-purpose programs designed to make particular applications easy to program even at a complex level. It is certainly feasible to have many different subroutines or programs that can do complex tasks but can be called upon by simple programs.

**8.1.6
Definitions
of Terms**

Since this book is to be used by many different people who may not be familiar with computer terminology, we will define here some of the more common terms used in robot control and programming. Many other terms are defined in the Glossary in Appendix C.

1. LOAD. Put a program or command into the computer registers or in memory. A common assembly language command, LOAD, is used to load a particular piece of information in a particular computer register. We also speak of loading a program into memory when it is to be run.
2. STORE. Put information into memory from a computer register.
3. SAVE. Copy a program from memory into secondary storage such as a disk file. The original program is still in the computer memory.
4. START. Start a program running. Another command frequently used to cause a program to run or execute is RUN.
5. CLEAR. Clear or erase a memory area or set a value in a register to 0.
6. ENABLE. Make operational so it can be used. A function is turned on so that it can be used, but it does not actually start until another command such as RUN or GO is entered.
7. DISABLE. Make non-operational so that it cannot be used even though a signal is applied to start or run. Often used as a safety precaution.
8. DOWNLOAD. Move a program from a higher-level memory or computer to a lower-level memory or computer. Often it is necessary to compile a program in a larger computer so that it can be run on a smaller computer. It is transferred to the smaller computer by downloading.
9. DEFAULT. A path or command that is normally followed unless

specifically changed by the operator or programmer. By default we may use the Cartesian coordinate system unless specifically told to use another coordinate system. Alternatively, the default speed is "20 inches per second." Default commands are often built into the program or into the monitor to simplify operation for unskilled users.

10. BRANCHING. A change in the path of a program due to a decision made by the program. Every computer language has some means to change its program. As an example, "IF force GREATER THAN 60 THEN STOP" might be used as a protective branch to prevent overloading of a robot arm. Simpler programs may say "WHILE A > B MOVE JOINT 6" to control the movement of Joint 6 as long as value A is greater than value B. Older programs used GO TO commands to transfer control of the program from one statement to another statement that might be anywhere else in memory. Such commands were a source of much confusion and error. Branching could be absolute or conditional. The examples just used are conditional. GO TO 12345 is an absolute command to transfer control to memory location 12345. Since memory names can be used, we could call the location 12345 by the name JOE and say GO TO JOE to get the same effect.

11. SUBROUTINES. A program designed to perform a specific task. Subroutines can be called by other programs, other subroutines , or even by themselves to perform their specific task. Subroutines that call themselves are said to be operating recursively. Some major programming languages implement all of their functions as subroutines. The operating system or other control program has a list of subroutines and their locations in memory. When the subroutine name is recognized in a program, the subroutine is called into a particular place in memory and given control of the computer system. It operates until it completes its task and returns control to the control program, which then reads the next statement in the program and continues.

12. PARAMETER PASSING. When a subroutine is called, it must have certain data to operate on. This data is supplied by the calling program through parameter passing. For example, a subroutine may be used to calculate the square root of a quantity. In the program, it may be written

$$B = SQRT(A)$$

The control program will recognize the SQRT as the call for the square root routine and the A in parentheses as the parameter or value whose square root must be taken. It will then call the SQRT routine, give it the value of A, and get back from the square root routine value B, which it will place in the proper memory location before going to the next step in the program.

Complex subroutines may have several values such as (X, Y, Z) as parameters. It is also possible to tell the subroutine to go to a certain place in memory and read a certain number of values in sequence to obtain the information needed to carry out a command. For example, a subroutine could be used to move an arm through the next 10 points as stored, beginning at memory location BEGIN if the programming language was set up to do this. In the AML language developed by IBM, it is possible to write a MOVE command using aggregates that specify the joints to be moved and the positions to which they should be moved.

8.2 **Levels of Programming**

8.2.1
Sequence
Control

All control in a sequence–controlled machine is done by mechanical or electrical sequencing devices such as timer motors, stepping switches, and relays. By our definition, there is no programming required. Sequence control has very little flexibility since all of the operational steps are wired in or controlled by mechanical stops or other fixed means. However, there are a large number of automatic machines operating under sequence control. This method has the major advantages of being inexpensive and easy for shop personnel to set up and operate.

8.2.2
Teach-by-
Guiding
(or Teach-by-
Showing)

At present, over 90% of robots are still controlled or programmed by being led through a sequence of operations, as described in Section 6.2.2. The term teach-by-guiding is preferred to teach-by-showing because the robot is guided through a path rather than shown a path visually.

Teach-by-guiding is a proven technique, easily performed by shop people who understand the task. It can be done with simple equipment and controls. Also, the teaching is performed quickly and is immediately useful. In the playback mode, the robot repeats the path and operations that were stored in memory during the teach operation. The operation can be repeated as many times as needed. In some systems, it is possible to play back at different speeds than the speed that was used in the teach mode.

It is possible to use teach-by-guiding to work with a single conveyor, or other moving device, if a signal is available from the conveyor to synchronize the operation of the robot and the conveyor. Several paint conveyors have been used in this way. The robot is taught a spray pattern with the conveyor moving at a constant speed. Then in operation, as the conveyor reaches a prespecified point, it closes a switch that starts the robot timer. At a specified later time, the robot goes through its learned pattern and sprays the objects on the conveyor. This has been a valuable training method easily used by skilled spray painters.

However, there are some disadvantages to teach-by-guiding:

1. It can be done only at human speed.
2. It is difficult to incorporate sensory information.
3. It cannot be used in some hazardous situations.
4. It is not practical in handling large robots.
5. High accuracy and straight-line paths are difficult to achieve.
6. Synchronization with other operations may be difficult.

Some of these difficulties can be overcome by using a Teach Box.

8.2.3
Teach by
Teach Box

Push button controls mounted on a control box can be used to drive the robot through a desired sequence of operations and record the result as also described in Section 6.2.2. In one teach box, for example, there is a pair of buttons for each joint. Pressing the left button causes the joint to move in one direction at full speed; pressing the right button causes the joint to move in the opposite direction at full speed. An auxiliary control is often provided to control the maximum allowable speed during the training operation. It is difficult to move more than one joint at a time in this teaching mode, although it is usually desirable to provide coordinated moves for maximum robot efficiency. Joysticks, such as those used on video games, can be used to provide controllable joint velocities in several directions, but they too have drawbacks. (Encoders or potentiometers in the control box are moved by the joysticks to control the velocity and direction of the individual joints.)

Teach boxes can be used to control large robots and robots operating under hazardous conditions, but it is still difficult to obtain high accuracy, synchronization with other devices, or appropriate reactions to sensory information.

The next improvement in control, therefore, was the move to pre-programming or off-line programming of robots as discussed in the next two sections.

**8.2.4
Off-Line
Programming or
Preprogramming
—Single Robot
Arm**

Preprogramming and offline programming are synonymous, and indicate that the programming is done in advance by the use of a robot programming language rather than being done in a teaching mode. The advantages of preprogramming are:

1. It can prepare programs without using the robot, so that the robot is available for other uses.
2. Layout and cycle time of the operations can be optimized in advance.
3. Previously worked-out procedures and routines can be incorporated in the program. It is not necessary to redesign each operation every time it is used.
4. Sensors can be used to detect external stimuli, and appropriate action can be taken in response. This action would increase the complexity of programming, and the robot would be considered to be in the adaptive mode of operation, as described in Section 8.2.7.
5. Existing Computer Aided Design (CAD) and Computer Aided Manufacturing (CAM) information can be incorporated into the control functions.
6. Programs can be run in advance to simulate the movements actually programmed without incurring the risk of damage. There are computer display techniques, for example, that simulate the movements of the robot on a screen to assist in program development.
7. Robots can be used to manufacture individual units by calling on previously developed routines. It would be impractical to do this by guiding and attempting to match different paths.
8. Engineering changes can be incorporated quickly by substituting only the part of the program that controls a particular design characteristic.

Controlling a single robot arm to carry out a task allows the program to be greatly simplified. In a nonadaptive system, there is no sensory feedback from the environment, so that the only inputs are the position sensors in the joints. Therefore, a simple program design can be used. The VAL program, described in Section 8.5.1, as well as several others, have been found to be very useful in this mode of operation.

**8.2.5
Programming—
Integrated
System**

An integrated system is one that has multiple arms and/or auxiliary equipment operating together to perform a task or series of tasks. It is difficult to imagine an integrated system using multiple robot arms in which teaching-by-guiding or even use of a teach box could be implemented successfully. Synchronizing the operations of the arms requires a complex series of interactions. If one robot is to pick something from a conveyor and place it in a machine, and then the second robot is to take the object from the machine after a specified time and place it on a take-away conveyor, we need a series of signals: item on the conveyor ready to be picked up, machine empty, robot has picked up the part, and so on. Programming is required to coordinate these tasks and send control commands at the correct times. A programming language to control an integrated system must have the necessary commands available.

More complex integrated systems would require some type of sensory feedback, which leads to the next level of programming: adaptive control.

8.2.6
Adaptive
Systems

An adaptive system is one that modifies its operations in response to sensory information from its environment. Sensors may measure force, torque, or proximity, or provide complete vision capability, as discussed in Chapters 5 and 9. Programs in the system control computer accept sensory input and control the robot's operations through appropriate commands.

More complex sensing action could involve multiple sensors of force, torque, and so on. A useful adaptive action would be <u>active compliance</u>, in which a robot arm or gripper may slide along a surface, maintaining a controlled force against the surface, until in encounters a hole in the surface or runs into a vertical wall. Much work has been done by researchers on active compliance, as discussed by Lozano-Perez [21].

8.2.7
Intelligent
Systems

Adding a knowledge base, expert system capability, and task planning to an adaptive system brings us to the level of intelligent systems, in which the system itself determines how to carry out actions. At this level of operation, the human controller may give a command such as "Put the red block in the middle of the yellow table," and the intelligent system would carry out the command. It would know the concepts <u>red</u>, <u>block</u>, <u>yellow</u>, <u>table</u>, and <u>middle</u>, and have sufficient sensory capability to recognize the objects. Then it would plan the sequence of moves necessary to carry out the operations required and perform the task. Similar tasks have been performed by Shakey, as described in Section 2.3.1, Part 1. Task planning is discussed in Lozano-Perez [20], and to some extent in Lozano-Perez [21].

8.3 Functions Performed by Programming

Robot systems perform complex, coordinated tasks requiring the planning of movements, sensing of internal and external conditions, and continuous control of the motions of individual elements. Coordination of movements is required to ensure that elements work together properly. It is also necessary to make frequent checks to ensure that operations are performed properly and safely. In advanced systems, there is a need for <u>world modeling</u> to provide the system with a means for planning operations. Data processing functions are necessary to maintain suitable records and communicate with the control hierarchy in a factory or other organization. Interfacing is required between system elements such as manipulator arms, vision cameras, conveyors, and devices external to the system.

Suitable programming and control languages must have provisions to carry out all of these tasks accurately and efficiently.

These functions have been categorized for purposes of exposition. The reader should keep in mind that there are many other possible ways to organize this information. Different systems and applications will undoubtedly be structured in a variety of ways.

Note that it is definitely not necessary to use one language for all functions. It is only necessary that the individual controllers used perform their tasks well and communicate with other controllers as necessary. Current practice is for individual elements to have their own computers with a specific language defined for a particular task. A good example of this specialization is in the vision systems that provide identification and position information to the associated manipulator controller.

In a parts-handling application using the VAL language, as described in Raymond [30], the VAL program uses a single command, PICTURE, to command the associated vision system (a Machine Intelligence Corporation V-100) to scan an area, analyze the features of objects in the area, and prepare a table identifying the objects and their locations. This table is then used by the VAL program to carry out its task of directing the PUMA robot in parts handling. VAL is a relatively simple language used for manipulator control; the vision system uses a sophisticated programming language specialized for this use.

**8.3.1
Motion
Control**

In the simplest robot system, there is a manipulator with some type of end effector, a workpiece or object to be manipulated, and an environment (workspace) in which the work is performed. A robot controller program must control the motions of the manipulator and end effector to carry out a task in the specific workspace. Control of the manipulator involves computing the motion path or trajectory, opening and closing the gripper of the end effector, and avoiding collision with objects in the workspace.

1. Types of Motion

Robot motions can be classified into four types: sequential joint control, slewing or uncoordinated joint control, interpolated joint control, and straight-line motion.

1. Sequential joint control means that the joints are moved separately, in a prescribed sequence. There is no interaction between movements, but the move time is usually unacceptably long.

2. Slewing all joints at the same time is faster, but the path of the end effector is not controlled and depends on the relative velocity of the individual joints. The possibility of damage and collisions is too great when long moves are to be made.

3. Interpolated joint control ensures that all of the joints start and finish together. Thus, the path is contained within a small volume, but the motion is erratic and not suitable for some tasks.

4. Straight-line control is slower because of the need to move the arm at the speed limited by the slowest joint, but is required for tasks such as welding along a seam. Controlling a straight-line path requires that the robot controller do a series of complex calculations at a rate of 20 to 80 times per second to ensure that

the path is straight within the required tolerance. Straight-line motion is so valuable that many systems have now made it available.

Programming languages often provide separate commands for interpolated joint control and straight-line control so that the faster move can be used when straight-line control is not necessary. The uncoordinated move and sequential moves are seldom used.

2. Trajectory Planning

Trajectory information input to a program is in the form of starting coordinates, coordinates of the intermediate points on the path or trajectory, and ending coordinates. The intermediate points, called vias are those through which the trajectory must pass to meet the requirements of the task. These vias may be arbitrary and it may only be necessary to come within a short distance of them in order to prevent collision with external objects. Therefore, it is desirable that the language have a way to specify allowable tolerances at these points. The mathematical problem of trajectory calculation is increased whenever high accuracy is required.

Homogeneous matrices are used, as described in Section 6.6, to state the kinematic equations that describe the trajectory of the robot arm. The programming language is required to provide a way to supply the information necessary to perform the calculations. The physical shape of the robot manipulator and other equipment can be specified by constructs in the language. Stating positions and orientations of each joint in reference to a base, by use of homogeneous matrices, is the most common way of specifying these items.

Actual calculation of the trajectory requires complex matrix manipulation, matrix inversions, and the solution of simultaneous differential equations either exactly or approximately. This complexity must be provided in the language, but it must be hidden from the user. A MOVE command may be a complex subroutine, but to the programmer it must look simple and easy to use. Examples of these commands are given in Sections 8.4 and 8.5.

3. Velocity and Acceleration Control

Often it is desirable for the robot arm to move at a specified velocity (for example, to match the velocity of a conveyor) or to accelerate and decelerate at controlled rates. These capabilities can be provided in the program in several ways: as overall (global) specifications that apply to all joints or limits set for each joint, or as specified starting and ending accelerations and maximum velocities. In the IBM robot language AML, there is a provision in each type of MOVE command to specify speed, acceleration, deceleration, and settling time, so that the velocity profile of each point or the overall velocity profile of the end effector may be specified. Acceleration control may be necessary to reduce the force on the arms themselves or on the objects being carried.

4. Force and Torque Control

When an unexpected obstacle is encountered or when a part is being fitted to another part, it is desirable that the amount of force exerted by the robot arm be controllable. Torque control is required for tightening bolts and similar tasks. The programming language should be able to specify forces and torques in three orthogonal directions to provide the necessary capability.

5. End Effector Control

Measurement of the opening in the gripper of the end effector is necessary so that it can be opened to pick up an object. If a vacuum pickup is being used, it is desirable at times to control the effective vacuum in order to pick up objects selectively. Other end effectors may have similar control requirements; magnetic pickups require the current to be set at a particular value, for example.

6. Obstacle Avoidance

A major problem in the control of robots is obstacle avoidance. Either analysis or simulation can be used to determine possible collisions and ways to avoid them. Both require some kind of modeling of a predicted path of the robot or its appendages with reference to either stationary or moving objects in its environment.

Stationary objects are, of course, easier to avoid. A straight-forward way is to draw (or extend by analysis) a clearance volume about each object that is sufficient to allow the robot to pass. This volume is dependent on the protrusion of the robot from known centerlines and reference points. Sometimes, in predicting the path of the robot, it is necessary to backtrack and try a new path because the modeling indicates that there will be a collision along the current path.

Moving objects require more complex modeling. Both the robot's movements and that of surrounding objects must be predicted. Except for a few special cases, little work has been done on this problem.

If a model of the workspace is stored in the computer memory, it is possible to plot a trajectory to move the arm without encountering any of the known obstacles in the workspace volume. Alternative collision avoidance algorithms are discussed in Section 6.3.2. Modeling of the workspace can be done by using lists, tables, or other means. The programming language should provide modeling capability to make obstacle modeling and avoidance possible.

Vision, acoustic, and tactile sensors may also be used to detect obstacles, as discussed in Section 8.3.2, so that a workspace model is not necessary. However, efficient ways of programming obstacle avoidance are still required.

7. Motion Coordination

Multiple robot arms, conveyors, feeders, and so on must operate together to carry out a task or many related tasks. Each robot arm may have a separate computer, with an overall computer to coordinate the operations. Signals are required to allow the multiple devices to report status and coordinate functions. The programming language can provide these signals in many ways: storing a word in a particular memory location, setting a bit in a common control register, or other means. Such controls are built into the robot languages AL, ADA and AML, for example.

Increasing sensory capability is being provided for robots and robot systems. Early robots had simple tactile switches to indicate the presence or absence of an object. Proximity sensors soon followed, and now there are many systems with full vision capability. These sensors, discussed in Chapters 5 and 9, make possible adaptive control of the robot system. It can change its operation in response to the conditions of its environment.

In order to use sensors effectively, the sensory input must be converted from raw data to useful information. The first task of the control computer is to gather data from the sensors, convert it to a standard form, analyse and record it for later use. This operation can be done with the usual computer commands and is not a difficult problem.

An important provision, needed for sensory control, is a powerful interrupt system that can monitor the sensors for particular events and be programmed to take necessary action. Older systems used a polling technique that periodically, perhaps once every 100 milliseconds, asked each sensor what it had to report. This was a slow and inefficient process. An interrupt-driven system allows each sensor to interrupt the computer program when an important event takes place. If the force on the gripper exceeds a specified value set by the program, for example, the program is interrupted and branches to a prepared subroutine automatically. The subroutine can then take action to stop the gripper drive before it crushes the object being handled. (Interrupts are described in Section 7.5.3 in more detail.)

Four major functions are provided by the sensing system. They are discussed in the next four sections.

1. Initiating and Terminating Actions

When sensory information is used for control, the program must specify the conditions of the sensors required to initiate and terminate an operation. These may be logical conditions or analog conditions that have been converted to digital values. The programming language must be able to handle these conditions in a simple and powerful way. Some languages provide a full range of logical commands for this use. In many languages, it is difficult to write a statement such as:

MOVE to Position 6.5 IF
Sensor $A > 2.5$ and Sensor B is On.

Clearly, in larger systems, these statements may be very complex and may need to be executed as subroutines.

2. Choosing Among Alternative Actions

Again, the ability to consider both logical conditions and values is required to write statements such as:

IF GRIPSENSE 5 = TRUE AND FORCE on
JOINT $5 < 65$ CONTINUE MOVE ELSE STOP,
PRINT ERROR MESSAGE, "FORCE ON JOINT 5 TOO LARGE".

This statement may be used to test the conditions repeatedly on joint 5 when the gripper is closed. Alternatively, it could be set up as an interrupt condition depending on the expected frequency of use.

3. Determining the Identity and Position of Objects

In working with a vision system, for example, the robot controller must tell the vision system what to look for and receive back from the vision system the necessary description of the scene and the objects in it. The robot's task may be to pick up an object with specific features from a conveyor and place it in an assembly fixture. It can then ask the vision system to look at a particular area of the conveyor and report what it sees. When the object with the desired features appears, the vision system will report the features and the location of the object, along with its orientation on the conveyor. Then the robot controller can direct the robot gripper to pick up the object and carry out its task.

Rates of data flow between the robot and the vision system may be high, so that a parallel data path may be necessary. Output from the vision system may be stored in a particular location in memory space accessible both to the vision computer and to the control computer. In a possible sequence of operations, the vision system would locate the object, store the required information in memory, and then interrupt the robot controller to inform it that the object had been found and that its coordinates were stored at a particular memory location. Then the robot controller could take over and complete its task.

4. Constraint Compliance

To illustrate the problem of handling constraints, let's consider the insertion of a round peg in a hole, as illustrated in Figure 8.2. The peg is held in a gripper mounted on the wrist assembly of a robot arm. We wish to move the peg in such a way that it will go into the hole in the block.

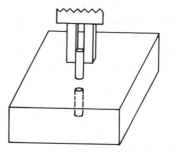

Figure 8.2 Peg-in-hole insertion.

There are two types of compliance considered in robot applications. _Passive_ compliance is purely mechanical. Section 5.3.7 describes the Remote Center Compliance device developed by Draper Laboratories. In this case, the mechanical structure of the device becomes deformed, so that the part being handled moves appropriately to ease the task being performed. This is the fastest way to implement compliance, but it does not work for all types of objects.

Active compliance requires that a sensor detect the presence of a constraint and signal the robot controller. Multiple sensors may be used to detect the forces and torques on the part being handled, so that the robot controller can move the part to minimize the forces being applied. A wrist sensor, as shown in Figure 5.20 provides six values: three forces and three torques. This sensory information can be used by the robot controller to determine the best path to follow in order to insert the peg in the hole. This is a difficult programming task in general but can be done in special cases.   Note that the program has to work with forces and torques as input to make decisions as to what to do at each step of the process.

Some languages do not provide for compliance operations, VAL, the Unimation language, performs many industrial tasks without having compliance capability. Even the advanced IBM language, AML, can handle only simple compliance control problems. The Stanford language, AL, has the most complete capability for compliance control. Compliance, however, is essential for a complete programming language. Robots must be able to maintain controllable forces and torques while operating at different velocities. This capability requires that force and torque information from sensors be monitored and be explicitly or implicitly provided for in the program.

An example of a portion of a compliance control program in AL (from Lozano-Perez [21]) used to test whether there is a hole beneath a peg being carried by the <u>barm</u> (blue arm) is

> MOVE barm to α -1 * inches
> (the " α " sign indicates current location)
> ON FORCE (zhat) > 10 * ounces
> DO ABORT ("No Hole")

The arm is moved down 1 inch from its current location, while the force sensor is interrogated for force in the Z direction (zhat). The arm continues to press on the peg until it either goes into a hole or the force exceeds 10 ounces. When the force is greater than 10 ounces, the error message "No Hole" is returned. Force information is measured in three orthogonal directions. Action in the program is taken on the basis of the forces measured, if required.

Another example indicates how downward force can be exerted while allowing movement of the object with respect to side forces (Lozano-Perez [21]):

> MOVE peg-bottom to hole-bottom DIRECTLY
>   WITH FORCE-FRAME = station IN WORLD
>   WITH FORCE (zhat) = -10 * ounces
>   WITH FORCE (xhat) = 0 * ounces
>   WITH FORCE (yhat) = 0 * ounces
>   SLOWLY;

In this case, the program is requiring a downward force of 10 ounces, but is allowing the peg to move since no x or y forces are specified. SLOWLY is a velocity control parameter that can be set by the program. The term station IN WORLD refers to the world coordinate system in which the model is embedded.

Compliance is required in turning cranks, tightening bolts, opening doors, assembly, and similar functions. Several mathematical approaches are discussed by Mason [23].

8.3.3 System Supervision and Control

In simple robotic systems, one computer may be used to control all functions, including monitoring. Most systems today use multiple computers, one or more for each joint, and a supervisory computer for a complete robot arm. Integrated robotic systems containing multiple arms and other equipment often have a separate computer to monitor and control the entire system.

The control computer provides monitoring and error checking of the operations of individual units, coordinates the activities of subsystems in performing various tasks, and interfaces with external devices and computers. In addition, it processes the data resulting from the system's activities and stores it in usable form for later retrieval. Part of the processing may be to keep track of work performed, time required, difficulties encountered, or errors found.

1. Supervision of Motion Control and Adaptive Control

Motion control functions are specified in general terms by the supervisory computer in an integrated system. The motion control computer provides the necessary computations and control functions. The system control checks to see that the operation has been performed as requested.

Performance of the supervisory functions means that the robot language requires error control and test commands. These commands should be able to find and report the error and its cause.

2. Interfacing to External Devices

Input and output from and to external devices may be in a different digital code than that used internally. Conversion commands are needed in the programming language to perform the necessary conversions. Typically, these commands should include binary-to-BCD conversion, ASCII to EBCDIC conversion, and others.

3. Coordination Functions

Multiple tasks, multiple robot arms, multiple processors, and auxiliary equipment must be coordinated in integrated systems. Flags, semaphores, and concurrency control capability are needed in the language to provide this coordination. Some processes must follow each other sequentially; others must start together and end together.

Concurrency of operations is provided for in AML, ADA, AL, and some versions of PASCAL. If two robot arms are to work together on the same task, they must be able to signal to each other. AL, for example, has COBEGIN and COEND commands, which can be used to tell two arms to start at the same time and carry out a task jointly (see Section 8.5.2).

4. Data Processing and Storage

Robotic systems must process data of many kinds if they are to be an effective part of an industrial hierarchy, as described in Section 1.6.3. In addition, a data base of information must be available to guide the system in making decisions. In Section 8.3.4, we discuss the need for knowledge representation and world modeling to support intelligent activities. Therefore, in addition to the usual data processing to maintain parts count, work done, time spent on each activity and so on, it may be necessary to store a detailed model of the environment for later use. In order to meet these requirements, the robotic programming language needs all the usual capabilities of a business data processing language, as well as the additional capabilities of a robotic language just discussed. It may be possible, however, to use separate languages, each optimized for its own task and capable of communicating data effectively to the other language. Since it is likely that there will be many computers in the control hierarchy, the use of multiple computers and multiple languages may be the ideal solution.

Tomas Lozano-Perez [20] has developed the task planning approach to robot systems and has shown the need for a way to allow the computer to plan the detailed activities of the robot system to alleviate the load on human programmers. The AUTOPASS Language developed by IBM is another attempt to provide task-level control of robot activities (Lieberman and Wesley [19]). As yet, neither of these approaches has been proven, but there is great promise for future research to solve the problems that still exist.

Intelligent activity or task planning, which is a simpler subset, requires that a world model be available to represent the environment, the robotic system, and the tasks to be carried out. Integration of this information requires that a means of knowledge representation be used to structure the voluminous amounts of information required. The subject of knowledge representation is complex and is being researched actively in the United States, Japan, and many European countries. McCalla and Cercone [24] introduce a special issue of IEEE Computer magazine on "Knowledge Representation." Torrero [39] introduces a special issue of IEEE Spectrum on "Next Generation Computers." Both discuss Artificial Intelligence and knowledge representation developments and applications.)

1. Knowledge Representation

Four general approaches to knowledge representation are discussed by McCalla and Cercone [24]: semantic networks, first-order logic, frames, and production systems. Analogic representations are added as an additional technique that does not fit well in any of the first four.

Semantic networks are described in Section 9.9.1. First-order logic, as exemplified by a language called PROLOG, developed in 1972, uses logic expressions to state relationships and then manipulates the expressions to determine new relationships implicit in the previous ones. Frames are taxonomic or classification structures that maintain hierarchical networks of relationships that can be manipulated by specific rules. They can be represented by linked lists or the trees of conventional programming. Production systems use a large number of individual rule-based relationships that can be manipulated in sequence to discover new relations. All of these techniques and several variants are described and discussed in the special issue edited by G. McCalla and N. Cercone [24]. Direct or analogic representation, semantic networks, and production systems are discussed by Arthur Sanderson and George Perry [32] who have applied knowledge representation techniques to sensor-based robotic assembly systems.

2. World Modeling

The following information is required to describe a suitable world model for task planning:

1. Geometric descriptions of the objects and their coordinate relationships in the robot workspace or environment.
2. Mass, inertia, and other physical characteristics of all objects in the environment if they will be moved during the operation.

3. Kinematic descriptions of robot linkages and other moving items.
4. Descriptions of robot characteristics such as joint ranges, velocity, acceleration capabilities, and sensor capabilities.

3. Task Specification

Specification of the task to the planner by the programmer requires that appropriate commands be provided in the programming language to do the specification unambiguously. Since unambiguous specification is difficult, it is desirable to provide a means for interactive communication between the programmer and the task planner program. This has been done by T. Winograd in his Blocks World program, as described by Winston [43]. An example of a blocks world situation is given in Figure 8.3.

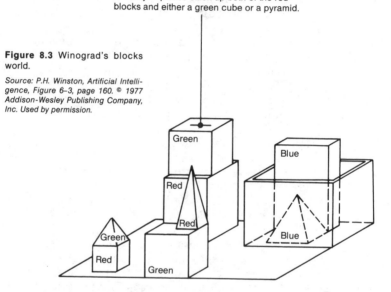

Will you please stack up both of the red blocks and either a green cube or a pyramid.

Figure 8.3 Winograd's blocks world.

Source: P.H. Winston, Artificial Intelligence, Figure 6–3, page 160. © 1977 Addison-Wesley Publishing Company, Inc. Used by permission.

In this figure, there are two green blocks. If the programmer commands "Pick up the green block and put it on the blue block," for example, the program planner would not know which green block to get. In an interactive program, it might ask, "which green block?" just as a human would do in the same situation. If the answer was "the green block under the red pyramid," the planner could then carry out its task. Winograd has written programs that show good understanding of the block world and can carry out complex orders regarding it. However, his program is not a general–purpose one since it does not have a model of any other environment.

4. Decision Making

Given the task specification, the task planner uses various approaches to decide what to do. These programs are similar to the expert systems used by Artificial Intelligence researchers. Many of the most successful planners use production systems to do the actual planning. Production systems have a global data base, production rules, and a control system. The global data base represents the knowledge in the system about its environment. Production rules are of the form

IF a condition exists THEN do the following

The control system tests the conditions in the data base against the production rules and performs the action specified.

In the blocks world of Figure 8.3, the data base would contain a description of all of the blocks and their coordinates. Some production rules might be

IF a block is to be moved THEN test to see if it has another block on top of it.

IF a block has another block on top of it THEN move the top block.

IF a block is to be moved THEN find a place to move it to.

Note that these rules interact. Each block to be moved may have another block on top of it. Thus, the control program has to have procedures that will examine the production rules in the correct logical order depending on the situation it finds.

8.4 Present Robot Languages—Features and Problems

This section will review the evolution of robot languages and discuss the features of some of the languages in use in the United States. In addition, some of the problems that have been found in current languages will be pointed out. No attempt will be made to list all of the languages that have been developed and used. Several review articles have evaluated and tabulated the characteristics of robot languages. Bonner and Shin [2], Gruver et al. [13], and Lozano-Perez [21] provide the most comprehensive reviews.

8.4.1 Evolution of Robot Languages

Robot programming languages have been developed in three ways: as completely new languages, as adaptations of existing programming languages by making modifications and additions to the syntax or rules of the language; and by adding subroutines to an existing language to provide useful additional capabilities. In some cases, both syntax modifications and new subroutines have been added.

Some entirely new languages developed for robot control are MHI, AL, and AML. A new data processing language, ADA, is now being adapted to robot control even though it is hardly active in other applications as yet.

Robot languages still in common use that were produced by modification and extension of existing data processing languages are: WAVE, VAL, PAL, RAIL, HELP, JARS, RPL, and MCL. Many other languages have been developed and used but are not widely used at the present time. In addition, some manufacturers have developed their own versions of languages similar to those discussed. Several of these languages—AL, AML, HELP, JARS, MCL, RAIL, RPL, and VAL— are described and evaluated in Gruver et al. [13].

We will discuss each of these languages briefly in this section. In Section 8.5, more complete descriptions of VAL, AL, and AML are provided. These languages are representative of three types of existing languages and provide an excellent review of the capabilities required and the ways in which these capabilities can be used.

1. MHI (1960–1961)

The first robot language capable of describing operations with programmable commands was MHI, developed at MIT by H. A. Ernst in 1960–1961 [7]. This new language was developed to control the Mechanical Hand One (MH1), which is described in Section 2.3.3 and the beginning of Chapter 5. Since the MH1 had several binary sensors, the language provided for moves controlled by sensory operations. Available language commands were:

MOVE—indicating a direction and speed
UNTIL—operate until a specific sensor condition occurs
IFGOTO   branch to a new location in the program on a specific condition
IFCONT—branch to continue action on a specific condition

These commands were adequate to perform several operations, including reaching for an object, grasping it, moving to a new location, and placing the object as directed. There was no provision in the language for arithmetic or control commands other than sensor-based ones.

2. WAVE (1970–1975)

WAVE, developed at Stanford University, was the first general-purpose robot programming system (Paul [27]. It was a new language modeled somewhat after the assembly language of the Digital Equipment Corporation (DEC) PDP-10 computer in use at Stanford University at that time. During the 1970–1975 period, WAVE ran off-line and produced programs for on-line use on the PDP-8. Algorithms developed for control of the robot were complex and "computationally intense," so that they could not be run on-line in real time. This was a pioneering effort. Much has been learned since that time. The present algorithms are more efficient, so that the computation problem is simpler today. In addition, today's computers are considerably faster.

WAVE had several important features that have been adopted, with some changes, by later languages. The most important features were:

1. Specification of compliance capability in Cartesian coordinates.
2. Coordination of joint motions to provide for continuity of velocities and accelerations through trajectory turning points or via points.
3. Description of end effector positions in Cartesian coordinates.
4. Guarded move capability, so that sensory information could be used to terminate a move when the end effector touched something.

3. AL

"AL is a high-level programming system for specification of manipulatory tasks such as assembly of an object from parts" (Finkel et al. [8]). It has an ALGOL-like source language, a translator to convert programs into runnable code, and a runtime system for controlling manipulators and other devices. The compiler is written in an HLL, SAIL, and the runtime system is generated on a powerful minicomputer. Trajectory calculations are done at compile time and are modified during runtime as necessary.

AL has had a considerable effect on other languages and has emerged as one of the leading contenders for a common robotic language. It was developed at Stanford University in 1974 and has been improved and tested since then (Finkel et al. [8], Binford et al. [1], Mujtaba and Goldman [25]).

Force sensing and compliance are implemented by a number of subroutines and by condition monitor statements in the syntax of the language. There are signal statements and wait statements available when one process must wait for the completion of another process. These and other statements make possible the coordinated operation of two or more robot arms. Arm and hand movement commands are available to control moves, velocities, forces, and torques. Two or more objects can be handled as one by the use of AFFIX commands that cause them to appear as one object. Data types are available to define scalars, vectors, rotations, frames, and other characteristics of the robot and its environment.

An example of a robot program in AL is (from [21]):

```
BEGIN
    FRAME PICK UP; {block is presumed here}
    FRAME PUT DOWN; {block is desired here}
    (Frame definitions are inserted here as textually produced by
        POINTY, the program for defining positions interactively.)
    OPEN bhand to 3.5* INCHES
    MOVE barm TO pick up
        VIA pick up + 4*ZHAT*INCHES;
    CENTER bhand; {close the fingers, grasping the block}
    MOVE barm to put down
        VIA pick up + 4*ZHAT*INCHES
        VIA put down + 4*ZHAT*INCHES;
    OPEN bhand to 3.5*INCHES; {release the block}
    MOVE barm to bpark
        VIA put-down + 4*ZHAT*INCHES;


END;
```

A COBEGIN and a COEND construct are available so that two operations can be controlled at one time. A WAIT command is used to allow one arm, for example, to wait for another arm that is handing something to it. There are also commands for definition of DURATION, STIFFNESS, and FORCE FRAME identifiers.

A lengthy description of AL, including AL programming examples, is provided in Section 8.5.2.

4. POINTY

POINTY is an interactive system for programming AL data structures to work directly with a compiled program to allow interactive operation. It can be considered part of the AL system and is described with AL in Section 8.5.2.

5. VAL (1975–Present)

Another language developed by augmentation of an existing language (BASIC) is VAL, developed by Victor Sheinman for the Unimation Corporation. VAL uses simple words to describe operations to be performed by the robot. When used with the PUMA robot manufactured by Unimation, the language looks like this:

```
;
; VAL example:  PUMA hands a block to BARM
;
    OPEN 100            ;open the hand 100 mm.
    APPRO PICK, 50      ;approach the block, location name is PICK
    SPEED 30            ;slow down for next motion
    MOVES PICK          ;go straight for the block
    GRASP 20            ;close hand to 20 mm
    MOVE HOLD           ;hold out the block to the position, HOLD
    SIGNAL 5            ;tell BARM we're ready
    WAIT 6              ;wait tell BARM's got it (event no. 6)
    OPEN 100            ;release the block
    DEPART 100          ;pull back out of the way
    SIGNAL -5           ;tell BARM we're clear
    MOVEPARK            ;alldone, go to the parking position
```

This example is easy to understand. Once we know that BARM refers to blue arm, the whole routine seems clear. VAL provides for speed control, grasping, arm movement, and signaling in a simple, easy-to-understand framework. Of course, all of these routines are implemented by subroutines written in BASIC and translated by an interpreter. Higher efficiency could be obtained by the use of a compiled BASIC if necessary, but usually it is sufficient to call precompiled subroutines. The need for a signal to other robots and machines is handled effectively by the use of the WAIT and SIGNAL commands, which take action on the occurrence of specific events.

The PUMA system uses six microcomputers, Motorola 6502s, to control the six separate motions of the arm plus a DEC LSI-11 16-bit microcomputer to control the overall system. A disk file, a keyboard, and a teach box provide storage, input, and teaching capability for the PUMA robot. The teach box has push button or joystick controls to move the arm through a sequence of steps. When the sequence of steps has been worked out to the operator's satisfaction, he or she can store the digital representation of the steps in computer memory and in the disk file. This is a powerful method of programming. Many users in the industry prefer the teach box to writing programs. In VAL, both teaching and programming can be used in an effective way.

Considerably more information is provided on VAL in Section 8.5.1, including a description of commands and additional programming examples.

6. AML (1977-Present)

AML was designed by IBM software engineers specifically for control of manufacturing processes, including robots. AML provides a systems environment in which different user robot programming interfaces may be built. It supports joint space trajectory planning, subject to position and velocity constraints. Relative and absolute motion can be handled, and sensor monitoring can interrupt motions as necessary.

The most unique capability of AML is its operations on data aggregates, so that many operations on vectors, rotations, and coordinate frames can be handled as multiple operations in one command. This capability makes the language more difficult to understand but simplifies programming and control.

Since AML is expected to be one of the most important robotic languages, a detailed description of it is provided in Section 8.5.3.

7. PASCAL

PASCAL is a standard programming language for data processing that uses <u>structured</u> statements to organize the program in a modular and efficient way. It also has strong <u>data typing</u> to protect the programmer from errors. The structured statements force the program into separate modules to reduce the number of branches. Each module is self-contained and can be entered only at the top and exited only at the bottom, so that errors due to branching mistakes are easier to catch and eliminate. Data types are the descriptors that identify elements in the language such as INTEGERS, REAL NUMBERS, CHARACTERS. It is a well-known and widely used language, and was chosen as the basis for several robot languages. There are several textbooks available in addition to the original manual by Jensen and Wirth [15]. One of the easiest books to use is by Welsh and Elder [42]. A more comprehensive book is by Tenenbaum and Augenstein [38].

8. PAL

Richard Paul, who was instrumental in developing WAVE and the AL language at Stanford, decided to add capability to PASCAL to provide a new type of programming language. He and his associates at Purdue developed PAL by extending the syntax, or set of rules, of PASCAL to include MOV, TOL, ARM, and other commands. This allowed them to take advantage of the proven, well-known PASCAL language and yet provide new capability (Takase et al. [35].

In the PAL data structure of Figure 8.4, the relationship between the arm and the object can be given explicitly by the use of homogeneous matrix transforms. Every motion statement in PAL causes the manipulator position and orientation to be specified in an equation representing a closed kinematic chain. (See Section 6.6 to review homogeneous matrices and kinematics.)

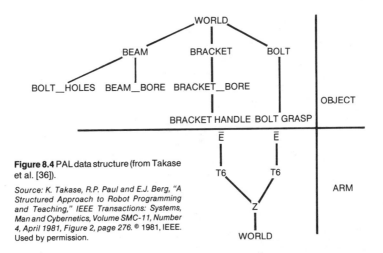

Figure 8.4 PAL data structure (from Takase et al. [36]).

Source: K. Takase, R.P. Paul and E.J. Berg, "A Structured Approach to Robot Programming and Teaching," IEEE Transactions: Systems, Man and Cybernetics, Volume SMC-11, Number 4, April 1981, Figure 2, page 276. © 1981, IEEE. Used by permission.

An interesting and potentially valuable capability of PAL is the ease with which unknown points can be picked up from a task and incorporated into the program. Essentially, the robot can be used interactively to make measurements. Then a program can be run to incorporate these measurements and carry out a task using them.

Every motion statement in PAL causes the solution of homogeneous equations that represent the position of the kinematic chain of links and joints in the arm. As shown in Figure 8.4, values E, T6, Z, and WORLD represent coordinate transforms in the manipulator. These values are related to the coordinate transforms for the object, as shown in the top part of Figure 8.4, which ultimately link back to the WORLD coordinate system. The position of any object or joint is obtained, and can be specified, by multiplying together the coordinate transforms represented by homogeneous equations. For example, the statement MOVE ARM TO BRACKET-HANDLE is equivalent to the equation

$$Z + T6 + E = BRACKET + BRACKET-HANDLE$$

where the plus sign indicates matrix multiplication and a minus sign would indicate matrix multiplication by the matrix inverse. The position of the base of the manipulator is specified in world coordinates by the transform, Z. The position of the end of the manipulator with respect to the base of the manipulator is specified by T6 where this value is obtained by the matrix multiplication of all of the joint transforms of the manipulator. E is the transform relating the end effector grasping point to the end of the manipulator.

Starting again from world coordinates, we require BRACKET + BRACKET-BORE + BRACKET-HANDLE to get to the same position so that this set of terms is equated with the left side of the equation.

ARM and TOL define the locations of the end of the ARM and the center of the tool, respectively. Both are defined in terms of matrices. It is important to remember that the plus signs represent matrix multiplication in Figure 8.4 and the minus signs represent matrix inversion. One current problem with PAL is the difficulty some programmers have in handling the representation used. This issue is discussed in Bonner and Shin [2].

9. MCL (1978-Present)

MCL was developed in 1978 by McDonnell-Douglas in conjunction with Cincinnati Milicron for an Air Force ICAM Project. This language is based on APT, the control language for numerically controlled machine tools used in CAM (Wood and Fugelso [44]). One of its advantages is that a large number of programmers already use APT in CAM applications. APT has specific constructs to specify POINTS, LINES, PATTERNS, and so on, which are useful in enabling robots to communicate and cooperate with machine tools. These capabilities have been carried over to MCL, with some specific new constructs added.

An example of a subroutine in MCL is:

```
FORM/@PRTNAM,@PROJ,@PA,@PG,@OV,@RV

    SEND/ATTACH, 'FGRIP'
    CONECT/ROBOT,FGRIP

    DEVICE/CVSS
    LOCATE/PROJEC,@PROJ

    USEFRM/FGRIP
    WORKPT/0,0,0,1,0,0,ROLL,0,0,1
```

SEND / FGRIP, 'OPEN'

GOTO /@PF,@OV, ROLL,@RV

SEND / FGRIP,'CLOSE'
CONECT / FGRIP/,@PRTNAM

ENDOF / EXTEN

Some advantages of MCL can be seen in this example. It uses recognizable abbreviations that a shop worker can learn quickly, and it has a clearly defined geometry. Also, since it is based on APT, it follows an international standard well supported in the industrial environment. Its chief disadvantage is the use of an older and less flexible syntax than the new languages designed for robotic use. MCL is described by Wood and Fugelso [44] and discussed by Lozano-Perez [21].

10. RAIL

RAIL was developed by Automatix, Inc., for the control of both vision and manipulation. It contains a large subset of PASCAL commands and can handle a variety of data types. An interpeter is used to convert the language constructs to machine language commands. Inspection, arc welding, and vision operations are supported by language capabilities. It is run on a powerful Motorola 68000 system. Use of the RAIL language in the Robovision welding system is described in Chapter 10. (Robovision is a trademark of Automatix, Inc.)

11. HELP

HELP was developed by the General Electric Company for its robot products. It is a PASCAL-like language and supports concurrent processes to control two robot arms at one time. It is comparable to RAIL.

12. JARS

JARS was developed by the Robotics and Teleoperators Group of NASA's Jet Propulsion Laboratory in 1979. The language base is PASCAL, with many specific types, variables, and subroutines added to make it useful for robot control. It provides a high-level programming capability to support basic research in robot control and manipulation. Currently, it runs on a DEC PDP-11/34 using floating point and 28K bytes of memory for robotic programs. It is used to control a modified Stanford Scheinman arm and has been interfaced to PUMA 600.

13. RPL

RPL was designed at SRI International specifically for use by people who are not skilled programmers: factory production engineers, line foremen,and so on. Some of the ideas of the LISP language are used but are organized into a FORTRAN-like syntax. (LISP is a language designed for artificial intelligence use. It is primarily recursive and is especially well adapted to handling lists of information.) The language is implemented as subroutine calls, so that it is modular and flexible in use. It has been used to control a Unimation PUMA 500 arm and the Machine Intelligence Corporation (MIC) vision module.   It has a compiler to produce interpretable code from programs and an interpreter for that code.

14. ADA

ADA was developed for the Department of Defense (DOD) in 1978. It has been specified by DOD as the sole system implementation language for the programming of real-time embedded systems. Chief advantages of ADA are the powerful data structures provided, the ability to separate control operations from data operations, and the inherent capability for concurrent operation of many tasks.   Concurrent operation is an important requirement for the efficient use of multiple robots and other equipment.

ADA is currently becoming available on several microcomputers. Zilog, Inc., for example, has ADA compilers for the Z8001 and Z8002, which are 16-bit microcomputers. However, the ADA compilers must run on a Zilog System 8000, VAX 11 VMS, or UNIX. Then the compiled program may be downloaded to run on the Z8002.   ADA is also available for the Motorola 68000 and several minicomputers.

Volz et al. [41] describe the use of ADA for the control of manufacturing cells using robots. They conclude that ADA is well suited to this task and that efficient programs can result from its use. In addition, ADA is expected to provide code that is easily transferred between computers and can be easily maintained.

**8.4.2
Language
Evaluation
Criteria**

In section 8.3, we discussed the functions to be performed by programming. Now we can ask, "How well do the various languages perform these functions?" This question is one the reader should ask in reviewing each of the languages considered in detail in the next section. Important criteria to be considered are now discussed.

1. Available Control Structures

Can the language provide for structured modular programming? Usually this means that commands such as IF-THEN, IF-THEN-ELSE, WHILE-DO, DO-UNTIL, CASE, FOR, BEGIN-END, and procedure, function, and subroutine capability are available. PASCAL-based languages have these capabilities or alternative means to achieve the same result.   ADA, AL, AML, RPL, and JARS are rich in these capabilities.   VAL, HELP, and MCL have less capability but are adequate for the simpler tasks they are designed to perform. VAL II is expected to provide the necessary capabilities when it becomes commercially available.

2. Multiple Tasking Capability

Robot languages of the future must be able to handle multiple tasks or operations at the same time in order to coordinate the activities of two or more robot arms and/or associated equipment. This means that such commands as COBEGIN, COEND must be available to signal the beginning and end of two or more operations or processes performed at the same time and in a coordinated fashion. Also, the operating system is required to have capabilities for semaphores, or monitors, or mailboxes to allow communication between operations. Some computers have built-in semaphore capabilities, as described in Section 7.6.1. However, the robot language must provide the necessary software commands if this synchronizing and control capability is to be used effectively. Only AL and ADA have COBEGIN and COEND, but HELP and MCL have usable alternatives and AML has recently added multiple arm control capability (Taylor and Grossman [36]). It is expected that many other languages will soon add this capability.

3. Rotation, Translation, and Vector Specifications

Geometric capabilities required for robotic control should be provided in the language for maximum efficiency. Coordinate frame rotation and translation, the handling of vector quantities, and simple means for expressing joint angles are important features. Rotation can be specified by homogeneous matrices, as described in Section 6.6, or by Euler angles, quaternions, or the standard roll–pitch–yaw angles. AL and JARS provide the most complete set of transformations. AL uses homogenous matrix manipulation internally to provide very flexible specification and control. PAL also uses homogeneous matrices but is not a complete language in other respects. AML, RAIL, and VAL use Euler angles, which are simpler to implement than homogenous matrices but less powerful.

4. Motion Control

Guarded moves, constraint compliance, and controllable bias force are desirable features of robot languages. Only AL, AML, and HELP provide guarded move capability, and only AL provides all three capabilities. Guarded moves are a special case of constraint compliance in which the end effector is moved slowly until a sensor detects an object and signals the robot arm to stop. This capability is essential in many robot tasks, including most assembly operations.

Of the types of motion discussed in Section 8.3.1, the most common is slewing, also called coordinated-joint control. Straight-line control is also widely available. These capabilities can be achieved with subroutines and need not be an integral part of the language, although the efficiency of operation is usually improved thereby.

5. Software Engineering Features

A number of capabilities are needed to make robot programming efficient, flexible, and maintainable. Desirable features are a text editor, an interpreter, a compiler, a simulator, a file system, and error logging. Debugging features are also required, such as single stepping, breakpoints, traces, and dumps. AL and AML have the most complete set of such aids. Some of them are described in Sections 8.5.2 and 8.5.3. Each of these categories of commands is described, for specific languages, in the studies by Bonner and Shin [2] and Gruver et al.[13].

8.5 Programming Examples for Existing Languages

In this section, we will give examples of the way in which existing languages provide the functional capabilities described in Section 8.3. Three levels of language will be discussed: those that describe the motion at the end effector level, at the level of the object being moved, and at the task level. Examples to be described at each level are selected from programming languages in daily use, either in industry or research organizations, except for AUTOPASS, the only available example of a task-level language, which is no longer in daily use.

End effector level languages are VAL, RAIL, HELP, JARS, and PAL. They specify what the end effector motion is to be. Object-level languages are AL, AML, and ADA. They specify the motion required of objects being manipulated. The language converts these specifications to the required manipulator motions. A task-level language is AUTOPASS. A task is specified in general terms. The language is required to issue commands in order to carry out the task.

8.5.1
VAL

"VAL is a computer-based control system and language designed specifically for use with the Unimation Inc. industrial robots." It is described more completely in <u>User's Guide to VAL</u> [40], from which this description is quoted. A more advanced version, VAL II, has also been developed and is now in use in some applications.

Since VAL is based on the BASIC language, it uses many of the commands and the general format of BASIC, such as the GOSUB and IF-THEN commands and the REMARKS code word for comments and remarks in the text of the program. Many additional commands have been added to make VAL suitable for robot control. Subroutines can be addressed by GOSUB, and unlike most BASIC languages, the subroutine may be given a name rather than a line number only. RETURN is used in the subroutines to return information to the calling program.

VAL is a real-time system, with trajectory computation being done continuously to permit complex motions to be executed quickly. An interpreter is used to generate commands continuously, so that it can interact with a human operator to provide on-line program generation and monitoring.

1. Types of Arguments Used

Several types of numerical arguments can appear in commands and instructions, subject to the following rules:

1. Distances, measured in millimeters, are used to define locations to which the robot is to move. Since only millimeters are permitted, it is not necessary to state units explicitly. Distance values can be expressed in increments of 0.01 millimeter, either positive or negative, up to the limits of reach of the individual robots—1024 millimeters for the PUMA 500, for example.
2. Angles are entered in degrees to specify orientations at named locations and to describe angular positions of robot joints. Values may be negative or positive, limited by 180 or 360 degrees depending on usage, and specified in 0.01 degree increments.
3. Integer variables can be used with ranges from −32,768 to +32,768. (This is the usual limit for microprocessors using 8-bit words, since integer variables are expressed by two words, providing 16 bits. However, 1 bit is used to express the sign, so 15 bits are used for the number value and 2^{15} has the value 32,768.
4. Joint numbers are integers starting from 1 for base rotation and extending to the total number of joints, including the hand, if it can be opened and closed.
5. Channel numbers identify the external signal lines. They are integers ranging from 1 to the total number of lines.

6. Other integers can be used in commands or instruction and can range from 0 to 327.67. (Effectively, these are integers that are used as decimal numbers in special programming.) They are suitable to express speeds and times, however.

Available modes of operation are JOINT, WORLD, and TOOL modes, corresponding to the available coordinate systems, and FREE mode, which allows the arm to be moved manually. In JOINT mode, the robot can be operated by use of the manual control unit, and each joint can be moved singly if desired or several joints can be moved at once. In WORLD mode, the coordinate system is fixed to the base of the robot. In TOOL mode, the coordinate system axes are fixed to the tool. FREE can be used for a single joint if desired or for multiple joints, as can the JOINT mode. All of these modes can be used under manual control, so that the Unimate programs can be guided through a path by using a teach box or by manual guiding.

VAL programs are entered by using an editor, a program that allows an operator to enter information directly into memory from the keyboard and display it on the control terminal. In the VAL system, this program is called .EDIT. Suppose the task for the robot is to pick up a part from a chute and place it in a box. An example of the use of .EDIT to enter a program called DEMO.1 to do this is the following (underlined characters are entered by the operator):

.EDIT DEMO. 1
.PROGRAM DEMO. 1
1.?APPRO PART,50
2.?

In this example, the .EDIT command was used to call the editor program and assign a name DEMO.1 to the program. The .PROGRAM command was entered to start work on the program called DEMO.1. Each entry is terminated by using a carriage return after each line of text. After the second line, the editor program automatically assigns a line number 1. followed by a ? mark to ask what the command is to be. The operator has filled in the APPRO PART, 50 response, followed by a carriage return, and the editor has responded with 2.? to ask for the second line. This program can be continued to become a full program as follows:

.EDIT DEMO. 1
.PROGRAM DEMO.1
1.?APPRO PART,50
2.?MOVES PART
3.?CLOSEI
4.?DEPARTS 150
5.?APPROS BOX,200
6.?MOVE BOX
7.?OPENI
8.?DEPART 75
9.?

A carriage return after the last question mark tells the editor that this program is complete and ready to execute. The meaning of the program is

1. Move to a location 50 millimeters above the part in the chute.

2. Move along a <u>straight</u> line to the part ("straight" is indicated by the "S" at the end of the command in this and other commands).
3. Close the hand immediately ("immediately" is signaled by the "I").
4. Withdraw the part 150 mm. from the chute along a straight-line path.
5. Move along a straight line to a location of 200 mm. above the box.
6. Put the part into the box.
7. Open the hand ("immediately").
8. Withdraw 75 mm. from the box.

PART has to be defined as a location to tell the system where it is. There are several ways to do this. The simplest is to move the hand to the desired location by using the manual controls and then read the location values into memory by using the .HERE command. It reads X-Y-Z coordinates of the hand and the three orientation angles (0,A,T) of the hand. An example is

<u>.HERE</u> PART

X/JT1	Y/JT2	Z/JT3	0/JT4	A/JT5	T/JT6
34.75	530.47	-23.06	-167.459	57.473	114.247

CHANGE?

(Underlined characters are those entered by the user.)

This is a little confusing because there are two ways to identify locations: either as coordinates and orientations or as the positions of the individual joints. Both outputs are provided for by the column headings. For the .HERE PART command, we read the left side of each heading and ignore the / and the joint name.

After the .HERE command is entered, the coordinates are read by the program, stored in memory as the location of PART, and printed out for the operator's information. Then the program asks CHANGE? to allow the operator to change any coordinate to a new value. Since no change was desired in this example, a carriage return was used to terminate the command.

Location variables can be identified in two ways; as <u>precision points</u> or as Cartesian coordinates and orientations, as previously mentioned. Variables specifying Cartesian coordinates and orientations are called <u>transformations</u>, and are implemented lby the use of homogeneous matrices, as described in Section 6.6.1. Precision points specify the positions of individual joints and provide the greatest precision of location. These variables are identified by a # sign in front of the name; for example, #PART would cause the joint values to be provided if .HERE #PART was used in the preceding location example. As given, the Cartesian coordinates were obtained and used in the program.

Other commands are also available. .STATUS can be used to determine the current speed setting, and .SPEED can be used to change the speed to a new value. The command .EXECUTE DEMO.1 will cause the program called DEMO.1 to be executed to perform all of the commands listed in the program. Programs can be saved on a floppy disk with the .STORE command and can be listed from the disk to the display with the .LISTF command. In both cases, the name of the file is required after the command. VAL has a good set of commands for handling programs, editing, setting formats, and so on.

2. Trajectory Control Methods

VAL has two methods to control the motion path of the robot from one location to another. The first method is Joint Interpolation. This is the fastest motion. All of the joint variables are interpolated in steps between their initial and final points, and each is moved about the same fraction of its total move at each step, so that the path taken does not vary excessively from a smooth path between the initial and final points. The path is complex, and the tip of the hand does not move at a constant speed.

Commands specifying joint interpolation are:

MOVE <location>]

moves the robot arm to the position and orientation specified by either a precision point or Cartesian coordinate-type variable name.

APPRO <location> , <distance>

moves the tool to the position and orientation defined by the variable <location> and an offset along the tool Z axis of the distance given.

DEPART <distance>

moves the tool the distance given along the current Z axis of the tool.

The second method used to control the path of the robot is Straight-Line Motion. Movement in a straight line at a controlled speed is accomplished by interpolating in the world coordinates in small increments that maintain the desired accuracy. The tip of the robot accelerates to a constant speed, travels at that speed, and decelerates smoothly to its final position. To accomplish this, the individual joints may move at different speeds, and accelerations and decelerations are required. Additional computation is required to provide straight-line motion, and the maximum speed is that of the slowest joint or the joint that travels the greatest distance, so that straight-line motion is slower than joint-interpolated motion.

MOVES, APPROS, and DEPARTS are the same as the respective MOVE, APPRO, and DEPART commands, except that the terminal S on each command specifies straight-line motion.

VAL has the ability to move through via points or approach points smoothly if desired. This capability prevents abrupt stopping at unnecessary times.

Hand control is provided by use of the following commands:

OPEN [<hand opening>] or CLOSE [<hand opening>]

open or close the hand <u>during</u> the next motion. The amount of motion is optional, as indicated by the use of the square brackets [, and] around the hand-opening specification. (In many programming languages, square brackets are used to indicate optional actions.) If a servo-controlled hand is used, the hand opening, in millimeters, will be the amount given. Nonservo hands, such as pneumatic grippers, open or close all the way.

OPENI [<hand opening>]or CLOSEI [<hand opening>]

cause the open and close operations to take place immediately, as signaled by the I at the end of each command.

GRASP <hand opening>, [<label>]

allows the amount of hand opening to be specified and tested. The hand will close until it cannot close further because of an obstruction in the path. If the final opening is less than the specified amount, the program can branch to the optional label and continue the program. If not, an error message is displayed. This is a convenient way to determine whether an object has been picked up or not.
        Location assignment and modification are done by a set of commands that use either transformations or precision points to specify the location.

HERE <location>

is equivalent to the monitor command .HERE used in the program example. The location is defined by the use of a name that must have been previously defined.

SET<transformation>=<transformation 2>[:trans 3>] ...
[:trans n>] or SET< precision point> = < precision point>

sets the value of the variable on the left to that on the right of the equal sign.

The transformation may be a simple one, with only one value on the right, or a compound one with optional multiple values on the right. This corresponds to multiple matrix multiplications indicated by the colon (:) and allows specifying locations relative to other locaitons by specifying the series of transformations required to convert to the desired reference base. Since there is low accuracy in the data representation in VAL, the error builds up with each transformation used due to computational error. VAL II corrects this error with the use of floating point representation and more digits of accuracy.

SHIFT <transformation> BY [dx], [dy], [dz]

X, Y, and Z components of the transformation are modified by adding the distances indicated by [dx], [dy], and [dz], respectively.

SHIFT PICK BY 100.8, -35.1

Causes the transformation PICK to be shifted 100.8 in the X direction and -35.1 in the Y direction.

INVERSE<transformation>=<trans 2> [:<trans 3>]...[:<trans n>]

The matrix inverse of the transformation on the right is calculated and becomes the value on the left.

FRAME <transformation>=<trans 2>, <trans 3>,<trans 4>

Note that the values on the right are not optional. All of them are required. The new value on the left is a reference frame with its origin at the point defined by <trans2>, its positive Y axis passing through the point defined by <trans 3>, and its X-Y plane containing the point defined by <trans 4>. A base transformation may be defined as

FRAME BASE=A1,A2,A3

Continuous-path motion is possible using VAL. A command used for welding operations is

WEAVE <distance>, [<cycle time>], [<dwell>]

This instruction sets up the parameters for a sawtooth weaving motion to be superimposed on the subsequent straight-line motion. It continues until terminated by a WEAVE 0 instruction or in certain other situations.

WEAVE 55, 10

initializes parameters for a weaving motion with a peak-to-peak amplitude of 55 millimeters, a 10-second cycle time, and no delays at the ends of the motion. A dwell could be specified at the top and bottom of each sawtooth if desired. The VAL program given in Section 8.4 illustrates the use of many of these commands.

**8.5.2
The AL System
at the Stanford
Artificial
Intelligence
Laboratory
(SAIL)**

Starting in 1973 with the WAVE system for manipulator control developed by Richard Lou Paul, there has been continuous development at SAIL of improved hardware and software for use in robotics. In the WAVE system and the initial AL system, all trajectory calculation and motion control was done by the compiler, and only minor modifications were made at runtime. In the latest version of AL, trajectory calculation is done at runtime since it is now possible, by use of improved hardware and software, to provide real-time operation. The description of AL in this section is based on the 1981 AL Users' Manual, Third Edition, written by Shahid Mujtaba and Ron Goldman [25] except as noted.

1. Design Philosophy of AL

AL was initially written in a symbolic language that was compiled into a runtime program by a set of compiler and system programs. (The latest version, as described in Goldman [25], has an interpreter and no longer needs the POINTY system to be described.) It was based on ALGOL initially but has more data types than other conventional HLLs. Control structures like those in ALGOL or PASCAL are used to provide a block-structured language with no jumps or GOTOs.

As listed in Figure 8.5, there are SCALAR numbers and specifications for VECTOR, ROTATION, FRAME, TRANS, STRING, and EVENT data types. A VECTOR consists of three real numbers specifying the three components of the vector. The data types ROTATION (ROT), FRAME, and TRANS are based on the homogeneous matrices described in Section 6.6.1. A ROT consists of a direction vector and an angle to indicate the amount of rotation. A FRAME contains a position vector and an orientation to describe the location and orientation of a coordinate system. The orientation and position of one coordinate system relative to another are described by a TRANS. All of these data types may be used in the form of arrays.

BLOCKS:
```
BEGIN S; S; S; ... S END
COBEGIN S; S; S; ... S COEND
```

DECLARATIONS:
```
TIME SCALAR ts1, ts2;                LABEL 11,12;
DISTANCE VECTOR dv1,dv2;                STRING st1,st2;
ROT r1,r2;                           FRA,E f1.f2;
TRANS f1,f2;                         EVENT e1,e2;
FRAME ARRAY f1[s1:s2],f2[s3:s4,s5:s6,...];
```

PROCEDURES:
```
PROCEDURE p1; S;
SCALAR PROCEDURE sp1(VALUE SCALAR vs1,vs2;
        REFERENCE ROT rr1;
        SCALAR ARRAY as1[2:3]); S;
```

OPERATIONS:
```
scalar s:      s*s,s−s,s*s,s/s,s↑s,|v|,|,s|,v.v,s MAX s, s MIN s, s MOD s, s DIV s,
vector v:      VECTOR(s,s,s),(s,s,s),s*v,v*s,v/s,v*v,v−v,v*v,r*v,t*v,f*v,v WRT f,
               UNIT(v),POS(f),AXIS(r)
rot    r:      ROT(v,s),(v,s),r*r,ORIENT(f)
frame f:       FRAME(r,v),f*v,f−v,f*f,CONSTRUCT(v,v,v)
trans f:       TRANS(r,v),(r,v),f f,f*f,INV(t)
boolean b:     −b, NOT b, b b, b AND b,b b, b OR b, b*b, b EQV b, b b, b XOR b,
               s<s,s≤s,s=s,s=s,>s,s≥s
dimension d:   d*d,d/d,INV(d)
```

FUNCTIONS
```
scalar         INT(s),SQRT(s),SIN(s),COS(s),TAN(s),ASIN(s),ACOS(s),ATAN2(s,s),
               EXP(s),INSCALAR,RUNTIME,RUNTIME(s)
boolean        QUERY(print list)
```

AL   s: π,Pi,BHAND,YHAND,GHAND,RHAND,DRIVER_TURNS
CONSTANTS   v: XHAT,YHAT,ZHAT,NILVECT
AND   r: NILROT
VARIABLES   f: STATION,BPARK,YPARK,GPARK,RPARK, (valid only in MOVE)
                    BARM,YARM,GARM,RARM,DRIVER_GRASP,DRIVE_TIP
            f: NILTRANS
            b: TRUE,FALSE
      strings: CRLF,NULL
        units: CM,INCH,INCHES,OUNCES,OZ,GM,LBS,SEC,
                    SECONDS,DEG,DEGREES,RADIANS,RPM
   dimensions: DISTANCE,TIME,FORCE,ANGLE,TORQUE,
                    ANGULAR_VELOCITY,VELOCITY, DIMENSIONLESS

STATEMENTS:

comment
```
COMMENT <any text without semicolon>;
{ <any text> }
```

control
```
FOR s  <scalar> STEP <scalar> DO <statement>;
IF <condition> THEN <statement> ELSE <statement>;
IF <condition> THEN <statement>;
WHILE <condition> DO <statement>;
DO <statement> UNTIL <condition>;
CASE <scalar> OF BEGIN S;S;... S END;
CASE <scalar> OF BEGIN [i1] S; [i2] S; ...ELSE j  0; [i3][i4] S END;
```

affix
```
AFFIX f1 TO f2 AT t1 RIGIDLY;
AFFIX f3 TO f4 BY t2 NONRIGIDLY;
AFFIX f3 TO f4 BY t2 NONRIGIDLY;
```

unfix
```
UNFIX f5 FROM f6;
```

condition   ON FORCE (<vector>) <rel> <force scalar> DO <statement>;
monitor   ON TORQUE <rel> <torque scalar> ABOUT <vector> DO <statement>;
statement   ON |FORCE| <rel> <force scalar> ALONG <axis vect> OF f1 DO ...;
and   ON TORQUE <rel> <torque scalar> ABOUT <axis vect> OF f1 IN HAND DO ...;
clauses   ON DURATION ≥ <time scalar> DO <statement>;
              ON ARRIVAL DO <statement>;
              ON DEPARTMENT DO <statement>;
              ON ERROR = <scalar> DO <statement>;
              <label>: DEFER ON <event> DO <statement>;
                    <rel> is ≥ or <

with   FORCE,TORQUE,DURATION similar to condition monitor
clauses   WITH FORCE_FRAME = <frame> IN <co-ord system>
              WITH SPEED_FACTOR = <scalar>
              WITH APPROACH = <distance scalar> or <distance vector> or <frame>
              WITH DEPARTMENT = ...
              WITH WOBBLE = <scalar>
              WITH NULLING or NO_NULLING
              WITH FORCE_WRIST ZEROED or NOT ZEROED
              WITH STIFFNESS = (v,v) ABOUT f
              WITH STIFFNESS = (s,s,s,s,s,s) ABOUT f IN WORLD
              WITH GATHER = (FX,FY,FZ,MX,MY,MZ,T1,T2,T3,T4,T5,T6,TBL)

enable   ENABLE <label>
disable   DISABLE <label>

motion   MOVE f1 TO <fval>;
              MOVE f1 TO <frame> VIA <frame>,<frame>,<frame>;
              MOVE f1 TO <fval>
              VIA <frame> WHERE DURATION = <time scalar>,
                          VELOCITY = <velocity vector>
                          THEN <statement>
                  <more clauses>
              MOVE f1 TO <fval> <more clauses>;
              OPEN <hand> TO <distance scalar>;
              CLOSE <hand> TO <distance scalar>;
              CENTER <arm>;
              OPERATE <device> <clauses>
              STOP <device>;
              RETRY;

print   PRINT(<e>,<e>,...<e>)   e is an expression, variable or string constant
abort   ABORT(<e>,<e>,...)<e>) similar to print
prompt   PROMPT(<e>,<e>,...<e>) similar to print
pause   PAUSE <time scalar>;

write   WRIST(fv,fv);
              SETBASE;

signal   SIGNAL e1;
wait   WAIT e1;

assignment   <var>   <expression>

return   RETURN
              RETURN(<expression>)

require   REQUIRE SOURCE_FILE "DSK:FILE.EXT";
              REQUIRE COMPILER_SWITCHES "LSK";
              REQUIRE ERROR_MODES "LAMF";
              REQUIRE MESSAGE "<message>";

macro   DEFINE <macro_name> = ⊂ <macro_body> ⊃;

MACROS: DIRECTLY   WITH APPROACH=NILDPROACH
                                WITH DEPARTURE=NILDPROACH

              CAUTIOUS   SPEED_FACTOR 6.0
              SLOW   SPEED_FACTOR 4.0
              QUICK   SPEED_FACTOR 1.0
              CAUTIOUSLY   WITH SPEED_FACTOR = 6.0
              SLOWLY   WITH SPEED_FACTOR = 4.0
              NORMALLY   WITH SPEED_FACTOR = 2.0
              QUICKLY   WITH SPEED_FACTOR = 1.0
              PRECISELY   WITH NULLING
              APPROXIMATELY   WITH NO_NULLING

Figure 8.5 Summary of AL commands.

Source: S. Mujtaba and R. Goldman, AL Users' Manual, Third Edition, STAN-CS-81-889, December, 1981. © 1981 Stanford University Computer Science Department. Used by permission.

PROCEDURES:	PROCEDURE p2(VALUE SCALAR vs1,vs2(23)); S;

OPERATIONS:

vector v:	v REL F
frame f:	CONSTRUCT(f,f,f),f1 REL f2

FUNCTIONS:

boolean	ARMREACH(arm,f),ISAFFIXED(f,f)

STATEMENTS:

affix	AFFIX f3 TO f4 +;
	AFFIX f3 TO f4 AT t1 *;
motion	MOVE f1 BY <vector exp>;
	MOVEX or MOVEY or MOVEZ f1 BY <scalar>;
	OPEN <hand> BY <scalar>;
	CLOSE <hand> BY <scalar>;
	DRIVE BJT(<jt no>) TO <scalar>;
	DRIVE YJT (<jt no>) BY <scalar>;
assignment	POS(<var>) <vect exppression>;
	ORIENT(<var>) <rot expression>;
	XCOORD(<var>) <scalar exppression>;
input/output	READ: READ <file>; QREAD <file>;
(file)	WRITE; WRITE INTO <file>; WRITE <id__lista3[6];
	WRITE ALL INTO <file>; WRITE <id__list> INTO <file>;
	PHOTO <file>;
edit	RENAME <var>;
rename	EDIT <var>;
display	DISPLAY SCALAR;
(terminal)	REDISPLAY; NODISPLAY;
	SHOW var1,var2,var3,...varn;
(vt05)	VT05__ON; VT05__BLUE; VT05__GREEN;
	VT05__OFF; VT05__YELLOW; VT05__RED;
deletion	DELETE s1,s2,v1,v2,...;
	DELETE;
	QDELETE ALL;
macro	DEFINE m2(mm1,mm2,...,mmx(mdx),mmy(mdy),mmz(mdz)) =⊃;
help	HELP; HELP <keyword>;

Figure 8.6 Summary of POINTY commands.

Source: S. Mujtaba and R. Goldman, AL Users' Manual, Third Edition, STAN-CS-81-889, December 1981. © 1981 Stanford University Computer Science Department. Used by permission.

AL can show the relationship between objects by using the affixment mechanism. A FRAME representing one object can be AFFIXED TO another FRAME so that the relationship between the objects will be automatically maintained by AL. AL has a complete set of controls for sensory input, condition monitoring, and input/output conversion.

Debugging aids, system software, and other similar features are provided in AL at the level of other complete programming languages. In addition, there is an interactive AL system or source code interpreter, POINTY, which can be used to create portions of AL programs and check them out on a statement-by-statement basis. POINTY can execute any AL command but has numerous additional commands of its own that are useful in interactive operation. An idea of the capabilities of POINTY can be obtained from Figure 8.6.

2. AL System Hardware

Understanding the operation of the AL system requires that some background knowledge of the hardware be included. As illustrated in Figure 8.7, there are two DEC computers in the AL system, a PDP KL10 (or PDP-10) and a PDP-11/45. There are also two

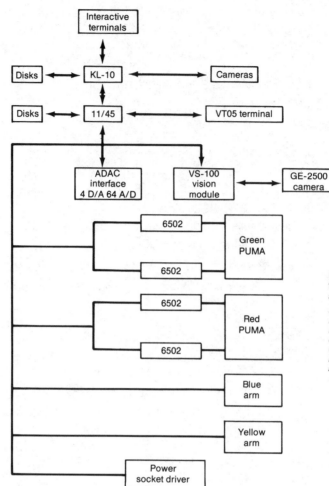

Figure 8.7 Hardware setup for AL at SAIL (from Majtaba and Goldman [25]).

Source: S. Mujtaba and R. Goldman, AL Users' Manual, Third Edition, STAN-CS-81-889, December, 1981. © 1981 Stanford University Computer Science Department. Used by permission.

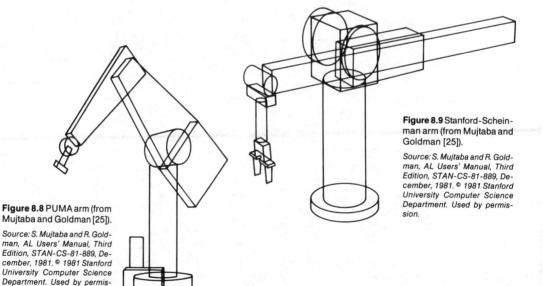

Figure 8.8 PUMA arm (from Mujtaba and Goldman [25]).

Source: S. Mujtaba and R. Goldman, AL Users' Manual, Third Edition, STAN-CS-81-889, December, 1981. © 1981 Stanford University Computer Science Department. Used by permission.

Figure 8.9 Stanford-Scheinman arm (from Mujtaba and Goldman [25]).

Source: S. Mujtaba and R. Goldman, AL Users' Manual, Third Edition, STAN-CS-81-889, December, 1981. © 1981 Stanford University Computer Science Department. Used by permission.

stanford model Scheinman arms, two unimate PUMA 600 arms, the power screwdriver, the ADAC (analog-to-digital and digital-to-analog) interfaces, the Machine intelligence VS-100 Vision Module, and associated disks, cameras, and a control terminal.

In this system, the PDP-10 is the supervisory computer and compiles the programs. The PDP-11 directly controls each joint of the stanford arms and sends commands to the 6502 microcomputers that control the joints of the PUMA arms. There are six microprocessors for each PUMA, one for each joint. An outline drawing of the PUMA arms is provided in Figure 8.8, and the Scheinman arm is shown in Figure 8.9.

3. Current AL Software Organization

Figure 8.10 shows the organization of the AL system software. Three types of programs can provide input to the AL parser, as shown. File names generated by the programs are listed below the program block. The parser accepts the program prepared by the user and checks it for syntax, generating error messages as necessary. It then generates output files as requested by the user.

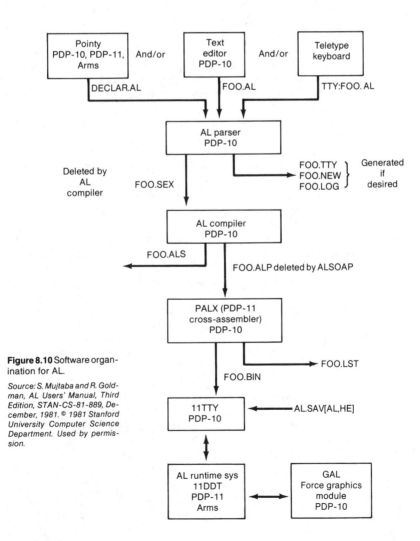

Figure 8.10 Software organination for AL.

Source: S. Mujtaba and R. Goldman, AL Users' Manual, Third Edition, STAN-CS-81-889, December, 1981. © 1981 Stanford University Computer Science Department. Used by permission.

Optional files are .LOG, which lists the errors; .NEW, which is a corrected copy of the source file; and .TTY, which is a disk file copy of programs entered through the teletype.

At least the file with the extension name .SEX must be generated for input to the AL compiler, which runs on the PDP-10 computer only. Output from the compiler is run on the PALX cross-assembler to generate the assembly language program for the PDP-11. This program is generated in binary coded form and therefore is labeled with the extension name .BIN. It must still be run with the AL code program AL.SAV[AL,HE] on the PDP-10 to generate the AL Runtime System, which is actually run on the PDP-11 computer. The GAL graphics program is used optionally to display forces encountered by the arms.

One of the problems with the 1981 AL system is the need for a large computer, the PDP-10, to run the compiler and the cross-assembler to provide the AL runtime system for the PDP-11. However, the same PDP-10 could support many PDP-11s since compilation is required only when major changes are made in the program. Also, the availability of low-cost, powerful computers could make this a minor problem.

8.5.3
IBM's
AML System

A completely new language and control system has been developed by IBM for robot control. It has been used successfully for several years and has many of the features desired in a complete robot language. The description here is a summary of the information provided in Taylor et al. [37] and Taylor and Grossman [36]. Much additional information is available from the referenced articles. R. Brooks [4] has written the complete AML reference manual published by IBM.

After several years of robotic system experience, IBM system planners started the development, in 1978, of a second-generation research robot system. For this system, the planners set out to provide an integrated system architecture, including both hardware and software, that would provide complete robotic systems control. Manipulation, sensing, intelligence, and data processing were considered to be the fundamental capabilities required.

Design objectives were based on previous experience and included the ability to link to a factory control system, use multiple arms, provide good operator interfaces, provide sensing capabilities, and provide a powerful programming language with suitable debugging tools. In addition, there was a need for flexibility in specifying system components, an easy-to-learn language for novices that could also be powerful and flexible enough for expert programmers, and the need for a reliable system that could continually monitor its own performance and shut down if a malfunction occurred. Finally, it was desirable to be able to add artificial intelligence and task-planning capabilities later as they became available.

1. AML System Software

Major system components in the AML system are a Programming System, the Workstation Interface Services, and the Supervisor Services, organized as shown in Figure 8.11.

Included in the Code Segment of the Programming System are the language interpreter, storage management, and built-in subroutines, as shown in Figure 8.12. Use of the interpreter allows the command expressions to be analyzed and applied line by line, so that it is easy to write an expression and determine its effects in an interactive way. This is similar, in principle, to the use of a language like BASIC for playing games on a home computer. Of course, the whole system is more complex and more powerful than a game–playing computer system. Storage management provides the capability for automatic control of files and memory, so that the programmer can call for information by name without having to understand what is happening or where the information is actually stored. Built–in sub-routines and user–programmed subroutines are treated in exactly the same way and may be called from any part of the program at any time. This feature allows system capability to grow as additional subroutines are added and tested. It is often possible to provide a new capability by using old techniques in a new way.

The Data Segment stores the programs, the STATIC variables, the stack, and the interface buffers. STATIC variables are assigned once each time the program is run and thus can be stored. The stack is used to control program execution. Values are put on the top of the stack and taken off the stack in a sequential order, which is useful for control of certain types of program sequencing and subroutine opera-tions. Interface buffers are used to store data for communication to the peripherals and to the robot arm controllers.

Workstation Interface Services provide sensor input/output (I/O) monitoring, motion planning, device interface output, and safety checking. This workstation allows interactive programming, on–line debugging, and application improvement through minor changes as needed.

As in other systems, the supervisory services provide for input and output from the data processing peripherals such as disks and terminals.

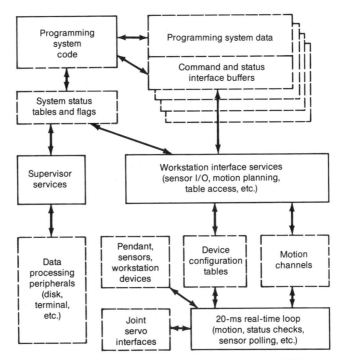

Figure 8.11 System software overview (from Taylor and Grossman [37]).

Source: R.H. Taylor and D.D. Grossman, "Integrated Robot System Architecture," Proceed-ings of the IEEE, Volume 71, Number 7, July 1983, Figure 12, page 853. © 1983 IEEE. Used by permission.

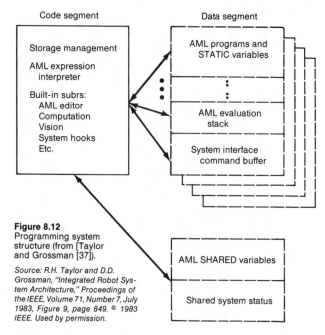

Figure 8.12
Programming system
structure (from [Taylor
and Grossman [37]).

*Source: R.H. Taylor and D.D.
Grossman, "Integrated Robot System Architecture," Proceedings of
the IEEE, Volume 71, Number 7, July
1983, Figure 9, page 849. © 1983
IEEE. Used by permission.*

Workstation Interface Services provide the many important functions of motion control through the use of several data tables stored in
memory. Some of the tables used are:

a. The Physical I/O Table contains the device control blocks used to
    transfer 16–bit control words from the Series/1 computer to the I/O
    interface registers. Control blocks provide information concerning
    the characteristics of a device such as size, speed, type of
    operation, location, and so on.
b. The Logical input/output Table defines logical input/output devices
    as subfields of the physical input/output devices. It includes
    scaling information between integer and floating point engineering
    units.
c. The Joint Table contains information regarding the status of each
    joint and has pointers (references) that tell where additonal infor-
    mation is located in memory. Joint position, feedback inputs,
    commands being executed, scaling ratios, acceptable error toler-
    ances, and similar information are maintained in memory. They
    can be found by looking up the pointer reference in this table.
d. The Motion Channel Table contains information required to de-
    scribe a coordinated motion of several joints. Coordinated motion
    is used for all joint moves in AML to ensure that the movement of
    the tip of the end effector in controlled.
e. The Monitor Table contains information used to monitor sensor
    threshold values.
f. The System Log Table contains accumulated statistics on such
    items as total joint travel and power–on time. These data are used
    by maintenance and diagnostic programs to evaluate system reli-
    ability and serviceability.

In the AML system, as in many other programming systems,
tables are used extensively to keep track of operations. Use of tables
provides an efficient and simple means of control.

Motion control commands are described in more detail later, but an example here will clarify some of our concepts.

The basic MOVE command in AML is expressed as

MOVE(joints, goals)

where joints is a list of the joints to be moved and goals is a list of the positions to which the joints are to be moved. This command can be used to move one or more joints to a specified position. An example is

MOVE ($<1,2,4>$, $<5.2,20,45>$)

which specifies that joint 1 is to be moved 5.2 inches in the positive direction, joint 2 moved 20 inches, and joint 4 moved 45 inches. Note that there is a one-to-one correspondence between the lists of numbers, as is often the case in programming languages. The numbers must be applied in the exact sequence specified. The carets $<$ and $>$ separate the lists of numbers into categories and are used to indicate to the intepreter that there are two separate lists to be handled in a particular way, as identified by the MOVE command subroutine. The list of numbers between the carets is termed an aggregate and is operated on by well-defined rules.

MOVE is recognized as a built-in subroutine because the interpreter takes any set of letters in a command format and compares it to a list of commands. If it finds the command, it branches to a memory location specified with the command and carries out the subroutine found at that location. This subroutine will look for the carets and will do one operation if it finds them and a different operation if they are not present. A MOVE for only one joint would not require the carets. Only a comma would be required between the joint identifier and the position information.

After determining that a particular set of joints is to be moved, the interpreter will call another function to provide motion planning in detail. This function will then issue as many additional commands as necessary to tell the hardware what operations to carry out. In a real sense, the MOVE command is just the tip of the control pyramid. Other functions, activated by the interpreter, will monitor the motion until it is complete and report back the complete information to the MOVE command, which returns it to the user. A typical trajectory, or motion path, is shown in Figure 8.13. It consists of an initial acceleration, travel at constant velocity, deceleration, and a period allowed for settling to the desired position. One or more sensors can be monitored to change the velocity, travel distance, or other characteristic of the trajectory under program control.

Figure 8.13 Motion control. Normal (a) and interrupted (b) trajectories are shown (from Taylor and Grossman [37]).

Source: R. H. Taylor and D.D. Grossman, "Integrated Robot System Architecture," Proceedings of the IEEE, Volume 71, Number 7, July 1983, Figure 8, page 848. © 1983 IEEE. Used by permission.

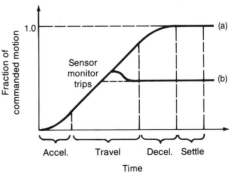

Guarded motions are needed in a robot system to detect obstacles and interference. In guarded motion the robot arm moves more slowly than usual and stops when one of its sensors encounters something. At that time, another segment of the program takes over and carries out an appropriate action determined by the sensor information received. In AML the subroutine call

monlist = MONITOR (sensors, tests,
threshold-parms, intervals)

causes the real-time system to monitor the specified sensors at the intervals stated.

There is a real time loop, as shown in Figure 8.11, which is executed every 20 milliseconds (ms). It scans the parameters stored in status indicators, monitors, sensors and so on, and returns this information to the status tables, where it can be used in motion planning and control. During each of these cycles, the commands in the motion channel are interpreted and the appropriate action is carried out. Therefore, the computation of trajectories must be done at least 50 times per second to provide the required accuracy for real-time control. This is a busy program since it also monitors several safety conditions in addition to status conditions and takes action if some safety criterion is not met.

2. AML Language

AML has been designed as a general-purpose programming language in recognition of the many tasks to be performed in a complete robotic system. The initials AML stand for A Manufacturing Language since one of the major goals is to use the language in manufacturing automation systems.

In many ways, AML is like a standard data processing language. It supports data types and scalar objects, and has the normal structured programming constructs much like ALGOL or PASCAL. Simple arithmetic expressions are available, plus the usual comparison and logical operators. However, there are a number of special constructs that add to the flexibility and power of the language. (They also make it more difficult to program.) It is possible to do simple tasks using a subset of the language, so that beginners can use it. Experienced programmers can use the more sophisticated commands to create package programs and subroutines that other programmers can use.

Other features of AML are as follows:

a. Aggregate expressions allow scalar operators (such as the arithmetic, comparison, and logical operators) to refer to more than one value. As in the MOVE command, one expression can carry out the same operation on multiple values. Examples are:

$$< 1,2,3 > *6 \text{ is } < 6,12,18 >$$

$$< 1,2,3 > * < 4,5,6 > \text{ is } < 4,10,18 >$$

A uniform set of rules allows all the usual operators to be used with aggregates, much like the set of rules in the older language, APL. If we define o as any operator, the expression

$$< x,y,z > o < a,b,c >$$
becomes
$$< xoa, yob, zoc >$$

so that

$$<1,2,3>+<2,3,4> \text{ is } <3,5,7>$$

It is also possible to use generalized indexing for referencing substructures in aggregates, again much like APL. More detail on these features is provided in Taylor et al. [37].

b. Referencing is done by the expression & followed by a variable. It acts like a pointer in PASCAL to allow one variable to point to or reference another. Dereferencing cancels referencing and is indicated by the ! sign.

c. Subroutine definitions and calls are fairly standard in concept although somewhat different in implementation. A subroutine may be defined as

```
subrname: SUBR(formall,formal2,....formaln);
    statementl;
    statement2;   — —Comments are identified with double dash
                  — —preceeding them, as demonstrated here.

        .
        .

    statementk;
END;
```

Formal parameters are identified by <u>formall</u> and so on in parentheses, as shown. Statements can be of any type and can be preceded by optional labels (much as in many other languages).

Subroutine calls have the general form

subrname(expressl,...expressn)

The values <u>expressl</u> and so on are passed as formal parameters to the subroutine, which is called by name. There is a one-to-one correspondence in subroutine operation. The value of <u>expressl</u> replaces the value <u>formall</u> wherever it appears in the subroutine, for example.

3. Commands in AML

Commands in AML are simply subroutines that have been provided in the language. Any command has the subroutine structure previously described and is called as a subroutine. There are several categories of subroutines when classified by application.

a. Fundamental subroutines are built in as part of the language definition. Mostly they affect the flow of control in the interpreter. RETURN handles the return of a result from a subroutine. APPLY is used to call a subroutine and to supply it with a program-generated aggregate of values or arguments. MAP is used to apply a subroutine, element by element, to aggregates.

b. Calculational subroutines perform data manipulations such as SQRT, to compute square roots; LENGTH, to find the length of a string; and DOT, to compute matrix dot products (also called <u>inner products</u>).

c. Workspace interface subroutines provide motion and sensing functions such as MOVE, SENSIO, which reads sensor values or outputs control values, and MONITOR, which provides for monitoring of sensor values.

d. Vision subroutines provide interface and calculational capability to implement a simple binary vision system. These subroutines can operate on a whole picture or pixel by pixel.

e. Editing and debugging subroutines include LOAD, which reads programs from a text file into program memory, and EFILE, which edits a text file. BREAK operates through the interpreter to call the BREAK subroutine if a statement does not meet program standards.

f. Supervisor service subroutines are used to provide access to data processing peripherals, among other uses. For example, OPEN will open and prepare a disk file for reading or writing, DISPLAY controls display at a terminal.

g. Multitasking subroutines are a special class of subroutines that create, terminate, and synchronize multiple AML programs. Examples are FORK, which creates a new task; ENDFORK, which terminates a task; and TESTSEM, which tests a semaphore.

The flexibility of using subroutines for all commands is illustrated by the previous examples. Multitasking has been added recently and implemented by writing new subroutines. This capability is described in Taylor and Grossman [36] and was not mentioned in Taylor et al. [37].

4. Examples of AML Application

Assembly of a print chain for an IBM printer is a typical application of AML to robot control. A print chain is used in a mechanical printer and contains 80 type slugs each with three alphabetic or numeric characters. During printer operation, this chain moves continuously across the printer above the ribbon and paper. Precision hammers behind the paper are actuated rapidly and precisely when a selected letter is in the correct printing position. Therefore, the positioning of the print characters must be quite accurate. Manual assembly of the print chain is tedious and prone to error, so that it is a good task for a robot assembly system.

Figure 8.14 Print chain assembly application (from Taylor and Grossman [36].

Source: R. H. Taylor and D.D. Grossman, "Integrated Robot System Architecture," Proceedings of the IEEE, Volume 71, Number 7, July 1983, Figure 8, page 848.
© *1983 IEEE. Used by permission.*

After the operator has placed an empty print chain cartridge in a passive transport mechanism, the robot grasps the mechanism and prepares to carry out its task. The operator designates the type of print chain to be assembled by pushing a single control key. Then the robot takes type slugs in the proper sequence from 42 gravity feeders and inserts them into the cartridge. A linear actuator indexes the cartridge to the next position after each slug insertion. The robot sensors are used extensively to ensure that space is available in the cartridge and to center its grasp on the type slug. The robot uses sensors to calibrate its grasp on the type slug. The robot uses sensors to calibrate its position and to find the zero point of its pitch, yaw, and roll motions. The print chain assembly operation is shown in Figure 8.14.

The AML program for the print chain assembly application is too complex to be discussed here in detail. However, the subroutine used to pick up a type slug is given in Figure 8.15. We will discuss some of the commands not previously described so that the reader can follow this sequence of program steps.

```
pick_up_slug: SUBR(fdr,tries);
  cc: NEW STRING(8);
  t: NEW 0;


step_1:                                                    —Move to grasping position.
  CMOVE(<feeder_loc(fdr),feeder_orient,.5>);
  IF DCMOVE(<<0,0,−.75>,ANY_FORCE(2*OZS) ,<.5>) THEN
  BEGIN                                                    —Hit something on way in
  DCMOVE(<<0,0,2>>)                                        —Back out
  RETURN('jammed');                                       —Return error
  END;


step_2:                                                    —Attempt to grasp slug.
  cc = GRASP(0.1,<−.04,.04>,PINCH_FORCE(1*LBS));
  IF cc NE 'ok' THEN
  BEGIN                                                    —Hit something on way in
  MOVE(GRIPPER,.5);                                        —Readjust gripper
  DCMOVE(<<0,0,1>>);                                       —Back out
  RETURN(IF cc EQ 'toosmall' THEN 'empty'
    ELSE if cc EQ 'toobig' THEN 'jammed'
    ELSE cc);
  END;


step_3:                                                    —Update feed location.
  fy = HAND GOAL()(1,2);                                   —"Y" position at grasp
  IF ABS(fy-feeder_loc(fdr,2)) GT 0.04 THEN
  feeder_loc(f,2) = fy;


step_4:                                                    —Pull out slug and reverify hold.
  DCMOVE(<0,0,1>);
  IF ABS(SENSIO(<LPINCH,RPINCH>)) LT 15*OZ THEN
  BEGIN                                                    —Dropped part.
  IF tries GE t = t + 1 THEN                               —If not too many,
    BRANCH(step_1);                                        —then try again.
  ELSE                                                     —Otherwise,
    RETURN('dropped');                                    —give up.
  END;


  RETURN('ok');
  END
```

Figure 8.15 AML subroutine to pick up a type slug (from Taylor and Grossman [36]).

Courtesy of IBM Corporation. Used by permission.

Robot position goals are represented as

$$<position,orientation,gripper>$$

sequences. Position is an aggregate of three real numbers giving the desired Cartesian coordinates of the fingertips. Orientation is an optional element with three real numbers giving the Euler angles of the hand orientation. Gripper is also optional and corresponds to a desired hand opening.

The subroutine call

$$goal = HAND\_GOAL(joint\_goals)$$

translates the joint_goals into AML goal aggregates. If joint_goals is omitted, the default values are the currently stored robot joint goals.

Another AML subroutine

$$CMOVE(goal, tests, cntl)$$

causes the robot to move to the coordinates of position, orientation, and gripper opening specified by goal. Position is specified in Cartesian coordinates. Tests is an optional parameter describing sensory monitoring required during the motion. The motion is terminated if any of the sensor thresholds are exceeded. Cntl is an optional parameter giving speed and acceleration parameters to be used. If cntl is omitted, the currently set default values are used. CMOVE returns an aggregate of TRUE and FALSE values to the user if tests is specified.

To displace the robot relative to the current Cartesian coordinates by an amount offset, the AML subroutine call is

$$DCMOVE(offset,tests,cntl)$$

Tests and cntl have the same meaning as in the CMOVE command.

Another subroutine needed to understand the program example is

$$GRASP(size,tolerances,force)$$

which causes the robot to center the gripper jaws on an object of the specified size and grasp it with the specified force. If the object is within the size range specified by the tolerances, the subroutine returns the AML string 'OK'; otherwise it returns either 'too big' or 'too small' as appropriate. Note that in step 2 of the program in Figure 8.15, the opening size is "0.1," separated by a comma from the aggregate $<-.04,.04>$ that specifies the two values of the tolerances. A valuable attribute of AML is the simple way in which multiple values can be indicated when needed but omitted if not required.

The subroutine pick_up_slug, shown in Figure 8.15, requires two parameters as input to the subroutine. These are fdr, the feeder number, and tries, the number of attempts the robot will make to pick up a slug from the feeder before giving up and signaling a failure. The variable cc is designated as type NEW STRING(8). NEW denotes a variable for which memory space is assigned on each entry into the subroutine and memory is released when the subroutine is completed. STRING(8) denotes a character string with up to eight characters in it. The variable t is NEW, is an integer and starts out with a value of 0.

This subroutine has been simplified for illustration; the actual subroutine used provides for many more error contingencies and does more bookkeeping. Steps in the algorithm are approximately the following. (This information is taken almost verbatim from Taylor and Grossman [36] except for comments in parentheseses.)

Step 1. Move the robot gripper to the approach position for the specified feeder and open the gripper to 0.5 inch. Then make a half-speed guarded move down to the grasping position. If the gripper hits anything, assume that a part is jammed in the feeder. Back the gripper off and report an error indication to the calling program. (Apparently there is a typographic error in the DCMOVE command since the number of right carets, >, do not match the number of left carets, <, causing an ambiguity in the meaning of the expression.)

Step 2. Center the gripper on the type slug and grasp it with 1 pound of pinching force. If the GRASP subroutine is unsuccesful, reopen the gripper, back out, and return an appropriate error code. (In this case, there are at least two levels of error return since the error return from the GRASP routine is passed twice, first to the pick_up_slug routine and then to the routine that called it. Many levels of error return can be expected in the total task.)

Step 3. If the centering performed in Step 2 moved the gripper more than 0.04 inch, update the Y coordinate of the feeder location. This action tends to improve efficiency by reducing the amount of time required on the next feed cycle. In effect, the robot tracks drifts in its own calibration and in how parts are presented from the feeder.

Step 4. Pull the slug out from the feeder. Then test the pinch force sensors to verify that it is still being held. If the part is dropped, check to see if the retry limit has been exhausted. If not, go back to Step 1. Otherwise, return an error code.

Further examples of programs are available in Taylor and Grossman [36] and in Taylor et al. [37] that describe several other commands and subroutines.

AML has been in use for 4 years at the IBM Research Center and at selected IBM manufacturing sites around the world. The product version, which is less complete than the research version described here, has also been installed and used in non-IBM sites. It has been used successfully for many applications, including assembly, inspection and test, light fabrication, and intelligent materials handling.

Experience has confirmed the value of a robot language that is also a good general-purpose programming language. The interactive capability of the language has also been proven to be valuable. It has been possible for relatively unskilled programmers to apply the language successfully. Some simple programs have been written in a few hours. Sophisticated applications require a few days to a few months to implement. The principal limitation of the language for research and prototype work has been the lack of a compiler, although in many cases the interpreter has been used successfully in this type of work.

AML is currently one of the most powerful systems and languages. It can be expected to be used for many years and to grow in capability as experience is gained and new subroutines are added.

Exercises

These exercises are provided to aid you in reviewing the information covered in this chapter.

1. Most robots today are programmed by teaching. Why is this type of programming likely to be replaced by preprogramming (off-line programming)?
2. How do operating systems and monitors differ?
3. What are some of the advantages of HLLs ? Under what conditions are other types of languages advantageous?
4. How do compilers and interpreters differ? If you were programming a robot for the first time, would you use a compiled language or an interpreted language? Why?
5. What are some of the disadvantages of teaching-by-guiding?
6. How do adaptive systems and intelligent systems differ?
7. What motion control functions must be performed by programming in a modern robot?
8. Describe the four main types of motion control used in robot programming. Which provides the best control? Why is this method not used all the time?
9. What are the main functions provided in an adaptive system by sensory information? Why are they needed?
10. What is the difference between active and passive compliance?
11. Intelligent activities are made possible in robot systems by the use of artificial intelligence methods. A basic requirement for intelligent activity is the use of a knowledge base. What other techniques are used to improve the capability of an intelligent system?
12. Review Sections 8.4.1 and 8.4.2 and discuss the main requirements of a robotic language that were provided by early computer languages. What new capabilities have been added by languages like AML and AL?
13. What joint control methods are available in VAL? Why is each type of control made available in the language?
14. Describe the relationships between the WORLD, TOOL, and JOINT modes in the VAL language.
15. How is the AFFIX mechanism used in the AL language?
16. Why does the AL language have both interpretive and compiled versions?
17. What is the function of storage management in a computer system such as AML?
18. What functions are provided by the Workstation Interface Services in AML?
19. What data are stored in the Physical I/O Table of AML?
20. Aggregates are used in many AML commands. What purpose do they serve, and how do they simplify program writing?
21. Describe how the AML MOVE command is interpreted to carry out a series of operations. Is this an efficient means of programming?
22. What is the principal limitation of AML as a robot language?

References

[1] T. O. Binford, C. R. Lui, G. Gini, M. Gini, I. Glaser, T. Ishida, M. S. Mujtaba, E. Nakano, H. Nabari, E. Panofsky, B. E. Shimano, R. Goldman, V. D. Scheinman, D. Schmeling, and T. A. Gafford.

[2] S. Bonner and K. G. Shin, "A Comparative Study of Robot Languages," IEEE Computer, Vol. 15, No. 12, December 1982, pp. 82-96.

[3] M. Brady et al. Robot Motion: Planning and Control. M. I. T. Press, Cambridge, Mass., 1982.

[4] R. Brooks, IBM Manufacturing System: A Manufacturing Language Reference Manual. Pub. No. 8509015, IBM Corporation, 1983.

[5] R. Cassinis, "Hierarchical Control of Integrated Manufacturing Systems," In [6], 1983, pp. 12-9 to 12-20.

[6] Conference Proceedings, 13th International Symposium on Industrial Robots and Robots 7, April 17-21, 1983, Chicago, Ill. Robotics International of SME, Dearborn, Mich. 1983.

[7] H. A. Ernst "MH-1, A Computer Operated Mechanical Hand," Doctoral thesis submitted December, 1961. Massachusetts Institute of Technology.

[8] R. Finkel, R. Taylor, R. Bolles, R. Paul, and J. Feldman, "An Overview of AL, A Programming System for Automation," Fourth IJCAI (International Joint Conference on Artificial Intelligence), September 3-8, 1975, pp. 758-765.

[9] J. W. Franklin and G. J. VanderBrug, "Programming Vision and Robotics Systems with RAIL." In [31], 1982, pp. 392-406.

[10] C. C. Geschke, "A System for Programming and Controlling Sensor-Based Robot Manipulators," IEEE Transactions of Pattern Analysis and Machine Intelligence, Vol. PAMI-5, No. 1, January 1983, pp. 1-7.

[11] R. Goldman, "Design of an Interactive Manipulator Programming Environment," STAN-CS-82-955, Department of Computer Science, Stanford University, Stanford, Calif. 1972.

[12] D. D. Grossman, "A Decade of Automation Research at IBM." In [31], 1982. pp. 535-543.

[13] W. A. Gruver, B. I. Soroka, J. J. Craig, and T. L. Turner, "Evaluation of Commercially Available Robot Programming Languages." In [6], 1983, pp. 12-58 to 12-68.

[14] A. Haurat and M. C. Thomas, "LMAC: A Language Generator System for the Command of Industrial Robots." In [6], 1982, pp. 12-69 to 12-78.

[15] K. Jensen and N. Wirth, PASCAL User Manual and Report. Springer-Verlag, New York, 1974.

[16] K. G. Kempf, "Robot Command Languages and Artificial Intelligence." In [31], 1982, pp. 369-391.

[17] H. Lechtman, R. N. Nagel, L. E. Piomann, T. Sutton, and N. Webber, "Connecting the PUMA Robot With the MIC Vision System and Other Sensors." In [31], 1982, pp. 447-466.

[18] A. Levas and M. Selfridge, "Voice Communication with Robots." In [6], 1983, pp. 12-79 to 12-83.

[19] L. I. Lieberman and M. A. Wesley, "AUTOPASS: An Automatic Programming System for Computer Controlled Mechanical Assembly," IBM Journal of Research and Development, July 1977, pp. 321-333.

[20] T. Lozano-Perez, "Task Planning." In [3], pp. 473-498.

[21] T. Lozano-Perez, "Robot Programming," Proceedings of the IEEE, Vol. 71, No. 7, July 1983, pp. 821-841. (Special Issue on Robotics and the Factory of the Future)

[22] T. Lozano-Perez and M. A. Wesley, "An Algorithm for Planning Collision-Free Paths Among Polyhedral Obstacles," Communications of ACM, Vol. 22, No. 10, October 1979, pp. 560-570.

[23] M. Mason, "Compliance." In [3], pp. 305-322.

[24] G. McCalla and N. Cercone, "Approaches to Knowledge Representation," IEEE Computer, Vol. 16, No. 10, October 1983. (Special Issue on Knowledge Representation).

[25] S. Mujtaba and R. Goldman, AL Users' Manual," 3rd Ed. STAN-CS-81-889, Stanford Computer Science Department, Stanford University, Stanford, Calif., December 1981.

[26] W. T. Park, "The SRI Robot Programming System (RPS)" In [6], 1983, pp. 12-21 to 12-41.

[27] R. P. C. Paul, "WAVE: A Model-based Language for Manipulator Control". Industrial Robot, vol. 4, 1977, pp. 10-17.

[28] V. Pavone, "User Friendly Welding Robots." In [6], 1983, pp. 6-89 to 6-102.

[29] C. Raymond, M. Donath, and W. R. Olson, "Problems of Vision Directed Robots in an Unstructured Parts Handling Environment." In [6], 1983, pp. 14-1 to 14-18.

[30] C. Rieger, J. Rosenberg, and H. Samet, "Artificial Intelligence Programming Languages for Computer Aided Manufacturing," IEEE Transactions on Systems Man. and Cybernetics, Vol. SMC-9, No. 4, April 1979, pp. 205-226.

[31] Robots 6 Conference Proceedings, March 2-4, 1982, Detroit, Mich. Robotics International of SME, Dearborn, Mich. 1982.

[32] A. C. Sanderson and G. Perry, "Sensor-Based Robotic Assembly Systems: Research and Applications in Electronic Manufacturing," IEEE, Proceedings, Vol. 71, No. 7, July 1983, pp. 856-871.

[33] B. I. Soroka, "Artificial Intelligence and Robot Software," presented at Future of Robotics: Research and Applications Seminar, University of California at Davis Extension, February 10-11, 1983. Palo Alto, Calif.

[34] B. I. Soroka, "What Can't Robot Languages Do?" In [6], 1983, pp. 12-1 to 12-8.

[35] K. Takase, R. P. Paul, and E. J. Berg, "A Structured Approach to Robot Programming and Teaching," IEEE Transactions on Systems, Man. and Cybernetics, Vol. SMC-11, No. 4, April 1981, pp. 274-289.

[36] R. H. Taylor and D. D. Grossman, "Integrated Robot System Architecture," IEEE Proceedings, Vol. 71, No. 7, July 1983, pp. 842-856.

[37] R. H. Taylor, P. D. Summers, and J. M. Meyer, "AML: A Manufacturing Language," Journal of Robotics Research, Vol. 1, No. 3, Fall 1982, pp. 19-41.

[38] A. M. Tenenbaum and M. J. Augenstein, Data Structures Using Pascal. Prentice-Hall, Englewood Cliffs, N.J. 1981.

[39] E. A. Torrero, "Tomorrow's Computers," IEEE Spectrum, November 1983,(Special Issue on Next-Generation Computers).

[40] "User's Guide to VAL(tm)—A Robot Programming and Control System." CONDEC, Unimation Robots, Unimation, Inc., 1980.

[41] R. A. Volz, T. N. Mudge, and D. A. Gal, "Using ADA as a Robot System Programming Language." In [6], 1983, pp. 12-42 to 12-57.

[42] J. Welsh and J. Elder, Introduction to PASCAL. Prentice-Hall, Englewood Cliffs, N.J., 1979.

[43] P. H. Winston, Artificial Intelligence. Addison-Wesley, Reading, Mass., 1977.

[44] B. O. Wood and M. A. Fugelso, MCL, The Manufacturing Control Language. In [6], 1983, pp. 12-84 to 12-96.

Chapter 9

Robot Vision

For many tasks that a robot performs, vision is the most important source of information about its environment. The recognition of surrounding objects, the perception of certain relations among these objects, and appropriate responses to a given scene lie at the foundation of much robot activity. Robot vision has been present in varying degrees since the late 1960s. Although there have been remarkable advances since then, the field must still be considered relatively undeveloped.

Part of the problem lies in the fact that the sophistication of the human visual system makes robot vision systems pale by comparison. Still, even a rudimentary vision system can augment a robot's capabilities manyfold, and may permit activities that would otherwise be difficult if not impossible to perform.

In this chapter, we will describe the hardware and software required to gather, process, and act on visual information. Hardware consists of cameras, scanners, preprocessors, computers, and interface equipment. It will be covered in Sections 9.1 to 9.4. Software includes the algorithms, procedures, and programs required to convert the digital image into useful information, and will be covered in Sections 9.5 to 9.9. Input to the hardware is a visual image; the output produced by the software is an action or decision by the robot.

Section 9.10 discusses some of the interface problems that depend on both hardware and software characteristics. Section 9.11 provides examples of two vision systems and their characteristics.

The automation of visual processes encompasses several levels of detail and conceptualization, and the sections of this chapter are arranged on this basis. At the lowest level, there must be a means of capturing and representing pictures in a robot's memory that is amenable to direct digital computation. Next, certain preprocessing occurs that is designed to enhance useful features of a picture and/or suppress noise and other unwanted aspects. Then, in many applications, edges and lines in the picture are detected. These edges are then synthesized into recognizable objects in the scene. Finally, the objects may relate to each other in a manner that invites a mechanistic "understanding" of a whole scene. The robot can then respond to the interpretation of this scene by carrying out appropriate tasks. It may alter the scene in this process, so that it is necessary to go through the preceding steps again.

Each level of the process just described is dependent on the one preceding it, so that careful processing at each level is necessary if the top level (that is, analysis of the entire scene) is to have meaning for the robot. The usual order of processing in this hierarchy is from the lowest level (the basic representational level) upward. However, it is sometimes possible to arrange things oppositely (top–down), so that the whole process is driven by objects and relations expected to be in the scene.

We shall take the bottom-up approach, beginning with the manner in which the picture is captured and stored in the robot's memory at the finest level of detail. The picture will be captured, converted into digital form (digitized) and stored in memory.

A digitized picture is shown in Figure 9.1. The first step of digitization is to partition the image into cells (pixels) addressed by row and column. An exaggerated image is shown on the left. Each pixel is assigned a number based on its light intensity or image brightness. This picture has been digitized into 256 levels requiring 8 bits per pixel location for storage of the light level. When the picture information is in this form, it is ready for processing by software.

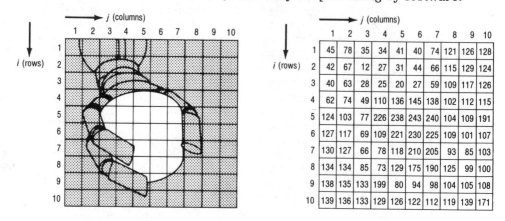

Figure 9.1 Digitized picture with 10 x 10 pixels and the digitized results (from Eskenazi [12]).

9.1 Capturing the Image

Scenes perceived by the robot's vision system consist of optical images of varying intensity. The first step in gathering information is to convert the light in the scene to useful electrical signals. This can be done by scanning the scene with various types of equipment and digitizing the resulting image.

9.1.1
Scanning a Scene
Scanning a scene consists of looking at it in a methodical way and recording the light intensity at each location or spot in the scene. We define a spot as the smallest area covered or observed by the viewing device at an instant of time. A well-known scanning method is the raster scan used in television. In a raster scan, the scene is scanned from left to right along the top edge; then the scanning point is stepped downward the height of the spot; and the left-to-right scan is repeated. (A short time is required for the scanning beam to be returned from the right edge to the left edge between scans.) The left-to-right scan sequence is repeated until the scanning spot has covered the whole scene from top to bottom.

Two general approaches may be used to scan a scene. We may generate a focused beam of light, such as a laser beam, move it across the scene in a controlled way, and record the light reflected from the scene as the video signal; this is the flying spot scanner approach. Or we may control which spot is looked at through a moving peephole or aperture and record at each instant what is seen through the aperture. This is the flying aperture technique. Note that in each case we are viewing only a small spot at each instant and recording the light intensity there. In the flying spot scanner approach, we assume that there is no other light source competing with the moving light beam. Methods for scanning the light beam or aperture will be described.

Controlled scanning provides a way to obtain the coordinates of a spot on the screen. As each spot intensity is recorded, it is stored in a time sequence that allows the location of the spot to be determined at a later time. In solid-state devices, individual photosensitive elements with known coordinate locations are scanned to obtain the light intensity and coordinates.

9.1.2 Standard Television Scanning— RS-170 Standard
In standard television operation, an optical image is formed on a photosensitive screen and is scanned by electrical means. Scanning is done at the television standard rate, prescribed by the RS-170 standard administered by the Electronics Industries Association (EIA). In this standard, the image has 480 horizontal lines and is divided into two fields, each consisting of 240 lines scanned from top to bottom. These fields overlap and alternate with each other; the odd lines of the image in one field are scanned in one pass, and the even lines in the other field are scanned in the next pass. This overlap improves the apparent resolution of the image. Each field is scanned in 1/60th of a second, so that the complete image is scanned in 1/30th second. Because of the need for time between fields for the scan to return to the top of the screen, there are actually 525 horizontal cycles per complete vertical scan. The useful spatial resolution of the image is 240 X 256 pixels. A pixel is a picture element that corresponds to the smallest spot of light or section of the image. In this case, there are 240 elements per horizontal row and 256 rows, so that there is a total of 61,440 pixels in the scene.

Specifications of the standard describe each part of the video signal and the necessary timing and control features required to provide a signal that is compatible with all television equipment used in broadcast television. Figure 9.2 illustrates the EIA RS-170 composite video signal. Video information is contained in the period shown as the horizontal interval. This is the varying analog signal that must be digitized for use by the computer. Two horizontal intervals are shown. The "intensity" of the signal, the height above the reference line, contains the useful information. Timing of the signal is controlled by the horizontal synchronization (Hor. Sync.) and vertical synchronization (Vert. Sync.) pulses.

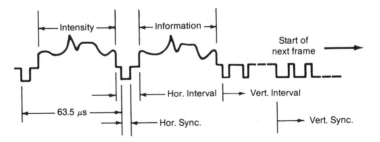

Figure 9.2 Composite video signal to the EIA RS-170 standard.

These signals are typical of most of the robot vision systems in use today, even those that use high-resolution scanners capable of as many as 4,000 x 4,000 pixels per image. Resolution is defined in various ways, but for image work we will use the number of discernible pixels per line as the measure of horizontal resolution and the number of lines as the measure of vertical resolution.

Robot vision cameras may have photosensitive screens that are scanned by electron beams, such as the vidicon or plumbicon, or the solid-state array devices that have individual sensing elements for every pixel location. There are several varieties of each type of camera.

Vidicon tubes are commonly used as television cameras because they are inexpensive and relatively high speed, and have acceptable photometric fidelity. The disadvantages of vidicons are their fixed scan pattern, limited spatial resolution, and the nonlinearity of the scan due to distortions in the electrical field in the scanner itself. In operation, an image is formed on the photosensitive screen of the vidicon tube and scanned by the electron beam. As the beam is scanned across the image, an electrical signal is picked up by an ancillary electrode. This is the video signal, but the location of information in the signal is controlled by the accuracy of beam deflection, which varies in a nonlinear way across the screen. When high accuracy of location is required, the vidicon is not adequate. It has the further disadvantage that the image drifts slightly back and forth across the screen due to voltage and temperature fluctuations.

Orthicon tubes have a flexible scan pattern. Image dissectors can have a very small aperture and obtain high resolution, but their light sensitivity is low because they do not integrate the signal as the vidicon tubes do. Neither of these tubes has found acceptance for robot vision applications.

Two types of solid-state devices are formed into arrays for use as solid-state cameras: charge-coupled devices (CCDs) and charge-injected devices (CIDs). Both of these devices are charge transfer devices (CTDs) and differ primarily in the way in which information is read out. Geometric accuracy of these arrays is extremely good, since they are formed by the photolithographic methods used for other solid-state devices. Commercially available linear arrays, one line wide, have as many as 4,096 elements. Two-dimensional arrays may have up to 512 × 512 elements per device; however, it is possible to butt multiple elements together optically. Arrays of 2,048 × 2,048 can be made available in this way but are quite expensive. These arrays are expected to replace the electron beam tubes for most vision applications.

1. Charge-Coupled Devices (CCDs)

CCDs work on an operating principle called charge coupling. Small amounts of electric charge called packets are created in specific locations in the silicon semiconductor material by the field of a pair of gate electrodes very close to the surface of the silicon at that location. These locations are arranged in rows in a linear array or in rows and columns in a matrix array, with connections between the elements of each row or line. At each location, there can be a different amount of charge. Charges represent an analog value. This analog value can be shifted from element to element along the row, so that a sequence of variable charges can appear at an output electrode at the end of each row. Shifting is done by applying control voltages along the array in a way similar to the action of shift registers used in computer logic circuits. The output charge can then be converted to a voltage or current level that represents the initial charge generated.

When light falls on a CCD element, either in a row of individual elements or at the row-column intersection of an array, the initial charge is increased by an amount proportional to the light level. This charge generation is due to the release of free electrons in the semiconductor by the light. Charges are stored and build up—the signal is <u>integrated</u>—at a storage element location until the shifting electrodes are actuated. Integration of the signal allows small amounts of light to build up relatively large charges, so that the CCDs are very sensitive to light and produce strong signals. (Since CCDs can store charge, they can also be used as memory elements in the absence of light.) Each charge packet represents the light intensity for one picture element or pixel. Shifting the charges to the end of the row and detecting them with an electrode produces a varying analog voltage. This voltage represents the video signal information in the image projected onto the array of CCDs. This signal can be converted to a series of digital signals by the sampling technique of digitizing.

CCD and CID image sensors share several desirable attributes:

a. They are sensitive over a wide spectral range, from 450 to 1,000 nanometers (corresponding to the range from blue light through the visible spectrum to the near infrared region).
b. They operate on low voltages and consume only a small amount of power.
c. They do not exhibit lag or memory, so that the traces of moving objects are not smeared.
d. They are not damaged by intense light. Present devices will over-saturate and "bloom" under intense light but are not permanently damaged (as a vidicon tube might be, for example).
e. Their positioning accuracy and therefore measurement accuracy are very good because of the accurate photolithography process used to form them.

As mentioned earlier, both linear imaging devices and area imaging devices are available in CCDs. Linear imaging devices are commercially available in packages of 256, 1,024, 1,728, and 4,096 elements. Area imaging devices, which produce a TV picture, are available in 488 × 380 elements, also in 100 × 100, 244 × 190, and other array sizes. Smaller packages are less expensive, of course, than the larger arrays. An example of an area imaging device is the Fairchild CCD211, a 244 × 190 element array, which dissipates 100 milliwatts when operated at a 7-megahertz data rate and a voltage range of 12 to 15 volts. This chip is only 0.245 inch square.

2. Charge-Injected Devices (CIDs)

CIDs are charge transfer devices like CCDs, but during sensing the charge is confined to the image site and must be read by a separate X-Y scanning addressing circuit like those used in computer memories. CIDs have greater tolerance to processing defects and less blooming than CCDs. However the control circuitry is more complex.

3. Area Camera Systems

An example of the use of a CCD array in a video camera system is given in Figure 9.3. Light is imaged from the scene through a lens and an infrared (IR) filter onto the CCD array, which has 448 × 380 elements. Included in the sense head of the camera are the horizontal

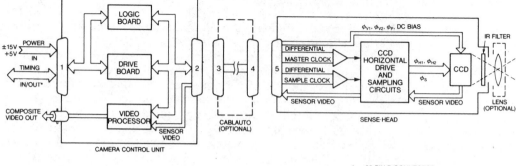

1 — 25 PIN D CONNECTOR
2, 3, 4, 5 — 31 PIN D CONNECTOR

Figure 9.3 Block diagram of a video camera using a CCD area array.

Courtesy of Fairchild, A Schlumberger Company–CCD Imaging. Used by permission.

drive and sampling circuits. This head is light and rugged enough to mount directly on a robot arm if desired.

As shown in Figure 9.4, the control electronics can be mounted separately from the sense head if desired. Output from the video processor is composite video to the NTSC or RS–170 standard. This

Figure 9.4 Separated sense head and camera control unit.

Courtesy Fairchild, A Schlumberger Company–CCD Imaging. Used by permission.

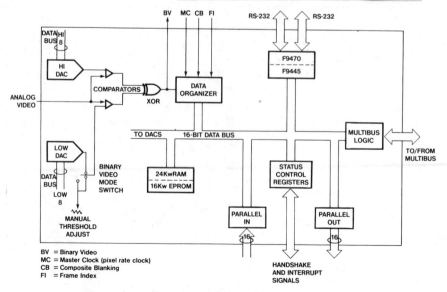

BV = Binary Video
MC = Master Clock (pixel rate clock)
CB = Composite Blanking
FI = Frame Index

Figure 9.5 Block diagram of the vision interface processor (VIP100), Fairchild Corporation.

Source: Fairchild, A Schlumberger Company–CCD Imaging. Used by permission.

can be input on the analog video line of the Vision Interface Processor (Figure 9.5) for further processing. The Fairchild VIP100 shown here is capable of performing the <u>frame grabbing</u> function, or complete shape and pattern recognition.

Analog video input from the camera is digitized by the two Digital-to-Analog Converters to convert the analog signal to digital signals that are stored in RAM and analyzed by the F9445 16-bit microcomputer. Note that in this case the video signal is thresholded and the video signal is converted to either a binary 1 or a binary 0, depending on whether it exceeds a certain value or not. This process is illustrated in Figure 9.6. Note that the 378 picture elements of one line of video are compressed and stored into 24 video data words of 16 bits each. Binary video requires only 1 bit per picture element. Gray scale video may have as many as 256 levels and require 8 bits per picture element. Using binary coding, the complete image can be stored in 488 × 24 or 11,712 words. For this application, the 24K of RAM memory is adequate. Much more memory may be required in other applications.

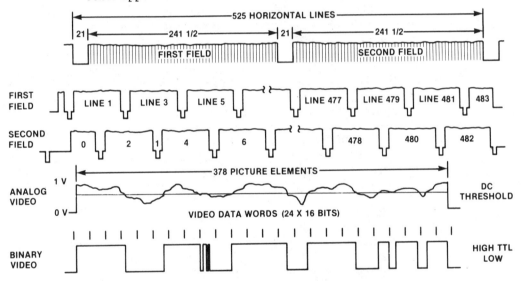

Figure 9.6 Video conversion of image data from CCD3000 into binary words.

Source: Fairchild, A Schlumberger Company–CCD Imaging. Used by permission.

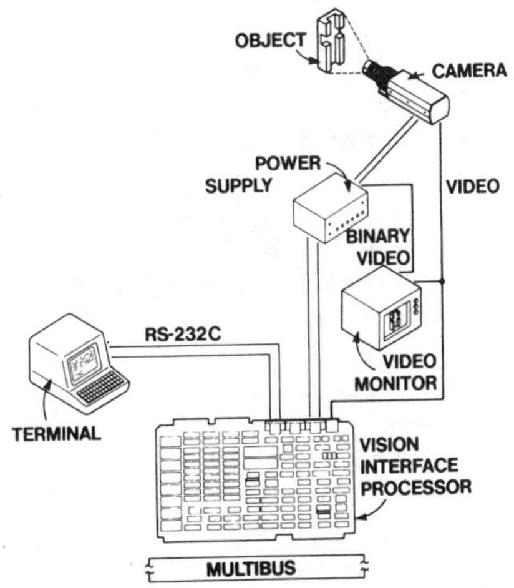

Figure 9.7 A single-board system for the video information processor showing component units.

Source: Fairchild, A Schlumberger Company–CCD Imaging. Used by permission.

The video processor is required to operate at a high instruction rate in order to perform the necessary operations in the time available. Clock rates of 16 to 24 megahertz (MHz) are used in the F9445. In the fastest system, an ADD, SUB, or COM operation takes place in 0.25 microsecond.

A complete video processing system, including the camera, video monitor, vision interface processor, terminal, and power supply, are shown in Figure 9.7.

**9.1.5
Scanning Laser
Camers**

Lasers can be scanned in either one or two dimensions by the use of rotating mirrors. These systems are most useful when the background is dark, so that the laser is the only source of light. Sensitive photodetectors can be used to pick up the reflected light as an input analog signal to a video processor. Position of the scanning mirrors must be available, usually by the use of encoders on the mirror drive axes, to provide the necessary information to locate the $X-Y$ position of the laser beam at any time.

**9.1.6
Linear Arrays**

Linear arrays of photocells, such as photodiode or linear CCD arrays, are valuable in determining the position of a line in an object. They can also be used for obtaining a picture in two dimensions either by moving the image past the linear array or by scanning the image by the use of a rotating or oscillating mirror. If a line perpendicular to a moving conveyor is imaged on the linear array, the movement of the conveyor will cause objects on the conveyor to be scanned by the linear array so that area images can be obtained. The resulting picture is obtained slowly, but the quality of the picture may be as good as that of a two-dimensional array.

The light imaged on the linear arrays can be filtered to take advantage of the different spectral ranges available. In an arc welding application, for example, the arc emits intense visible and ultraviolet light but relatively small amounts of infrared light.

However, the heating due to the arc causes the metal being welded to emit strong infrared light. Therefore, by using a scanner sensitive only to the infrared light, either intrinsically or because of filtering, the heating pattern of the welded area can be observed and used to guide the welding process.

**9.1.7
Random Access
Devices**

When only particular parts of an image are of interest or one wishes to track a boundary in the picture, it is desirable to have a camera that can be directed to follow a desired path in the $X-Y$ plane. The image dissector tube is capable of randomly addressing an image when the light source is external to the scene and the image dissector is used as the light detector. Conversely, if the scene is dark, a CRT can be used as a controllable light beam to scan the image. In this case, a photocell is used to pick up the light reflected off the image. By suitable filtering, it is possible to use both a random access scanner and a raster scanner, such as a CCD camera, at different light wavelengths to obtain two sets of information about the scene.

CID cameras, which can be scanned randomly, should also be useful in random access applications.

**9.1.8
Obtaining Depth
and Range Infor-
mation (in Three
Dimensions)**

Robot end effectors or hands are used to grasp and manipulate objects of many kinds. A basic requirement for grasping an object is to know its location in three dimensions. When a visual scanning system such as one of the area cameras or linear arrays just described is used, it is relatively easy to obtain the location of the object in the two dimensions perpendicular to the camera vertical; we will designate those dimensions as X and Y for this purpose. It is not easy to obtain accurate dimensional information, along the line parallel to the camera centerline, which we designate as the Z direction.

In human vision there are several methods for obtaining depth information, the most usual being the use of binocular vision, in which the two eyes see different images separated by a small angle. The separate images can be superimposed and analyzed to determine the third dimension by triangulation. There are several problems in implementing stereo vision in computer vision systems. One of the major difficulties is the problem of determining which points in the different views correspond to each other. Although considerable work has been done on stereo vision for robots, there is still no satisfactory, practical, cost-effective solution using this method.

1. Specialized Lighting (Structured or Contrived)

Considerable success has been achieved in obtaining depth and range information for a robot by using specialized lighting that is adapted to the problem. SRI was the first to develop a complete system using auxiliary lighting (Rosen et al. [42]). This work was based on earlier work (Will and Pennington [51], Shirai and Suwa [47], Popplestone et al. [37], and Agin and Binford [2]). Much of the early work was done by Artificial Intelligence researchers throughout the world, a good example of theoretical work that has been used in an important practical application.

In all of these methods, a selected light pattern is projected on the scene to be viewed. Usually these are regular patterns such as bars, grids, or circles. Light patterns are projected from an angle (30 to 45 degrees is typical) from the vertical to the scene, as illustrated in Figure 9.8.

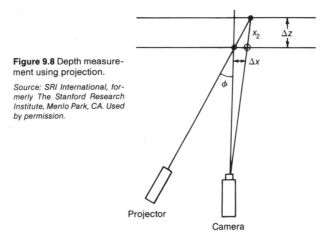

Figure 9.8 Depth measurement using projection.

Source: SRI International, formerly The Stanford Research Institute, Menlo Park, CA. Used by permission.

A displacement, Δx, is observed between the front edge of an object and the rear edge of an object a distance Δz away. By measuring the displacements along the projected beam, it is possible to obtain excellent depth and range information. The projected light pattern may be generated by a scanning laser, so that the location is controlled accurately, or by a rotating slit scanner, or more simply by a conventional slide projector. Greater displacement is observed when the camera is farther from the object or when the displacement angle is increased. However, there is a loss of light intensity at greater distances and possible obscuring of parts of the image if the displacement angle is too large. Lasers provide very intense light bars, which are an advantage in some instances. Jarvis [21] discusses some of the factors involved in selecting the type of light source and pattern. An example is given in Figure 9.9.

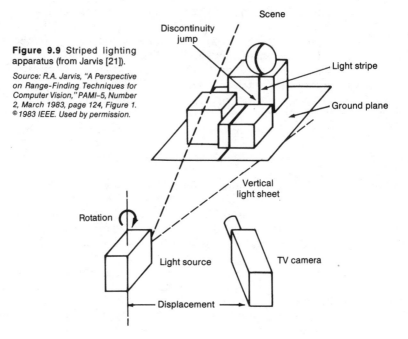

Figure 9.9 Striped lighting apparatus (from Jarvis [21]).

Source: R.A. Jarvis, "A Perspective on Range-Finding Techniques for Computer Vision," PAMI-5, Number 2, March 1983, page 124, Figure 1. © 1983 IEEE. Used by permission.

2. Projection Patterns

Different projection patterns can be used to obtain particular results. A projected dot or point is easy to detect, unless it is lost in the noise of the picture, because it produces a reference discontinuity from any scanning direction. Single lines are also easy to detect when they are perpendicular to the scanning direction. If the scanner is time based, the location of the discontinuity due to the line is directly obtainable from a time measurement that provides maximum accuracy.

Multiple lines can provide some of the same effect, and the

Figure 9.10 Various patterns for the projected image. Grid resolution differs for squares A, B, and C (from Hall et al. [18]).

Source: E.L. Hall, J.B.K. Tio, C.A. McPherson, F.A. Sadjaki, "Measuring Curved Surfaces for Robot Vision," Computer IEEE, December 1982, page 44, Figure 2. © 1982 IEEE. Used by permission.

projection can be analyzed to obtain depth or range information over the whole image. Grids can be used to obtain information along both the x and y axes of the image. These and other alternatives are discussed by Hall et al. [18]. Various patterns are shown in Figure 9.10, as suggested by these authors.

3. Alternative Ranging Methods

Although not yet in wide commercial use, several other techniques discussed by Jarvis [21] are potentially valuable in robot applications. Jarvis lists: range from occlusion cues, depth from texture gradients, range from focusing, surface orientation from image brightness, range from stereo disparity, range from camera motion, Moiré fringe range contours, and time-of-flight range finders as alternative methods. One technique mentioned by Jarvis, simple triangulation range finding, is used as illustrated in Figure 9.11.

An energy source, separated by a baseline distance, b, from an energy detector, can measure the distance to a nearby object. The distance, L, can be calculated from the simple equation

$$L = b \tan \theta$$

where θ is the angle between the horizontal baseline and the energy beam emitted from the source. The beam can be swept through an arc to allow ranging at different distances. An important limitation of this method is the possibility of loss of the return beam because of occlusion due to part of the object or another object blocking the field of view. This method also requires the object to be either a diffuse reflector, so that some light is reflected in all directions, or a specular (mirror like) reflector that is perpendicular to the incident rays. (Lambertian scattering describes the situation which the reflected energy is scattered with equal probability in all directions in a hemisphere). A specular reflector aimed away from the detector will not return any energy to the detector and will not be detected.

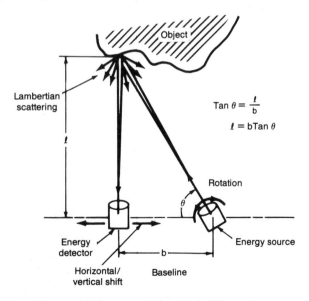

Figure 9.11 Simple triangulation range-finding geometry (from Jarvis [21]).

Source: R.A. Jarvis, "A Perspective on Range-Finding Techniques for Computer Vision," PAMI-5, Number 2, March 1983, page 135, Figure 11. ©1983, IEEE. Used by permission.

4. Time-of-Flight Range Finders

Time-of-flight range finders also appear to be applicable to use with robots. Ultrasonic ranging is described in Section 5.3.6 and can have accuracies on the order of 0.003 inch, but the effective beam width is more than 10 degrees so that good resolution cannot be obtained. Laser beams can provide high resolution and resolve details as small as about 0.5 cm. At present they are expensive, but costs should come down rapidly. Processing time is also long for present systems. The reader is referred to Jarvis [21] for more details on all of these systems.

9.2 Frame Grabbers

Each complete image generated by a television scan is called a frame. Conversion of the analog image in the video signal to a digital signal requires that each frame be converted individually from a varying analog signal to a series of digital representations. This conversion can be done in one frame time, 1/60th of a second, or over several frame times by converting only part of each image during each frame time. Slow conversion is feasible if the image is unchanging or slowly changing, and is much less expensive than flash digitizing, which converts a complete frame in one frame time.

Slow conversion is accomplished by setting up a simple timing circuit that converts a corresponding spot on each row during each frame cycle. On each horizontal sweep, a timer is started. It is set initially to sample the analog value of the first spot position of each row into a circuit that holds it temporarily. This action allows the digitizer a complete horizontal sweep time to convert the analog signal to a digital representation in an analog-to-digital converter so that a slow, inexpensive circuit may be used. After each frame cycle, the sampling point is moved one step along the row until the whole of each row and thus the whole image has been sampled and digitized. As each sample is digitized, it must be stored in memory to free the digitizer for the next sample. Memory space required depends on the number of pixels and the number of gray levels used. A minimum memory requirement for a 128×128 image is 2,048 bytes if only a binary image is desired.

When a horizontal resolution of 128 elements is used, a common value, the total time required to digitize an image is $128 \times 1/60$ or 2.15 seconds. This time delay is suitable for some inspection applications but is much too long for a robot guidance system.

9.2.1 Flash Digitizers

Flash digitizers convert each analog sample into digital form during the sampling period. Very fast analog-to-digital converters are required, especially if the full resolution of the image is taken as 488 lines by 380 horizontal elements. In this case, there are 185,440 picture elements or pixels, and the frame is repeating every 1/30th of a second. Therefore, the digitizing rate is $30 \times 185,440$ or 5,563,200 conversions per second.

Available commercial analog-to-digital converters are able to exceed this rate and generate 8-bit digital bytes from analog inputs at a rate of 20 million operations per second (20 Mhz). These digital outputs represent 256 levels of gray code, which seems to be ample for any planned robot application. Even faster digitizers are available: 100 Mhz at 8 bits/sample and 20 Mhz at 10 bits/sample. Extra bits in the digitized sample are used when full-color samples are to be represented. Each frame requires 185,440 bytes of memory for each image when the 488 × 380 image is digitized over 256 levels or 8 bits per pixel.

1. Analog-to-Digital Converters

There are four major types of analog-to-digital (A/D) converters: counting, successive approximation, parallel-comparator, and ratiometric. Only the parallel-comparator converter will be described here, since it is by far the fastest (Millman [26].) Counting converters take 2^N clock cycles for an N-bit digital output; successive approximation converters require about N clock cycles, and ratiometric converters are even slower.

As shown in Figure 9.12, the analog voltage, v_a, is applied simultaneously to a bank of comparators with equally spaced threshold voltages (the reference voltages: $V_{R1} = V/8$, $V_{R2} = 2V8$, etc.)

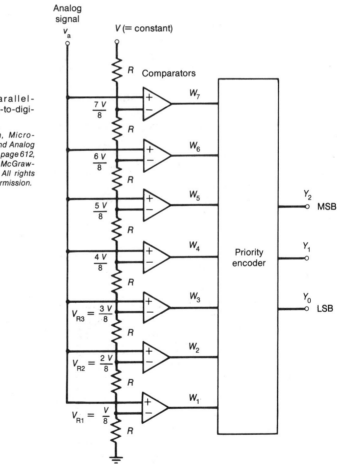

Figure 9.12 Parallel-comparator analog-to-digital converter.

Source: J. Millman, Micro-electronics–Digital and Analog Circuits and Systems, page 612, Figure 16-44. © 1979 McGraw-Hill Book Company. All rights reserved. Used by permission.

As the input analog signal is compared to each of the reference voltages, a digital bit is generated in the priority encoder circuit, so that the output of the priority encoder is the digital code desired to represent the input analog signal. There are 7 comparators to generate a 3-bit code, so that 255 comparators would be required to generate an 8-bit digital output. Essentially, the digital circuits in the priority encoder look at all of the inputs, W_1 to W_7, and note which values are 1 and which are 0. Then the encoder converts this information to the required digital output.

Since there is only one simultaneous comparison and only one encoding cycle, the total delay in the circuit is only the operation time of these two circuits, which can be on the order of 10 to 20 nanoseconds with fast circuits. These circuits are relatively expensive because of their great complexity.

Output of the digital circuit goes to the video processor and must conform to the interface standards. Digital information from a flash digitizer is at a high data rate of more than 5 million bytes per second. This rate requires a suitable high-speed interface bus, which may be serial or parallel. An alternative is to preprocess the data out of the digitizer in order to reduce the data rate. Differencing and other techniques are applicable for this purpose.

9.3 Vision Processors

The functions of the robot vision processor are to control the hardware of the vision system, accept input from the analog-to-digital converters in the camera subsystem, process vision information, determine significant parameters to be output, and transmit the output in suitable form to the control computer that is operating the robot system.

An example of one vision processing system is given in Figure 9.13. This system was developed by the National Bureau of Standards (Nagel, et al. [28] and illustrates the control equipment needed to carry out a robot vision task. A task is assigned by the robot control system computer, which provides a maximum range within which the vision system is to search for a part. Output of the vision system to the robot computer includes status information, the location of the part in the wrist coordinate system of the robot, and an edge characteristic.

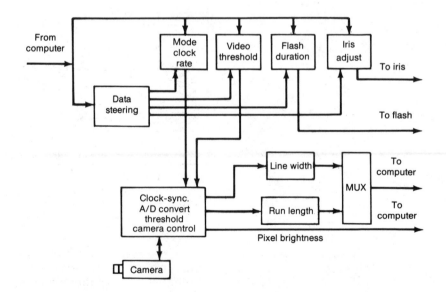

Figure 9.13 Block diagram of NBS vision system interface hardware.

Source: National Bureau of Standards, U.S.A

The camera interface is controlled by an 8-bit microprocessor that provides control functions and data reduction. Control functions are flash duration of the strobe light used for a light source, camera clock rate, and iris adjustment. Video threshold, run length representation of the binary image, and line width are also processed by the microprocessor, as shown in the block diagram.

9.3.1 Microprocessors for Vision Systems

Microprocessor architecture and operation are discussed in Chapter 7 in greater detail. In this chapter, we discuss only those characteristics of major importance to vision systems.

Multiple microprocessors used for different tasks are common in vision processing systems. Both 8-bit and 16-bit processors are used. Word size is the distinguishing characteristic used here. The 8-bit processors have 8-bit words in memory and contain internal registers that can handle 8 bits at time. Larger word sizes are valuable in most of these applications, so that the use of 16-bit words in memory and 16-bit storage registers has increased.

Addressable memory size for the 8-bit microprocessors is 64K bytes. Two memory words are sent to the addressing registers to control access to memory, so that there are 2^8 possibilities along each direction (X and Y) of the memory. Since $2^8 = 256$, the total number of addressable bytes of memory is 256×256 or 65,536 bytes. This is equivalent to 64K bytes because in computer nomenclature K stands for 1,024 which is the closest binary number to 1,000—the multiplier usually designated by the word <u>kilo</u> (K) in the metric system. (Those not familiar with this nomenclature are referred to the Glossary in Appendix C. Processors with 16-bit word sizes can address more than 4 billion bytes directly; by using some advanced techniques they can address up to 16 billion bytes or more.

Larger word sizes allow more complex instructions and more information to be handled in the 16-bit than in the 8-bit microprocessors. This ability provides more computing power, more flexibility, and higher processing speeds—important features in vision processing.

Some 8-bit processors in common use in vision systems are the Motorola 6809, Mostek 6502, Zilog Z-80, and the Intel 8080 and 8085. More powerful systems use the 16-bit Intel 8086, Intel 8088, Motorola 68000, and DEC's PDP-11 series, the PDP-11/40 and the smaller LSI-11 microcomputer. The PDP-11/40 is technically a miniprocessor but has the same instruction set as the newer LSI-11.

These same microprocessors are used for different parts of the vision system. They are also used for control of the robot joints and as the master control or host computer of the robot system.

Tasks that have been assigned to different processors are:

1. Capturing the Image. This includes the processors used to control multiple cameras, multiple analog-to-digital converters, and the lighting equipment used to illuminate the scene. Lighting equipment may include strobe flashes to stop motion during video scan, laser scanners to generate one or more reference scan lines for depth measurement, and perhaps an overall light level control. Multiple cameras may be used to obtain views of a scene or object from different angles at specified times. Each of these cameras generates video information that may be switched to one or more analog-to-digital converters. One processor may perform all of these functions, or a processor may be assigned to each function. If several processors are used, a control processor is needed to control and coordinate their outputs.

2. Preprocessing the Image. Functions such as thresholding, image compression, and noise removal may be done in the main vision processor or in a separate processor. Maintaining full gray scale capability requires large amounts of processing; an auxiliary computer is useful for these functions. Other functions may also be performed. The division of tasks is purely arbitrary.

3. Finding Edges and Recognizing Objects. Output of the preprocessor, if one is used, becomes input for the main vision processor, which does edge finding and object recognition, sometimes using very complex algorithms or programs. This same processor may then calculate the location of the object and feed that information to the master control computer for the robot system.

All of these functions have been done with one microprocessor. Some systems may use many microprocessors in parallel to speed up the processing.

9.3.2
Multiple
Processors and
Special-Purpose
Systems

Several high-speed image processing systems developed for other purposes could be used for robot vision applications. These can be classified as multiple processor systems or special-purpose systems.

In addition to the use of full systems, it is quite feasible to add specialized circuits to existing processors to perform a specialized task. An example is the use of floating point multipliers to speed up arithmetic operations. Specialized edge detectors, convolution circuits, and histogramming circuits are other examples. The greatest throughput improvement has come from designing complete systems for specialized tasks, which will be described.

1. Parallel Image Processing

Parallel image processing by the use of cellular arrays of processors has become a feasible way to improve the speed and capability of image processing. A cellular array is defined as a two-dimensional array of processors working on the same task. As discussed by Azriel Rosenfeld of the University of Maryland [45], cellular arrays can operate in parallel on a group of pixels in an image. Local operations can be performed over the whole array in the same period of time that a single processor would require for a single neighborhood. This capability allows very-high-speed processing of images.

A cellular array with dimensions of m processors in the X direction and m processors in the Y direction has a total of m^2 processors. Arrays of up to 128×128 processors (16,384 total) have been built, although they are still expensive even with the great reduction in microprocessor cost. Figure 9.14 is an example of a cellular array structure. These arrays can be built up from multiple silicon chips on a common structure.

One way to use these arrays is to process one $m \times m$ block of a larger image of size $n \times n$ pixels. This action allows each block of the image to be processed separately, but requires additional operations to provide communication across the boundaries between image blocks.

Another technique for using cellular arrays is to give each cell as input an $n/m \times n/m$ block of the image. Then a pair of neighboring cells must exchange information about n/m pairs of neighboring pixels along their common border.

Algorithms for the use of cellular arrays are discussed in Rosenfeld [45] and will not be covered in detail here. There is an increase in processing speed by the use of cellular arrays, but there is also a cost for input, communications between cells, and output that prevents the full potential speedup from being accomplished. Rosenfeld suggests that one-dimensional cellular arrays called <u>cellular strings</u> would be valuable in processing. An application is the use of cellular strings to derive information about curves or regions using as input the chain codes for the curves or the region's borders. Processing times nearly proportional to n, the number of pixels along one edge, instead of time proportional to n^2, can be achieved for some operations by the use of cellular strings of processors of length n.

K. Preston [39] has summarized the many types of cellular logic processors. He identifies four architectural types of processors and provides a diagram, Figure 9.15, which shows the growth in capability of cellular logic machines from 1960 to 1980. The data provided indicates that there is an increase in processing speed of a thousand-fold every 10 years. Not shown on this diagram, although discussed by Preston, is the Massively Parallel Processor (MPP) constructed by the Goodyear Aerospace Corporation. This machine, which has 128 × 128 processors, has a potential speed of 100 billion pixops (picture operations per second).

Figure 9.14 A two-dimensional cellular array (from Rosenfeld [45]).

Source: A. Rosenfeld, "Parallel Image Processing Using Cellular Arrays," Computer, IEEE, Volume 16, Number 1, January 1983, Figure 3, page 17. © 1983 IEEE. Used by permission.

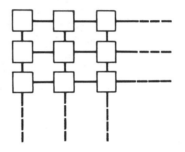

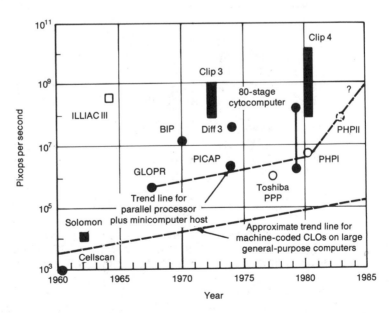

Figure 9.15 Operating rate of cellular logic machines versus time (from Preston [39]).

Source: K. Preston, "Cellular Logic Computers for Pattern Recognition," IEEE Computer, Vol. 16, No. 1, January 1983, pp. 36–50. © 1983 IEEE. Used by permission.

2. Architecture of the Massively Parallel Processor (MPP)

More detailed information on the MPP is provided by Potter [38], who describes the structure of the machine and suggests some ways in which it could provide new image processing capabilities. Processing on the MPP is done by 16,384 microprocessors working in parallel, each operating on the same instruction but on a different pixel or group of pixels. Data is transmitted between processors and the staging memory on a high speed I/0 bus capable of transferring 320 megabytes per second.

Sequential, parallel, and interface instruction types are provided. Sequential instructions are conventional: loads, stores, adds, subtracts, compares, branches, and logical operations. Parallel instructions are similar, but there are no branch instructions. When the sequential controller detects a parallel instruction in the program, it sends it to all of the 16K processors to be executed in parallel. Effectively, these instructions operate like serial instructions but at the tremendous speed provided by the parallel structure. Since there are no branch instructions, the programming of the parallel operations is not difficult. Interface instructions allow the movement of data between the sequential and parallel portions and to and from external storage or display.

The effective speed of the MPP is about 6 billion operations per second. Examples of some speeds of execution for typical image processing algorithms are provided in Table 9.1. [Potter 38].

Table 9.1 Algorithm execution speeds

ALGORITHM	Speed (Millions Of Pixels Per Second)
CONVOLUTION (3 X 3)	77.625
CONVOLUTION (7 X 7)	14.25
TEMPLATE MATCHING (7 X 7)	1783.25
PSEUDOMEDIAN FILTER (3 X 3)	925.00
HISTOGRAM COMPUTATION	0.125
GRAY-SCALE THRESHOLDING	7123.5
REGION GROWING	96.5
TWO-DIMENSIONAL CROSS CORRELATION (13 X 13)	2043.25
GRAY-SCALE AVERAGING (32 X 32)	20.48

Source: J.L. Potter, "Image Processing on the Massively Parallel Processor," Computer, IEEE, Volume 16, Number 1, January 1983, Table 3, page 65. © 1983, IEEE. Used by permission.

9.3.3 Vision Processors in Japan

It appears that the Japanese are also working diligently on aspects of advanced image processing. M. Kidode [24] describes several new systems now being developed in Japan. Macsym is a good example of a parallel processor that uses a master processor to control multiple slave processors to obtain speed increases nearly proportional to the number of processors used. Several processors, 4 at present, but with 16 planned, share a common data bus to provide the increased speed. Other systems in Japan are being developed using several different computer architectures. The general approach is similar to those being developed in the United States, although different in detail.

9.3.4 Special-Purpose Hardware

1. Real-Time Digital Image Enhancement

Image enhancement for robotic applications must operate in real time, that is, fast enough so that the robot operation is not delayed by waiting for information from the vision system. In most cases, this means that the processing of images must be done in less than the scanning cycle time of a TV image, which is 1/30th of a second. When binary vision is used (with just two values, 0 and 1, in the image), the processing is not difficult, so that many systems work on binary values, as discussed previously. When multiple gray levels of information are to be processed, processing needs increase rapidly, so that special hardware is required to process at the required rates.

R. E. Woods and R. C. Gonzalez at the University of Tennessee [52] have developed a hardware system to provide image enhancement of images with 256 gray levels. Their system is capable of automatic and interactive enhancement based on histogram equalization, function processing, and histogram specification techniques. This equipment is able to extract useful information from images that appear to be so dark as to be unreadable. Figure 9.16 is a block diagram of the basic system architecture of the system. A separate

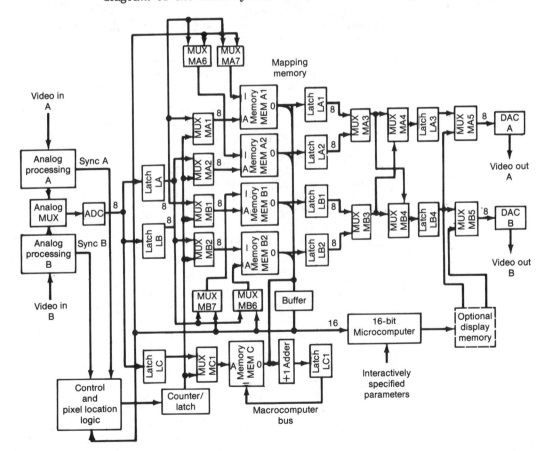

Figure 9.16 Basic image enhancement system architecture.

Source: R.E. Woods and R.C. Gonzalez, "Real Time Digital Enhancement," Proceedings of the IEEE, Volume 69, Number 5, May 1981, (Special Issue on Image Processing), Figure 11, (a,b,e,f), page 652. © 1981 IEEE. Used by permission.

16-bit microcomputer controls the operation of analog-to-digital and digital-to-analog converters, gray-level mapping logic, and a gray-level histogram computation stage. The results are shown in Figure 9.17.

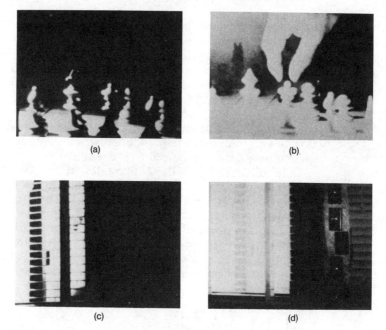

(a)   (b)

(c)   (d)

Figure 9.17 Examples of images before (left) and after (right) histogram equalization.

Source: R.E. Woods and R.C. Gonzalez, "Real Time Digital Enhancement," Proceedings of the IEEE, Volume 69, Number 5, May 1981, (Special Issue on Image Processing), Figure 6, (a,b,e,f), page 647. © 1981 IEEE. Used by permission.

Note in Figure 9.17 the greatly increased detail available after histogram equaliziation. In (b) the hand, which wasn't visible in (a), is seen. In (d) the details of the Venetian blinds have been brought out. These results were obtained by the use of spatial domain techniques that operated directly on the individual pixels of the image. Smoothing by neighborhood averaging, sharpening by gradient-type operators, and global enhancement by means of histogram modification techniques were used. Details of the techniques are discussed further in Woods and Gonzalez [52].

9.4 Image Storage and Retrieval

9.4.1
High-Speed
Memory

Images must be stored for immediate processing and also for possible future processing. Sometimes it is desirable to store the original image and the image after thresholding, differencing, and other processes have been completed on the original image. When extracting three-dimensional information from stereo pairs, it is necessary to store both images so that corresponding points in each image may be identified. All of these functions require large amounts of memory, and memory cost has been a limiting factor in the storage, retrieval, and processing of images. Examples of memory requirements are: for a 100 × 100 image, 10,000 bits; for a 256 × 256 image, 65,536 bits or 8,192 bytes; for a 512 × 512 image, 32,768 bytes. These are the requirements for each image and must be multiplied by the number of images to be stored.

Fortunately, the cost and size of RAM has been decreasing rapidly, so that the need for increased memory is being met. The new

256K semiconductor chips make large memories possible and are being produced in quantity. Costs are coming down rapidly, so that for robot use, in which considerable memory cost is acceptable, the availability of memory is not expected to be a limiting factor.

9.4.2
Disk Files

Disk files are already more than adequate for storage of any presently required information for image handling in robot application. This situation may change rapidly as more and more information is stored. At present, floppy disk storage units are capable of storing over 1 million bytes per disk, while the hard disks of the Winchester type are available in multimegabyte capacities, up to more than 1 billion bytes per disk system.

Future imaging systems may store multiple copies of images for processing using knowledge base approaches and artificial intelligence concepts. At that time, there may be some limitation on capability because of storage capacity. Even then, a solution is in sight with the use of perpendicular magnetic recording on disks, where densities of 20,000 bits per lineal inch and track densities of several hundred tracks per inch are already available.

A potential problem is the matching of disk data rates with the very high data rates of parallel processors. One solution, previously implemented, is the use of multiple recording and reading heads in parallel. Potentially, hundreds of heads could be used in parallel, each recording at multiple megabit rates, so that the problem is solvable in principle at least, although the solution may be expensive. With multiple heads, the average access time is equal to one-half of the disk rotation time or latency. Buffering of several tracks of data in high-speed memory would be required to implement this solution.

9.4.3
Magnetic Bubble
Memories

Several robot manufacturers use magnetic bubble memories instead of disks for local storage of information, primarily for programs. Magnetic bubble memories are used because of their greater reliability in electrically noisy industrial environments. As the costs of bubble memories are reduced, the importance of high reliability may offset the greater cost compared to that of disks even for image storage information.

9.5 Picture Representation

9.5.1
Digitization

In this book, all pictures or scenes to which a robot has access will be considered to be digital—that is, a picture is represented as a finite grid of squares or rectangles. Each small square is termed a pixel, which is a contraction of the phrase picture element. Each pixel takes on one of a finite number of values according to how white or black the picture is in that small area. For example, if white is represented by the binary value 1 and black by 0, then the letter *F* might look as shown in Figure 9.18. Actual pictures are much larger than the 7 × 5 matrix of Figure 9.18, and the pixels usually take on many values rather than just 0 and 1.

Entire pictures for robot vision purposes are frequently stored in matrices of size 256 × 256 or 512 × 512, which is approximately the

Figure 9.18 The letter *F* as seen by a robot vision system could be stored in a 7 × 5 matrix of 0s and 1s.

```
0  0  0  0  0
0  1  1  1  1
0  1  1  1  1
0  0  0  1  1
0  1  1  1  1
0  1  1  1  1
0  1  1  1  1
```

resolution of a standard television receiver. Furthermore, each pixel can usually take on a range of values, often anywhere from 16 to 256. For example, if there are 16 possible pixel values, then the bit configuration 0000 (decimal 0) represents <u>black</u> and the bit configuration 1111 (decimal 15) represents <u>white.</u> (Those not familiar with binary numbers and hexadecimal numbers should consult Appendix A.1) Numbers between these two extremes represent brightness intensities between black and white (that is, gray), and the whole black–white spectrum here is frequently called a <u>gray scale.</u> The possibility of color is not considered here, and is unnecessary for many robot applications. However, color is available, at greater expense, where it is needed.

There are a number of hardware devices that can convert a scene that is part of a robot's environment into a matrix of gray values in the robot's memory. For example, CCD systems convert the input analog light signal through a digitizer to a digital picture at high speed, typically with a resolution of 256 × 256 pixels. Pixels may take on as many as 256 different gray–level values.

In the remainder of this chapter, it is assumed that the robot's picture is digitized; it has been converted into a matrix of pixels of the type just described. In many applications, we are talking about a whole set of pictures that constitute the robot's environment; also, these pictures may be changing over time as a result of movement of the robot, natural changes in the environment, and/or changes the robot makes in the environment. It is important to keep in mind the dynamic situation that is usually present.

9.5.2 Image Compression

It is clear that storing images (pictures) in the memory of a robot requires much space. A picture may be worth a million bits. Memory can be easily consumed by just one or a few pictures. Can anything be done to reduce these requirements?

If the picture is random, in the sense that adjacent pixels show no correlation, and it is necessary to preserve every pixel, then probably nothing can be done. However, these two conditions are seldom true in practice. Furthermore, in many pictures (random or not), the actual range of pixel values may be much less than the theoretical maximum.

For example, if the nominal range of pixel values is from 0 to 255, but the actual range is from 101 to 164 due to natural low contrast in the images or characteristics of the image acquisition system, then 64 (2^6) rather than 256 (2^8) values are present. In this case, each pixel value can be represented by 6 bits rather than 8 bits.

It is also possible to cut down on the precision of each pixel value. For example, one can truncate the 2 low–order bits of an 8–bit value, and no significant degradation of the picture may occur. A human would probably notice no difference, although the loss in the robot imaging system might not be trivial. (One must be careful not to always apply the same metric in both human and robot systems.)

One may increase the size of each pixel, that is, decrease the resolution. For example, a 256 × 256 picture can be regarded as 128 × 128 pixels, where each new pixel here is the average of a 2 × 2 square of pixels in the original image. This new image can be created, stored, and used thereafter in place of the original. This procedure reduces the memory requirements to one–fourth of what they were, provided no additional precision is required in the values of the new pixels created. An example of one type of image compression, Haar transform coding, is provided in Figure 9.19. In this case, blocks of pixels of different sizes were transformed and then reconstructed to determine the effects of image compression using this advanced technique. Block sizes were $R \times R$ pixels, with R varying from 2 to 5. The results can be seen in Figure 9.19, with increased loss of detail at each step.

(a)

(b)

(c)

(d)

Figure 9.19 Haar transform coding. (a) $R = 2$. (b) $R = 3$. (c) $R = 4$. (d) $R = 5$ (from Chen and DeFigueiredo [9]).

Source: T.C. Chen and R.J.P. DeFigueiredo, "An Image Transform Coding Scheme Based on Spatial Domain Considerations," IEEE, PAMI-5, Number 3, May 1983. ©1983 IEEE. Used by permission.

Various techniques from coding theory can be applied to images that have relatively large areas of constant or near-constant pixel values. For example, if the top row (line) of an image consists entirely of 0's as the pixel values, one can simply record the line number and this zero value, without the necessity of delineating each 0. Similarly, if at some point in the line the value changes from 0 to 1 and remains at 1 thereafter, it is necessary only to record the location of the pixel where this change occurs. Related techniques can be used to record a rectangle or other region of constant-value pixels. (Of course, this operation can be done by a computer program as desired.)

1. Differencing

Differencing is a method whereby only the change in value from one pixel to the next is recorded. For example, if the top line of an image is

203 204 207 207 205 204...

then one might record only the first differences:

1 3 0 -2 1 ...

Only the first actual pixel value (203 in this example) would need to be retained, and then only if the image needed to be reconstructed. If the differences between adjacent pixels are small—which is often the case in practice—then considerable savings in storage requirements may result by storing the differences rather than the actual values. Furthermore, a number of algorithms for image processing use only differences in pixel values, or at least can be reformulated in this manner.

There are many other techniques for reducing the space required for an image. Some of them will be described in Section 9.6, but a detailed discussion is beyond the scope of this book. References to further reading are given at the end of this chapter.

9.5.3
Memory and
Speed
Considerations

The preceding section showed that considerable computer memory is involved in the storage of pictures, and that it is advisable to reduce the number of pixels and/or gray levels whenever possible. The fact that a robot is sensing and acting on information in or near real time adds another dimension to the problem: There may be multiple images that need to be stored, and these images often need to be processed within severe time constraints.

Multiple images may arise from the motion of the robot scanner or from the fact that the robot's environment is changing due to external factors or to changes effected by the robot's operation. Whatever the reason, it is necessary to hold and process these images. In some cases, it is possible to store just the differences in two images, for example, before and after a certain action of the robot. As we saw in the preceding section, storing differences is often more efficient than storing the pictures themselves, and in this case would be the very focus of our interest anyway.

Time constraints may also be a critical consideration. Many operations upon images must be done for every pixel or small group of pixels. Furthermore, some of these operations are quite complicated, leading to significant processing time when carried out over the whole picture. For example, a certain complex edge detection algorithm might take 1 millisecond per pixel; for a 256×256 picture this process would take 66 seconds, which would probably be unacceptably long. Thus, there is always an interest in efficient algorithms at every level.

The time problem is helped, potentially, by the fact that many picture-processing algorithms can be done in parallel over the image. That is, the algorithms involve only a small part of the image, and yet must be replicated over the entire area. Thus, if parallel processors are available, each can work on a different part of the image, taking care to interface the sub-images properly and to communicate results as necessary. In practice, these techniques are relatively undeveloped; progress should be made as the cost of interfacing parallel microprocessors decreases. Section 9.3.2 discusses the available types of parallel processors and their use.

Computing cost has dropped dramatically in recent years. Microprocessor chips are inexpensive, as are reasonably large RAMs. There are specialized chips for doing certain simple algorithms in parallel. Computing speeds are also improving, to the point where the most basic operations (such as a single addition) are measured in nanoseconds (billionths of a second). Still, the considerable requirements in both speed and memory imposed by picture processing for robots present a real challenge to the designer and user. Further hardware improvements can be expected to ameliorate these problems, but corresponding gains in software at all levels are also needed. Active research is proceeding in this area.

Time and space requirements for the various vision algorithms in this chapter cannot be discerned in all cases from the level of detail given here, but they are an appropriate consideration in any robotic application. The "best" techniques are useless if they effectively immobilize the robot for long periods of time or if they require more memory than is available.

9.6 Preprocessing

9.6.1
Local
Transformations

The word <u>preprocessing</u> covers a multitude of operations that may or may not be appropriate for any particular application. The term more usefully refers to any technique that will improve the picture early on in some way, or that will prepare the picture for easier later processing. Reducing noise in the picture is another common use of preprocessing. In this section, we look at a few preprocessing algorithms that can be applied at each point of the image, hence the use of the word <u>local</u> in the heading.

Preprocessing steps should be motivated by what is to be done with the image at higher levels. If lines in the picture are important and are typically only 1 pixel wide, then one normally would not want to do the smoothing about to be described. On the other hand, smoothing is appropriate for reducing certain kinds of high-frequency noise, and it may have no undesirable side effects in some applications. (Note that <u>high-frequency</u> as used here is not related to radio or circuit frequency. We use the term to indicate that there are frequent changes in the values of adjacent or nearby pixels.)

An example of one type of smoothing has been presented in Section 9.5.2, although it was not referred to in that way. There, the storage requirements of a picture were reduced by collapsing a 2×2 square of pixels into a single new pixel and assigning its value as the average of the four component values.

More typically, smoothing does not reduce the resolution of the picture; it reduces the magnitude of isolated, extremely high pixel values (or increases very low values). As an example, imagine a new picture of the same resolution that is created from an old one by replacing each pixel value by the average of the nine values in the 3×3 square of which that point is the center. Symbolically, this operation can be described by the following matrix or filter:

$$
\begin{array}{ccc}
\dfrac{1}{9} & \dfrac{1}{9} & \dfrac{1}{9} \\[2em]
\dfrac{1}{9} & \dfrac{1}{9} & \dfrac{1}{9} \\[2em]
\dfrac{1}{9} & \dfrac{1}{9} & \dfrac{1}{9}
\end{array}
\tag{1}
$$

This matrix is superimposed on the picture at every possible position, and the result is computed by multiplying each filter value by the underlying pixel value and summing the nine products to produce a pixel value in the new picture.

This matrix could just as well be written as:

$$
\begin{array}{ccc}
1 & 1 & 1 \\[1em]
1 & 1 & 1 \\[1em]
1 & 1 & 1
\end{array}
\tag{2}
$$

Here each pixel value is replaced by the sum of its eight neighbors and itself. Since this process greatly increases the average pixel value, the normalization to the 1/9th shown in the first matrix may be desirable. The reader can fill in this step as necessary in what follows.

It should be understood that a new "copy" of the picture is produced in this process, so that changing a pixel occurs only in the copy and does not affect subsequent computations involving neighboring pixels. However, after the new picture is completed, the old one may be unnecessary. In this case, the new one could then occupy the memory space where the old one had been. It should also be noted that many local transformations cause some loss of pixels at the edges. For example, the averaging operation just described maps a 256×246 image onto a 254×254 one, an insignificant loss in most applications.

The possibilities for smoothing filters are endless. For example, the reader might think the preceding filter "unfair" in that the neighboring points are not equidistant from the center. Furthermore, the middle point might be excluded entirely from the computation. (This is not unusual.) The following filter could result:

$$
\begin{array}{ccc}
1 & \sqrt{2} & 1 \\
\sqrt{2} & 0 & \sqrt{2} \\
1 & \sqrt{2} & 1
\end{array} \tag{3}
$$

Larger filters such as 5×5 or 7×7 can be used. Moreover, the filters need not be square, and the number of pixels on a side usually does not have to be odd. Making the number odd is a convenience in explaining some concepts and in certain types of programming, because there will then be a single pixel identified as the center.

Another type of local transformation is useful in finding edge elements and is described in Section 9.7.1. It is quite different from smoothing, and the reader is cautioned not to infer that all filters are used for smoothing or related processes. There are many filters (and some other local transformations) for many purposes. As noted before, the selection of any particular one depends upon the application and the later processing that is to be done.

**9.6.2
Global
Transformation**

A global transformation is one that uses information about the picture as a whole in tranforming each pixel or group of pixels. For example, if the nominal range of pixel values in a picture is 0 to 255 (8 bits), but the actual range is 3 to 57, then the first 2 bits can be dropped since they are 0 in every case. That this can be done is a result of the global knowledge of the pixel range.

Many types of global transformations are used in image processing. As always, the ones employed, if any, depend on the specific application. Some are computationally expensive. Two of general interest are presented here: histogram equalization and the Fourier transform.

1. Histogram Equalization

The histogram of an image is a bar chart in which the x-axis represents the various pixel intensities, and the y-values are the number of pixels at each intensity level. [Figure 9.23(b) is an example of a histogram.]   Intuitively, the more uniform the distribution (i.e., an equal number of pixels at each intensity), the greater the overall contrast in the picture. From information theory, it can also be shown that a uniform distribution of pixel values results in maximum information content of the picture.   Thus, in the absence of other criteria, a histogram of roughly equal numbers of pixels at each level is a desirable goal. We show this process in an example that combines histogram equalization with a reduction in the number of gray levels.

Consider an 8×8 picture in which each pixel can take on one of eight values, numbered in binary from 000 to 111. Suppose the number of pixels having each of these values is as follows:

Intensity	Number of pixels at this intensity
000	3
001	4
010	8
011	10
100	7
101	8
110	10
111	14
	64 (total number of pixels)

To economize on storage, suppose that it is desired to use only 2 bits per pixel, thus saving one-third of the memory space required. A naive approach would be to discard the low-order bit, forming a new 2-bit pixel value from each pair of old ones. The new distribution would be:

Old intensity	New intensity	New frequency (sum of two old ones
000		
	00	7
001		
010		
	01	18
011		
100		
	10	15
101		
110		
	11	24
111		

$$(5)$$

The new distribution (7, 18, 15, 24) is unequal. We can do better. Consider the following alternative grouping of old intensities.

Old intensity	New intensity	New frequency	
000			
001	00	15	
010			
011			
	01	17	(6)
100			
101			
	10	18	
110			
111	11	14	

Here, the frequencies are quite evenly distributed, and we cannot improve upon this result in such a small picture. For most purposes, this would be a superior new coding. It is left to the reader to see how a general algorithm could be developed to carry out histogram equalization on an image of any size. Sometimes histogram equalization is just a redistribution of pixel values; at other times, it is carried out in conjunction with another process, as we have just seen.

Although histogram equalization is conceptually a global process, it is possible to segment a picture along natural boundaries and then to apply histogram equalization to each section. For example, in a robot picture containing both an illuminated wall and a dark floor, separate histogram equalization could be done for each, resulting in potentially better contrast within each area.

Still another approach, which is a hybrid global-local technique, is to replace each point by what it would become if histogram equalization were done on a neighborhood of the point. For example, consider the following 5×5 neighborhood about the center pixel, where the entire picture consists of just four pixel values (represented here as 0 through 3):

					Intensity (in decimal)	Frequency	
3	3	2	3	1	0	2	
2	1	2	3	3	1	4	
3	3	1	0	2	2	7	(7)
3	0	2	3	3	3	12	
2	1	3	2	3		25 (total number of pixels)	

If the frequency distribution is divided into quartiles, both 0 and 1 belong to the lowest quartile, which consists of 6 pixels (2 + 4). In this local histogram equalization, this lowest quartile is recorded as 0, so that the 1 in the center is replaced by a 0. This process is repeated for every pixel, creating a new picture with improved local contrast. In practice, a neighborhood larger than 5×5 probably would be employed.

2. Fourier Transform

There are many global transformations of images. The Fourier transform is one of the best known. In it, the picture is converted to a frequency domain through the use of trigonometric functions. The result is a totally different image, representing periodicities in the original picture. It is more amenable to subsequent mathematical and pattern recognition procedures. It is not discussed further here, but the interested reader can easily find descriptions in many standard reference texts. The so-called Fast Fourier Transform is an especially efficient implementation.

9.6.3 Noise Removal

Broadly speaking, noise in an image is anything that degrades the image with respect to further processing or interpretation. There are many varieties and sources of noise. Some noise may be present in the scene itself, such as light specks or "snow"; some arises in the imaging system that converts the scene into a digital representation. Some noise is local, such as a brilliant reflecting digital spot in the scene. Other noise may be global, as when blurring occurs from movement in the scene or imaging system. This section describes some of the more common types of noise, together with approaches, used to remove or lessen their effects.

A robot imaging system, however good, will usually contribute some noise to the internal representation of a scene. For example, the image produced by a lens system always includes some distortion, usually more pronounced at the edges; an image of a rectangular grid will have many curved lines. A mathematical transformation of the image can restore it approximately to its "true" form. Different transformations may be needed over different parts of the image. In this case, there must be smoothly varying parameters or smooth piecing together of the parts of the image undergoing the various transformations.

The imaging system may be subject to occasional random errors that cause part of the image to be dropped. For example, if one line of a particular image were missing, a solution might be to replace each pixel value in the line with the average of a few of its neighbors in the vertical direction.

Another error of the system might be a systematic tendency to magnify pixel intensities on the right side of the image. If the degree of this magnification were known, an appropriate mathematical transformation could compensate for the error.

Note that in all of these cases, the nature of the noise is known, so that appropriate software can reduce or eliminate its effects. However, the source of the noise may be unknown or random, presenting a problem in processing. Examples include sporadic failure of a component in the imaging system, a power surge in the system, and a bit of dirt temporarily in the optics. It may be difficult to detect (and therefore compensate for) these and many other types of noise. If one knows what to expect in a picture, software may be used to see if the picture generally conforms to this expectation, but if one knew exactly what to expect, the image analysis would not be necessary in the first place.

The scene itself may have many kinds of inherent noise that will make proper analysis difficult. Shadows and illumination that are changing are two such examples. In some cases, the shadows exist where none should be, and they must be treated as noise. In other cases, the shadows provide clues to the height and depth of objects and must be processed carefully to extract the maximum information from them. The robot's environment may be different from time to time, even if the only change is in the illumination. Or the robot vision system may be looking at the scene from different angles. One approach to handling this problem is to build an explicit internal model of the three-dimensional scene in which the notions of illumination and shadows can be accurately simulated. Kanade [22] discusses some aspects of three-dimensional scenes.

Obscuration of one part of a scene is another difficulty. While this event might not technically be called noise, it is a problem in that it impedes correct interpretation and action. Again, a solution may lie in an internal three-dimensional model. Such a model can be developed by looking at the scene from different viewpoints, either at different times or as in stereo vision.

Abnormally bright or dark spots may be present. For example, a small bright area in a scene may result from the reflection of a light source on a shiny surface. It might be possible to model this situation and thereby compensate appropriately, but often precise determination and correction are not feasible. Rather, small bright spots can be processed out after the fact, either by ignoring them or filling them in with an average of their neighbors or with some other values that are thought to be representative. Of course, there is always the problem that a particular bright area may be a significant feature of a scene, and eliminating it is the wrong thing to do. Local context may assist in this matter, but writing programs to incorporate this feature is not necessarily an easy task.

There may be errors in the scene itself. For example, an edge of a machined part may have a nick in it, or a hole in a part may not have been cut to the desired shape. Again, expectations about the scene may permit a correction via appropriate software. Artificial intelligence techniques may offer guidance in such matters.

Noise in images is a fact of life in the real world. It is usually not possible to remove all noise, nor is it necessary. Certain kinds of noise pose no problem in specific applications; others can be more or less adequately compensated for. The designers and users of robot vision systems must learn to live with a certain amount of noise and recognize the degree to which this problem can be overcome in practical situations.

9.7 Edge Detection

9.7.1
Edges and Lines   Edge information in a scene is perhaps the most important single aspect of robot vision. In most applications, it is the knowledge of edges that enables higher-level software to recognize objects and to meaningfully separate parts of a scene. Much has been written about edge detection, but it still remains a fairly difficult problem in the presence of noise. Humans are superb edge detectors, bringing to bear on a given scene not only basic structural mechanisms of the retina and brain but also a full range of intelligence and a world model that materially assists in finding edges under degraded conditions of noise, low contrast, and partial obscuration. A robot vision system can only touch on these capabilities, although considerable progress has been made in recent years. In this section, only a few of the many edge-finding algorithms will be highlighted.

An <u>edge</u> is a local boundary within a scene such that the pixel values on one side of the boundary are significantly different from those on the other side. The four-level picture segment in Figure 9.20 has a well-defined vertical edge in the center. Of course, edges are not usually so precisely delineated.

The segment in Figure 9.21 has a "rough" edge running diagonally from upper left to lower right, which the reader is invited to check some simple computations. Also, note that there seem to be no other candidate edges in this 4 × 4 segment.

Figure 9.20

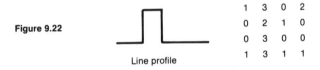

Edge profile

1	1	3	2
1	0	3	3
1	1	3	2
0	1	2	3

Figure 9.21

1	3	2	1
0	2	1	2
1	2	1	3
3	1	0	2

A <u>line</u> is a local boundary composed of relatively high pixel values that separates two regions of significantly lower and approximately equal values. The words <u>high</u> and <u>lower</u> in this definition may be replaced by <u>low</u> and <u>higher</u>, respectively. For example, the second column in the picture segment of Figure 9.22 is a well-defined line. (In larger pictures, lines may be several pixels wide.)

Figure 9.22

Line profile

1	3	0	2
0	2	1	0
0	3	0	0
1	3	1	1

Which is more important for subsequent processing, edges or lines? The answer depends upon the application, but for many purposes, lines are easier to handle, partly because of their (theoretical) binary nature—unlike edges, which depend on gradations over many sets of pixel values. In fact, the most common processing at this point is to reduce a picture with edges to one with lines that replace the edges.

There are many line-processing algorithms that can then be used to connect segments, recognize figures, and generally provide input for a higher level of processing.

**9.7.2
Local Edge
Elements**

Perhaps the simplest way to obtain a local edge (that is, a small edge segment) is to use a template or filter. A small template for finding a vertical edge is:

$$
\begin{array}{ccc}
-1 & 0 & 1 \\
-1 & 0 & 1 \\
-1 & 0 & 1
\end{array}
\tag{8}
$$

Consider the following 5 × 5 image, which appears to have a vertical edge at approximately the middle column:

$$
\begin{array}{ccccc}
1 & 1 & 2 & 3 & 2 \\
1 & 0 & 1 & 2 & 2 \\
1 & 1 & 2 & 3 & 3 \\
1 & 0 & 2 & 3 & 2 \\
0 & 1 & 1 & 2 & 2
\end{array}
\tag{9}
$$

This matrix is filtered by computing the sum of products of each point in the matrix with all points in the filter. Take the center matrix element, 2, as an example. We center the filter over that point so that the center 0 of the filter is over the 2. Now multiply each element of the matrix by the filter element superimposed on it and add them together: $0 \times -1 + 1 \times -1 + 0 \times -1 + 1 \times 0 + 2 \times 0 + 2 \times 0 + 1 \times 2 + 1 \times 3 + 1 \times 3 = -1 + 2 + 3 + 3 = 7$. However, the filter is not allowed to lap over the edges of the matrix, since the values beyond the edges are undefined. Therefore, we lose the values of all of the matrix edge values on every filter operation, so that the matrix shrinks each time. Remember that a matrix is usually at least 64×64, so that this loss is tolerable.

If we superimpose the template upon each of the possible positions on the image, and compute the sum of products at each position, the following matrix results:

$$
\begin{array}{ccc}
2 & 6 & 2 \\
2 & 7 & 2 \\
3 & 6 & 2
\end{array}
\qquad (10)
$$

Note that the vertical edge is much more sharply defined now than it was before.

Thresholding

Now threshold this matrix, that is, replace all elements above a certain threshold (say, 3 in this example) with 1s and the rest with 0s:

$$
\begin{array}{ccc}
0 & 1 & 0 \\
0 & 1 & 0 \\
0 & 1 & 0
\end{array}
\qquad (11)
$$

Center this matrix over the original 5×5 image. The 1s will lie along the middle column, showing where the vertical edge is in that image. By applying other templates, such as

$$
\begin{array}{ccc}
1 & 1 & 1 \\
0 & 0 & 0 \\
-1 & -1 & -1
\end{array}
\qquad
\begin{array}{ccc}
1 & 1 & 0 \\
1 & 0 & -1 \\
0 & -1 & -1
\end{array}
\qquad
\begin{array}{ccc}
0 & 1 & 1 \\
-1 & 0 & 1 \\
-1 & -1 & 0
\end{array}
$$

$$
\qquad (12) \qquad\qquad\qquad (13) \qquad\qquad\qquad (14)
$$

one can find one horizontal edge and two diagonal edges, respectively. (In these examples, the 0s in the templates locate the edges to be found.)

The foregoing transformations can be done over an entire picture. The picture can actually be replaced with appropriate threshold matrices, so that binary <u>line</u> elements take the place of what were originally <u>edges</u>. Note that directional information can be retained, since different templates were used to extract edges. A large picture can thus be reduced to a line drawing, although it may be quite rough and require much subsequent processing. Figure 9.23 shows some examples of the steps in processing a digitized picture of industrial parts.

Larger and more complex templates may be used; they may incorporate smoothing, for example. Different thresholds may be used in different parts of the picture, depending on local contrast and other factors. Various adapative processes may also be employed.

The <u>Laplacian</u> is another method of finding edges. It can also be implemented via a template; for example:

$$
\begin{array}{ccc}
-2 & -3 & -2 \\
-3 & 20 & -3 \\
-2 & -3 & -2
\end{array}
\tag{15}
$$

The weights in this template are approximately proportional to the sum of four directional numerical second derivatives, modeled after the mathematical Laplacian operator. It is used in essentially the same way as the previous templates, with thresholding above a certain value of the output. However, directional information is lost. This may not matter if one is simply connecting points that comprise an edge, and the only interest is in the finished line drawing. Also, the Laplacian operator tends to enhance noise (as operators based on derivatives usually do); larger Laplacian templates (for example, 9 × 9) can ameliorate this condition by performing smoothing as well. Again, the proper size Laplacian to use depends upon a number of

(a)

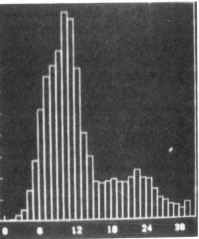

(b)

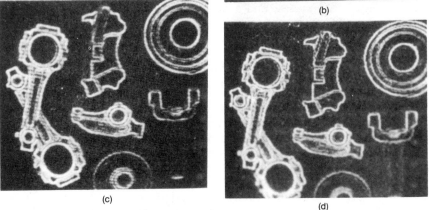

(c)

(d)

Figure 9.23 (a) Digitized picture of some industrial parts. (b) Intensity histogram. (c) Edge magnitude. (d) Thinned edge magnitude (from Perkins [36]).

Source: W.A. Perkins, "Area Segmentation of Images Using Edge Points," IEEE, PAMI-2, Number 1, January 1980. © 1980, IEEE. Used by permission.

factors, including the resolution of items of interest in the image and the kind and extent of noise present.

Once individual edge elements are located by any method, it is necessary for the robot vision system to attempt to connect these edge segments together into meaningful edges of the picture. Joining these pieces together is sometimes called segmenting, but this term could imply that we are cutting the pieces smaller. Actually, we are trying to find the outlines of boundaries of the objects in the picture by joining small edge pieces together to make larger pieces. Unless the picture is very noisy, this would be a fairly easy task for humans because expectations and experience can be brought to bear on the problem. For robots, the problem is often difficult. Artificial intelligence techniques can be used, but it is first necessary to do certain lower-level processing.

First, it may be desirable to have the program discard isolated edge elements (i.e., those with no immediate neighbors), since these are probably nothing more than noise in the scene. Isolated pairs of edge elements (two adjacent edge elements with no other immediate neighbors) may also be discarded, although as pairs and triples are considered, it is important to know whether there is a reasonable chance that these small groupings of edges might actually be meaningful parts of the scene. For example, if it is known that the scene contains only relatively large objects with no minute features, then small isolated sets of edge elements are probably insignificant and can be discarded with no further consideration. On the other hand, small features may be important, such as a set of small holes in a machined part. The actual algorithm used clearly depends on the specific application. The correct threshold for discarding sets of edge elements may be determined by experimentation in some situations.

Assuming that processing has been done to remove obviously insignificant edge elements, some kind of tracking algorithm is now necessary to put elements together into coherent edges. One such method is as follows: Find any two adjacent edge elements, and then draw a line through these 2 pixels to an adjacent pixel. If that pixel is an edge element, then continue the line in that direction. If it is not, then look at the immediate neighbors of the adjacent pixel. If any of these are edge elements, then continue the line in that direction. If no further edge elements are found in the vicinity, the local process terminates, and a new starting pair is found elsewhere in the picture, if any others exist.

Pixels are marked as they are processed, such as by appending an additional bit to each. A copy of the scene with its developing edges is kept. If directional edge element information was acquired in the first place, then the preceding algorithm can be modified to incorporate this knowledge. For example, adjacent elements that point in the same direction are evidence of a true edge in the scene. Likewise, adjacent elements that are nearly opposite in direction may be cause for discarding one or both, although in some scenes a sharp point on an object could cause this circumstance to occur legitimately.

When there is an intersection of lines or edges in a scene, edge-following algorithms will yield multiple choices for continuation at some points. At such a point, the algorithm must arbitrarily choose one direction and keep track of the other possibilities. When the processing in the first direction is exhausted, the algorithm backs up to another possibility. Of course, in following the first direction, there may be alternative choices at a later point; these too must be saved for later exploration. The whole process can be done in a systematic manner, and is referred to as backtrack programming in computer science (Nilson [31], Ballard and Brown [4]). The algorithm finally terminates, yielding a new picture that is basically a line drawing of the original scene.

Processing may still be necessary to yield a "clean" picture; if the original image was not too noisy, certain heuristics can be profitably employed to provide a still better picture. The word heuristic means a process that seems relevant to a particular situation but is not guaranteed to be effective. An example of a heuristic that might be useful at this point is one that looks for gaps in edges. Whenever there is a relatively long edge with one segment missing in the middle, fill in the gap with a segment whose direction is the average of the directions on each side of it. (Humans do this automatically.) Care must be exercised in programs that add segments to pictures, lest noise (i.e., an inappropriate segment) be added that degrades rather than enhances further processing.

If one knows what to expect in a scene that the robot vision system is processing, as is usually the case, then edge detection and later processing can be guided by these expectations, ultimately leading to a more accurate understanding of the scene and more appropriate robot action based on this comprehension. For example, if the scene contains circular pieces of known radii, then the edge detector can be broadened to string together segments that would collectively have the right amount of curvature. More complex shapes can also be anticipated in the processing here, but at some point it is useful to separate out these techniques as a higher-level process, namely, that of pattern recognition. This subject is briefly covered in the next section.

9.8 Recognizing Objects

9.8.1
Template
Matching

The problem of determining the identity of an object from knowledge of its shape and/or various other characteristics or measurements is the branch of applied science called pattern recognition. This subject embraces many other disciplines, including mathematics, statistics, and the particular domain in which the patterns of interest lie, such as music, linguistics, or biology.

Pattern recognition is a very extensive subject, and only a brief introduction is given here. It is restricted to the situation in which the items of interest are two-dimensional line representations of objects in a robot's environment. Furthermore, the number of different objects to be recognized is usually small, and even the relative locations of certain objects may be known fairly accurately. Thus, the constraints found in many robot environments assist in the recognition, although this effect may be offset by the fact that recognition by robots often needs to be more accurate than that in many other pattern recognition applications. One cannot afford to have a robot select the wrong part in an assembly process, for example.

The most obvious method for recognizing the outline of a figure by automated means is to use template matching, that is, to compare the edge-processed figure (as described in the previous section, assuming here that only a single object is present) with stored representations (templates) of various possible objects, and determine which is the best match. For example, consider the following unknown object (where 1s represent a line drawing of the object) and ask whether it is most like the L-shaped, C-shaped, or the O-shaped figure.

Out of a possible 16 matches, the following scores are obtained in comparing the unknown with each of the shapes: $L = 10$; $C = 13$; $O = 11$. (The reader should verify these numbers.) We thus conclude that, of the stored figures, the unknown figure is most like the C-shape, and should be so classified in the absence of further information. There is also the assumption that 13 out of 16 is a sufficiently high score to not reject the unknown figure as unclassifiable.

1	1	1	1		1	0	0	0		1	1	1	1		1	1	1	1
1	0	0	0		1	0	0	0		1	0	0	0		1	0	0	1
0	1	0	0		1	0	0	0		1	0	0	0		1	0	0	1
0	1	1	1		1	1	1	1		1	1	1	1		1	1	1	1

Unknown L Shape C Shape O Shape

There are many variations on the basic template-matching concept, such as more complicated similarity measures and transformations of the picture space. Real-world problems arise, such as rotation, scale change, and distinguishing between figures that have identical or closely matching scores. The reader is invited to consult almost any book on pattern recognition for further detail on these matters.

**9.8.2
The Hough
Transform**

There are many other ways of recognizing objects in a scene besides template matching. In this section, we give an example of another in wide use; the Hough transform (Ballard and Brown [4], Nevatia [30]). First, we illustrate its use when the object to be recognized can be described in its entirety by one set of parameters. Consider the problem of recognizing straight lines in a scene. (In earlier sections, we encountered the problem of finding short straight-line segments; here, we are concerned with finding longer straight lines in a scene by a more global technique.) The equation of a straight line in an (x,y) coordinate system is

$$y = mx + b$$

where $b =$ the y axis intercept of the straight line and $m =$ the slope of the line.

Suppose there are pixels in a scene that are edge elements, as identified earlier. Consider any one of these, and call its coordinates (x', y'). Then we have

$$y' = mx' + b$$

which can be written

$$b = -x'm + y'$$

where x' and y' are considered for the moment as fixed and m and b are variable. In (m,b) space this is a line with slope $-x'$ and intercept y'. Each point in the scene thus generates a line in (m,b) space. Points that lie on a straight line in (x,y) space generate lines in (m,b) space that intersect at the particular m and b that are the parameters of the original straight line. The algorithm then is to make a grid in (m,b) space, which can be regarded as a rectangular array initially containing all 0s. For each candidate point in the scene, determine the m and b from the preceding equation, and increment by 1 all of the grid cells that the line passes through. At the end, a threshold is used to extract those points in (m,b) space that have the largest counts. Each one corresponds to a line in the original space.

1. Application of the Hough Transform

An example will help make the Hough transform clear. Suppose the scene to be analyzed contains the following subscene, with five points (identified as edge elements by previous methods) located as shown:

$$
\begin{array}{llllll}
2 & \text{o} & . & . & . & . \\
  & . & \overset{3}{\text{o}} & . & . & . \\
  & . & . & \overset{4}{\text{o}} & . & . \\
1 & \text{o} & . & . & . & . \\
  & . & . & . & . & \text{o} \;\; 5
\end{array}
\tag{16}
$$

It is immediately obvious to the human eye that points 2 to 5 lie on a single straight line, while point 1 is outside this line (although it could lie on some other line in the larger scene). But the automation of this detection via the Hough transform is not a trivial process. First, record the points and their coordinates. (The origin is arbitrary: For this example, it can be considered to be located at the lower left.) Also, record the equation that results when the (x,y) coordinates of the points are substituted into the general straight-line equation $y = ax + b$:

Point number	Coordinates	Result of substituting into $y = ax + b$
1	$(0,1)$	$b = 1$
2	$(0,4)$	$b = 4$
3	$(1,3)$	$b = -a + 3$
4	$(2,2)$	$b = -2a + 2$
5	$(4,0)$	$b = -4a$

Then plot each of the five lines (whose equations are in the rightmost column) in (a,b) space. The points in (a,b) space with the most intersections correspond to the parameters of straight lines in the original scene. In this example, it is seen that four lines in (a,b) space intersect at the point $(-1,4)$. See Figure 9.24.

This point corresponds to the equation

$$y = -x + 4$$

in the original subscene, which is in fact the equation of the line through points 2 to 5. (The reader should check this result.)

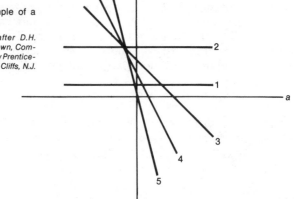

Figure 9.24 Example of a
Hough transform.

*Source: redrawn after D.H.
Ballard and C.M. Brown, Com-
puter Vision, © 1982 by Prentice-
Hall Inc., Englewood Cliffs, N.J.*

How does the computer actually know that four lines intersect at
the point (-1,4) in (*a,b*) space? As mentioned before, imagine a grid of
square cells, each 1 × 1, superimposed on (*a,b*) space such that the
center of each grid cell is a node in (*a,b*) space that has integer
coordinates. Every time a line is "drawn" in (*a,b*) space, corresponding
to a point in the original scene--increment by 1 all of the grid cells
through which that line passes. For example, line 3 in the previous
diagram will cause those grid cells to be incremented whose centers
lie at the points (-2,5), (-1,4), (0,3), (1,2), (2,1), (3,0), (4,-1) ... and so
on, that is all the cells the line passes through within the bounds of the
(*a,b*) diagram.   The computation of which grid cells are to be
incremented for any given point in (*x,y*) space is fairly straight-
forward, based on simple geometric considerations. After all of the
points in (*x,y*) space are processed by the preceding algorithm, one
looks for local maxima in (*a,b*) space, that is, grid cells that have
higher counts than their neighbors and are significantly different from
0, considering the scenes as a whole.   These values will represent
parameters of lines or line segments in the original scene. In actual
practice, each grid cell is usually smaller than 1 × 1 in order to get
more accurate determinations of *a* and *b* .   The best size to use
depends upon the particular application, and is controlled by such
factors as the number of edge elements in the scene and the kinds of
noise present in the image.

A problem with the preceding formulation is that *m* is potentially
infinite (in the case of a vertical line.) This situation can be handled
by using the normal form equation of the straight line, which is

$$x \cos \theta + y \sin \theta = r$$

Here we are looking for concentrations of points (intersections)
in (*r,θ*) space rather than in (*m,b*) space.   Otherwise, the process is
similar. If directional information about the edge elements is known,
then the Hough algorithm becomes a search in one-dimensional space,
which is inherently more efficient.

The Hough algorithm is especially useful if segments of a
straight line are missing for any reason, such as obscuration of one
object by another or general degradation of the scene.   Boundary-
following techniques generally cannot handle such situations.   However, this same feature can cause pieces of a line to be improperly
regarded as a single line in the real world—for example, two separate
but adjacent tables of the same height being interpreted as one.
Additional knowledge must be used to resolve such a situation. Also,

the Hough transform usually will have trouble with close parallel lines. Despite these limitations, the technique has proven useful in a variety of applications.   It is directly applicable where the object to be recognized can be described as a single function of several para- meters, such as an ellipse.   It can also handle a rotation and/or scale change simply by incorporating these parameters as additional vari- ables in the formal description of the object.

The Hough transform can be generalized to recognize objects of particular shapes that are not readily expressed as a single function of several variables, for example, a square.   An algorithm for doing this is as follows: Take a prototype (an idealized representation) of the figure and choose some interior reference point.   With this point as the center, imagine a radius that sweeps around one complete rotation of 360 degrees, variable in length and always just touching the boundary of the figure.   For any angle at the center, a particular angle is created with the edge segment.   This latter angle and the associated radius can be stored in a table for a number of different angles formed at the boundary.   For a given angle there may be more than one possible radius, due, for example, to concavity of the figure.   To determine whether an unknown figure is an instance of the stored one, trace its boundary, and for each boundary segment compute a possible center using the table information and incrementing accumulators of possible centers.   At the end of the process, high densities of points in (x,y) space indicate probable centers of instances of the located figure.

The preceding technique suffers from some of the same problems as the simpler version.   Also, two dissimilar objects that nevertheless share a significant common part may be recognized as the same object.   Of course, postprocessing may be used to resolve the ambiguity in this case by looking at other features of the object.   In fact, much of pattern recognition can be characterized by a feature extraction process that seeks to identify the distinguishing charac- teristics of an object and then to classify it on the basis of the presence or absence of certain of these features.   Again, the reader is invited to consult any of several pattern recognition books for details of this approach (Ballard and Brown [4], Duda and Hart [11].

9.8.3 Syntactic Methods

A different approach to the recognition of objects is afforded by the branch of pattern recognition called syntactic, structural, or linguistic (Gonzalez and Thomason [16], Fu [13, 14]).   The basic idea is that objects to be recognized are regarded as being composed of primitives, which are small picture segments (line segments, curves, etc.) that comprise the picture and are not further divisible.   Further- more, the relationships among the various segments are of great importance and can usually be described by some formal notation that is utilized in the actual recognition.

For example, the recognition of a table in a scene might proceed on the basis of its structural definition as a "horizontal rectangular surface supported at the corners by legs of equal length."   There must also be stated relationships among the various parts.   (An immediate problem, of course, is to devise the necessary three-dimensional information from two-dimensional pictures.)   In any scene, if a table may be present, virtually every word of the preceding definition and the accompanying relationships must be carefully checked for conformity.

9.8.4 Feature Recogni- tion Methods

After an image has been processed into lines and shapes, there are several ways in which further recognition steps may be applied. Template matching and syntactic methods are possibilities, but neither of them has been successfully applied to recognition of industrial parts such as covers, hinges, or yokes, where recognition is required. SRI International has developed several approaches to this type of recognition, as reported in a series of reports by Rosen et al. [41, 42] and Nitzan [32, 33, 34]. Two distinct methods will be discussed here. Although the second method is really an improvement on the first, the change in approach justifies considering it a new method.

1. Image Parameter Measurements

Computer vision systems can measure objects precisely, and can determine orientation and a number of mathematical relationships by suitable programming (Nitzan [33]).

Particular features of objects can be used to identify the objects uniquely in some cases. Sometimes the features can be used to classify the object for further processing. Before discussing the features used, we will define some useful terms in common use.

a. Blob. An arbitrary pattern of contiguous pixels. As a result of processing, the blob has a border of line segments but has not yet been identified. An example of a blob is given in Figure 9.25.

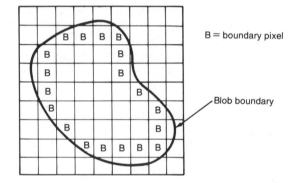

Figure 9.25 Pixels on the boundary of a blob.

Source: SRI International, formerly The Stanford Research Institute, Menlo Park, CA. Used by permission.

B = boundary pixel

Blob boundary

b. Area of a blob. The area is somewhat arbitrary because the border of the blob does not exactly coincide with the rectangular corners of the pixel elements. By counting the pixels inside the heavy border of straight lines, the area can be determined with sufficient accuracy. In this case, the area is 38 squares or pixels. On a repeated scan of the same blob, the area might vary by a few pixels.

c. Centroid of a blob. The X coordinate of the centroid is found by summing the X values of all the pixels and dividing by the total number of pixels. However, determination of the X value must be done with care. In Figure 9.25, we will assume that $X = 0$ and $Y = 0$ at a point up 1/2 pixel and right 1/2 pixel from the lower left corner. Then the center of each pixel is at an integral X value and integral Y value. There are then the following values: 1×4 pixels $+ 2 \times 6$ pixels $+ 3 \times 7 + 4 \times 7 + 5 \times 7 + 6 \times 4 + 7 \times 3 = 145$. So the X coordinate is at a distance $145/38 = 3.815$ from the starting point of the coordinate system we selected. The Y coordinate may be found in the same way. This problem will be left as an exercise. Note that this definition of a centroid is the same as the balanced moments definition of mechanics, which states that

$$x_c = \frac{W_i\, X_i}{W}$$

where X_c = the X coordinate of the centroid, X_i = the X location of the ith set of pixels, W = the total number of pixels (in mechanics, it is the total weight of the object), and W_i = the number of pixels at the X_i location. This corresponds to weight in the usual mechanical equation for the centroid. The location of the centroid is a function only of the shape of the area and is independent of the selection of coordinate axes with respect to the area.

d. Perimeter Length of the Object. The shape of an object or blob determines its perimeter length. In Figure 9.25 there are 28 pixel edges in the perimeter, but there are only 19 pixels labelled B as boundary pixels.

e. Area-to-Perimeter Ratio. A useful identifier is the ratio of the area, which is 38 in this case, to boundary pixels, which is 19, so that this measure is 38/19 or 2.0. A long object may have a different ratio, a circle would have a larger ratio.

f. Maximum Dimension. If the object is oriented along the x and y axes, we can measure the maximum x and y dimensions. Even if the object is rotated the maximum distance across, it can be measured and will not change with rotation. In Figure 9.25, the maximum dimension is along the diagonal and is about six units or pixels. The maximum x and y dimensions are seven pixels.

g. Maximum Edge Segment Length. By counting the pixels along the edges, we find that the longest segment is five pixels long.

h. Boundary List. A sequential list of the segments that make up the boundary of the object.

These are some of the parameters that can be obtained by measurement and mathematical analysis of the features of a simple blob. More complicated analysis can provide the second, third, and higher moments of a figure. Moment invariants—certain linear combinations of moments—are often useful in recognizing simple objects (Wilf [50]).

An example of the application of these and other features to part identification is summarized by Kruger and Thompson [25] based on the work at SRI International. Seven possible features were extracted from each discrete object. These were:

x_1 Perimeter of the object
x_2 Square root of the area
x_3 Total hole area
x_4 Minimum radius
x_5 Maximum radius
x_6 Average radius
x_7 Compactness (x_1/x_2)

These features were abstracted from the parts shown in Figure 9.26 and analyzed by the decision tree of Figure 9.27. Sometimes only two features were needed to identify a part uniquely, since there were only a few possible parts. The items shown were foundry castings placed on a conveyor for processing.

This general approach to visual processing has been incorporated into the SRI Vision Module. A description of the hardware for this Module is given in Section 9.11.1, along with a discussion of some of the system concepts that were important. The Machine Intelligence Corporation was formed by SRI personnel to make this system available commercially. Other companies produce industrial vision systems based on this method, although each of them has developed its own variations and improvements. The Automatix Corporation, for example, has developed a new programming language called RAIL, which is used with their vision system.

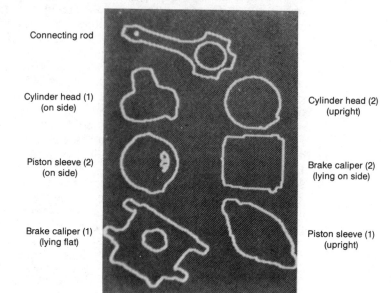

Connecting rod

Cylinder head (1)
(on side)

Cylinder head (2)
(upright)

Piston sleeve (2)
(on side)

Brake caliper (2)
(lying on side)

Brake caliper (1)
(lying flat)

Piston sleeve (1)
(upright)

Figure 9.26 Edge-detected parts on a conveyor.

Source: R.P. Kruger and W.B. Thompson, "A Technical and Economic Assessment of Computer Vision for Industrial Inspections and Robotic Assembly," Proceedings of the IEEE, Volume 69, Number 12, December 1982, Figure 10, page 1530. © 1982 IEEE. Used by permission.

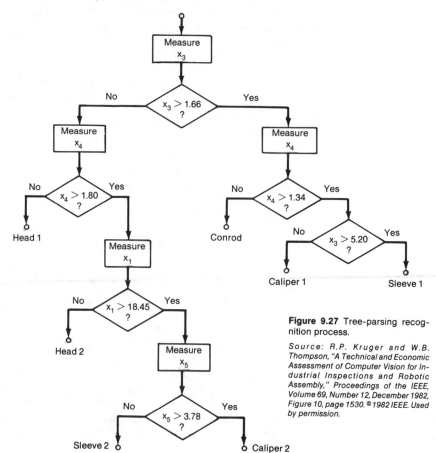

Figure 9.27 Tree-parsing recognition process.

Source: R.P. Kruger and W.B. Thompson, "A Technical and Economic Assessment of Computer Vision for Industrial Inspections and Robotic Assembly," Proceedings of the IEEE, Volume 69, Number 12, December 1982, Figure 10, page 1530. © 1982 IEEE. Used by permission.

2. Local Feature Focus System

SRI International has continued to work on improved vision systems and is currently doing development on the Local Feature Focus System (LFF). This work was started in 1978 and has been improved in several ways since then, as reported by Bolles [6,7] and Nitzan et al. [34].

LFF can recognize both individual and partially viewed objects that can be viewed in a two-dimensional image. Typical objects are cylinder heads, hinges, covers, and so on. There are three phases in the LFF system:

a. Model acquisition.
b. Feature selection.
c. Runtime processing.

During model acquisition, the system acquires pictures of randomly oriented objects by the usual scanning and preprocessing algorithms. It then locates the features considered most important for recognizing a particular part or group of parts. The features determined are the same as those listed in Figure 9.25, but additional features are used. Especially important in LFF are the boundary list of line segments, orientation of major and minor axes, corner locations, and the number and sizes of holes in the parts.

Usually only a few of these features are needed. An important capability of the LFF method is the ability of the recognition program to learn which features are most important for a particular application. Several parts can be shown to the vision system, and a feature selection program will decide which features are best able to differentiate between the parts. During the training phase, the parts are separated to provide the program with sufficient information so that feature selection is optimum. During operation the parts may be overlapped, so that a second set of features may be needed.

Runtime processing occurs in three major steps:

a. All local features are extracted from an image.
b. A hypothesis is made by the program as to which part is being observed. This hypothesis is based on the local feature considered most important by the feature selection program.
c. The hypothesis is verified, if possible, by finding the other features that are expected to be present if the part is the one hypothesized. If the hypothesis is not verified, a new one is tried.

Use of a primary feature and several secondary features is important in making the LFF method effective. Previous methods used either sequential testing of features, which was fast but not always effective, or parallel testing of all features, which used an excessive amount of processing time. LFF uses a minimum of processing time to find only those features most useful in identifying the part, and yet is capable of handing complex and even overlapping parts.

9.9 Scene Understanding

For many industrial tasks, it is sufficient to recognize parts without needing to understand what is going on in the image. Once a part is located and identified, it can be picked up and handled as a unit without regard to its relationship to other parts in the same image. However, for many applications, it is necessary to know the relationships between objects in the picture in order to determine the necessary action. Scene understanding is the ability to understand relationships between parts of the scene as a prelude to formulating an action plan.

In the assembly of small mechanisms, the vision system must locate the parts to be assembled, recognize them, pick them up in the correct orientation, and place them on the assembly correctly. Complex welding tasks depend on the ability to locate the seam to be welded even though it has a three-dimensional contour. Vision is valuable in tracking the seam and verifying that the weld pool is melted properly so that the weld will have the proper strength. In these cases, scene understanding can be useful.

Relationships between objects in a scene must be represented in memory so that information can be stored and retrieved as needed. In three-dimensional scenes, it is necessary to store depth relationships in addition to the normal two-dimensional relationships. Several techniques are available to organize information efficienty.

The purpose of the vision system is to provide information to aid the robot in carrying out its task. Therefore, the information in the scene must be analyzed and acted on in such a way that the robot is able to refer to it and use it. Achieving this ability to carry out a complex task, depending on vision and other sensory information, is the domain of Artificial Intelligence. As we understand better what is required, we turn more and more to the techniques and methods of Artificial Intelligence (AI). We will discuss only a few of the basic ideas of AI and the way in which robot vision systems can operate. The reader is referred to the literature on Artificial Intelligence for further information. Ballard and Brown [4] provide a thorough and enlightening discussion of the application of AI to vision and will be of great value to the interested reader. Barrow and Tenenbaum [5], Gonzalez and Safabakhsh [17], Kanade [22], Kelley et al. [23], and Chen and DeFigueiredo [9] are also of great value in consideration of some aspects of computer vision application.

9.9.1
Knowledge
Representation

Models of an object can be used to represent what we know about the object, but how do we represent the model? In CAD it is possible to represent arbitrarily complex objects by the use of a coordinate system and a listing of the points, intersections, surfaces, and so on in the object. A listing of the features of an object is actually a model of a particular kind used for a specific purpose. In Section 9.8.4 the feature method of representation was shown to be a useful way to recognize objects so that a robot could handle them.

Relationships between objects in a scene can be done with semantic nets. In Figure 9.28 the relations between an ordinary chair, an armchair, a highchair, and a stool are shown by means of a semantic net. This could just as well be the description of parts being assembled in a factory. Terms such as wide, left of, between, and similar relational terms can be used to aid in organizing information in useful ways.

Gonzalez and Safabakhsh describe the use of a hierarchical structure diagram to represent a simple scene. This structure has the same form as the organization chart for a company. At the top level is the scene identifier, at the next level is the name of an object, and each level is a more detailed breakdown of the preceding one.

Several other methods are discussed in the literature previously referenced.

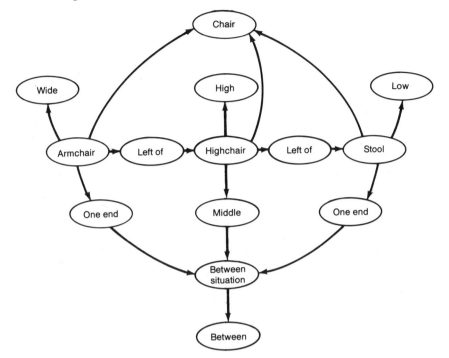

Figure 9.28 An example of a semantic net (from Ballard and Brown [4]).

Source: D.H. Ballard and C.M. Brown, Computer Vision, page 327, Figure 10. © 1982, Prentice-Hall Inc., Englewood Cliffs, N.J. Used by permission.

Specific information supplied to the robot control system from the vision system is:

a. The coordinate location of objects in a scene.
b. Pickup points that have the proper characteristics to allow the robot gripper to seize the part.
c. Identification of parts and objects in a scene. The robot performs different functions in regard to a conveyor belt than it does for a part on the conveyor, for example, and thus needs different information.
d. Orientation of parts and objects in a scene.
e. Measurement of the relative velocity of moving objects with respect to the robot. This is not being done often, as yet, by vision systems. Velocity is measured by other means for most uses.

The success of robot vision systems over the next decade will be determined by how well this basic information is determined and supplied to the robot to control its operations.

9.9.2
Three-Dimensional
(3D) Scenes

Hardware methods for obtaining depth and range information in an image are described in Section 9.1.8. Availability of this added information improves, in some cases, the ability of the software to recognize and understand scenes. In other cases, it is a source of additional complexity that must be dealt with.

In stereo views of images, it is necessary to match corresponding points in the two images of a scene in order to take full advantage of the depth information available. This task is difficult and so has not often been used for robotic tasks. Active measurements of range through triangulation, laser time-of-flight techniques, and projection of structured light patterns have been found to be more immediately useful. These methods are discussed by Jarvis [21].

When the depth of an object is known a priori as a result of other information, it is possible to simplify the matching of stereo images to obtain other information about the scene. Use of a known surface is especially valuable in matching points in the picture and extracting information. Kanade [22] makes use of the fact that buildings have vertical edges and are mostly block-shaped to simplify matching of stereo images. In this case, some of the 3D locations of junctions can be computed and used in analysis. This technique is used in the Carnegie-Mellon University Incremental 3D Mosaic system to build a scene description from stereo images. An advantage of the system is its ability to add new information incrementally as it becomes available as a result of additional image information.

An important task for robots is picking parts from bins, something that humans do well but robots do poorly. It is frequently uneconomical or even impossible to maintain the orientation of parts moving from one machine to another. Kelley et al. [23] discuss three vision algorithms that have been used for acquiring workpieces from bins. The algorithms do not identify the position and orientation of the part. Instead, they are used to recognize a section of a part where edges are parallel so that a parallel-jaw gripper may be used to grab it; in other cases, there is a flat area where a vacuum pickup can be used to pick up the object. Since the objects to be picked up are known in advance, the local recognition problem is simplified.

Much work has been done on 3D scene analysis, so that many useful tasks can now be performed. Considerable research will be required, however, before robot operation in an unstructured environment is possible.

**9.9.3
Feedback to the
Robot**

If it is to be useful, robot vision must supply information to guide the robot in performing its tasks. Success in achieving this goal has made possible the use of robots in assembly, welding, and handling tasks that would otherwise not be possible.

9.10 Interfacing and Control

Interfacing between various pieces of equipment is always an important consideration in any computer system. As discussed more fully in Chapter 7, it is necessary to provide physical, electrical, signal level, and timing controls so that information can be communicated across the boundary between pieces of equipment.

Physically, the electrical cable connectors must fit together. Since there are many different cables and plugs in use, this is an important requirement. There are subtleties of wiring that can cause utter confusion to both a novice and an experienced engineer or technician.

Digital signals are sent in coded form, with a particular voltage level representing a binary 0 and a different level representing a binary 1. These signals may be inverted in value or displaced along a voltage reference. Characters or bytes of information may be sent in communications form, with a start bit, several data bits, 1 or 2 stop bits, and a parity bit. These bits must all match to ensure that accurate communication is possible. Therefore the exact coding must be known and adjusted for by the interface electronics.

Some signals are sent in parallel, with 8 or 16 wires each transmitting 1 bit to another parallel interface. More common is the use of serial interfaces where two wires—the signal and ground,—or three wires—the transmitted signal, the received signal, and ground—are used. Parallel and serial interfaces may both have the same kind of plugs, but they are not compatible and no communication will take place if they are plugged together. In addition, there are other wires carrying handshaking signals that must be considered.

Data rates are important. Common communication data rates are 150, 300, 600, 1,200, 2,400 and 19,200 baud, where baud corresponds roughly to bit per second. Actually, the baud is the individual bit in a communication code, excluding start and stop bits, and thus averages out to a rate of less than a bit per second. Some circuits will automatically adjust the receiving baud rate to the transmitted baud rate; others will not. Usually it is possible to tinker with the baud rate switch on a piece of equipment until it works correctly if everything else is correct. But everything else may not be correct, and it takes considerable knowledge and some equipment to determine what is wrong.

Video signals from cameras are in analog form, which means that they appear as varying voltage levels, varying current levels, or, in special cases, as constant voltage or current levels of varying time duration. Signal magnitude, intensity, or duration carries the important information to be processed, so that considerable care is required to ensure that none of the signal is distorted or lost. This type of signal is familiar to anyone who uses a high-fidelity audio system that also uses analog signals that can be easily disorted by extraneous noise or electrical disturbances.

Digital signals carry information at standardized levels, so that a small variation in intensity is not critical. However, the timing of signals is critical and must be preserved. Noise pulses can be injected from high currents or voltages in the environment that can mask the digital signal and cause a string of data pulses to become invalid.

Control circuits in the computer and in associated pieces of equipment are required to ensure that each of the preceeding considerations is handled properly.

Software interfaces must also be considered. Internal data may be in ASCII, EBCDIC, pure binary, or other forms and must be converted to the correct form by suitable software before use. Data may be stored in packed form to save space but must be unpacked before use. Complex algorithms may be in use to improve the efficiency of data location and storage, so that several levels of software must operate properly to ensure that correct results are obtained.

Processing programs can be run under some operating systems or monitors, but not others. It is important that these characteristics be considered in evaluating system operation.

9.11 Examples of Vision Systems

**9.11.1
SRI Vision
Module**

Automatic computer vision is essential for flexible automation systems in an industrial environment. The SRI Vision Module developed by SRI International was the first complete vision system suitable for industrial use. (This vision system, known as the VS-100, is now available from the Machine Intelligence Corporation in Sunnyvale, California.)

The Vision Module has three major components: a GE Model TN-2200 TV camera with a 128 X 128 array of solid-state elements, a DEC LSI-11 microcomputer with a 28K memory, and an interface processor between the TV camera and the microcomputer. Vision library programs to operate the system are stored in the microcomputer memory. A block diagram of the system is given in Figure 9.29.

The key attributes of the SRI Vision Module are (Gleason and Agin [15], Nitzan et al. [32]):

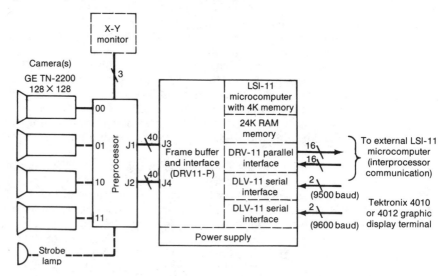

Figure 9.29 SRI Vision Module. This block diagram shows the components and their relationships.

Source: SRI International, formerly The Stanford Research Institute, Menlo Park, CA. Used by permission.

1. Emphasis on software that can be used to develop application programs by calling the vision library subroutines stored in the LSI-11 memory. This capability is important because it makes the Vision Module a general-purpose system.

2. The ability to recognize parts on the basis of their size and shape regardless of their position or orientation. Not only does this help reduce the need for jigging of parts to preserve orientation, it is also useful for qualitative in-process inspection. Position and location of parts may be calculated by the Vision Module so that a robot arm can be visually guided to acquire it. A part usually has several stable states that can be recognized by a set of shape descriptors, such as area, perimeter, number of holes, area of holes, length of major and minor axes, and other descriptors.

3. Training-by-showing is the third key attribute. A skilled programmer is not required to operate the equipment. Instead, an unskilled operator may show the Vision Module several parts in different orientations. These views are analyzed by the Vision Module, and the new image descriptors are computed from these images automatically.

One reason for the success of the Vision Module was the deliberate decision to limit all pixels to either black or white, thus eliminating the great amount of processing required to operate with gray-scale information. It also allowed the use of run length encoding, which provides greater computational efficiency than pixel-by-pixel analysis. For this reason, it was necessary to ensure that high-contrast illumination of the scene be provided, using either strong lighting or a dark background. For industrial applications, this was not a major problem.

The preprocessor, located between the cameras and the LSI-11 microcomputer, performs the thresholding, run length encoding, histogram generation, and storage of the 128 X 128 binary image. It also controls the flash lamp strobe used to obtain "frozen" images of moving objects.

Connectivity analysis, blob analysis, and extraction of features is done by the LSI-11 computer using the techniques discussed in the software sections of this chapter. Windowing, the ability to specify operations over a selected rectangular area of the image, can be used to reduce the computation required when only certain areas of the image are of interest. Threshold adjustment is possible under computer control to allow selection of the optimum level conversion of gray-scale information to binary values.

Initially, the total time required to read in and process an image was found to be about 1.4 seconds, depending on its complexity. This time has now been reduced by a factor of two or more by the use of a faster LSI-11 microprocessor and improved software.

Other hardware used in the system is a conventional video monitor (a TV set in principle), a control console or terminal, and a graphic display terminal.

Some examples of tasks that the SRI Vision Module has successfully accomplished are:

1. Part Recognition and Acquisition. Parts on a moving conveyor were located in position, orientation, and identity by the Vision Module, which then directed a Unimate® manipulator to pick up each part and transport it to a specified destination.
2. Stenciling Moving Boxes. Boxes were placed randomly on a moving conveyor, and the vision subsystem determined the position and orientation of each box. The Unimate® then placed a stencil on each box, sprayed the stencil with ink, and removed the stencil.
3. Visual Inspection. Lamp bases were inspected to verify that each base had two properly located electrical contact grommets. Washing machine water pumps were inspected to verify that the pump handle was present and in the proper position.

Several commercial manufacturers have adopted some of the general principles of the Vision Module and have made various improvements or modifications for specialized applications. This development has had a substantial impact on the generation of a computer vision industry, and especially on robot vision applications.

9.11.2
CONSIGHT—
General Motors
Vision System

CONSIGHT is a vision-based robot system suitable for operation in a manufacturing environment. It has the following features (Ward et al. [49]):

1. It determines the position and orientation of a wide range of manufactured parts including complex curved objects.
2. It provides easy reprogrammability by insertion of new part data.
3. It works on the visually noisy picture data typical of many plant environments.

Vision, robot, and monitor subsystems in CONSIGHT are logically separate but are integrated to work together, as shown in Figure 9.30.

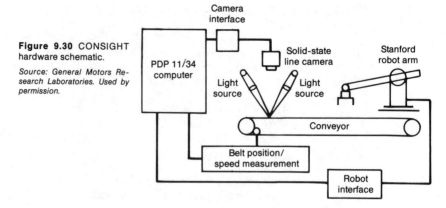

Figure 9.30 CONSIGHT hardware schematic.

Source: General Motors Research Laboratories. Used by permission.

Hardware of the experimental system consisted of a DEC PDP 11/34 computer, a Reticon RL256C line camera (256'X 1), a Stanford arm, and a belt conveyor. An encoder on the conveyor provided position and speed measurement. A production system has since replaced the experimental hardware. Design concepts followed the SRI Vision Module approach, with some variations and improvements.

The Reticon linear array camera imaged a narrow strip across the conveyor belt perpendicular to the belt's direction of motion. It viewed the moving belt and built up a two-dimensional image in memory. Only 128 of the 256 diodes in the Reticon camera were used.

As shown in Figure 9.31, the structured light approach was used. A slender tungsten bulb and cylindrical lens projected a narrow, intense line of light across the belt surface parallel with the line of the linear array camera. With no part on the belt, the camera saw the beam of light undeflected. When a part was present, the line of light hit the part at a different location along the belt at a point proportional to the height of the part. This effect can be seen in Figure 9.32.

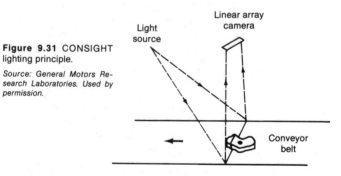

Figure 9.31 CONSIGHT lighting principle.

Source: General Motors Research Laboratories. Used by permission.

A shadowing effect occurred due to the height of the part, so that the light beam did not reach the conveyor at all points. The solution to this problem was to use a second light source (as in Figure 9.33) and adjust it so that its line of light coincided with the light from the first source. Using the structured light sources, the location and height of the part could be obtained accurately as the image built up in the memory of the vision computer. An ambiguity in determining

Figure 9.32 Computer's view of parts.

Source: General Motors Research Laboratories. Used by permission.

the inside and outside of some binary regions was resolved by using six-connected regions instead of the usual four- or eight-connected regions.

In the production model of the CONSIGHT system, the single control computer was replaced by a PDP 11/34 vision computer, an LSI-11 monitor computer, and a separate controller that is part of the Cincinnati-Milacron T3 robot used to replace the Stanford arm. These changes, plus a speedup in the T3 robot, reduced the total robot cycle time to 5 seconds for a simple pickup, transfer, and putdown operation under vision control. Belt speeds up to 20 cm/sec. are used.

Additional applications of vision systems are described in Chapter 10.

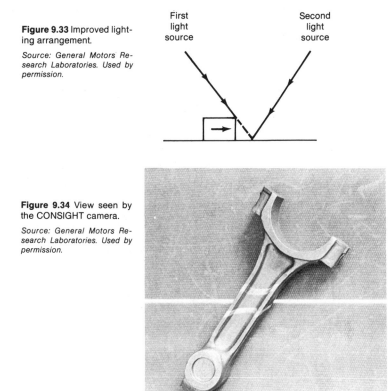

Figure 9.33 Improved lighting arrangement.

Source: General Motors Research Laboratories. Used by permission.

First light source   Second light source

Figure 9.34 View seen by the CONSIGHT camera.

Source: General Motors Research Laboratories. Used by permission.

Exercises

These exercises provide a review of the subjects covered in this chapter and emphasize some of the more important points.

1. What is the purpose of the digitization function? How is it accomplished? (See Section 9.1 for further detail.)
2. What important information does a robot need to pick up a part from a mechanical feeder and place it in an assembly? Can all of it be obtained from a vision system?
3. What is a raster scan?
4. Why are only 480 lines used for imaging out of the total 525 lines in the EIA RS-170 Standard?
5. Discuss the advantages of solid-state sensors as compared to vidicon tubes.
6. Can an image be obtained by using only one photosensitive ele-

ment? If so, how is it done and what precautions are necessary?

7. Describe current methods for obtaining depth and range information in an image. What method do you consider the most useful at this time?

8. How does a flash digitizer work? Why is it used?

9. What are some functions of a vision processor?

10. What is the purpose of image compression? Describe one way in which it can be done.

11. If the speed of image processing is not adequate, what possibilities exist for increasing it?

12. How can the amount of noise in a picture be reduced?

13. What is a global transformation?

14. What is meant by thresholding in picture processing?

15. Describe the essential difference between an edge and a line.

16. What two simple steps can be used to find and emphasize an edge segment in a picture?

17. How is a histogram generated? How can it be used? (See also Section 9.3.4.)

18. What is template matching? What are some of its features?

19. Given the following points in (x,y) coordinates — $(2,7)$, $(4,6)$, $(4,4)$, $(7,3)$, and $(8,2)$—determine the best segment to fit these points by using the Hough transform.

20. In Figure 9.25, find the Y coordinate of the blob centroid.

21. Follow the tree-parsing recognition process in Figure 9.27 and determine the features that must be possessed by the caliper 2 to be accepted by the tree-parsing algorithm.

22. Why is the LFF system an effective recognition method?

23. Draw a semantic net for the furniture in one of your rooms at home. How useful would it be in describing and recognizing the room? How could it be improved?

24. Locate and read the article on image enhancement by R. E. Woods and R. C. Gonzalez [52]. What processing is used to perform histogram equalization?

25. What were the key attributes that made the SRI Vision Module an effective vision system? What special provision was necessary to improve the image contrast?

26. Describe how the structured light approach is used to determine the heights of parts on a conveyor in the CONSIGHT system.

References

[1] J. K. Aggarwal and N. I. Badler, eds., "Special Issue on Motion and Time-Varying Imagery," Pattern Analysis and Machine Intelligence, PAMI-2, No. 6, November, 1980. (Several articles on the subject.)

[2] G. J. Agin and T. O. Binford, "Computer Descriptions of Curved Objects," IEEE Transactions on Computers, Vol. 25, No. 4 (April 1976).

[3] H. C. Andrews, "Digital Image Processing," IEEE Spectrum, April 1979, Vol. 12, no. 4, pp. 38-49.

[4] D. H. Ballard and C.M. Brown, Computer Vision, Prentice-Hall, Englewood Cliffs, N.J.,1982.

[5] H. G. Barrow and J. M. Tenenbaum, "Computational Vision," Proc. IEEE, Vol. 69, No. 5, May 1981, pp. 572-595.

[6] R. C. Bolles, "Robust Feature Matching Through Maximal Cliques," Proceedings of SPIE's Technical Symposium on Imaging Applications for Automated Industrial Inspection and Assembly, Washington, D.C., April 1979.

[7] R. C. Bolles, "Locating Partially Visible Objects: The Local Feature Focus Method," Proceedings of the First Annual Conference on Artificial Intelligence, Stanford University, Stanford, Calif. August 1980, pp. 41-43.

[8] F. H. Bower, "CCD Fundamentals", Military Electronics/ Countermeasures, February 1978, Reprinted in CCD, The Solid State Imaging Technology. Fairchild Corporation, Palo Alto, Calif.

[9] T. C. Chen and R. J. P. DeFigueiredo, "An Image Transform Coding Scheme Based on Spatial Domain Considerations," IEEE, Pattern Analysis and Machine Intelligence, Vol. 5, No. 3, May 1983, pp. 332-336.

[10] R. Cunningham, "Segmenting Binary Images," Robotics Age, July/August 1981, Vol. 3, No. 4, pp. 4-19.

[11] R. O. Duda and P. E. Hart, Pattern Recognition and Scene Analysis, Wiley, New York, 1981.

[12] R. Eskenazi, "Video Signal Input," Robotics Age, Vol. 3, No. 2, March-April 1981, pp. 2-11.

[13] K. S. Fu, Syntactic Pattern Recognition and Applications, Prentice-Hall, Englewood Cliffs, N.J., 1982.

[14] K. S. Fu, "Pattern Recognition for Automatic Visual Inspection," IEEE, Computer, Vol. 15, No. 12, December 1982, pp. 34-41.

[15] G. J. Gleason and G. J. Agin, "The Vision Module Set Its Sight on Sensor-Controlled Manipulation and Inspection," Robotics Today, pp. 36-40, Winter 1980/81.

[16] R. C. Gonzalez and G. Thomason, Synthetic Pattern Recognition—An Introduction. Addison-Wesley, Reading, Mass. 1978.

[17] R. C. Gonzalez and R. Safabakhsh, "Computer Vision Techniques for Industrial Applications and Robot Control," Computer IEEE, Vol. 15, No. 12, December 1982, pp. 17-33.

[18] E. L. Hall, J. B. K. Tio, C. A. McPherson, and F. A. Sadjaki, "Measuring Curved Surfaces for Robot Vision," Computer IEEE, December 1982, pp. 42-54.

[19] J. W. Hill and A. J. Sword, "SRI Reports on Its Programmable Parts Presenter," Robotics Today, Summer 1980, pp. 20-23.

[20] K. Hwang and K. Fu, "Integrated Computer Architectures for Image Processing and Database Management," Computer, Vol. 16, No. 1, January 1983, pp. 51-60.

[21] R. A. Jarvis, "A Perspective on Range Finding Techniques for Computer Vision," Pattern Analysis and Machine Intelligence, Vol. 5, No. 2, March 1983, pp.

[22] T. Kanade, "Geometrical Aspects of Interpreting Images as a Three-Dimensional Scene," IEEE, Proceedings, Vol. 71, No. 7, July 1983, pp. 789-802. (Special Issue on Robotics).

[23] R. B. Kelley, H. A. S. Martins, J. R. Birk, and J-D. Dessimoz, "Three Vision Algorithms for Acquiring Work Pieces from Bins," Proc. IEEE, Vol. 71, No. 7, July 1983, pp. 803-820.

[24] M. Kidode, "Image Processing Machines in Japan," Computer IEEE, Vol. 16, No. 1, January 1983, pp. 68-79.

[25] R. P. Kruger and W. B. Thompson, "A Technical and Economic Assessment of Computer Vision for Industrial Inspections and Robotic Assembly," Proc. IEEE, Vol. 69, No. 12, December 1981, pp. 1524-1538.

[26] J. Millman, Microelectronics: Digital and Analog Circuits and Systems, McGraw-Hill, Inc., New York, N.Y., 1979.

[27] W. Myers, "Industry Begins to Use Visual Pattern Recognition," IEEE Computer, May 1980, pp. 21-31.

[28] R. N. Nagel, G. J. Vanderbrug, J. S. Albus, and E. Lowenfeld,

"Experiments in Part Acquisition Using Robot Vision" 1979, National Bureau of Standards, Washington, D.C. MS79-784. Reprinted in [48].

[29] G. Nagy, "Optical Scanning Digitizers," Computer Magazine, IEEE, Vol. 16, No. 5, May, 1983, pp. 13-24.

[30] R. Nevatia, Machine Perception.   Prentice-Hall, Englewood Cliffs, N.J., 1982.

[31] N. J. Nilsson, Principles of Artificial Intelligence. Tioga Publishing Company, Palo Alto, Calif. 1980.

[32] D. Nitzan et al., "Machine Intelligence Research Applied to Industrial Automation," Ninth Report, NSF Grants APR75-13074 and DAR78-27128, SRI Projects 4391 and 8487, SRI International, Menlo Park, Calif., August 1979.

[33] See [32], Tenth Report, November 1980.

[34] D. Nitzan et al., "Machine Intelligence Research Applied to Industrial Automation," Eleventh Report, January 1982. (There are now 12 reports from SRI on this subject, which report some of the advanced research done by SRI under NSF and industrial grants.)

[35] M. Oshima and Y. Shirai, "Object Recognition Using Three-Dimensional Information," IEEE, Pattern Analysis and Machine Intelligence, Vol. 5, No. 4, July 1983, pp. 353-361.

[36] W. A. Perkins, "Area Segmentation of Images Using Edge Points," IEEE, Pattern Analysis and Machine Intelligence, Vol. 2, No. 1, January 1980, pp. 8-15.

[37] R. J. Popplestone et al., "Forming Models of Plane-and-Cylinder Faceted Bodies from Light Stripes," Fourth International Joint Conference on Artificial Intelligence, Tbilisi, Georgia, USSR, 1975.

[38] J. L. Potter, "Image Processing on the Massively Parallel Processor," Computer IEEE, Vol. 16, No. 1, January 1983, pp. 62-67.

[39] K. Preston, "Cellular Logic Computers for Pattern Recognition," IEEE Computer, Vol. 16, No. 1, January 1983, pp. 36-50.

[40] C. A. Rosen and D. Nitzan, "Use of Sensors in Programmable Automation," IEEE Computer, Vol. 10, No. 12, December 1977, pp. 12-13. Reprinted in [48]

[41] C. A. Rosen and D. Nitzan, et. al., "Exploratory Research in Advanced Automation," Second Report, NSF Grant GI-38100X1, Stanford Research Institute, Stanford, Calif., August 1974.

[42] C. Rosen, and D. Nitzan, et. al., "Machine Intelligence Research Applied to Industrial Automation," Eighth Report, NSF Grant APR75-13074, SRI Project 4391, SRI International, Menlo Park, Calif., August 1978.

[43] A. Rosenfeld, "Image Processing and Recognition," in Advances in Computers, Vol. 18, Academic Press, New York, 1979.

[44] A. Rosenfeld, "Image Pattern Recognition," IEEE Proceedings, Vol. 69, No. 5, May 1981, pp. 596-605.

[45] A. Rosenfeld, "Parallel Image Processing Using Cellular Arrays," Computer IEEE, Vol. 16, No. 1, January 1983, pp. 14-20.

[46] Y. Shirai, "Image Processing for Data Capture," IEEE Computer, November 1982, Vol. 15, no. 11, pp. 21-34.

[47] Y. Shirai and Motoi Suwa, "Recognition of Polyhedrons with a Range Finder," Second International Joint Conference on Artificial Intelligence, London 1971.

[48] W. R. Tanner, Ed., Industrial Robots, 2nd ed., Vol. 1, Fundamentals. Robotics International of SME, Dearborn, Mich. 1981.

[49] M. R. Ward, L. Rossol and S. W. Holland, "CONSIGHT: A

Practical Vision-Based Robot Guidance System," Ninth International Symposium on Industrial Robots Conference, March 1979. Reprinted in [48], 1981, pp. 337-353.

[50] J. M. Wilf, "Chain Code," Robotics Age, Vol. 3, No. 2, March-April, 1981, pp. 12-19.

[51] P. M. Will and K. S. Pennington, "Grid Coding: A Preprocessing Technique for Robot and Machine Vision," Second International Joint Conference on Artificial Intelligence, London, 1971.

[52] R. E. Woods and R. C. Gonzalez, "Real Time Digital Image Enhancement," Proc. IEEE, Vol. 69, No. 5, May 1981, pp. 643-654. (special issue on image processing).

Chapter 10

Applications of Robots

In this chapter, some of the many applications of robots will be described. Problems affecting applications and some known solutions will be discussed. Examples of working systems will be given, along with references to further information. Each section describes a major category of applications. Simpler systems are described first in each section, with the more sophisticated systems following.

Some applications of robots to be described in this chapter are listed in Figure 10.1. Applications listed in the left column are justified by the robotic capabilities and benefits listed in the headings on the right side of the figure.

Since machine loading and unloading was the first application of robots, it will be discussed first, followed by examples of some of the other applications listed in Figure 10.1.

Figure 10.1 Major categories for robot applications and rationale for using robots (from Ottinger [31], Figure 1, p. 19).

Source: Tech Tran Corporation, Naperville, Illinois. Used by permission.

Application	Examples	Robot capabilities justifying use				Primary benefits of using robots			
		Transport	Manipulation	Sensing		Improved product quality	Increased productivity	Reduced costs	Elimination of hazardous/unpleasant work
Material handling	Parts handling Palletizing Transporting Heat treating	●						●	●
Machine loading	Die cast machines Automatic presses NC milling machines Lathes	●	●				●	●	
Spraying	Spray painting Resin application		●			●		●	●
Welding	Spot welding Arc welding		●				●	●	●
Machining	Drilling Deburring Grinding Routing Cutting Forming	●	●			●	●		
Assembly	Mating parts Fastening	●	●				●	●	
Inspection	Position control Tolerance			●		●			

10.1 Machine Loading and Unloading

Many kinds of machines are being loaded and unloaded by robots. The first application of a robot was the use of the Unimate robot in

1961 to run a die-casting machine in a General Motors Plant in Trenton, New Jersey. This application is thoroughly described by Engelberger [15]. Operations performed by this robot were: unloading of one or two die-casting machines, quenching parts in a tank of water or oil, trimming parts in a trimming die, and inserting the parts into a loading fixture.

Other machine loading and unloading applications described by Engelberger are as follows:

1. Forging. Holding parts in a drop forge, upsetter, roll forge, or presses.
2. Stamping Presses. Loading and unloading presses, and performing press-to-press transfer.
3. Injection and Compression Molding. Unloading molds, trimming and inserting parts, palletizing and packaging parts.
4. Machine Tool Operations. Loading, unloading, palletizing, transferring between machine tools, and so on.

All of these applications depend on the use of a suitable gripper or end effector to allow the part to be picked up and held for operation or transferred to and from a machine. Much effort goes into designing suitable end effectors and planning the timing and control of the robot and the related machines. Commercially available robots have differing load capacity, reach, and speed characteristics that must be taken into account in designing these systems. Detailed listings of the characteristics of available robots are to be found in Tanner [44] and in Flora [18].

More advanced machine loading and unloading systems work in flexible manufacturing cells where one or more robots service several machines at a time.

Justification of the use of robots has been discussed in Chapters 1 and 2, but we will repeat some of the major arguments here as discussed by Sullivan [43]. Human operators loading and unloading production machines get tired and slow down, take breaks—both scheduled and unscheduled—and do not operate the machine at its full capability. As a result of delays in machine loading and unloading, the machine may be idle as much as 70 to 80 percent of the total shift period.

Robots are tireless and do not take breaks. Also, they can be programmed to load and unload parts fast enough to take full advantage of the productive potential of the machine. Other compelling reasons for using robots are:

1. Robots in hot, noisy, or hazardous jobs not only prevent exposure of humans to these environments but will usually perform better than human operators.
2. Batch processes can be converted to continuous automatic processes, since the robots can be programmed to work more closely and efficiently with the machines than human operators could.
3. Robots can perform nearly any loading and unloading operation now being done by human operators.

10.1.1
Press Tending

An example of the use of robots for machine loading is shown in Figure 10.2. A PRAB Model 5800 hydraulic robot swings 90 degrees, after picking up a part from a stamping press, to deposit the finished part on a pallet.

Figure 10.2 Prab Model 5800 hydraulic robot picking up a part from a press and depositing it on a pallet. *Courtesy of PRAB Robots, Inc. Used by permission.*

The part being picked up is an oven liner for a kitchen range manufactured by the Grand Rapids Manufacturing Company in Grand Rapids, Michigan. It weighs about 15 pounds and is fabricated from enameling iron sheet stock. Four identical 200-ton presses are used to form the liners. Cut sheets are manually extracted from an upline blanking press and placed on a roller conveyor. An operator loads and orients the blanks, one at a time, into the stamping press's first die. He then presses a pair of palm buttons (a safety factor) to actuate the press. The ram descends and closes the die, forming and punching the blank.

An interlock signal from the press signals the robot when the press has cycled. The robot reaches into the die opening with a special end effector, using two separate sets of vacuum grippers with independent vacuum cups. Each set of grippers can handle a separate part. The stamping press has two dies: one for piercing and trimming and the other for forming tabs on the edge of the part. The robot picks up a semifinished part from the first die and a finished part from the second die (after the initial loading).

Retracting its arm a short distance, the robot deposits the semifinished part on the second die. It then pivots 90 degrees to the left and places the finished part on one of two stacks on the pallet by controlling its arm extension.

Figure 10.3 shows the robot gripper with one set of vacuum cups loaded and one set empty after it has deposited the semifinished part in the second die and is placing the finished part on the pallet.

As the pallet is loaded with 60 liners on each of two stacks, the full pallet is automatically indexed by the conveyor to the next station and an empty pallet is moved into place. This operation is shown in Figure 10.4; A and B are two dies in the press, C and D are the two stacking areas on the pallet.

Vacuum cups on the robot's end effector are spring loaded and have a simple search–until–find limit switch to provide rudimentary tactile sensing. This switch signals the robot controller when the part is on the pallet stack so that the robot gripper will release the part to the pallet. Parts are counted by the robot controller, and a red light is

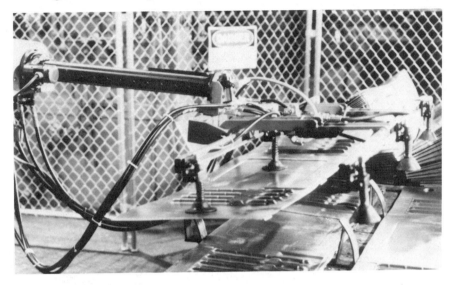

Figure 10.3 Robot arm about to release a finished part to one of two stacks on a pallet. Vertical guides on the pallet to ensure straight stacking.

Courtesy of PRAB Robots, Inc. Used by permission.

Figure 10.4 Plan view of robotic press system (from T & P, p. 105).

Source: PRAB Robots, Inc. Used by permission.

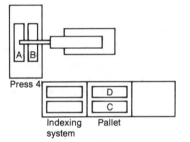

lighted on top of a nearby post to notify a fork lift operator to pick up the full pallet and move another empty pallet into position on the pallet conveyor.

Peak output of the system is 244 liners per hour, or two every 27 seconds. The plant engineer at Grand Rapids Manufacturing calculated that the system had paid for itself in less than a year. In addition, the system is much safer than removing parts from the press by hand and easier for the operators. At 244 liners per hour, 15 pounds each, there would be 1.83 tons of parts to handle per hour, quite a job for human operators.

10.2 Material Handling

Moving parts from a storage warehouse to machines, transfering assemblies from one conveyor to another or between a conveyor and a machine, and stacking completed parts or assemblies and other transfer operations are thought of as material handling. Packaging operations are also included in material handling in most organizations. There is actually little difference in operation between machine loading and material handling, except that in material handling the parts or assemblies are larger and multiple parts or assemblies may be handled at one time. We could consider machine loading and unloading as a special case of material handling.

Robots can be used with conveyors and automated guided vehicles (AGVs) to mechanize completely material-handling functions. Conveyors may be of the overhead rail type or the better-known belt type. Automated guided vehicles (AGVs) are used to carry large parts and pallet loads of parts or assemblies. AGVs are low platforms, about 1 or 2 feet high, 4 or 5 feet wide, and 6 to 8 feet long. They are usually driven by battery-powered electric motors and follow a guide path or wire in the floor. Loads of 2,000 to 6,000 pounds are commonly moved from one location of the factory or warehouse to another.

**10.2.1
Palletizing**

Robots are frequently used for palletizing and depalletizing. A Cincinnati Milacron T^3 robot is shown picking up a cardboard carton from a palletized load in Figure 10.5. Both palletizing and depalletizing are done under computer control. All four major robot configurations— rectangular, cylindrical, spherical, and revolute—can be used for palletizing. Each configuration must be operated differently, and each has particular advantages. Jointed-arm robots (revolute) lose horizontal reach at the bottom and top of their stroke. A good solution is to raise the pallet up from the floor level so that the pallet volume is centered in the work envelope of the robot arm. Rectangular robots, such as the gantry type, can cover a large volume and can therefore handle larger palletizing tasks. Stauffer [41] describes some alternative palletizing operations using different robots on AGV's.

Figure 10.5 Palletizing and depalletizing with a Cincinnati Milacron T^3-566 robot (from Stauffer [41], p. 43).
Courtesy of Cincinnati Milacron. Used by permission.

Programming the robot for palletizing may be done by teaching each individual move, but it is more effective to use a programming language to off-line program the operation. Using a language such as VAL, developed by Unimation, a robot can be instructed to place cartons in a particular sequence at a particular spacing by the use of a SHIFT command. This action causes the spacing parameters to be automatically incremented or decremented by the dimensions of the object being palletized, so that it is only necessary to specify the starting points on the pallet and the desired spacing of the objects in three dimensions. Complex pallet patterns may be used by suitable programming and specification of the pattern relationships.

Placing of <u>dunnage</u> is often part of the palletizing task. Dunnage is the industry term for the packing material used to separate and protect objects while being transported. It may consist of formed cardboard sheets, plastic forms, or inflated rubber cushions and is placed at appropriate locations during the palletizing operations.

End effectors used for palletizing may be vacuum pickups, grippers, magnetic pickups, or even specialized tooling designed to handle a particular part. Palletizing of automobile engine blocks is done by the use of a gantry crane at one company. A special clamp is used to grip the sides of the block to avoid damage to the upper part of the block. The weight of heavy tooling of this type must be considered when analyzing the cycle time of the robot. Typically, a gantry robot handling an engine block may have a cycle time of 10 to 15 seconds in moving a block from one conveyor to another.

**10.2.2
Packaging**

Placing small assemblies of parts into packages for shipment is a tiring and repetitious task. But it requires careful attention to detail to ensure that all the required items are included and protected for shipment. Robots can be used effectively in this type of operation to reduce error, eliminate handling damage, and assure proper labeling of the shipment.

Figure 10.6 Diskette writer system with a PUMA 560 robot (from Bloom [4]).

Photo courtesy of IBM Corporation and Paul Nelson. Used by permission of the Society of Manufacturing Engineers, Dearborn, Michigan, U.S.A. Copyright 1984 from the Robots 8 Conference Proceedings.

IBM ships thousands of diskettes (floppy disks) from its Pough-keepsie, New York plant to supply information about its large computer systems to customer installations. These particular disks contain engineering change information and features for use in operating the systems in the field. Writing information on the diskettes and packaging the 8-inch disks for shipment was a tedious, unpleasant task when done manually. Errors in labeling were frequent.

After considerable study of alternatives, including the use of hard automation, it was decided that robots provided the best solution to the problem. The shipping system was put together as part of the overall integrated manufacturing system at the Poughkeepsie plant.

Major components of the system are two diskette writers, a box storage carousel, two envelope feeders, two label printers, a Unimation PUMA 560 robot, and an IBM Series/1 computer with associated terminals, printers, and hard disk drives. In addition, four Z80 microprocessors are used to control the diskette drives, since there are two drives per writer station. These components are arranged as shown in Figure 10.6 and diagrammed in Figure 10.7.

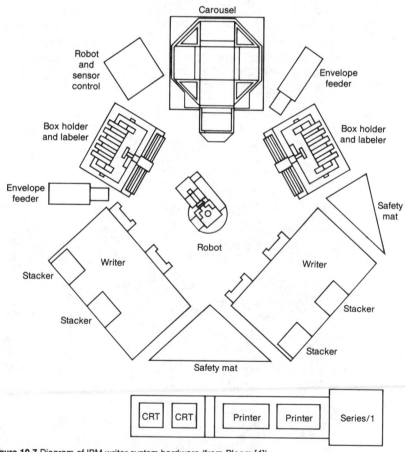

Figure 10.7 Diagram of IBM writer system hardware (from Bloom [4]).

Source: Robots 8 Conference Proceedings, Copyright 1984 by the Society of Manufacturing Engineers, Dearborn, Michigan, U.S.A. Used by permission of SME and Paul Nelson, IBM Corporation.

A PUMA 560 robot was selected because of its ease of programming (using the VAL language), its electric servo drive, large work envelope, and physical configuration. The PUMA has a reach of 1 meter, can rotate 360 degrees, carry a load of 5.5 pounds, and has

0.1-millimeter (0.004 inch) repeatability. It can pick up a full box of diskettes and transfer them to the top of the box carousel as required.

The box carousel is a four-sided revolving device that both dispenses new boxes and accepts full boxes of disks. Empty boxes are manually loaded and dropped by gravity feed to the robot as needed. Boxes are picked up by the robot and placed in the adjacent box holder. In each box holder are two assembly stations so that four orders can be processed simultaneously. This arrangement also provides redundancy in case of failure of one of the stations.

Each writer frame has two independent 8-inch diskette drives modified to feed up to 200 diskettes to each drive. Writer operation is under the control of the Z80 computers. Disks are released from the stacker units, fed to the writer, written on by computer control, and released to an output chute. Pneumatic cams and motor-driven rollers move the diskettes through this sequence of operations.

A complex set of sensors and computer-controlled interlocks ensures that every operation is completed properly before the next is started. When a disk is detected in the output chute of a writer, the robot is signaled to pick up the completed disk and place it in an envelope. The Series/1 computer maintains a queue of operations to be performed by the robot and determines which operation it will execute next. When the robot has completed one task and is ready to pick up the disk at a particular output chute, the associated label printer is directed to print a label for that diskette. Then the robot picks up the diskette in its gripper, moves it to the designated labeler, and holds it in place while the label is applied by a blast of compressed air. An optical sensor verifies that the label is applied properly. The robot then moves the labeled disk to the assigned box and uses compressed air to force the diskette into the box. Gripper sensors signal to the Series/1 computer that the diskette is picked up and released properly. An envelope is fed from the envelope feeder at a signal from the robot, picked up by the robot, and placed on top of the diskette in the box. These envelopes act as separators during shipment and are used for diskette protection at the field site.

After the box of diskettes is full, a lid holder supplies a lid to the robot. This lid is placed on the box by the robot after careful checking to ensure that no envelopes or disks are sticking out of the box. The box is then labeled and sealed by the robot.

This whole complex operation is described in detail in Bloom [4]. Included in the system are several important safety measures and error recovery techniques. Operational problems, especially labeling problems, were resolved after considerable effort. Over 150 possible error considerations are provided for in several ingenious procedures, including boxes improperly loaded by the operator, dropped envelopes, warped diskette covers, and similar circumstances.

The stated main goal of the system was improved product quality. This was achieved. "Not a single handling, labeling, sorting or packaging error has escaped since May, 1983," according to Bloom [4]. The system handles between 1,000 and 2,000 diskettes per two-shift day.

10.3 Fabrication

Fabrication describes the use of robots in handling tools and production items in manufacturing operations. Instead of loading a machine tool, the robot carries out the operation much as a human would.   There are several applications of robots where the robot is actually doing the work. Some of these are now discussed.

10.3.1
Investment Casting

In investment casting, the robot picks up a "tree" of plastic parts and dips the assembly into a slurry of ceramic material to form a coating on the plastic. After several dipping operations, the plastic is baked out by high heat and the cured ceramic mold is filled with molten metal to form several parts at one time. This process is used to make small metal castings and even metal heads for golf clubs. This is the famous "lost wax" process in modern form. Plastic parts are formed by plastic molding. Figure 10.8 shows the use of a Shell-O-Matic robot at the Waukesha Foundry in Watertown, Wisconsin. Waukesha is a division of the Abex Corporation. Shell-0-Matic has

Figure 10.8 Investment casting—immersing a mold cluster into a ceramic slurry (from Schreiber [37], p. 42).
Courtesy of Waukesha Foundry. Used by permission.

Figure 10.9 Removing the parting line flash with a grinding wheel (from ASEA pamphlet YB 11–106E, Figure 1).

Courtesy of ASEA. Used by permission.

installed 130 robot systems especially for investment casting. These robots are designed to be well sealed and resistant to the dust and high temperatures that may occur in the investment casting foundry. Several other companies, Unimation and Prab among them, have installed robot systems for investment castings in many factories.

Three basic dipping operations are performed by the robots— prerinse, dipping, and seal coat. Cycle times average 2 to 2.5 minutes. Mold clusters average $2.5 \times 3.5 \times 4$ feet in size (Schreiber [37]).

10.3.2
Grinding

In this operation, the robot picks up a heavy grinder and uses it to grind burrs off castings and other parts, or picks up the part and holds it against the grinder, as in Figure 10.9, which shows the use of an ASEA robot for this application.

In more sophisticated applications, the robot follows a complex contour and removes metal as needed to make the part conform to dimensions stored in the robot controller's memory. Accuracies of a few thousandths of an inch can be achieved depending on the accuracy of the robot.

10.3.3
Deburring

Deburring is similar to grinding, except that the robot removes only a small amount of material to clean up the edges of parts after machine operations such as sawing, drilling, and routing.

Figure 10.10 shows the deburring of a forged steel blank after machining. A tungsten-carbide file is being used. The drive motor is mounted in a resilient tool holder to allow for slight part variations. Movement of the robot is controlled by a deburring program. Deburring is a monotonous, risky and laborious job well suited to a robot.

Figure 10.10 Deburring of steel item against a tungsten-carbide file (from ASEA pamphlet YB 11–108E, Figure 1).

Courtesy of ASEA. Used by permission.

10.3.4
Fettling

Removing excess metal in foundries is known as <u>fettling</u>. It can be done by sawing or grinding and is dangerous, dirty work where robots can be and are being used effectively. This is another situation in which the tools are heavy and difficult for a human operator to handle.

10.3.5
Routing

The robot moves a routing machine around a template or guide in order to cut parts to shape. Routing machines can be awkward and heavy for a human to handle. The robot has an advantage in both speed and accuracy for this operation. In some operations there are flying particles of metal, so that human operators must wear goggles and protective clothing. Use of a robot provides a safety advantage in this case.

**10.3.6
Drilling**

Printed circuit boards for electronics are drilled by a Numerically Controlled or computer-controlled machine installed at the ASEA Group's Electronic Division in Vasteras, Sweden (Stauffer [42]). A robot is used to load and unload the drilling machines, assemble lamination stacks in preparation for drilling, disassemble the stacks after drilling, and change drill bits as they become worn. In this case, the robot is acting as an assistant to the machine and is not doing the drilling directly.

Robots directly handle the drilling motor at the General Dynamics aircraft plant in Fort Worth, Texas. Large drill templates are mounted on aircraft skins, and the robot is trained to drill the skins using the drill bushings in the drill templates as guides. Since the skins are contoured, the robot must drill from many angles to a predetermined depth. This has been found to be an effective labor-saving arrangement. In addition, the quality of the drilled holes is better than that achieved by human operators. There are 10 robotic work cells in operation at the Fort Worth plant. Over 80 different end effectors are in use to provide the necessary flexibility and spare end effectors needed. (Heywood [21]). Drilling is done on graphite-epoxy composite skin and on aluminum skins.

T. L. Welch [47] has simulated the drilling operation for an aircraft windscreen and has studied the effect of drilling holes in different sequences on the total production achieved.

**10.3.7
Water Jet Cutting**

A spectacular application of robots that has recently become more popular is the use of a high-pressure (50,000 psi) water jet to cut soft materials at high speeds. The part is held by the robot and moved in a horizontal plane through the vertical path of the water jet, or the robot gripper, carrying the jet nozzle, is moved around the periphery of the part. Cut edges are clean and smooth, and intricate cuts can be made. Plastic, rubber, cardboard, and molded foam products are now being cut in this way for the automobile industry. Typical parts are noise insulation for car doors and frames, floor mats, and so on. Since the water jet is only 0.6 millimeter (0.024 inch) in diameter, there is little material waste. Very little water is actually used, and most of it disappears through the output orifice so that parts are not damaged by water. The nozzle has a sapphire cyrstal pierced with a small hole to withstand the abrasive effect of the water.

Thin metals are now being cut by water jets containing abrasive materials. This is an excellent application for robots because of the speed and accuracy attainable, considerably better than a human operator could maintain.

**10.3.8
Electrical Wire
Harness
Manufacture**

An overhead mounted PUMA robot used to form complex wire harnesses is shown in Figure 10.11. The robot picks up a precut wire, wraps one end around a pin at one end of the harness, and then follows a programmed path around pins set in the harness pattern board. At the end of each travel, the wire is wrapped around a pin to hold it in place. Operating speed and accuracy are superior to those of a human operator. This type of operation is suitable when there are as many as 100 similar wire harnesses to fabricate.

Robots with six degrees of freedom are desirable for this type of operation so that the robot wrist has the flexibility to wrap wires around pins in various ways. Loads are small so that a robot capable of carrying 5 pounds is adequate.

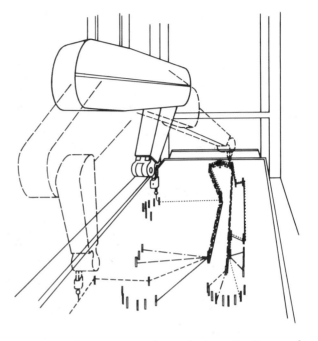

Figure 10.11 Wire harness lay-out and wrapping using a PUMA robot (from Unimation, Inc., 600–T2–382, p. 9.)

Courtesy of UNIMATION Incorporated. A Westinghouse Company. Danbury, Conn. USA. Used by permission.

**10.3.9
Application of
Glues, Sealers,
Putty, and
Caulking Materials**

Many semifluid materials are applied using robots. Caulking of car bodies, gluing of sound–deadening material to skins, and similar tasks are routine in the automobile industry. Greater flexibility and accuracy of application are major reasons for robot use. Other reasons are provision of a better working environment, since some of these materials are volatile and toxic; more uniform flow of materials toward the production line, since the robot always completes the job in the same time period; and better utilization of investment, since the robot is faster and an assembly station is not tied up as long as it would be with human operators.

In Figure 10.12, car body items are fed into the glueing section by a conveyor and are then placed securely under the robot. Glue strings are applied from the dispenser carried by the robot according

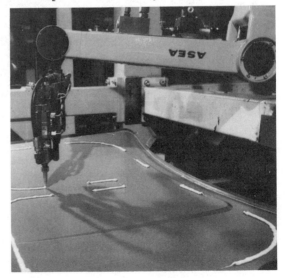

Figure 10.12 Glue application using an ESEA robot (from ASEA Pamphlet YB 11–107 E, Figure 1).

Courtesy of ASEA. Used by permission.

to a pattern preprogrammed into the robot controller. The thickness of the glue string can be varied by changing the speed of application or by a flow regulator in the dispenser head. In this case, the glue is used to join stiffener plates to the body part at the next assembly stage. Accuracy of placement of the glue, in both position and width, ensures that a good joint will be made.

There are several other fabrication applications being done by robots. Polishing operations use a steel brush instead of a grinder. Thermal sprays lay on a coat of molten metal from a spraying device carried by the robot. Cleaning is done with detergent jet sprayers carried by a robot. Other robots pick up parts and move them into and out of heat treatment ovens or plating baths. Robots can do nearly any simple handling task that human operators can do, and usually more rapidly, more safely, and with higher accuracy and lower cost.

10.4 Spray Painting and Finishing

Robots are now being used in many painting and finishing applications worldwide. Labor replacement, consistent quality, energy savings, removal of workers from toxic or harmful substances, material savings, and reduced finishing booth maintenance provide substantial advantages to the industrial user. There are also many regulations issued by government agencies with which the user must comply and which are difficult or expensive to comply with when the work is done by humans but much easier to meet using robots. All of these considerations make the use of robots attractive. Figure 10.13 shows the use of a robot for spraying on an automobile assembly line as an example of one application.

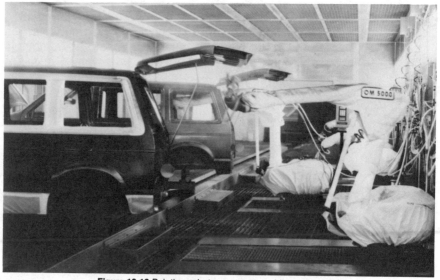

Figure 10.13 Painting robots on an automobile production line.
Courtesy of Chrysler Corporation. Used by permission.

10.4.1
Application
Methods

Major application methods for spray painting and finishing materials are:

1. Air Spraying. Air under pressure causes the paint to atomize and be propelled to the article to be painted.

2. Airless Spraying. Finishing materials, such as paint, are sprayed under considerable hydraulic pressure through a fixed orifice, which causes the paint to be atomized directly without the need for air.

3. Electrostatic Spraying. Atomized particles from either air or airless spray are electrically charged as they emerge from the sprayer and are attracted to the object being sprayed by the large electrostatic field which is applied. Considerable material savings is achieved, since very little of the sprayed material bypasses the object and is lost. Objects being sprayed are kept at a ground potential to achieve a large electrostatic field.

4. Heating of Materials. Paint decreases in viscosity when heated and can be sprayed with lower pressures. Less solvent is required, and there is less overspray of paint. Heating may be used with any of the preceding systems.

Air spraying and electrostatic air spraying are the most common methods of applying paint and other materials (enamels, powders, sound-deadening coatings), but some finishes are applied by other means. Robots used for electrostatic spray painting have an advantage over humans because it is possible to use higher voltages (in the 150,000-volt range) to direct the paint spray onto the workpiece, thus providing a better coating and less paint spray loss. Electrostatic sprayers used by humans are restricted to 70,000 volts for safety reasons.

Watching the way the paint spray curls around and coats the reverse side of parts handing from an overhead conveyor dramatically demonstrates the advantage of electrostatic spraying to even the critical skeptics among skilled painters. However, there is a problem called the Faraday cage effect. Electric fields are characterized by strong concentrations of field force along edges. When the electrostatic voltage is high, the edges of deep objects such as boxes or deeply formed covers gather an excess of paint, with an insufficient amount reaching the bottom of the box or depression. Control of the electrostatic field voltage during the spray operation reduces or eliminates this problem, but it requires additional programming either during training or in preprogramming operations. During the period when the bottom of a depression is to be painted, the voltage is reduced to a low value so that the spray will not be deflected toward the edges of the object.

10.4.2 Advantages of Robots

In all of the finishing systems, robots can serve a useful purpose at considerable savings in labor and materials. Finishing materials are injurious to the worker's health. Commonly, it is necessary for workers to wear tight-fitting masks and goggles to protect their respiratory system and eyes. Masks are hot and uncomfortable, and do not always protect the worker. Goggles become coated with paint and are difficult to see through. Robots do not have these problems and are capable of more accurate and consistently high-quality painting. However, even robots are sometimes covered with plastic covers to protect the joints and working parts from the paint spray, as illustrated in Figure 10.13.

A detailed list of the advantages of robots was compiled by one of the major automobile manufacturer and reported by Engelberger

[15]. This list is reproduced here. It helps explain why the spray painting application is one of the major users of robots:

1. Reduce the hostile environment factors
   a. Noise.
   b. Carcinogenic Materials.
   c. Particulate Matter.

2. Reduce Energy Requirements for Finishing
   a. Less fresh air required. Protection of operators requires much more air than necessary just to remove solvent and particulates to control the coating process.
   b. Reduced energy use for heating and cooling.
   c. Possibly less energy needed for lighting the spray area.
3. Improved Finish Quality
   a. Less dirt since less air flow.
   b. Uniform build-up of coating due to controlled spraying by the robot.
   c. Quality level more consistent.
   d. Specialized spray techniques may be used with better control.
   e. Quality not dependent on the fitness of the operator who may be fatigued.
4. Quality Improvement Results
   a. Reduced warranty problems.
   b. Improved customer acceptance.
5. Cost Reduction
   a. Reduced material costs.
   b. Reduced direct labor.
   c. Reduced health insurance and compensation costs.
6. Improved Employee Relations
   a. Positive response to improved working conditions.

10.4.3
Robot Character-
istics Required
      Robots used for spray painting and finishing must work in paint booths that are often small since they were initially designed for human use. This was a problem for early robots because they were designed with long arms and could not operate in these booths. One company, Trallfa Niles Underhaug, in Norway, designed a robot just for this purpose and gained an early market advantage. The Trallfa robots are sold in this country by the DeVilbiss Company. Over 1,000 paint-spraying robots have been sold.

      A more important problem in spray painting is the large concentration of volatile solvents in the air. These mixtures can become concentrated enough to explode and are easily ignited by an electric spark or by an open flame. Therefore, there are strict safety regulations laid down and enforced by OSHA and widely supported by spray painting equipment manufacturers and users. OSHA regulations require that the voltage be less than that required to cause a spark in the air or in the hazardous atmosphere, or that the current be low enough so that a hot spark cannot occur. This means that objects capable of producing high currents are restricted to about 9 volts or less and that high-voltage electrostatic spray systems are internally limited to currents of only a few microamperes so that they cannot heat the air enough to cause ignition of the volatile gases.

      In practice, all robot painting systems (known to this author, at least) use hydraulic or pneumatic drives rather than electric drive motors. Hydraulics is used because it is intrinsically safe; there is nothing to cause a spark. Sensors and controls that might carry an unsafe voltage or current are isolated from the spray-painting area by a sealed wall through which the low-voltage, low-current conductors pass to the robot. Usually, a sealed or separately ventilated chamber is used to house the electronic control system for the robot.

Spray painting requires the use of continuous-path robots. To achieve the required flexibility, speed, and accuracy, five, six, and even seven-axis servo-controlled robots are used. Continuous-path rather than point-to-point operation is needed because robots must often enter the cavities of parts to be painted. Accuracy is required to ensure that edges are painted without overspraying or wasting paint. Robots are also used to paint thin lines for decoration. Load capacity at full speed is in the 18-20-pound range to carry the spray gun and the attached hoses. Typical ranges and speeds are (this data is for the Binks Model 88-800 industrial finishing robot):

Horizontal stroke/speed—48 inches/45 inches per second
Vertical stroke/speed—84 inches/32 inches per second
Rotary stroke/speed—135/60 degrees per second
Wrist roll/travel speed—270/150 degrees per second
Wrist movement—180 degrees

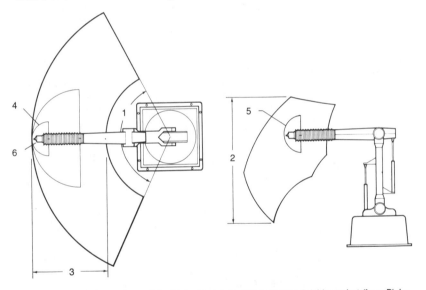

Figure 10.14 Work envelope of the Binks Model 88-800 industrial finishing robot (from Binks Bulletin A97-7R-1, p. 4).

Courtesy of Binks Manufacturing Company. Used by permission.

Repeatability of plus or minus 0.125 inch or better is common.

In spray painting of many items, the robot is required to bend back nearly 180 degrees at the wrist to spray the back surface as well as the front surface of the object being painted. This requirement has led to the development of additional wrist joints such as the one illustrated in Figure 10.15, which can bend back to spray paint into otherwise inaccessible places.

A <u>flipper</u> solenoid is used to drive the wrist through an additional 90 degrees as required. This movement, added to the normal wrist motion, provides the necessary flexibility.

Selection of a robot for a particular finishing task requires analysis of the complete system to be used, including the conveyor or other transport means used to bring the workpiece items to the robot. In some cases, the robot is mounted on a rail so that it can move along a production line or reach all areas of a large object such as an airplane wing. It is important that the work envelope be adequate to reach all items to be painted. Vertical reach and depth measurements are especially important. For conveyorized items, the horizontal reach is not as critical in most cases, since the conveyor will bring the parts to the robot. In all cases, the paint-spraying speed must be known to calculate the rate at which objects of specific sizes can be finished.

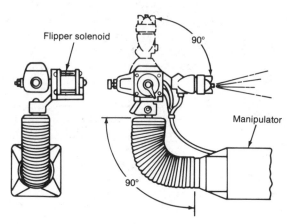

Figure 10.15 Flipper solenoid accessory provides flexibility (from the DeVilbiss Corporation).

Source: DeVilbiss Corporation. Used by permission.

10.4.4 Programming of Finishing Robots

Usually the robots are programmed by a skilled painter who guides the robot through the complex painting path. Two handles are provided on the robot. One of these handles has a trigger to control the spraying operation, while the other has controls to control recording of the robot path. Large memories are used in the robot controllers of the spray robots to record the large number of points required for accurate spray painting along a complex three-dimensional path. Path information may be stored directly in large solid-state memories, on tape cassettes, and in floppy disks or diskettes. Tape cassettes were used exclusively in early robots before floppy disks were available, but most systems now store the spray path on floppy disks. Some advanced systems use nonvolatile magnetic bubble memory for path storage.

Teaching with a teach pendant is possible for stationary parts but is not always suitable for use with moving conveyors. Off-line programming is potentially possible through the use of simulators or CAD information, but successful cases have not been reported in the literature.

10.4.5 General Motors Finishing System

A successful, fast, but expensive system has been developed by General Motors for its own use. This system was developed under the direction of Dr. Hadi A. Akeel and is now being produced commercially by the GMF Corporation (General Motors–Fanuc). This system has completely eliminated manual spraying in order to attain the many advantages of high-volume production, quality, cleanliness, and cost savings inherent in the use of robots.

New robots are designed to move at 1,200 millimeters per second. Auxiliary robots are provided as assistants to open and close automobile doors and trunk lids as necessary to allow the painting robots to function. A complete system has been set up to paint complete automobile body shells at high speed. Eight robots, four on each side of the painting line, do the actual painting. These robots have seven axes each and are mounted on rails so that they can follow the car along the line. Door operators are simple two-axis robots. There is a vision camera on the production line that recognizes the body style approaching and selects from its memory the prestored programs to control the finishing robots and the door operators. In addition to the individual controllers on each robot, a supervisory computer coordinates their joint operations.

GMF has produced a video recording of this system illustrating its abilities. The system will paint a car every 90 seconds, inside and out, with two coats of paint in the wet-on-wet process now being used for automobile finishing. Viewing this video tape is an exciting experience. Some details of the system are described in Akeel [2].

10.5 Spot Welding

Two types of welding are commonly done by robots: spot welding and arc welding. Arc welding is discussed in the next section. Spot welding, discussed in this section, was one of the first applications of robots because the required accuracy and speed were available in early hydraulic robots and the robot's reach and load-carrying capacity were far superior to those of a human operator. In spot welding, two pieces of conductive metal are joined together by passing a large electric current through them. This high current heats the contact point sufficiently to cause local melting so that molten metal is formed momentarily. The current is on for a short time only (pulsed),so that the molten metal quickly solidifies, forming a strong joint.

Heating occurs at the joint because of the resistance of the metal to the passage of current through it. This is often called the

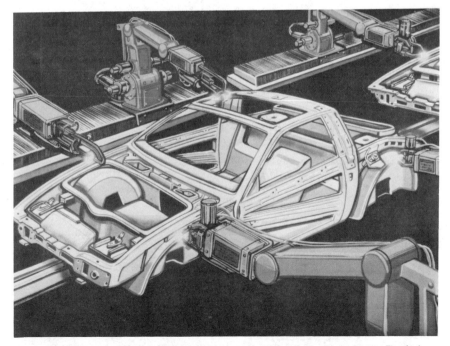

Figure 10.16 An artist's rendition of spot welding on an automobile production line for Pontiac Fiero's shows how integral transformers could be used (from Cecil [9], p. 18).

Courtesy of Robotics World magazine, April, 1984. © 1984 Communication Channels, Inc. Used by permission.

I^2R heating effect. The usual symbol for current is I and current is measured in amperes. Resistance in electrical circuits has the symbol R and is measured in ohms. Power in a resistive circuit is proportional to the square of the current multiplied by the resistance, hence the use of the term I^2R for the heating effect.

An artist's rendition of an automobile assembly line with spot-welding robots at work appears in Figure 10.16. On an actual assembly, it would be difficult to see the spot-welding action because of all of the necessary support beams, cables, and other equipment. Nearly all of the sheet metal and formed metal welding on automobiles is done by spot welding. The robot end effector is the welding gun, which has powerful pincer electrodes to force the metal together and provide good electrical contact during the welding operation. Power for forcing the electrodes together is usually pneumatic or hydraulic from a separate power source.

Currents as high as 1,500 amperes are used for spot welding at voltages of about 5 to 15 volts. Older spot-welding guns weighed as much as 200 pounds, including the heavy cables used to carry the high currents. It was necessary to counterbalance the guns from an overhead suspension so that human operators could handle them. In addition, water cooling was required to keep the copper electrodes of the gun from overheating during operation at high currents. This made the gun assemblies heavy and awkward to handle. During the spot-welding operation, there is a spray of sparks and molten metal, requiring the operator to wear protective clothing and a hood. Removing the operators from spot-welding area was therefore an important safety factor. Because of these conditions, factory workers welcomed the use of robots for spot-welding tasks. It was also found that the robots repeated the weld sequence with great precision, so that the quality of welds and the resulting frame structure were improved. Operator fatigue affected the accuracy of welds after a long day spent handling the heavy equipment. This problem was also eliminated by the use of robots.

Engelberger [15] describes a typical spot-welding operation in considerable detail:

1. Squeeze. The two surfaces are held together in contact with the electrodes which exert a force of 800 to 1000 lbs per square inch.

2. Weld. Current is turned on and flows through the material. Heat is generated in the vicinity of the electrodes.

3. Hold. The tips of the gun are kept closed long enough for the weld to cool, usually assisted by water circulation through the electrodes.

4. Off. The machine is rested until the next operation. In order to avoid overheating of the welding equipment, the 'off' cycle is governed by the welding control unit which programs an interval long enough for cooling to occur."

Computer-controlled sequencing of the robot and the spot-welding operation is used today to improve the quality of welds by accurately controlling the duration and intensity of the current pulses used.

Copper and aluminum are such good conductors that it is usually impractical to spot weld them. Ferrous metals are most suitable for spot welding. Fabrication of automobile bodies and domestic appliances are the most common applications of spot welding. Over 1,200 spot-welding robots are used in the automobile application at the present time.

Integral transformer weld gun systems have recently been developed as described in Cecil [9]. Older spot-welding guns were carried by powerful but bulky hydraulic robots such as the Unimate® 4000. As cars were reduced in size and lighter metals were used, it was no longer necessary to use such heavy currents to perform the weld. Also, the smaller cars did not have the space to accomodate the large robots, so it was desirable to design smaller welding guns and to control them with smaller electric-drive robots. As a result of considerable redesign, it was found possible to carry all of the cables and cooling tubing along with or inside the robot arm, thus providing a more compact design. Also, an integral transformer on the welding gun allowed higher-voltage, lightweight cabling to bring power to the gun itself where the voltage was reduced and the current increased by the transformer to eliminate the need for heavy cables carrying large currents to the spotweld gun. The weight of the spot-weld gun and cable system was reduced by about half, so that smaller robots could be used and the improved accuracy of electric drives could be obtained. Cecil [9] describes these improvements and discusses the considerable cost savings obtained. Figure 10.17 illustrates the integral transformer weld gun.

Figure 10.17 Integral transformer weld gun developed by Milco Manufacturing Company. This gun is mounted on the end of an electric welding robot's arm without the use of overhead cabling.

Courtesy of Milco Manufacturing Company. Used by permission.

10.6 Arc Welding

There are two major types of electric welding: gas metal arc welding (GMAW), also known as metal inert-gas welding (MIG) and gas tungsten arc welding (GTAW), also called tungsten- inert-gas welding) (TIG). In MIG welding, the electrode is consumed and laid down as filler material in the weld. In TIG welding, the tungsten electrode is not consumed, so a separate filler rod or wire is supplied to be melted down by the high temperature of the electric arc. Both MIG and TIG welding use an inert gas such as helium, argon, or carbon dioxide to protect the weld from oxidizing due to oxygen in the air. This operation is related to the Heliarc welding technique used in some portable welders. In practice today, the mixture of inert gases may be 90% argon and 10% carbon dioxide since helium is more expensive. Arc welding can be used to join stainless steel, carbon steel, aluminum and aluminum alloys, copper, and magnesium.

Arc welding is done at about $6,500^{\circ}$ F. This temperature, generated by the electric discharge between two electrodes, causes the materials being welded to fuse together. Direct currents of 100 to 200 amperes are used at 10 to 30 volts, depending on the length of the arc, the material being welded, and other factors. Normally, the workpiece is grounded to the welding power supply, while the positive side of the power supply is attached to the welding wire. There is a large body of knowledge on the science and technology of arc welding, but until recently, a highly skilled human operator was required to perform good welds.

There are five general steps in welding: cleaning the workpiece, set-up or fixturing the part or assembly to hold it in place, arc welding, tear-down or disassembly of the fixtured part, and grinding the welded part to remove excess metal. During the welding process, the filler metal is supplied from a reel of wire that is automatically fed through the gas tubing line to the welding gun.

All of these functions can be and have been done by robots. However, as an initial step, it is preferable to use a robot to perform the high-precision welding function that would otherwise require a highly skilled welding technician. Other functions can be performed by welding assistants, who are less skilled and are paid at a lower hourly rate.

Robots can weld at a rate of at least 40 inches per minute (300 inches per minute have been achieved) on simple straight-line welds of the type performed in the average welding operation.

Several hundred robots are now being used in arc-welding applications, so the use of robots in arc welding is well understood and of proven value. Recent developments in computer-controlled robots, computer-controllable welding equipment and rotary tables, and in software especially developed for robot arc welding have greatly improved the quality of robot welding and made it possible for experienced welders to use their existing knowledge and skills to program robots. As a result, the productivity of robots is two to five times that of human welders, the quality is better, and the welder is protected from the intense light, heat, and poor atmosphere of manual welding.

There are three main types of joints between parts to be welded, as illustrated in Figure 10.18. When two plates or sheets with straight edges are laid flat and pushed together until the straight edges touch,

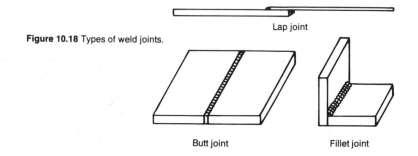

Figure 10.18 Types of weld joints.

Lap joint

Butt joint   Fillet joint

a butt weld is formed. Plates at an angle to each other, such as the two sides of a box, form a fillet weld. Lap welds are formed when two sheets are overlapped. This is a good approach when the sheets to be joined are different in thickness. All of these joints are called seams at the point where the actual welding is performed.

Robots may be taught to perform the weld, as discussed in the next section, or more sophisticated control techniques may be used. When robots are used for complex welds, it is often not practical to depend on teaching to perform the weld. In other cases, it is less expensive to use control equipment to guide the robot rather than depend on teaching. Therefore, several methods have been developed and are used for following the seam, depending on the weld operation and type of seam. Seams may be contoured in two or three dimensions and the width of the seam may vary, so that the amount of filler metal required to join the two parts together will vary.

10.6.1
Simple Welding

In some applications of robots for arc welding, the well-known teach-by-showing method is used. The skilled welder guides the robot end effector holding the welding gun through a welding path while maintaining the gun at the optimum distance from the workpiece, at the correct angle to the workpiece, and at a speed that will form a good weld. This optimum path is recorded and can be repeated as necessary to form good welds on a particular part. As part of the teaching operation, the voltage and current supplied and the rate of wire feeding are also controlled. Teaching works well for straight-line welds, corners, and simple curves where there is only a small variation in the width of the seam between the pieces being welded. A skilled welder controls the weld by observing the size and color of the weld puddle, the pool of molten metal formed by the heat from the electric arc. The welder knows from experience how large the pool should be and how fast to move the weld gun, depending on the type of weld being performed.

10.6.2
Tactile Seam Following

The simplest control method is the use of a probe or tactile sensor to follow the seam and feed back control information to the robot. On long, straight seams, the probe is placed an inch or so ahead of the weld gun and follows the groove between the two plates to be joined. It can be mechanically linked to the weld gun to guide it down the seam, or the probe motion can be sensed electrically, with the resulting signal used to control the robot motion. This system works well for straight and slowly curving seams but does not work well in most other cases.

10.6.3
Through-the-Arc Sensing

Noncontacting sensing can be achieved by oscillating the gun or

Figure 10.19 Through-the-arc sensing (from Stauffer [40], Figure 4, p. 33).

Source: CRC Automatic Welding, Inc., Houston, Texas. Used by permission.

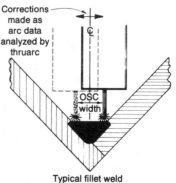

Typical fillet weld

torch back and forth across a seam and measuring the resulting current and voltage variation, as shown in Figure 10.19. Analysis of the information obtained allows the center of the seam to be determined. This method was developed and analyzed in detail by George E. Cook [10] and is available from CRC Welding Systems, Inc. in Houston, Texas. It can be used for MIG, flux-cored, submerged arc, TIG, and plasma arc processes.

This and other through-the-arc methods are discussed in Stauffer [40].

10.6.4
Vision Guidance
Systems

Several companies have developed robot welding systems based on the vision systems described in Chapter 9. In fact, welding systems are one of the biggest users of vision systems. Some companies and the systems offered are:

1. General Electric—WeldVision System

The vision sensor is built into the welding torch. A lens system focuses the image on one end of a fiberoptic bundle, which transmits the image to a GE TN2500 solid-state TV camera. Two parallel laser stripes are positioned across the weld joint and detected by the vision system. Deviations in the apparent path of the laser beams provide three-dimensional information to the tracking computer.

2. Unimation, Inc.—UNIVISION System

The UNIVISION system also uses a fiberoptic cable to transmit the image to the camera. Two passes are made across the joint. On the first pass, the whole length of the joint is scanned at a high rate, about a meter per second. On the second pass, the actual welding is done, using the stored information from the first pass.

3. Adapative Technologies, Inc.—Adaptavision 3-D

Adaptive Technologies is a small company in Sacramento, California that has demonstrated the ability to follow three-dimensional paths along the seam to be welded. They use a simple structured light system with a planar beam projected across the seam to outline the shape of the seam. This is also a two-pass system. It has been used to weld such irregular shapes as the two castings that make up an oil well drill-bit in which the seam between parts is of varying width and depth.

4. Automatix, Inc.—Robovision Welding System

Since there is considerable information available on the Robovision system, it will be explained in more detail as an illustration of the use of vision in robotic welding systems. The Automatix system is also one of the earliest and best-selling commercial systems.

Automatix, Inc., in Billerica, Massachusetts, was one of the early developers of vision systems for robots. They have developed three generations of systems using vision for inspection, visual servoing of arc welding operations, and robot guidance in bin picking of small parts. Autovision 4 is the third-generation system. Early generations were modifications of the feature extraction system developed by SRI as described in Chapter 9. Villers [46] describes some of the development process and system characteristics.

Computer usage in the latest system is quite sophisticated. Two powerful Motorola 68000 microcomputers are operated at 12 megahertz to control the vision system. Since the 68000s are capable of addressing up to 4 megabytes directly, they can handle up to 16 cameras with up to 512×256 picture elements and 64 level gray-scale capability on each camera.

Vision software provides display or output of 58 geometric features such as angular and linear displacement, area, and moments of inertia. The RAIL language, described in Chapter 8, is used for command and control of the vision system. Input/output is handled with multiple RS-232 lines. The total memory available to the vision and control processors is 512K bytes.

The Robovision II arc-welding system is shown in Figure 10.20. A laser beam is used to illuminate the seam during the welding operation. Optical tracking of the seam under computer control allows correction for part mislocation, part inaccuracy, and thermal distortions of the part due to the heat of welding.

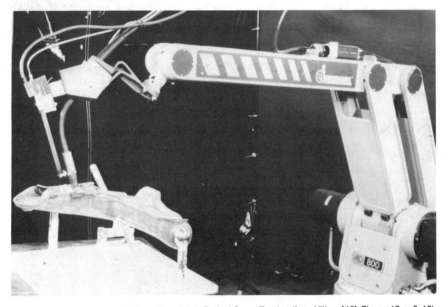

Figure 10.20 Robovision IIA with an Autovision Optical Seam Tracker (from Villers [46], Figure 19, p. 3–18).
Courtesy of Automatix. Used by permission.

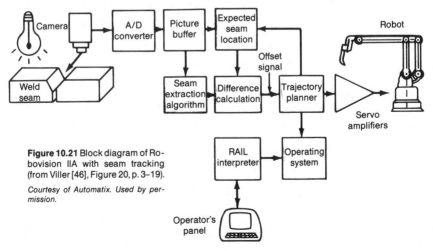

Figure 10.21 Block diagram of Robovision IIA with seam tracking (from Viller [46], Figure 20, p. 3–19).

Courtesy of Automatix. Used by permission.

Figure 10.21 is a block diagram of the visual servo used to control the welding robot. A typical part, as shown in Figure 10.22, can be welded in about 5 minutes along 80 inches of weld. The tolerance on the 0.25-inch lap weld is 0.05 inch and the parts vary in dimensions up to 0.25 inch. Using the vision system, the position error is reduced to about 0.03 inch.

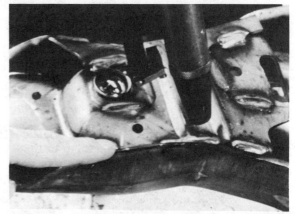

Figure 10.22 Complex customer part welded with an Autovision Optical Seam Tracker.

Courtesy of Automatix Inc. Billerica, MA. Used by permission.

10.6.5 Payoff for a Simple Robotic System

The justification analysis in Section 10.12 is based on straight-line welding using a robot taught by an experienced welder. An allowance is made for the time required for teaching. Economic calculations are given in that section . Here, however, we will discuss some of the background information used.

Experience with robot welding has demonstrated that, for the average operation, the robot can keep the welding gun active 80% of the time rather than the average 30% attained by the average human welder. This time saving occurs because the robot welder carries out a pretaught or controlled task, while the human welder must continually monitor the weld itself. Because of the intense light from the electric arc and the molten metal surrounding it, the human operator has to wear a heavy hood and heavy protective clothing. Ultraviolet light emission from the arc is so strong that the glass window in the welding hood must be thick and opaque to normal light. When the arc is off, the welder wearing a welding hood cannot see. It is therefore necessary to throw back the hood, locate the point to be welded, pull down the hood, and strike the arc. Then the light from the arc, passing through the glass of the hood is bright enough to view the seam being welded. If the welding torch or gun is lifted too high, the arc is extinguished and the welder cannot see the seam. It is therefore difficult to weld continuously even along a straight line. A robot can follow a path within 0.008 inch. A man cannot consistently weld with this accuracy.

Guiding a welding torch accurately is difficult. On straight-line welds, a man can move accurately at about 10 inches per minute. A robot can easily weld at 40 inches per minute under the same circumstances. In analyzing the output of a man versus a robot on straight-line welds, we can safely assume a welding throughput of four times as much for a robot as for a man. In the analysis of Section 10.12, a speed ratio of three is used to be conservative.

10.7 Assembly Applications

Another applications area of growing importance is the assembly of products using robots. A good example of assembly, the Westinghouse APAS project, is described in Section 10.7.2.   Many other applications can be found in the technical literature.   We will list a few of them to illustrate the type of work now being done before describing the APAS project.   Assembly of automotive engines and other units in Italy is described in Section 10.7.3 to complete this short survey of assembly applications.

**10.7.1
Electronics
Assembly**

Electronics assembly is the fastest-growing application of robots at the present time.   Assembly in the electronics industry is labor intensive and expensive, yet many of the operations performed are well within the capabilities of a small robot.   One forecaster, Predicasts, Inc. of Cleveland, Ohio, predicts that sales of assembly robots will increase at a rate of 74% annually through 1985 and continue at 41% per year, to hit $500 million sales annually by 1990.

Some assembly examples follow (the first four were reported by Harris [20]).

1. Apple Computer, Inc. in Carrollton, Texas, uses an Intelledex model 605, six-axis computer to assemble power supplies on base plates for the Apple IIe personal computer.   With the help of hard-automation fixtures, the 605 is building 250 assemblies per hour, about the equivalent output of eight people.   However, the quality of the parts supplied to the robot is critical; some parts can jam the automated feeded system.

2. The Digital Equipment Corporation uses an Automatix, Inc., AID 600 system with an Automatix vision system to place keys in terminal keyboards.   Production of the system is about eight times as fast as that of humans.   The vision system inspects, counts, and orients the keys.

3. Shugart Associates uses a Unimate PUMA 550 to assemble the mechanical portion of the SA 410 and SA 450 floppy disk drives.   It is turning out about 500 drives per day.

4. General Electric uses an Anorad robot on the radio assembly line to tune radio circuit elements.   There are 19 tuning points in commercial two-way radios, which are tuned by the use of multiple turning tools mounted on a turret carried by the robot.   These tools are rotated into place and used to adjust the position of the tuning coils in the radio while the frequency of the coil output is monitored.   The robot used costs $80,000 and is a Cartesian coordinate type mounted over a conveyor that brings radios to it for test and adjustment.   A Hewlett-Packard 9825 computer directs the robot's operations, test instruments, and turret with its tools.

5. Many electronics companies have used insertion machines for several years to insert standard-size components into printed circuit boards and then added the "odd-form" components by hand. Now robots are being used extensively to insert the odd forms such as potentiometers, relays, transformers, and special circuits.

    Dedicated machines handle up to 70% of the components on a board, leaving 30% to be handled by the robot.   Another factor influencing the use of robots is the tendency to make lower quantities of some special boards and a greater variety of boards. In this case, the use of a robot becomes more competitive with dedicated tooling.

    Figure 10.23 illustrates the use of a robot in placing a 64-pin grid array in a printed circuit board.   Insertion position for this and other components is repeatable within 0.001 inch (0.02 millimeter).

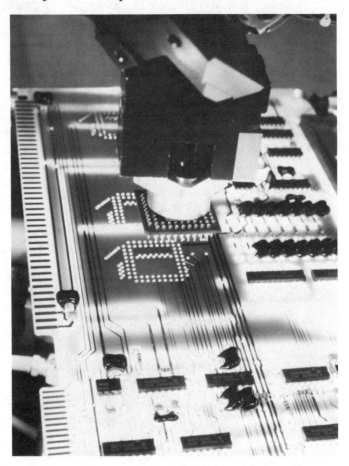

Figure 10.23 A Control Automation Mini-Sembler robot used to insert a 64-pin grid array in a printed circuit board (from Stauffer [42], p. 45).

Courtesy of Control Automation. Used by permission.

6. Another example of a special use of a robot is insertion of pins in military-type connectors. The connector is held in a fixture while the pins are picked up and inserted by the robot. Scheduling of connector type and pin insertion is done off-line and then down-loaded to the robot controller memory under supervisory control from another computer. Often in this type of operation, the CAD database already contains the information needed, so there is an advantage in using the information to control the robot directly rather than printing it out for human use. At the same time, an associated vision system can inspect the component to verify that it agrees with the CAD information. Stauffer [42] discusses these and other applications of robots in electronic manufacturing.

7. A total of 12 robots and 28 employees work together at 19 automated stations to assemble the new Touch-a-Matic telephone at the AT & T Shreveport, Louisiana, plant. This type of assembly is expected to make U.S. operations competitive with low-cost foreign labor and to produce a high-quality, reliable product at a reasonable price. Robots will place and solder parts on circuit boards and do final assembly. Employees will monitor the line and oversee repairs on phones not properly assembled.

10.7.2
Westinghouse
APAS System

Westinghouse recently announced the Adaptable Programmable Assembly System (APAS), partially funded by the National Science Foundation. Development of the system took about 5 years. APAS applies automated programmable assembly technology to batch assembly operations. This approach is expected to lead to significant productivity improvement in low-volume manufacturing.

Many products were considered as candidates for the program; fractional horsepower electric motors were finally settled on for the initial system. Westinghouse builds 450 different types of fractional horsepower motors, mostly in small lots and with frequent style changes on the assembly line. Two good articles on APAS by Stauffer [39] and Bortz[5] describe the operation of six end-bell assembly stations in detail.

The functions and lineup of the six stations are as follows:

Station 101 Inspect end bells.
Station 102 Pallet transfer
Station 103 Add a steel thrust washer, felt washer, bearing cap, and two lubricants
Station 104 Insert a plastic plug and a contact point, and drive three screws.
Station 105 Add an oil flinger, felt washer, dustcap (six styles), and mounting ring (two styles).
Station 106 Final inspection of completed end bells.

Some of the problems encountered in handling parts are discused in Stauffer [39]. Vibratory bowl feeders, frequently used in assembly, were considered impractical, so Westinghouse engineers developed new types of part feeders using concepts generally credited to Prof. G. Boothroyd of the University of Massachusetts. Feeders used two endless belts with parts fed from a hopper. Felt washers were a tougher problem. "A paddle wheel with a cutout rotates against the flow of parts letting only the flat, round washers pass through in single file" (Stauffer [39]). Other ingenious techniques were developed to handle other types of washers and parts.

As shown in Figure 10.24, there are four robots assigned to the six stations of the closed-loop assembly line for motor end bells. They perform simple parts handling, inspection, and assembly tasks. The modified Auto-Place robot at station 103 has a gripper adapted to pick up three different parts in sequence (Figure 10.25). A PUMA robot at station 105 (Figure 10.26) can handle a variety of parts for seven different end bell styles. Final inspection is done by a vision camera at station 106.

Control of the complete APAS is provided by a hierarchical system of computers. The block diagram of the system in Figure 10.27 shows the PDP 11/34 master computer, the four PDP 11/23 computers, and the individual PC microprocessors in the three-layer control hierarchy. Three PUMA robots and one AUTO-PLACE robot are controlled by the DEC PDP 11/23, which also controls the PC microprocessors handling sensory information. One PDP 11/23 provides interface facilities to the separate vision system.

Experience in developing and using the APAS has shown the importance of designing suitable feeders and end effectors to accomplish the system goals. Also, it was found that the cost of programming was underestimated and grew to be larger than the hardware cost.

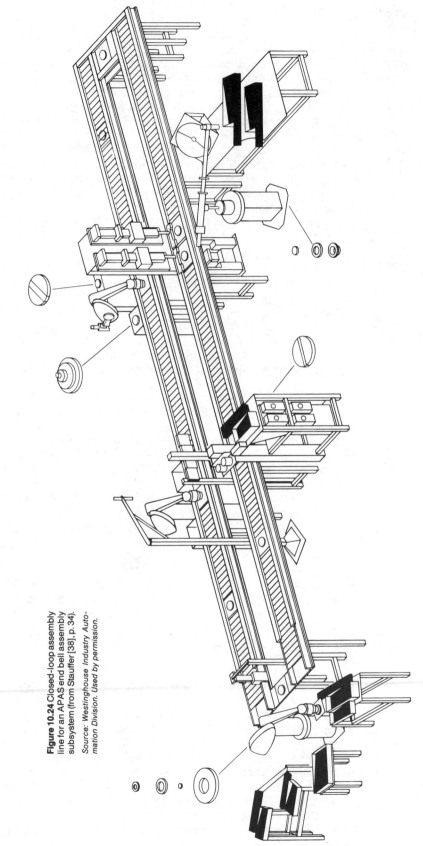

Figure 10.24 Closed-loop assembly line for an APAS end bell assembly subsystem (from Stauffer [38], p. 34).

Source: Westinghouse Industry Automation Division. Used by permission.

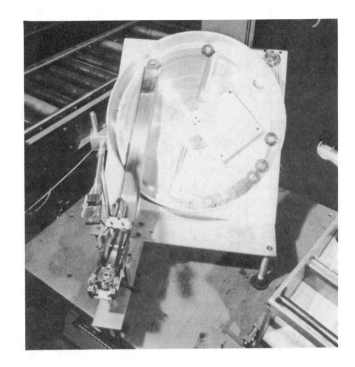

Figure 10.25 Modified Auto-Place robot.

Source: Westinghouse Industry Automation Division. Used by permission.

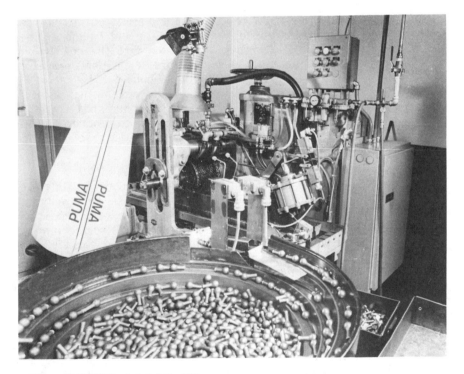

Figure 10.26 PUMA robot at station 105.

Courtesy of UNIMATION, Incorporated, A Westinghouse Company, Danbury, Conn., USA. Used by permission.

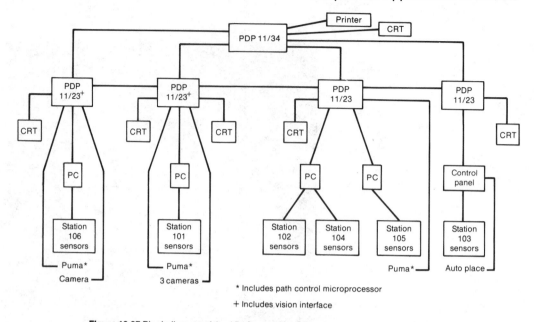

Figure 10.27 Block diagram of the APAS computer, sensor, and control arrangement (from Bortz [5], p. 16).
Source: Westinghouse Industry Automation Division. Used by permission.

**10.7.3
Robotic Assembly
in Italy**

An Italian robot manufacturer, Digital Electronic Automation S.p.A. (DEA), is building robotic systems for use in automotive assembly. Examples are (See Krauskopf [25]):

1. Fourteen DEA robot arms are arranged in five assembly cells to assemble cylinder heads for Fiat and Lancia engines. The robots assemble all valve components into the head, along with a camshaft cover and studs. A total of 147 parts are assembled on each head at a rate of 180 units per hour.
2. Six DEA robots are used in a system that assembles rack-and-pinion gear and housing components and tie rod subassemblies. Five tie rod parts are assembled to produce 500 assembled units per hour. A total of 285 steering gear assemblies are put together from 12 separate parts.

Other parts being assembled by robots in Italy are refrigeration compressors. Installation of bearings on the compressor crankshaft requires maintenance of critical tolerances. Therefore, cranks and bearings are accurately premeasured and sorted into five classes for selective mating. A total diameter tolerance of 0.02 mm. is permitted over the five classes; the fit between each bearing and crank must be within 20 micrometers. Parts are sorted by robots into classes and then placed into feeder mechanisms to be assembled by either the same or other robots. Tactile sensors on the robots detect part size, thickness, and position. Dual grippers are used on the robots, when possible, to perform multiple functions and speed up the assembly. A total of nine workers had been used with previous assembly machines. Now only three persons are required to load parts for robot handling.

Some of the same types of assembly are now being done in the United States. However, it is believed that the Italian automotive manufacturers are more advanced in this type of automated assembly.

10.8 Inspection and Test

A growing application of robotics and associated vision systems is in inspection and testing. Robots are used alone and with vision systems. There are also many cases in which the vision systems are used alone or as part of another automation system. Since the vision systems grew out of robot system work, they will be included as part of robot applications.

**10.8.1
Three-Dimensional
Measurement**

Some small robots can position with a repeatability of 50 millionths of an inch over a volume of 2 × 3.3 × 5.25 inches, and thus can be used to inspect directly parts that will fit within this volume or can be accurately stepped to position them within that volume. Robots with this capability, such as the Digital Automation Corporation's Mach-1 (M-1) model, can be purchased at a typical price of $12,000. Since they can move 0.25 inch and reposition within 40 milliseconds, they are quite cost competitive with human inspectors for inspection of machined or fabricated parts. A useful application of this capability would be to classify machined parts for later assembly by robots, as described in Section 10.7.3.

Many robots can position repeatably within 0.001 inch and so can be used where larger components are to be measured within this accuracy. The Everett/Charles Automations Systems, Inc., ERMAC 6-100 robot has a Cartesian coordinate system that will cover 2 × 6 × 6 inches, with a repeatability of ± 0.001 inch and a resolution of 0.0005 inch. By preprogramming the robot, the inspection operation can be made almost completely automatic.

**10.8.2
Robot Plus Vision
(or Other Sensors)**

Factory floor applications of robots with vision include the inspection of machined, formed, and fabricated parts. These systems are being applied in heavy equipment manufacturing, power generation systems, inspection of large flat sheets, and multiple-bend tubing assemblies. They are also used for the inspection of small assemblies and small parts. An interesting application is the use of robot "sniffers" to locate water leaks in vehicles. Some examples are now described.

1. A 3-D Robo Sensor is shown mounted on a Cincinnati Milacron T^3 robot in Figure 10.28. (Robo Sensor is made by Robot Vision Systems Inc.) This design is an extension of some of the designs described in Chapter 9 in using active optical triangulation between a projected light pattern and the location of its image on the surface to be viewed. The image is sensed by an offset camera. It is used to do automated inspection of engine casting and robotic arc welding. Surface measurement data, dimensions, position, contours, edges, radii, and concentricity are obtained, as well as object recognition. A Motorola 68000 microcomputer is used to process the raw vision data into a digital format. Accurate measurements can be obtained over a field of view of ± 1 × ±4 inches.

2. Ford Motor Company, at its Milan, Michigan, plant, uses a robot to pick up a thermoplastic bumper and hold it in place for inspection by a thermovision scanning system. Davis and Crawley [13] describe this system, which was used to examine bumpers for the 1984 Escort automobile. Thermoplastic bumpers are made up of two molded plastic parts that are welded together by vibratory

Figure 10.28 A 3–D Robo Sensor mounted on a Cincinnati Milacron T³ robot (from Nelson [28], and Crawley [13], p. 28).

Courtesy of Robotic Vision Systems, Inc. Used by permission.

friction welding. Heat from the friction causes melting at the weld joint. When the vibration stops, the weld joint quickly resolidifies and must be examined within 60 seconds before completely cooling to determine the continuity and strength of the weld by infrared light (thermovision).

This inspection is aided by a robot that picks up the bumper assembly and holds it vertically between two large vertical mirrors to be viewed by the thermovision camera. (The mirrors allow both sides of the bumper to be scanned at the same time.) In the Phase II operation of this inspection system, the thermovision inspection is done entirely by computer linked to the robot, although in Phase I the infrared image was presented on a CRT for approval by a human operator. Advantages of the robotized system include the following:

    a. Robot handling reduces manual part–handling defects and allows the operator to perform additional tasks.

    b. Thermovision computerized testing eliminates human judgment errors and reduces direct labor cost.

3. An English company, B. I. Technology, Ltd., has developed a unique method for testing for water leaks in vehicles by using sniffing robots. Understandably, vehicle manufacturers wish to seal their vehicles from penetration by water from outside. Previously, testing for leaks was done by spraying the vehicle with large quantities of water and examining the inside on the vehicle afterward. Damage to trim and other areas was easy to see, but it was often difficult to locate the actual source of the leak. In addition, this was a slow and expensive process.

Now the interior of the vehicle is filled with a mixture of helium and other gases, called the <u>tracer gas</u>, while moving down the assembly line. A flexible hose in a sealing frame is fitted through a car window to maintain a flow of gas while the vehicle moves down the line. The car is carried on a track to hold it in place with the required accuracy.

Robots carry sensor heads that can detect small amounts of leaking helium and relay this information to a computer test controller. As the vehicle moves down the assembly line, the robots track the vehicle and pass the sensor heads over the outside of the vehicle. Whenever a leak is detected, the computer records its position and displays it on a schematic drawing prepared by a test printer in the test station. Since the robots and sensor heads never contact the vehicle, there is no damage to the paint or external accessories. The sensing head on the robot channels the helium into a mass spectrometer type of gas analyzer to determine the amount of helium and therefore the severity of the leak. This system was carefully planned to take into account all possible mechanical, testing, and control problems. As a result, it has worked very well and produced the results expected. A schematic view of the system is shown in Figure 10.29.

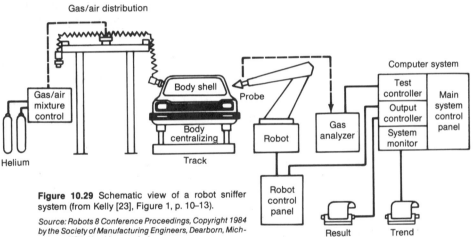

Figure 10.29 Schematic view of a robot sniffer system (from Kelly [23], Figure 1, p. 10–13).

Source: Robots 8 Conference Proceedings, Copyright 1984 by the Society of Manufacturing Engineers, Dearborn, Michigan, USA. Used by permission.

4. Large parts with many dimensions to be checked can be inspected by a robot–carried vision camera. Parts inspected in this way are gears, rings, and bulkheads. In this case, the robot can carry a camera so that the interiors of objects may be inspected. Accuracy is limited to the positioning accuracy of the robot unless there is some characteristic of the part that can provide a position reference. Suitable position references are usually available, however, so that the accuracy is limited only by the resolution of the camera and the geometry of its location.

For example, a robot camera with a resolution of 400 X 256 can be moved close enough so that the image to be measured is less than 0.25 inch across. Then the accuracy of measurement is within 0.001 inch because the 0.25 inch field of view is divided into 256 parts in the minimum direction. If the reference point is within the image, high accuracy can be obtained. Cameras are available with as many as 4,096 pixels across the field of view so that corresponding accuracies can be obtained. (Accuracy of the solid–state camera is seldom a limitation because the sensor is fabricated by extremely accurate equipment and used with high–precision lenses.)

Some examples of the growing use of vision for inspection in manufacturing plants are:

1. Currently, one of the biggest uses of vision systems is in the inspection of printed circuit boards. Inspection of etched boards before the components have been added is done routinely in many factories, although it was a new technique just 4 years ago. In 1984, at the Robots 8 Symposium, there were many exhibits demonstrating the use of computer vision for inspecting complete printed circuit boards at rates far exceeding the capability of human inspectors. Boards were mounted on X-Y drives with two cameras in use, one above and one below. It was possible to inspect boards as large as 18 × 24 inches in less than a minute with the high-speed inspection stations. Step-and-repeat scanning was used. The board was positioned for a few seconds in one position and scanned top and bottom by the vision system. Then the board was moved to a new position. These systems used microcomputers, for control and vision processing. Some of the newer systems had built-in systolic processors, as described in Chapter 9, to speed up processing. This application of vision can be expected to grow rapidly because of its high return on investment in terms of labor saving and improved quality control.

2. General Motors will use the standard model ScanSystem 200 manufactured by Object Recognition Systems, Inc., for inspection of automobile wheel bearing units. This system verifies correct assembly orientation, retaining ring crimps and lubrication within the wheel bearing units. Statistical quality control information is provided by the computer associated with the system.

3. An automated inspection system (ACOMS) is operating at the Cummins Engine Company in Indiana. The system consists of a main column supporting an array of four three-dimensional sensors. Scanning is done by structured invisible laser light. Measurements are obtained by optical triangulation techniques. Although millions of points can be measured, the system records and reports approximately 1,250 dimensions. This and other similar systems are reported by Nelson [28].

4. The Cognex Corporation makes the Checkpoint vision system for industrial inspection. It has been used for label, print quality, keycap, and printed circuit board inspection.

5. Several manufacturers now provide vision systems for gauging, label verification, part counting, sorting, and identification. It should be mentioned that all of the robot vision systems used for control of welding and other operations are also capable of providing inspection capability.

10.9 Auxiliary Equipment

Various types of auxiliary equipment are required to work with robots in application systems depending on the functions to be performed. End effectors are usually considered part of the robot itself but may become quite elaborate pieces of auxiliary equipment when the many possibilities for changeable end effectors are considered. Other types of auxiliary equipment most often used are conveyors, automatic guided vehicles, positioners, feeders, and separate sensor systems. There is also a need for additional equipment to provide a safe environment for the operators, the workers, and the robot itself.

10.9.1
Conveyors

Two types of conveyors are commonly used to bring work to the robot and to take away the work completed at a particular station. Overhead conveyors have chain-driven rollers on rails with supporting hooks for transporting objects that can be hooked easily. Flat conveyors, usually of the belt type, are used to carry small objects or assemblies to the robot workplace.

A particular kind of flat conveyor, the CARTRAC conveyor, carries multiple carts that roll or slide on horizontal guides. This type of conveyor is useful for positioning parts in front of a robot, as seen in Figure 10.30. Although it is quite expensive, it is justified for some tasks because it has built-in rotators and can position multiple objects under computer control. A Barrett Cravens AGV, used to bring pallet loads to the CARTRAC, can be seen in the background of the photo.

Synchronizing the robot to pick up objects from conveyors is done by computer controllers that calculate the robot motions necessary to coordinate the pickup task. Conveyor following is described in Section 6.7.3.

Figure 10.30 CARTRAC conveyor with robot and AGV. Taken at the Robots 8 Symposium, June, 1984 (from SI Handling Systems, Inc.).

Courtesy of Arthur Critchlow.

10.9.2
Automatic Guided
Vehicles (AGVs)

Automatic guided vehicles (AGVs) are being used more frequently in bringing large objects to the robot. They are capable of carrying several thousand pounds and thus can transport a complete pallet load of boxes or parts between points. Positioning accuracy of an AGV is usually about 1.5 inches but ± 0.25 = inch accuracy is available. A Barrett-Cravens Model 140/240 Unicar is shown in Figure 10.31. These vehicles follow a guideline in the floor and are under full computer control. Typical vehicles with associated equipment cost about $90,000. Although guidance is obtained from a wire buried in the floor, the vehicle is powered by heavy-duty industrial batteries driving DC motors.

10.9.3
Positioners

Robotic welding requires positioners for many tasks. The positioner can hold the part or assembly to be welded in the optimum

Figure 10.31 AGVs.

Courtesy of Barrett Electronics Corporation. Used by permssion.

position for welding and can be moved, under program control, as the welding operation progresses. Turntables, headstocks, and gear-drivern positioners are in use. Turntables rotate the workpiece horizontally; headstocks rotate the workpiece in a vertical plane (as on a lathe); and gear–driven positioners provide one, two, or three additional degrees of freedom. Positioners can present the workpiece, or weldment, so that weld seams are downward from the robot.

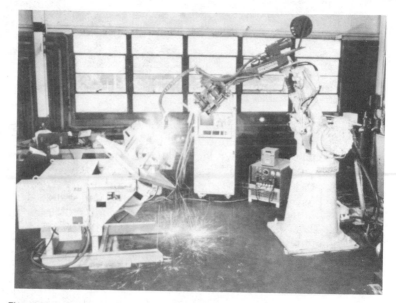

Figure 10.32 Positioner used in a welding operation (from Aronson Machine Company Bulletin CM82).

Courtesy of Cincinnati Milacron. Used by permission.

Gravity assists in controlling the flow of the molten metal and simplifies welding. Geared positioners made by the Aronson Machine Company have adjustable backlash control so positioning accuracy can be held to ±0.0005 inch per inch of gear radius. Typical positioners range from $10,000 to $50,000 per installation. Figure 10.32 illustrates the use of a positioner in a welding operation.

**10.9.4
Feeders**

Some alternative parts-feeding methods for use in robotic assembly are discussed by J. L. Nevins and D. E. Whitney [29]:

1. Conventional feeder tracks, slides, and vibratory bowl feeders may be used to provide parts that are oriented for pickup. However, the supply of parts must be used in the sequence presented.
2. Parts can be dumped onto a vision station and selectively picked up by the robot by using information supplied by an associated vision system. Since any part may be picked up at random, the choice of part is programmable.
3. Palletized kits of identical parts can be supplied to the robot already oriented in a supporting box. This operation simplifies the robot selection task at the cost of some preprocessing.
4. Kits of parts can be presented with all parts available for a product unit. Again, the load on the robot is simplified.
5. Kits can be supplied as just noted, but with the necessary tools also provided. Now the robot has everything it needs to carry out an assembly task.

Feeder tracks are simple slides with roller bearings on each side of the slide to support the parts and allow them to feed by gravity into position for pickup. Slides are adjusted to the right width for the parts being fed. Only one type of unit is supplied with each slide, so that there must be a separate feeder for each part. Some feeders are shown in Figure 2.8 with the HI-T-HAND.

Vibratory feeder bowls are usually about 12 to 18 inches in diameter. All the parts in the bowl are identical and may consist of such items as screws, nuts, levers, or caps, usually less than 2 inches long. An inclined helical track winds from the bottom of the bowl, around the sides of the bowl, to the feed position. Mechanical vibration of the bowl causes small parts to march up the bowl along the helical track. Parts with the desired orientation are allowed to continue to advance up the helical track. Ingenious detents, knock-offs, and scrapers are set up so that parts not properly oriented fall off or are knocked off the track. As parts reach the top, they are picked up by the robot or steered into a holding track for further use. Parts that fall back in are randomly reoriented by the vibration and again try to ascend the track. This type of feeder is widely used for orientation of small parts.

**10.9.5
Compliant Wrists**

In assembly, drilling, deburring, and routing applications, among others, there is a need for the robot and effector to give way to applied forces in order to allow itself to be guided by the task. Inserting a part into a hole or attaching it to another part requires that the part tilt slightly to fit into the hole or mate with the adjacent part. This is achieved by the techniques of Remote Center Compliance (RCC) described in Section 5.3.7. In other cases there is a need to measure the forces in three directions and modify the robot's actions in response. Force measurement is done with strain gages or other sensors in a way similar to that described in Section 5.3.5.

Examples of RCC devices used for robotic applications are shown in Figure 10.33.   Controlled flexibility in these devices is accomplished with laminated elastomer and metal shim elements.   In compression these elements are much stiffer than in shear.   By selecting elastomers, the number of shims, and part geometry, the performance of RCC devices can be altered to meet many application needs.   The devices shown, manufactured by the Lord Corporation, have a typical cost of $450 each. They are mounted on the end of the wrist mechanism of a robot. In turn, the end effector is mounted on the compliant mechanism.

Figure 10.33 RCC devices (from Flora [18], p. 348, column 1).

Courtesy of Lord Industrial Products. A Division of Lord Corporation. Used by permission.

10.9.6
End Effectors

As discussed in Section 3.4, end effectors are mounted on the wrist of the robot and perform the actual manipulation of objects. In this section, some of the advanced end effectors will be described. These end effectors make possible adaptive operation of the end effector itself. Many of them contain one or more microcomputers to control the detailed operation of the end effector.

Some recently developed end effector assemblies are being produced by EOA Systems, Inc. (End-of-Arm tooling). Examples of these end effectors, their characteristics and applications, are now discussed (Figure 10.34).

1. Complex Drill. This sophisticated drill can detect broken and dull drill bits and drill breakthrough into a vacant or less dense space. It has the ability to change up to 16 drill bits automatically. Each drill assembly can perform functions such as countersinking, drill- ing to a specified depth, and with a specified feed rate.   A computer provides control of these functions through closed-loop control of the end-of-arm tooling.
2. Parts Measurement Gripper.   This gripper utilizes a VA High- Resolution Linear Encoder to gauge machined or fabricated parts to less than 0.001 inch. Two pairs of infrared scanners are used for part detection in the fingers. The parts measurement gripper has a lifting capacity of 50 pounds and uses nonservoed, air-actuated fingers with a jaw travel of 4 inches.
3. Tactile Gripper.   This is a lightweight, force-sensing gripper with a 10-pound lifting capacity and 3.25 inches of travel. The tactile

gripper can grasp at various speeds until a certain position, grasping force, or part is detected. Fingertip position, grasping force, or part presence can be reported by the preprocessor to a robot controller or other device.

4. Transducerized Nutrunner. A variety of nutrunners are available to provide the user with a large degree of flexibility. Through the preprocessor, the user can format four sets of metric or English measurement units, specify torque or turn of the nut mode, and specify 16 different tensions or torque control actions. The preprocessor can detect cross-threading or other fastener abnormalities through comparison of the "joint signature" of a nut with the detected characteristics.

Other advanced end-of-arm tooling devices are under development at EOA, including a transducerized screwdriver and a deburr tool. These units are expensive but are justified in several special cases.

Several manufacturers provide different types of standard end effectors, as described in the Robotics Industry Directory (Flora [18]).

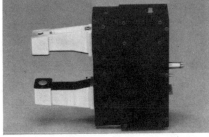

Figure 10.34 (a) Complex drill. (b) Parts measurement gripper. (c) Tactile gripper. (d) Transducerized nutrunner (from EOA Systems, Inc.).

Courtesy of EOA Systems Incorporated. Used by permission.

10.9.7 Fences and Safety Shield

All robots should be surrounded with fences and safety protection to protect the operators, the workers, and the robot itself from possible damage. The usual practice is to surround the area with a chain link fence and install interlocks on entry gates so that the robot is powered down if any gate is open. In addition, there are often special safety requirements specified by OSHA. Safety shields should separate the operator from any robot welding operations. In assembly operations in which an operator is assisting the robot, it is desirable to install protective means to stop the robot if the operator intrudes on the robot space. A simple light curtain is available commercially, which covers a vertical area of about 18 inches high x 24 inches wide, with multiple beams of light so that the robot will not operate if the operator's hands and arms pierce the curtain. In other cases, acoustic beams have been used successfully for protection, as discussed in Section 5.1.8.

10.10 Optimizing Robotic Systems

Robotic systems go through four major phases between conception and full operation. These phases or steps are:

1. Investigation. Determining what the goals of the system are and educating those involved in the necessary technology and operational requirements.
2. Planning. Identifying alternative ways to meet system goals, analyzing costs and benefits, and selecting the optimum system configuration. Planning includes hardware, software, and the effects on the people involved in the system. During this stage, detailed specifications for all types of equipment, procedures, and operator training should be developed.
3. Implementation. Designing the system to meet specifications, making the necessary tradeoffs and compromises, purchasing or building the system elements, and installing the system are parts of the implementation phase.
4. Evaluation and Follow-Up. Verifying that the system met its goals or the reasons why it did not is an important task. It is necessary to learn as much as possible from each system so that it can be improved on and the next system made better.

10.10.1 Investigation

Steps to carry out the system investigation may include the following:

1. Educate all personnel and staff on the available capabilities of equipment and related systems to some basic level. There are several ways to do this. One of the best ways to get started is to attend a comprehensive, well-organized seminar on robotics arranged by a competent group. The next step is wide reading in the field. After you have read this book thoroughly, for example, you should have a good overall understanding of the field. Now concentrate on the type of work you expect to do at your facility. Take advantage of the wealth of information available. Detailed accounts of the way in which systems have been planned and implemented are available. Some good summaries are given in Rosato [36], Abair and Logan [1], Klipstein [24], and Newman and Harder [30]. Much good information is available in the proceedings of conferences such as Robots 7 [45], and Robots 8 [35]. Many other sources are listed in the references for this and other chapters of this book.
2. Identify opportunities for economic benefits, improved quality, better working conditions, and so on in your organization based on the knowledge you have acquired. If your present staff is small, it may be valuable to call in an experienced robotic consultant to help you in identifying opportunities and planning the investigation. A long-term rather than a short-term approach should be taken. At this stage, do not ask, "what can we use a robot on?", but consider where the use of robots will lead and what results can be obtained over the next 5 years.
3. Survey the products to be made or the operations to be performed and investigate possible ways to peform the desired functions. Determine whether you are ready for robots and what steps will be necessary if you are not. Keep in mind that the first application should be one that has a high probability of working. One way to ensure success is to choose a task that robots have performed

before.   Also, this first application should show an obvious and easily understood payoff, either in economic savings, quality, flexibility, throughput, removal of personnel from a hazardous or boring environment, or some combination of these results.

4. Survey the existing plant and document the present capabilities and costs.   Good "before" and "after" data will be needed to show the advantages of using robots in this application.

10.10.2
Planning

Some of the steps in good planning are:

1. Establish system goals. Know what results you expect to achieve.
2. Determine system selection criteria.   Some useful criteria are these:
   a. Economic Benefits—Measurable in terms of Payback period; Return on Investment, or Discounted Cash Flow
   b. Quality Improvement in Products
   c. Improved Working Conditions
   d. Increased Output
   e. Flexibility and Longevity of the System
   f. Life Cycle Costs—Consider total maintenance, supplies, modification, programming, and operation costs over the life of the system.
   g. Human Factors (Ergonomic Considerations)

3. Determine system requirements: throughput, accuracy, reliability, work force characteristics, and so on.
4. Determine specifications: weight capacity, peak and average velocity, number of axes, wrist motion, type of end effectors, auxiliary equipment needed, memory required, control required, accuracy and repeatability, and so on.
5. Consider Integration with CAD/CAM, Computer-Integrated Manufacturing (CIM), and Flexible Manufacturing Work Cells (FMC).
6. Encourage specification review by all concerned or affected parties.   Those who are involved are more likely to understand and cooperate than those who do not.   Besides, they will have many useful ideas to contribute.
7. Determine documentation requirements: Operating manuals, programming manuals, maintenance manuals, layout drawings, and so on.
8. Set-up training and education plans for all those who must learn something about the system.   Keep everyone involved as fully informed as possible. There should be no secrets. If a position is to be eliminated, tell the workers involved and arrange for training, transfer, or other solutions to the problem.
9. Plan the organizational structure required to implement the planned operation and operate the robot system most effectively.
10. Determine acceptance testing procedures and specifications. Define what you mean by successful operation and devise means to test it.

In a small system not all of these steps may be necessary, but most of them will be. It does not take long to decide what has to be done, and it is less expensive and more effective in the long run.

10.10.3
Implementation

During the actual implementation of the system, there will be a frequent need to modify the plans previously made. This action will usually improve the system. But it is essential that a log book be kept of changes made and the reasons for them. Documentation should be kept completely up-to-date. An easy way to ruin a system is to get lost and not know the status of a system. This applies to hardware, software, procedures, and training especially. Of course, it is essential that a procedure be available to review and approve changes.

All affected personnel must be involved as much as possible in the implementation. Rosato [36] suggests that the following be kept involved and informed. Plant/Manufacturing/Process Engineers; Maintenance, Production, and Safety departments, Technical Training Manager, Human Resource Manager, Quality Manager, Employee Representatives, Programmers, Operators, and Skilled Tradesmen.

A partial check list of the functions to be performed during implementation is: (see also Stauffer [39], Laney [27], and Evans [17]):

1. Site Planning. Determine space and facilities requirements and ensure that they are made available.
2. Detailed design to meet the specifications outlined during the planning phase. Determine which items will be purchased and which ones fabricated or programmed in house.
3. Equipment selection, including vendor selection and evaluation of capability. Be sure to include spare parts, test equipment, and maintenance equipment as needed.
4. Operational procedures and programs, including training and programming. Also, include means for system monitoring and evaluation. How well does the system work?
5. Purchase of equipment and software to arrive on schedule and up to specifications, or fabrication and programming in house as required. Careful scheduling and schedule monitoring are necessary in both situations.
6. Programming and control planning: Choosing the programming technique (point-to-point or continuous path) and choosing between teach-by-lead-through, preprogramming, and others. How will auxiliary equipment be controlled?
7. Personnel selection and training. Some operators should be sent to vendor's schools. (Operators and supervisors at the Nissan truck plant were sent to Japan for training. See Drozda [14].)
8. Maintenance and reliability planning. Some in-house capability is required for maintenance. Space, tools, instruments, and spare parts are required, along with trained maintenance people.
9. Safety and health planning. Safety fences, pollution control, and so on. OSHA regulations must be observed, as well as worker considerations.
10. Installation and testing. Each piece of equipment should be tested separately before trying to test the system as a whole.
11. Education and training should be a continuing operation. What happens if an operator or maintenance person is off the job or quits?
12. Acceptance testing to specifications and procedures set down during the planning phase.

**10.10.4
Evaluation and
Improvement**

It is unlikely that the first robotic system will be the last. It is also unlikely that the first system will be perfect. Therefore, it is desirable to evaluate all aspects of the system and try to determine what its good features are and what aspects can be improved.

Robotic systems are evolving. Hardware, software, and techniques are being developed at a rapid rate. As time passes, there will be changes required or desirable in the system. Thus, performance data should be acquired on as much of the system as is economically feasible. Fortunately, new systems will be controlled by computer in most cases. The computer can gather the necessary data as a by-product of the daily operation.

Now is the time to begin planning for the improvements in the existing system and the addition of new systems. Perhaps the next step is FMCs, integration with CAD/CAM, CIM, or even starting to plan for advanced systems such as expert systems or intelligent systems.

10.11 The Future Factory

Many people have talked and written about a hypothetical Future Factory. This would be an almost unmanned, automated factory, with all of its operations controlled and integrated by computer. A visualization of one system approach, developed by Integrated Automation, is given in Figure 10.35.

In Figure 10.35, we see that Corporate Headquarters, the Raw Materials Warehouse, Manufacturing, and the Finished Goods Warehouse are all connected together by an integrated material handling and reporting system. This concept is similar to the hierarchical design described in Section 1.6.3 based on work at the NBS.

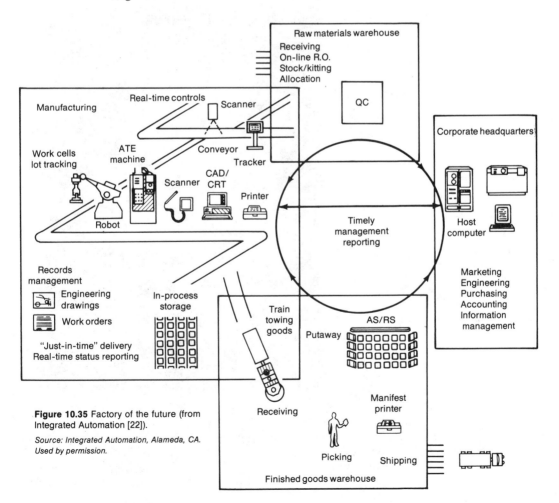

Figure 10.35 Factory of the future (from Integrated Automation [22]).

Source: Integrated Automation, Alameda, CA. Used by permission.

Some of the terms used in Figure 10.35 may be new to some readers. We have mentioned Computer-Aided Design (CAD), Computer-Aided Manufacturing (CAM), and the combination usually written CAD/CAM. This approach is in wide use, with estimated sales of $20 billion per year in these two areas.

In CAD, an engineer or other planner, sitting at a workstation connected to a powerful computer, designs a part or assembly directly without the need for a drafting table, paper, or the usual design equipment. The designer can draw on stored information in the computer for standard parts designs and dimensions, material characteristics, and analysis programs to aid in the design. An actual design drawing is presented to the designer on the face of a CRT. This drawing can be viewed in three orthographic views, or in isometric or other three-dimensional form. Under the designer's control, the object can be rotated about any axis, local details enlarged, and the whole drawing changed in size, as with a zoom lens. Usually, the drawings are in outline form, wire frames, which show the outline of parts and assemblies. However, some more expensive systems show the design information in apparently solid form and even in color. Figure 10.37 shows a robot arm in various positions. Figure 10.36 shows a robot arm in several positions using three-dimensional CAD display. Gunn [19] has a good discussion of CAD/CAM and the automated factory.

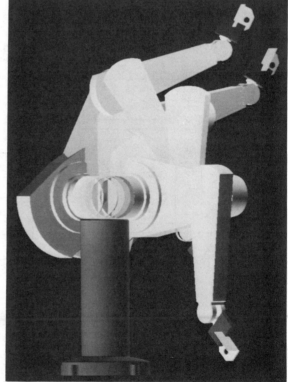

Figure 10.36 Three-dimensional CAD display showing various positions of robot-arm.

Source: Computervision Corporation, Bedford, MA.

Output from the CAD system is in the form of digital data stored in the computer memory, usually on a large disk file. This information may be transmitted to other locations or can be manipulated in many ways by a computer.

CAM was first applied to the use of computers to control machine tools such as lathes, milling machines, and drills. Now it refs to the control of the whole complex of equipment used in a modern factory. To the purist in CAD/CAM, robots are merely another tool used in a factory, and there is some justification for this point of view.

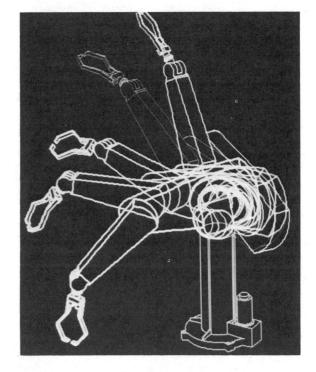

Figure 10.37 Wire frame form showing various positions of robot arm.

Source: Computervision Corporation, Bedford, MA.

An important term, not mentioned in Figure 10.34, is CIM, which stands for <u>Computer-Integrated Manufacturing</u>. This is envisioned as taking CAD/CAM one step further. In this approach, the design information obtained from CAD is used to control machine tools directly through CAM, without the need for human intervention. In addition, the robots are considered to be part of the CIM system and operate with other equipment in an integrated way to produce parts and complete assemblies. It is entirely conceivable that complete large assemblies can ultimately be made without human intervention, directly from the CAD information.

In Figure 10.35, in the Finished Goods Warehouse, we see a block labeled AS/RS; this is the Automatic Storage and Retrieval System. Such systems are in wide use; at least several hundred large warehouses are fully automated at the present time. Pallet loads of parts, equipment, supplies, and so on, are delivered into the AS/RS and are put away automatically by the computer-controlled equipment. AS/RS systems use large, specialized, electrically driven vehicles much like enlarged fork lift trucks. Most of these vehicles follow a track on the floor, move down the aisles of a warehouse with full pallet loads of merchandise (2,000 to 8,000 pounds per load), and store them on pallet racks as much as 100 feet high. They are expensive but provide a rapid, cost-effective way to handle enormous amounts of material. Detailed records are kept in the AS/RS computer so that any item may be quickly retrieved on demand.

Pulling into the Finished Goods Warehouse, at the top left of the warehouse block in Figure 10.35, is the train towing goods. This can be a mobile vehicle such as AGV, or it may be a vehicle on a track pulled by a tow chain in the floor. Both types are in use. These vehicles can be used in warehouses, in the manufacturing area, and in other areas of the factory.

Some vehicles are used to transfer bulky parts between machine tools automatically. For example, a large casting may be mounted on a pallet, carried to a milling machine for the first operation, and automatically off-loaded. AGVs can raise and lower the loads slightly, align them with the machine tool bed, and push them into place. While that operation takes place, the AGV can go for another load, or it can wait for the completion of an operation, pick up the machined casting, and take it to the next operation. Flexible operation is possible by controlling the path of the AGV.

In the Manufacturing area of Figure 10.35, we see the entities we have become familir with in this book: optical scanner, the CAD/CRT just described, the robot used in a work cell, the real-time controls provided by the computer, and the conveyor connecting all of these together. Combinations of conveyors and guided vehicles are used in manufacturing, depending on the products made, production rates, and many other factors.

Just-in-Time (JIT) delivery was first used extensively by Japanese manufacturers and has now become important for American manufacturers. JIT means that there is little or no stock kept in the manufacturing plant or even in a warehouse nearby. Parts are purchased or manufactured and scheduled to arrive at the factory or assembly line as needed. This is a tremendous cost saver. Expensive space is saved along the assembly line, parts are not damaged by excessive handling, and the cost of inventory is greatly reduced. JIT is possible only where very tight control is kept of all parts in shipment, on order, or in production. It is possible with the tight control available when a well-planned computer system is used. Major companies such as Chrysler, American Motors, Hewlett-Packard, Apple, Burroughs, Deere & Co., General Electric, General Motors, and Ford are using JIT and finding that it has a large payoff for them.

JIT requires improved discipline by suppliers in packaging, transportation, quality, and communication. If there are plenty of spare parts, for example, the quality of incoming parts is less critical. When there are no spare parts in stock, delivery of bad parts shuts down the assembly line. Thus, much better process control, etc. are necessary to attempt to obtain zero defects in delivered goods. Burgam [8] describes the problems and results of JIT.

Another term picked up from the Japanese is Kanban to describe the Toyota Motor Company, Ltd.'s order system. In a Kanban system the order for a part to be used on the production line is generated only by the requirements of the next station on the line. A chain of orders is set in motion by a single order for finished products at the end of the line. Unlike the manufacturing-resource planning system, which depends on detailed, centralized planning of all components and subassemblies, the Kanban system depends only on centralized planning of the output of finished products. It has the advantages of improved planning and reduced waste, but requires high standards of quality since there are no surplus parts. Gunn [19] discusses this concept in more depth.

Flexible Manufacturing Work Cells (FMC) are part of the Future Factory concept but have already been extensively implemented. The goal of an FMC is to group work together and do as much as possible at one time. It is especially suitable when a relatively small number of items are to be made and a small group of robots and associated machines can be used to carry out the complex task.

Jack Bradt, chairman of SI Handling Systems, Inc. Bradt [6] states, "My contention is that we are at the front end of a virtual revolution—not an evolution, but a revolution—in batch manufacturing operations, and that an understanding of that revolution is essential for the success and, indeed, the survival of United States manufacturing companies in the not-too-distant future." He and others believe that it is necessary to use JIT and other techniques—robots, precision tooling, and interfacing transportation systems—to provide higher-quality products, savings through inventory reduction, and the flexibility to fabricate parts as needed. It is predicted that there will be an increasing demand for small quantities of a large number of different models of items, and that only through the flexibility of the Future Factory concept will it be possible to produce them economically and of high quality.

Flexible Manufacturing Systems (FMS) is essential in the automated factory. Several simple FMSs can be integrated into the automatic factory. Each FMS will incorporate an in-process storage system and a high-speed, flexible transportation system to move materials between the storage areas and the work stations, all under computer control.

All of these capabilities—FMC, FMS, robots, sensory controls, CAD/CAM, AS/RS, AGVs, and computer control—are expected to make possible an automated factory that has people only for supervision, control, and maintenance. The first of these factories is expected to be available within the next decade to produce large quantities of reliable, well-designed, attractive products at low cost so that all members of society can have access to them.

At least one factory, the Fujitsu Fanuc plant in Japan, now has an unmanned third shift, the "ghost shift," where robots and other machines work unattended to build robot parts and motors. It is an eerie sight to see long lines of robots and machines operating efficiently with no humans around. Actually, there are people on the first and second shifts who set up the machines, and even on the third shift there are some maintenance people who watch over the machines. But it is a step toward the automatic factory.

10.12 Management Considerations

Major management considerations in the adoption and use of any system or technology are economic justification, improvement of product quality, effect on people, compliance with legal requirements, efficiency of operations, flexibility and growth potential, the effects on organizational structure, and the effects on the future of the organization. We will discuss each of these briefly, and refer to other sources for more detailed analysis and consideration of these topics.

This presentation is made from the point of view of the engineers and middle management people who make recommendations to higher management on system planning and adoption. Engelberger [15], Stauffer [39], Evans [17], and Petrock and Grumman [33] discuss some of these topics in more detail.

**10.12.1
Economic
Justification**

Economic justification is a very broad and important subject. In the long run, every decision made and every operation in the organization is an economic decision. Short-term optimization is possible in any company—cut out research and development, customer service, quality control, and almost anything else, and there will be a short-term economic advantage—but the long-term effect may be disastrous.

Economic benefit comes about from such items as these:

1. Reduced labor cost.
2. Higher productivity.
3. Greater throughput in the same facilities.
4. Less scrap and rework.
5. Material savings.
6. Power savings.
7. Lower handling cost.
8. Decreased inspection cost.
9. Improved management control.
10. Lower inventory cost.
11. Decreased insurance costs by providing a safer workplace.
12. More cooperation from workers who are better satisfied.

These are only some of the benefits that are potentially available from the use of robots and automated equipment. They should all be considered in justifying the use of a robot or other new system.

Justification is defined here as that set of circumstances that motivates a decision in favor of a particular alternative action. We justify the decision to install a welding robot because it welds faster, better, and more safely, and pays for itself through cost savings in 1.5 years. If it didn't do these things, it might not be justified and we would not choose to install it. In deciding among multiple alternatives, we would choose the one that best met our objectives or achieved the highest score by some decision process.

In determining economic justification, we can quantify in some way each of the factors just listed. All of these factors produce tangible savings (or losses), except improved management control and improved worker cooperation. These intangibles may be very important, but will not be considered here because of the difficulty of quantifying them. In a specific organization, they may be easy to quantify, however.

Although many ways of justifying expenditures are used by accountants, we will describe only three simple approaches. In any organization there is an accounting department that can provide the standard approach used by that organization. If the proposed expenditure meets the criteria established in the following discussion, there is a good possibility that it will meet stringent accounting rules.

Payback (or payoff) period in years, return on investment, and discounted cash flow are the three most common simple methods of economic analysis and justification.

1. Payback Period

The simple payback period equation is

$$\text{Payback} = \frac{\text{capital cost of system}}{\text{yearly savings} - \text{operations cost}}$$

Defining

$P =$ payback period in years, $C =$ capital cost of system, $S =$ yearly savings from all sources, and $O =$ operations cost (maintenance, spare parts, operators, etc.),

We have

$$P = \frac{C}{S - O}$$

Capital cost is made up of several items, including the robot itself, auxiliary equipment, end effectors, safety equipment, installation cost, training cost, and others. It should include all the costs incurred in setting up the original installation, programming and testing it, and training people to operate and maintain it. Planning and education costs for the first robot may be considered as overhead if more robots are likely to be installed. This initial cost may be so high that the robot would not be justified otherwise.

Savings can result from labor savings, material savings, space savings, reduced scrap and rework, faster turnaround so that inventory can be reduced, and other factors.

Operations costs include all costs incurred in the day–by–day operation of the system, such as daily operation and maintenance including an allocated overhead charge for each activity. Wages and salaries paid to operators who supervise the robot, perform set–up and change–over between jobs, and teach new jobs should be included. Maintenance includes the labor time spent on maintenance, the cost of spare parts, and perhaps a charge for the amortization of tools and test equipment used in maintenance. Actual charges for space, power, and other costs may also be included. In addition, the cost of insurance for the capital equipment involved should be included.

A sample set of data can be used to illustrate how this analysis might work using some assumed values. This data is for a simple arc welding system.

Capital Investment

1.	Robot, including shipment and insurance	$69,500
2.	Training fee—two operators	1,000
3.	Positioner—geared, two positions	18,500
4.	Safety fence—installation and switches	1,200
5.	Installation and testing—80 hours at $15 per hour	1,200
6.	Engineering and design time	10,000

Total Capital—$101,400

Savings

1. A robot welds at three times the speed of the human operator. It replaces three operators working 2,000 hours per year at $15 per hour, including fringe benefits.
Total labor saving: 6,000 hours $\times$ $15 per hour = $90,000 yearly.
2. Material savings—not easily measurable, small amount. Neglect.
3. Quality—better quality, fewer rejects, less grinding required. Savings, 10% of grinding time (1,000 hours), or 100 hours at $15 per hour.

Quality Savings = $1,500 per year

Total Savings—$91,500 per year

This analysis does not take into account the greater working time of the robot. Human welders usually work only about 6.5 hours per shift due to personal and fatigue allowance. Robots can work 95–98% of the time if a two–sided positioner is used, with a human operator preparing set–ups on one side while the robot is welding on the other. It is assumed that set-ups are prepared manually for both human and robot welders and that the same amount of time is required in each case. Also, grinding of finished welds is assumed to be done separately.

Operating Costs

1. Half-time for one operator for set-ups and training—1,000 hours at $15 per hour $15,000
2. Maintenance—third shift, 200 hours per year at $15 per hour   3,000
3. Depreciation per year on tooling and test equipment   3,000
4. Spare parts—Electric robot   2,000
5. Power—2 kilowatts per hour for the robot, welding power decreased. Net cost nearly zero. Neglect.
6. Insurance   1,100

          Total Yearly Operating Cost $24,100

Payback Calculation

$$P = \frac{C}{S-O} = \frac{101,400}{91,500 - 24,100} = 1.5 \text{ years}$$

The usual rule of thumb when using the payoff formula is that anything less than 3 years, which has a high probability of success, is a good application. If the project is risky, the payoff should occur sooner, perhaps in 2 years. When the payoff rises to over 3 years, this method is not sufficiently accurate because it does not take into account the cost of money (interest), capital equipment depreciation, tax credits, salvage value, and other factors. It is the simplest way to get a useful answer for justification purposes.

We have assumed that three operators are replaced by one-half of an operator's time and the robotic system. This is possible only if there is enough work to keep the robot busy. When there is an uneven flow of work, only two welders might be required at times, so that the other welder would work on some other task. Although this system might still have a good payoff, it is important to consider the actual work flow. The robot's costs go on whether it is working or not.

2. Return on Investment

Another way to calculate the value of a capital expenditure is by the return-on investment method. We will use the same values as in the payback analysis in this calculation.

The total initial investment is $101,400. We are interested in determining the cash returned yearly by this investment. It is assumed that the robot and associated equipment is depreciated over its useful life, assumed to be 8 years if it is well maintained. Then the costs of operation are the depreciation and operation costs determined previously. There are several depreciation methods used by accountants. We will use the simplest: straight-line, in which the total investment is depreciated evenly over the life of the equipment. This gives $101,400/8 or $12,675 as the yearly depreciation.

Total Robot System Costs

Depreciation	$12,675
Operating Costs	24,100
	$36,775 per year

Labor Savings   91,500 per year

Net Savings = Labor Savings - System Costs

$$= \$91,500 - 36,775 = \$54,725$$

$$\text{Return on Investment} = \frac{\text{Net Savings} \times 100}{\text{Investment}}$$
(percentage)

$$= (54,725 \times 100)/101,400$$

$$= 53.97\% \text{ per year}$$

This rate of return is quite good and would justify borrowing the money to purchase the robot system at an interest rate of 20% per year, since the net return on investment would still be nearly 34% per year.

One of the advantages of robots over fixed automation is the greater flexibility of the robot to do other tasks. In practice, it has been found that many robot systems have had a life of 8 years or more, so this type of analysis is reasonable.

If a faster depreciation period is used, say 5 years, the return on investment would decrease. However, if the robot could be kept busy on two shifts, a much higher rate of return would be possible.

3. Discounted Cash Flow (Internal Rate of Return)

A more accurate way to measure the value of an investment is to compare it to other possible investments. If we invest $100,000 today in a savings account or treasury bond, we could expect at least 10% interest compounded yearly. At the end of 5 years at this rate, we would receive $161,051 for our initial investment. Since we can invest in this way, it is clear that a dollar today is worth more than a dollar received 1 year or 5 years from now. When we purchase a robot system, there is a period during which it does not return any income because it is being installed. Then, after it starts to work, there may be a period of low activity due to startup problems. Finally, after 6 months to a year, the robot system is producing as expected, and we start to obtain labor savings or other savings in return for the investment made. The discounted cash flow method is a way to compute the expected value today of money returned at some time in the future. It is calculated by using interest rates in reverse. Table 10.1 gives an example of this calculation if we assume interest rates of 10, 15, 20, or 30% per year.

From Table 10.1, the present value, PV, of money received after n years is given as a ratio of the amount received. For example, at 20% interest the value of $10,000 received in the fifth year ($n=5$) is worth only $4,020 today, because you could invest $4,020 today at 20% and receive a total of $10,000 in 5 years with low risk. This return is calculated by the formula $PV = (1 + r)^n$ with $r = .20$ and $n = 5$. A more striking example is the present value of $10,000 in 10 years at an interest rate of 30%. Then the present value has dropped to only $720.

The discounted cash flow method is also known as the <u>internal rate of return</u>. It is too complex to cover here in detail but is described in Blois [3] as it applies to robots. In general, the effective return on investment will be lower. Blois also discusses in detail the effect of tax laws, alternative ways to calculate depreciation, and other topics.

Table 10.1 Present value of money received in the future

	Interest Rate			
Year	10% Ratio	15% Ratio	20% Ratio	30% Ratio
0	1.000	1.000	1.000	1.000
1	.909	.870	.835	.770
2	.826	.757	.695	.590
3	.751	.660	.578	.455
4	.683	.572	.483	.362
5	.621	.498	.402	.270
6	.564	.433	.334	.208
7	.513	.376	.279	.160
8	.467	.327	.232	.123
9	.424	.284	.193	.094
10	.386	.247	.162	.072

The ratio, R, is calculated from the equation $R = 1/(1 + r)^n$ where: r = the yearly interest rate in percent and n = the number of years.

10.12.2
Improvement of
Product Quality

Today, robots are making parts to higher accuracies than ever achieved by humans. Welds that are difficult or impossible to make reliably are routinely made by vision-equipped robots at speeds three to five times those of humans. General Motors is painting automboiles with a multiple-robot system with no people involved. High and consistent quality of finish is obtained due to the tight control possible using robots and the elimination of error caused by human fatigue.

10.12.3
Effect on People

People have been removed from tedious and hazardous jobs and given jobs supervising robots and other functions. Training has been found to be essential in enabling people to do their jobs. More important, perhaps, is that training people and allowing them to increase their value to the company has motivated many workers to commit themselves more fully to company goals.

Worker attitude at the new Nissan truck plant at Smyrna is excellent, and so far there is no union at the plant because the employees are happy without one. Productivity is high and the quality is excellent as a result. Drozda [14] describes the plant in some detail and reports that technicians are encouraged to take cross-training in as many skills as possible. An allowance is made in the production schedule for on-the-job training, and additional pay is granted to those with multiple skills.

Keeping workers informed is also considered to be important by those who have successfully introduced robots. Several companies have had "name-the-robot" contests to get people involved with the robot—with excellent results. Evans [17] discusses experience with involving workers and keeping them informed. If jobs are going to be eliminated, the workers are informed ahead of time and prepared for the necessary changes.

It appears that a new approach to workers, involving them in the task, is paying off for both management and labor. Any company considering the use of robots or other automation should become familiar with the new techniques being developed.

**10.12.4
Compliance
with Legal
Requirements**

Safety is mandated by OSHA in the form of regulations governing the use of all types of equipment. Robots, for example, must be surrounded by safety fences, and each gate through the fence must be interlocked with the robot so that opening the gate will stop the robot. Special consideration must be provided while an operator is training the robot in the form of an override switch. It is good practice to ensure that the override switch is normally locked and that a supervisor is responsible for the key and the use of the switch.

Robots allow people to be removed from hazardous environments which in some cases simplifies compliance with legal requirements. Air flow in painting booths can be reduced at considerable savings if the painting is done by a robot and no humans are permitted in the area.

It is important that legal requirements be well understood by any company introducing robots in order to meet the requirements and take advantage of opportunities in the use of robots.

**10.12.5
Efficiency of
Operations**

Productivity can be increased by using robots in many different ways. In Sections 10.1 through 10.8, we discussed some of the many applications of robots. In all cases, the use of robots has increased productivity. In addition, management control of the system has been improved due to the availability of immediate information from the computers in the system.

Less waste and rework is another manifestation of efficiency. Robots perform the same operations repeatedly. Once they have been properly programmed, they can work tirelessly up to 20 hours per day, essentially without error. Part of this gain is due to the high reliability of robots. Uptime of 98% is routinely achieved in well-maintained robots.

**10.12.6
Flexibility
and Growth
Potential**

Unlike hard automation, robots can be quickly and easily configured for different tasks. As programming has become easier due to the use of preprogramming and improved teaching techniques, it has become possible for a robot to change jobs every few hours if necessary. One General Electric advertisement lists the following schedule of activities for the GE A/3 robot:

Monday—Palletizing-300 crates of selector switches within 2 hours.
Tuesday morning—materials handling over an 8.62-square-foot area.
Tuesday afternoon—inspection/qualification.
Wednesday—parts loading/unloading   .
Thursday—electro/mechanical assembly of disc drives, VCRs, electric
          fans, motors, and so on.
Friday—PC board stuffing, with a vision sensor and ± 0.002 inch
          repeatability. The robot can handle almost any odd-shaped
          component in use.
Saturday—sealant/adhesive joining.

All of these activities can be preprogrammed and stored in low-cost memory so that the robot can switch from job to job at any time.

Since there can be many similar robots in a plant, they can all be used on one job or any combination of jobs. Only one robot has to be trained for a new task; then all the other robots can operate from the same program. Humans must learn each new task individually.

Flexibility also leads to a capacity for growth. Robots can work three shifts, 7 days per week, except for a short time (about 2 hours, per week) for preventive maintenance. If additional capacity is required, additional robots can be added.

**10.12.7
Effects on Organizational Structure**

It is clear that robots are affecting work practices in the factory and that a different organizational structure is required. In larger organizations, robot engineering and application departments are being set up to develop and supervise robot applications. Maintenance will become a more specialized task as robots are used in increasing numbers. Many robots today are controlled by programmable controllers that can be maintained by skilled electricians, but more complex systems will require people with electronics and computer knowledge.

Vision systems, adaptive robots, and integrated flexible manufacturing systems will require maintenance people and operators with many new skills. Again, it is important that management recognize the importance of training and set up training courses to upgrade the skills of employees. Many large companies have already set up training courses in house for their employees.

Each organization will find it necessary to adjust its organization to meet its own needs. This will be an important part of using robots and the future factory ideas that are rapidly being developed.

**10.12.8
Effects on the
Future of the
Organization**

In Section 10.11, Future Factory ideas were discussed. Many of them are in place now, others are being implemented, and perhaps some of them will never be used. Large organizations will use robots in increasing numbers. The new Nissan plant at Smyrna now uses 231 robots (Drozda[14]) in the following applications:

    127 robots for spot welding
     54 robots for arc welding
     44 robots for painting and anticorrosion spraying
      4 robots for sealing
      2 robots for tire handling

The use of robots is expected to increase at this plant because of the increased productivity and quality improvement that have occurred. It is clear that competitors of those companies that use robots will either have to use robots, find a better way, or go out of business. That is the challenge to today's management.

As robots become more adaptive and flexible, we can expect to see them used in smaller organizations. A 10-man machine stop today may become a 2-robot, 3-man machine shop in a few years and still provide greater productivity and improved quality. There are a large number of automobile body repair shops and paint shops in the United States. They perform many of the same types of operations that automobile manufacturers perform. Now that vision is available at a reasonable price, it should be possible to use robots to reshape fenders, and to do welding, grinding and painting in these shops.

Agriculture has not used robots appreciably, but why not? Many tasks in the fields have been automated and could be done better with robots. Here, too, we will see robots in use within a few years.

Clothing manufacturers are already trying to use robots to handle cloth, cut cloth, and sew large pieces together. With vision available, a robot can do these tasks cost competitively and bring back some of the clothing manufacturing that is now being done in less developed countries.

Management today has a great opportunity and a great challenge. It is up to all of those involved to take advantage of this opportunity and meet this challenge.

Exercises

Review the exercises below to test your knowledge of robot applications and refresh your memory.

1. What are the three most important reasons for using robots?
2. What was the first application of robots in a factory? When did it occur?
3. Why can robots do more work than human operators in some kinds of tasks?
4. Describe the sequence of operations involved in press tending with a robot?
5. The acronym AGV is frequently used in discussions of material handling equipment. What does it stand for?
6. What types of end effectors are used for palletizing?
7. Describe the IBM packaging system using a robot. What were the major components of the system? What was the major goal of the system? Was this goal achieved?
8. How are robots used in the investment casting process?
9. What is the meaning of the term <u>fettling</u>?
10. Water jet cutting is a new use of robots. How does it work, and what kinds of materials can be cut in this way?
11. How many degrees of freedom are needed for electrical wire harness manufacturing with a robot?
12. How does electrostatic spray painting work?
13. Spray painting is one of the major users of robots. Why is this so?
14. Are spray-painting robots operated as point-to-point or continuous-path robots? Why?
15. What are some of the handling problems in doing spot welding with robots? How are these problems being solved?
16. There are five general steps in arc welding. Which of them must be done manually, and which can be done by a robot?
17. Describe the alternative sensing methods for arc welding. In your opinion, which is the best sensing method for high-precision work?
18. How is the image transmitted from the sensing head to the TV camera in the WeldVision and UNIVISION systems?
19. When a vision system is used, what accuracy of weld can be expected?
20. Market researchers have stated (Section 1.7.3) that assembly will be one of the largest applications for robots by 1990. What are some of the reasons for this statement?
21. What are the functions performed at the six stations of the Westinghouse APAS system? What were some of the major problems in implementing this system?
22. Describe the way robots are used with sniffers to detect leaks in

automobiles. Describe other possible applications of this type of detection using robots.

23. What are some of the inspection applications using vision systems?

24. Several types of auxiliary equipment are used with robots. In your opinion, which type of equipment will increase most in use by 1990?

25. Section 10.10 discusses Optimizing Robotic Systems. Discuss the important characteristics of a good robot system plan.

26. What technological developments can you foresee that would make possible major improvements in the Future Factory concept?

27. How would the payback period change in the economic analysis of Section 10.12 if the robot worked two shifts instead of one? Make reasonable assumptions about the effect of two shift operation on the cost of maintenance and other costs.

28. Whose responsibility is it to implement robot systems effectively? State your own opinion, not necessarily that of the author.

References

[1] D. Abair and J. C. Logan, "The Road to a Successful Robot Project—It's a Two Way Street." In [45], pp. 4-32 to 4-49.

[2] A. Akeel, "Expanding the Capabilities of Spray Painting Robots," Robotics Today, April 1982, vol. 4, no. 2, pp. 50-53.

[3] J. F. Van Blois, "Robotic Justification Considerations." In [34], 1982, pp. 51-83.

[4] H. J. Bloom, "Robotic Diskette Writer System." In [35], 1984, pp. 3-14 to 3-28.

[5] A. B. Bortz, "Robots in Batch Manufacturing—The Westinghouse APAS Project," Robotics Age, January 1984, vol. 6, no. 1, pp. 12-17.

[6] L. J. Bradt, "The Automated Factory," Robotics World, March 1984, vol. 2, no. 3, pp. 18-19.

[7] T. J. Bublick, "Guidelines for Applying Finishing Robots," Robotics Today, April 1984, vol. 6, no. 2, pp. 61-64.

[8] P. M. Burgam, "JIT: On the Move and Out of the Aisles," Manufacturing Engineering, June 1984, Vol. 92, No. 6, pp. 65-71.

[9] J. Cecil, "Integral Transformer Weld Gun Systems," Robotics World, April 1984, vol. 2, no. 4, pp. 18-21.

[10] G. E. Cook, "Position Sensing with an Electric Arc." In [45], 1983, pp. 6-46 to 6-69.

[11] K. Crawford, "Interchangeable End Effectors." Robotics World, May 1984, vol. 2, no. 5, pp. 30-32.

[12] B. L. Dawson, "Moving Line Applications with a Computer Controlled Robot," (Cincinnati Milacron), Robots 2 Conference, October 1977. Reprinted in [44], Vol. 1, pp. 294-308.

[13] R. C. Davis and R. L. Crawley, "A Robotic Workcell for Linear Welding/Thermovision Scanning of Thermoplastic Bumpers." In [35], 1984, pp. 1-16 to 1-25.

[14] T. J. Drozda, "The Nissan Truck Plant—One Year Up and Running," Manufacturing Engineering, June 1984, vol. 92, no. 6, pp. 55-59.

[15] J. F. Engelberger, Robotics in Practice, AMACOM, Division of the American Management Associations, New York, 1980.

[16] EOA Systems Inc., 15054 Beltway Drive, Dallas, Texas. Phone: (214) 458-9886.

[17] T. P. Evans III, "A Robotics Learning Experience," Robotics World, April 1984, vol. 2, no. 4, pp. 38-41.

[18] P. C. Flora, ed., Robotics Industry Directory, 4th Ed. Technical Data Base Corporation, Conroe, Texas, 1984.

[19] T. G. Gunn, "The Mechanization of Design and Manufacturing," Scientific American, September 1982, pp. 115-130.

[20] M. A. Harris, "Electronics Assembly Draws Spotlight at Robot Show," Electronics, April 7, 1983, Vol. 55, No. 7, pp. 47-49.

[21] J. M. Heywood, "Robotic Drilling End Effector Research and Development and Production Support." In [45], 1983, pp. 18-14 to 18-22.

[22] Brochure, Integrated Automation, Alameda, Calif.

[23] M. P. Kelly, F. R. Piper, and D. Bray, "Testing for Water Leaks Using Sniffing Robots." In [35], 1984, pp. 10-1 to 10-15.

[24] D. H. Klipstein, "Practical Considerations in Quantifying the Purchase of Automated Vision Systems." In [35], 1984, pp. 2-63 to 2-73.

[25] B. Krauskopf, "Italians Move to Robotic Assembly," Manufacturing Engineering, April 1984, vol. 92, no. 4, pp. 98-99.

[26] A. Kusiak, "Robot Applications in Flexible Manufacturing Systems. In [35], 1984, pp. 3-1 to 3-13.

[27] J. D. Laney, "A Suggested Standard for Robots Comparisons." In [35], 1984, pp. 4-67 to 4-79.

[28] F. Nelson, Jr. "3-D Robot Vision," Robotics World, March 1984, vol. 2, no. 3, pp. 28-29.

[29] J. L. Nevins and D. E. Whitney, "Assembly Research," Automatica (Great Britain), vol. 16, 1980, pp. 595-613.

[30] D. J. Newman and M. J. Harder, "Comprehensive Calculation of Life Cycle Costs for Robotic Systems." In [35], 1984, pp. 2-83 to 2-96.

[31] L. V. Ottinger, "Applications for Robots—The Next Wave," Robotics World, vol. 1, no. 3, March 1983, pp. 18-21.

[32] V. J. Pavone, "User Friendly Welding Robots." In [45], 1983, pp. 6-89 to 6-102.

[33] F. Petrock and P. T. H. Grumman, "The Adoption of Technological Change: Management Issues and Practices." In [35], 1984, pp. 15-34 to 15-52.

[34] Robots 6 Conference Proceedings, March 2-4, 1982, Detroit, Mich. Robotics International of SME, Dearborn, Michigan, 1982.

[35] Robots 8 Conference Proceedings, June 4-7, 1984, Detroit, Mich. Robotics International of SME, Dearborn, Mich. 1984.

[36] P. J. Rosato, "Robotic Implementation--Do It Right." In [45], pp. 4-32 to 4-49.

[37] R. R. Schreiber, "Waukesha Welcomes Robots to Its Foundry," Robotics Today, April 1984, vol. 6, no. 2, p. 42.

[38] R. N. Stauffer (1983) See [39]

[39] R. N. Stauffer, "Equipment Acquisition For the Automatic Factory," Robotics Today, April 1983, Vol. 5, No. 2, pp. 37-40.

[40] R. N. Stauffer, "Update on Noncontact Seam Tracking Systems," Robotics Today, August 1983, vol. 5, no. 4, pp. 29-34.

[41] R. N. Stauffer, "Palletizing and Depalletizing: Robots Make It Easy," Robotics Today, February 1984, vol. 6, no. 1, pp. 43-46.

[42] R. N. Stauffer, "Robots Assume Key Role in Electronics Manufacturing," Manufacturing Engineering, vol. 92, no. 4, April 1984, pp. 84-87.

[43] M. J. Sullivan, "Putting Robots to Work: Loading/Unloading," Modern Machine Shop, November 1982, pp. 69-77.

[44] W. R. Tanner, ed. Industrial Robots, 2nd ed., Vol. 1, Funda-
     mentals, Vol. 2, Applications. Robotics International of SME,
     Dearborn, Mich. 1981.

[45] Thirteenth International Symposium on Industrial Robots and
     Robots 7, Conference Proceedings, "Applications Worldwide,"
     April 17-21, 1983. Robotics International of SME, Dearborn,
     Mich. 1983.

[46] P. Villers, "Recent Proliferation of Industrial Artificial Vision
     Applications." In [45], 1983, pp. 3-1 to 3-20.

[47] T. L. Welch, "Evaluating Robotic Systems Through Simulation."
     In [45], 1983, pp. 7-55 to 7-72.

Appendix A
Computer Notation and Codes

A.1 Binary, Octal, and Hexadecimal Notation

Computers operate on the binary number system (number system to the base 2) since most of the available semiconductor devices can have either of two values: on or off (or high or low), as discussed in Section 7.2. Since there are only two values in the binary number system, a method was developed to do computation using these values.

Conventional number systems used in our society are based on the base 10 number system. Since people have 10 fingers, it is convenient to count up to 10 and then start over. Numbers are represented in a positional notation. Each numeral in a number, going from right to left, is worth 10 times as much as the number on its right, so that the number 1093 for example, means

$$
\begin{array}{rcccr}
1 & \times & 1000 & = & 1000 \\
0 & \times & 100 & = & 0 \\
9 & \times & 10 & = & 90 \\
3 & \times & 1 & = & 3 \\
\hline
& \text{Total} & & = & 1093
\end{array}
$$

The positions of the numerals in the preceding sequence give the number a value, as in the number 1093. The same numerals in a different sequence would have a different value, such as 9310 or 9031.

In the binary number system, the same idea is used but there are only two possibilities, 1 and 0. We can still assign a positional value to each place in the binary number. In the binary number system, each place has a value twice as great as the position to its right. We can make a table of these decimal equivalent values as follows:

$$64 \quad 32 \quad 16 \quad 8 \quad 4 \quad 2 \quad 1$$

As an example, the binary number 1101011, when translated to decimal form so we can readily understand it, really means

$$
\begin{array}{rcccr}
1 & \times & 64 & = & 64 \\
1 & \times & 32 & = & 32 \\
0 & \times & 16 & = & 0 \\
1 & \times & 8 & = & 8 \\
0 & \times & 4 & = & 0 \\
1 & \times & 2 & = & 2 \\
1 & \times & 1 & = & 1 \\
\hline
& \text{Total} & & = & 107 \text{ in decimal}
\end{array}
$$

A useful table can be made up relating the first five binary digits to their decimal, octal, and hexadecimal equivalents as follows:

	Decimal Number		Binary Number					Octal Number			Hexa-decimal Number	
Position values	10	1	16	8	4	2	1	16	8	1	16	1
	0	0	0	0	0	0	0	0	0	0	0	0
	0	1	0	0	0	0	1	0	0	1	0	1
	0	2	0	0	0	1	0	0	0	2	0	2
	0	3	0	0	0	1	1	0	0	3	0	3
	0	4	0	0	1	0	0	0	0	4	0	4
	0	5	0	0	1	0	1	0	0	5	0	5
	0	6	0	0	1	1	0	0	0	6	0	6
	0	7	0	0	1	1	1	0	0	7	0	7
	0	8	0	1	0	0	0	0	1	0	0	8
	0	9	0	1	0	0	1	0	1	1	0	9
	1	0	0	1	0	1	0	0	1	2	0	A
	1	1	0	1	0	1	1	0	1	3	0	B
	1	2	0	1	1	0	0	0	1	4	0	C
	1	3	0	1	1	0	1	0	1	5	0	D
	1	4	0	1	1	1	0	0	1	6	0	E
	1	5	0	1	1	1	1	0	1	7	0	F
	1	6	1	0	0	0	0	0	2	0	1	0

Since there are several ways to write a particular value, depending on the number system employed, a subscript is often used to identify the number system being used. The number 11_8, for example, represents 11 in the octal number system, which is equivalent to 9 in the decimal system.

Four binary digits are required to represent the value 15_{10}, while only one hexadecimal digit is required, since one hexadecimal digit can have 16 different values. In order to make it possible to represent hexadecimal digits with one printer position, the capital letters A through F have been assigned values as follows:

Hexa-decimal		Decimal	Hexa-decimal		Decimal
A	=	10	D	=	13
B	=	11	E	=	14
C	=	12	F	=	15

Frequently, computer output will contain a series of hexadecimal digits such as D56F. These digits can be converted to decimal form by assigning the positional values and adding them up.

In hexadecimal, each position is worth 16 times as much as the position to its right, so the position values for the first four digits are:

4096   256   16   1

Therefore, the equivalent decimal value of D56F is

13	$\times$	4096	=	53248	since D = 13
5	$\times$	256	=	1280	
6	$\times$	16	=	96	
15	$\times$	1	=	15	since F = 15

Decimal value	=	54639

Converting a decimal value to an equivalent binary number is easily done by repeatedly dividing by 2 and keeping track of the remainder, and then reading the remainders in reverse order, that is, bottom up. As an example, convert 107 decimal back to binary form.

Remainder

2	107	1	2	$\times$	53 +	1	=	107
2	53	1	2	$\times$	26 +	1	=	53
2	26	0	2	$\times$	13 +	0	=	26
2	13	1	2	$\times$	6 +	1	=	13
2	6	0	2	$\times$	3 +	0	=	6
2	3	1	2	$\times$	1 +	1	=	3
2	1	1	2	$\times$	0 +	1	=	1
	0							

Reading the remainders bottom up, we get 1101011, which is the same binary number with which we started. Decimal numbers can be converted to octal numbers in the same way, except that the division is by 8 each time. For hexadecimal conversion, the decimal number is divided by 16.

A.2 ASCII Code

ASCII stands for American Standard Code for Information Interchange. ASCII code is the most widely used code in microcomputers and robotics, and has been standardized and approved by the American National Standards Institute as Standard No. X3.4-1977. This standard is also accepted by the Institute of Electronic and Electrical Engineers (IEEE) and by the Association for Computing Machinery (ACM) and several other groups.

In Figure A.1, the columns across the top specify bits 7 to 5 of the 7-bit code, while the rows down the left side specify bits 4 to 1. Normally, the code is written in the order 7,6,5,4,3,2,1,with the 7-bit called the high—order bit. This 7 bit code is used to display 96 printable and 32 nonprintable characters, for a total of 128 characters. This group of characters uses all of the 2^7 or 128 possible combinations of 7 bits. In some applications, an eighth bit is tacked on as a high-order bit (in the 8 position); it is used for parity and sometimes for indicating special characteristics of a code.

Bits	$b_7 \longrightarrow$ $b_6 \longrightarrow$ $b_5 \longrightarrow$	0 0 0	0 0 1	0 1 0	0 1 1	1 0 0	1 0 1	1 1 0	1 1 1
$b_4\,b_3\,b_2\,b_1$ $\downarrow\,\downarrow\,\downarrow\,\downarrow$	Column→ Row ↓	0	1	2	3	4	5	6	7
0 0 0 0	0	NUL	DLE	SP	0	@	P	`	p
0 0 0 1	1	SOH	DC1	!	1	A	Q	a	q
0 0 1 0	2	STX	DC2	"	2	B	R	b	r
0 0 1 1	3	ETX	DC3	#	3	C	S	c	s
0 1 0 0	4	EOT	DC4	$	4	D	T	d	t
0 1 0 1	5	ENQ	NAK	%	5	E	U	e	u
0 1 1 0	6	ACK	SYN	&	6	F	V	f	v
0 1 1 1	7	BEL	ETB	'	7	G	W	g	w
1 0 0 0	8	BS	CAN	(	8	H	X	h	x
1 0 0 1	9	HT	EM	)	9	I	Y	i	y
1 0 1 0	10	LF	SUB	*	:	J	Z	j	z
1 0 1 1	11	VT	ESC	+	;	K	[	k	{
1 1 0 0	12	FF	FS	,	<	L	\	l	:
1 1 0 1	13	CR	GS	−	=	M	]	m	}
1 1 1 0	14	SO	RS	.	>	N	^	n	~
1 1 1 1	15	SI	US	/	?	O	_	o	DEL

Figure A.1 ASCII code.

All characters with bits 7 and 6 set to 0 (hex 0 and 1) are nonprinting characters used for control functions. The DEL character, made up of all 1s, is used to mean "DELETE" but is also used as a logical symbol and prints the logical NOT symbol in some systems. The equivalent hexadecimal (hex) codes can be obtained for each of the ASCII codes by converting the column and row numbers to hex numbers and noting that the column number is the first hex digit. For example, the SP, or "space," character is designated by the hex value 20. DEL has the hex value 7F.

**A.2.1
Control Key
Usage of the
ASCII Code**

Many keyboards use the control key followed by a letter as a special control character. The effect of the control key is to invert the value of bit 7, so that the combination becomes one of the control characters in the first two columns of Figure A.1. For example, control-*K* (the control key held down while the letter *K* is also pressed) causes the 100 1011 code for a *K* to be converted to a 000 1011, so that the VT code is sent to the computer. The VT control character can then be used for any purpose desired in the computer by suitable design, although it is usually used as a vertical move control for a display. Control characters are often written with an upward caret^to indicate that the control key is to be pressed, so that this character could be written "^*K*".

Appendix B
Mathematical Operations

Some of the more common mathematical techniques used in the control, analysis, and description of robotic systems are described in this appendix. Vector operations and notation are covered briefly in Appendix B.1, matrix theory and notation are described in Appendix B.2. In both appendixes, only the main ideas are given, those needed to understand the control theory and homogeneous matrix operations described in Chapter 6.

B.1 Vector Operations and Notation

Vector analysis is another commonly used name for vector operations and notation. It is a systematic way of handling operations on vector quantities. Vector quantities are those such as force, velocity, and acceleration, which have both magnitude and direction. Vector quantities can be added, subtracted, and their product determined to obtain a resultant magnitude and direction. However, there are specific manipulation rules that must be followed in order to obtain the correct resulting values.

Figure B.1 illustrates the addition of two vectors. The direction of the vector is given by the angle made by the vector with a reference line. The magnitude is given by the length of the line. Addition is done by placing the base of one vector to the tip of the other vector and determining the new sum of the magnitude and direction (a + b). Any number of vectors can be added in this way by placing the vectors end to end. Figure B.2 illustrates the resultant with three vectors.

In order to develop a mathematical notation for vectors, the magnitude and direction information is converted to a specialized type of algebraic notation based on a cartesian coordinate system, as illustrated in Figure B.3. Here, the i, j, k terms are the unit vectors in the three mutually perpendicular (orthogonal) directions x, y, and z. Unit vectors have a magnitude of 1 or unity and are aligned along a coordinate axis to provide a direction specification. In Figure B.3 the magnitudes of the vectors are represented in terms of the components a_x, a_y, and a_z along the three orthogonal axes. The resultant vector is a, as shown in Figure B.3 and is given by the equation

$$a = a_x i + a_y j + a_z k \qquad (B.1)$$

Note that vector values are given in boldface type, while scalar values are given in regular type. Scalar values are quantities only; vectors have both magnitude and direction. The component values a_x, a_y, and a_z, are the components of the vector magnitude along the x,

y, and z axes and are equal to the magnitude of the vector multiplied by the cosine of the angle between the vector and the axis under consideration.

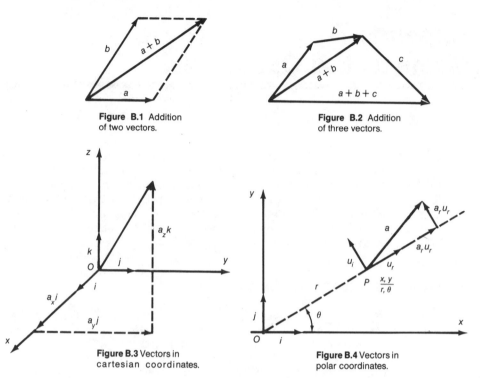

Figure B.1 Addition of two vectors.

Figure B.2 Addition of three vectors.

Figure B.3 Vectors in cartesian coordinates.

Figure B.4 Vectors in polar coordinates.

Other coordinate systems can be used, as in Figure B.4, which shows the vectors in a polar coordinate system. Only two coordinates are shown, with unit vectors in the radial direction and perpendicular to the radial direction. Only the cartesian vectors will be considered further in this section.

There are two types of vector multiplication, the dot product or scalar result and the cross product or vector result.

B.1.1
Scalar or
Dot Production
of Two Vectors

A scalar value (magnitude only) is found when the dot product of two vectors, **a** and **b**, is calculated. The resulting value has a magnitude or scalar value equal to that obtained by multiplying the magnitudes of the two vectors together and then multiplying that result by the cosine of the angle between them, as shown in Figure B.5. If **c** is the scalar value, it is determined by ("•" signifies the dot product)

$$c = a \bullet b = |a| \bullet |b| \cos \theta \qquad (B.2)$$

The following relations hold:

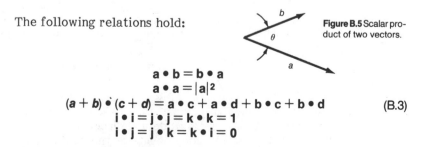

Figure B.5 Scalar product of two vectors.

$$a \bullet b = b \bullet a$$
$$a \bullet a = |a|^2$$
$$(a + b) \bullet (c + d) = a \bullet c + a \bullet d + b \bullet c + b \bullet d \qquad (B.3)$$
$$i \bullet i = j \bullet j = k \bullet k = 1$$
$$i \bullet j = j \bullet k = k \bullet i = 0$$

Since the angle between two unit vectors along the same axis is 0, the cosine has the value 1. Also, the angle between two unit vectors along different axes is 90 degrees, so the cosine is 0.

Given two vectors:

$$A = A_x\mathbf{i} + A_y\mathbf{j} + A_z\mathbf{k} \text{ and } B = B_x\mathbf{i} + B_y\mathbf{j} + B_z\mathbf{k} \tag{B.4}$$

The dot product is obtained by multiplying term by term, and canceling out those terms where the unit vectors are different, to obtain

$$A \bullet B = A_xB_x + A_yB_y + A_zB_z \tag{B.5}$$

Determining the component of a force, moment, or velocity along an arbitrary direction is a major use of the dot product in robotics.

B.1.2
Cross Product
of a Vector

When the cross product of two vectors is taken, the result is a vector. In robotics applications, it is common for vectors to be oriented in three-dimensional space. If two vectors are not parallel, their cross product may be formed. Since two vectors with the same origin define a plane, the cross product resultant will be perpendicular to this plane, as shown in Figure B.6.

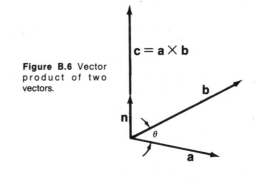

Figure B.6 Vector product of two vectors.

$$\mathbf{c} = \mathbf{a} \times \mathbf{b} = |a||b| \sin\theta\, \mathbf{n} \tag{B.6}$$

Note that the new unit vector, **n**, is perpendicular to the plane of **a** and **b** and in the direction taken by a right-hand screw rotated from **a** to **b**. If the angle between the vectors is 0 the value of $\sin\theta$ is zero and **c** is 0.

When using the cross product as defined in equation (B.6), the following relations hold:

$$\mathbf{a} \times \mathbf{b} = -(\mathbf{b} \times \mathbf{a})$$
$$(\mathbf{a} + \mathbf{b}) \times (\mathbf{c} + \mathbf{d}) = \mathbf{a} \times \mathbf{b} + \mathbf{a} \times \mathbf{d} + \mathbf{b} \times \mathbf{c} + \mathbf{b} \times \mathbf{d}$$
$$\mathbf{i} \times \mathbf{i} = \mathbf{j} \times \mathbf{j} = \mathbf{k} \times \mathbf{k} = 0 \tag{B.7}$$
$$\mathbf{i} \times \mathbf{j} = \mathbf{k}, \mathbf{j} \times \mathbf{k} = \mathbf{i}, \mathbf{k} \times \mathbf{i} = \mathbf{j}$$
$$\mathbf{j} \times \mathbf{i} = -\mathbf{k}, \mathbf{k} \times \mathbf{j} = -\mathbf{i}, \mathbf{i} \times \mathbf{k} = -\mathbf{j}$$

These relations can be verified by using equation (B.6) and noting that rotation opposite the right-hand screw rule generates a negative value.

Taking two vectors where

$$A = A_x\mathbf{i} + A_y\mathbf{j} + A_z\mathbf{k} \quad \text{and} \quad B = B_x\mathbf{i} + B_y\mathbf{j} + B_z\mathbf{k} \tag{B.8}$$

the cross product is

$$A \times B = (A_x\mathbf{i} + A_y\mathbf{j} + A_z\mathbf{k}) \times (B_x\mathbf{i} + B_y\mathbf{j} + B_z\mathbf{k}) \tag{B.9}$$

Multiplying these together and substituting values from equation (B.7), the usual form of the cross product is obtained.

$$A \times B = (A_yB_z - A_zB_y)\mathbf{i} + (A_zB_x - A_xB_z)\mathbf{j} \tag{B.10}$$
$$+ (A_xB_y - A_yB_x)\mathbf{k}$$

This relation is conveniently expressed by the 3×3 determinant

$$A \times B = \begin{vmatrix} \mathbf{i} & \mathbf{j} & \mathbf{k} \\ A_x & A_y & A_z \\ B_x & B_y & B_z \end{vmatrix} \tag{B.11}$$

Multiplying terms along the diagonals and adding the results by using the usual rules for evaluation of a determinant generates the same result as equation (B.10). Note that starting with a value along the bottom row and moving upward introduces a negative sign in the' term. Also, the reverse sequence, $B \times A$, does not give the same value as $A \times B$.

The cross product may be used to find the moment due to a force. Other applications and some illustrative problems are given in Shelley [3] and other textbooks on statics and dynamics.

B.2 Matrix Theory and Notation

A <u>matrix</u> is a rectangular array of values, numbers, or objects. In matrix theory we use an array of variables to represent values in a set of equations according to some specified conventions. Matrixes (the preferred term specified by the IEEE) conventionally have m rows and n columns. Such a matrix is called an $m \times n$ <u>matrix.</u> Rows of a matrix are numbered from the top down, and the columns are numbered from left to right. Subscripts on the names of the variables identify the row and column; the first subscript is the row and the second subscript is the column. Therefore, the element a_{ij} is in the ith row and the jth column of the matrix. (This same convention is also used in computer systems to specify arrays.) Equation (B.11) is an example of an array.

$$A = \begin{vmatrix} a_{11} & a_{12} & \cdots\cdots\cdots & a_{1n} \\ a_{21} & a_{12} & \cdots\cdots\cdots & a_{1n} \\ \cdots & \cdots\cdots\cdots\cdots\cdots\cdots \\ a_{m1} & a_{m2} & \cdots\cdots\cdots & a_{mn} \end{vmatrix} \tag{B.11}$$

One application of matrixes is in the manipulation and solution of a linear set of equations such as

$$\begin{aligned} 5x + 2y + 2z &= 15 \\ -2x + 3y + \;\; z &= 7 \\ x + \;\; y - z &= 0 \end{aligned} \tag{B.12}$$

Since the variables x, y, and z occur in each equation, we can represent the set of equations as

$$\begin{matrix} 5 & 2 & 2 & 15 \\ -2 & 3 & 1 & 7 \\ 1 & 1 & 1 & 0 \end{matrix} \tag{B.13}$$

These equations can be manipulated by matrix algebra to obtain solutions for the values x, y, and z. This is similar to the use of determinants when numerical values are used. However, determinants operate only with square matrixes and always have a value. Matrixes can operate with different numbers of columns and rows.

Another example of a set of equations is as follows:

$$x = x' \cos \theta - y' \sin \theta$$
$$y = x' \sin \theta + y' \cos \theta$$

(B.14)

which becomes, in matrix form,

$$\begin{vmatrix} \cos \theta & -\sin \theta \\ \sin \theta & \cos \theta \end{vmatrix}$$

(B.15)

Equations (B.14) and (B.15) represent the rotation of axes of a cartesian coordinate system. Note the similarity to the rotation equations in Section 6.6.1.1.

B.2.1 Matrix Definitions and Terminology

Some matrix terms are defined here for reference:

1. Order of a Matrix. The order of a matrix specifies the total number of rows and columns in the matrix. The matrix in equation (B.13) has three rows and four columns, and is thus a 3×4 matrix.
2. Square Matrix. A square matrix has the same number of rows as columns. Note that homogeneous matrixes are square matrixes (See Section 6.6.1).
3. Diagonal Matrix. A diagonal matrix is a square matrix with all values equal to 0 except those along the diagonal, where the value of i and j are equal.
4. Unity Matrix. A unity matrix has all diagonal values equal to 1 and all other values equal to 0.
5. Transpose of a Matrix. The transpose of a matrix is defined as the matrix obtained by interchanging the corresponding rows and columns of the matrix. It may be visualized as a rotation of the elements around the diagonal of the matrix so that all elements a_{ij} become the corresponding elements a_{ji}. The transpose of A is written A'. Note that an $n \times m$ matrix becomes an $m \times n$ matrix due to the transposition of elements.
6. Equality of Matrixes. Two matrixes are equal when they are of the same order and every element in one is equal to the corresponding element in the other. That is, if matrix $A = B$, then $a_{ij} = b_{ij}$ for all values of i and j.

B.2.2 Matrix Algebra

1. Addition of Matrixes

Two matrixes, A and B, can be added to form a new matrix, C, only if they are of the same order ($m = m'$ and $n = n'$). Then the prime values of each term may be added individually to obtain a new term such that

$$C_{ij} = A_{ij} + B_{ij}$$

(B.16)

2. Matrix Subtraction

If two matrixes are of the same order, each item can be subtracted from the corresponding term of the other matrix so that

$$C_{ij} = A_{ij} - B_{ij}$$

(B.17)

3. Matrix Multiplication

Two matrixes, A and B, can be multiplied together to form the product AB if they are conformable. This means that the number of columns of A must equal the number of rows of B when B is the second value in the product. This is necessary because we multiply the rows of A, term by term, by the columns of B, term by term, and add these products to get the desired result. Therefore, the number of terms in each row of A must equal the number of terms in each column of B. This arrangement can be seen in Equation (B.18).

$$A = [a_{ij}]^{2,3} \quad \text{and} \quad B = [b_{ij}]^{3,1}$$

then the two matrixes are conformable: there are three columns in and three rows in B, as shown by the subscripts.

$$AB = \begin{vmatrix} a_{11} & a_{12} & a_{13} \\ a_{21} & a_{22} & a_{23} \end{vmatrix} \begin{vmatrix} b_{11} \\ b_{21} \\ b_{31} \end{vmatrix} . \tag{B.18}$$

$$= \begin{vmatrix} a_{11}b_{11} + a_{12}b_{21} + a_{13}b_{31} \\ a_{11}b_{11} + a_{22}b_{21} + a_{23}b_{31} \end{vmatrix}$$

An entry is generated in the product matrix for every intersection of the rows and columns in the two matrixes being multiplied together. If matrix B had two columns of three rows each, it would still be conformable and the resultant matrix would have two columns of two rows each, instead of just one column of two rows with three terms per entry.

Determination of the inverse of a matrix is more complex and will not be covered. The reader is referred to other texts such as [2] for this information.

References

[1] HyTerm Communications Terminal, Model 1610/1620, Product Description, 82332 Rev. G. December, 1978, Diablo Systems Inc. (A Xerox Company), Hayward, California 94545.

[2] B. C. Kuo, Automatic Control Systems, 4th ed., 1982. Prentice Hall, Englewood Cliffs, N.J.

[3] J. F. Shelley, Engineering Mechanics: Statics and Dynamics, 1980. McGraw-Hill, New York.

Appendix C
Glossary of Robotic Related Terms

Some of the most frequently used terms in robotics are listed here with their usage and meaning in the field of robotics. When the same word has multiple uses, the separate meanings are identified by numbers.

In order to avoid excessive duplication, many terms have not been listed in this glossary that are defined in the text. ·Those terms may be found on the pages listed in the index.

ACCESS TIME   The time required to obtain information from a memory or storage unit. Measured from the request time to the time it becomes available for use.

ACCELERATION The rate of change in velocity over time. It has the units of feet per second per second in English units of meters per second per second in the MKS or meter–kilogram–second system.

ACCUMULATOR A register in the Central Processing Unit used to store operands and the results of operations.

ACTUATOR   A device to convert electric, hydraulic or pneumatic energy to rotary or linear motion.

ADAPTIVE CONTROL   Control capability of a robot or machine to alter its output in response to sensory information from its environment.

ADDRESS A number specifying the location of a unit of information in computer memory.

AIR MOTOR   A device to convert pneumatic pressure and flow into motion. A pneumatic actuator.

ALGORITHM   A set of well–defined rules (including processes and mathematical equations) for the solution of a problem in a finite number of steps.

ALPHANUMERIC   All alphabetic, numeric, and special characters in a computer language. Control characters are excluded.

ANALOG   Values that can vary continuously over a range, such as rotation, translation, or resistance. Not digital or stepped in discrete values.

ANALOG BOARDS   Printed circuit boards that develop amplified analog signals modified by compensation circuits to drive the servo motors.

ANALOG-TO-DIGITAL CONVERSION   Conversion of analog information to digital form.

ANSI (American National Standards Institute) The organization that sets many electrical standards.

ANTHROPOMORPHIC Having humanlike shape or characteristics.

ARCHITECTURE In computers, the physical and logical structure of the hardware and software systems. The meaning is similar to that of the term used in building and construction.

ARM A general term for an interconnected set of links and joints in a robot (or a human). Robot arms are sometimes called manipulators.

ARTIFICIAL INTELLIGENCE Machine or computer capability to do humanlike functions such as learning, reasoning, and self-control. One step beyond adaptive control.

ASCII (American Standard Code for Information Interchange.) It specifies a 7-bit code (8 bits when parity is used) for representing 128 different alphanumerics, punctuation marks, and characters used for computer and communications control.

ASSEMBLER A program that converts ASSEMBLY LANGUAGE to machine language. One machine language instruction is produced for each assembly language instruction.

ASSEMBLY LANGUAGE A simple computer language in which each word or group of words represents a single instruction in machine language.

ASYNCHRONOUS Not occurring at the same time as another operation.

ASYNCHRONOUS DEVICE A device that runs at its own speed. It is not controlled by clock signals.

AUTOMATION Automatic operation and control of equipment, processes, or systems. This is the general term for operation and control independent of direct human action.

BACKLASH Free space or "play" in gears and other drive units so that there is a slight lag after the driver moves before the driven unit moves. It results in a form of hysteresis and nonlinearity.

BACKPLANE A printed circuit board that provides the interconnection path between other printed circuit cards.

BANG-BANG CONTROL A control unit that causes movement until the unit hits a mechanical stop. Used in point-to-point mechanisms.

BASE (1) The number of characters in a number system, also called the radix. For the decimal system the base is 10, since there are 10 numbers, 0 to 9, in the system. Binary systems have a base of 2. (2) In a mechanical unit, the platform or structure that supports the robot or other equipment.

BASIC (Beginner's All-purpose Symbolic Instruction Code) An easy-to-learn high-level language available for most microprocessors. There are many dialects, not all of which are compatible.

BATCH MANUFACTURING Production of parts or material in lots or discrete runs, or batches interspersed with other units. Not continuous production.

BAUD One unit of a communication signal. May be a start bit, signal bit, or stop bit. Therefore, a band per second corresponds approximately but not exactly to 1 bit per second.

BAUD RATE The rate, in bits per second (approximately—see "BAUD"), of data communication over a serial link such as a telephone.

BCD (Binary-Coded Decimal) A computer code that uses 4 binary bits for each decimal digit. Only 10 of the possible 16 combinations of bits are permitted.

BENCHMARK A test program for computers used to compare the speed or other characteristics of two different computers.

BIDIRECTIONAL LINES Communication lines, or bus lines, that can be switched to carry information in either direction, but not simultaneously in both directions.

BINARY The base-2 number system that uses only the numerals 0 and 1. This number system is used in all computers today.

BINARY PICTURE   A video picture that has only two levels of brightness, black or white. Other pictures may be represented by many levels (gray levels) of brightness.

BIPOLAR   A technology used for integrated circuits. It is faster than some other technologies but is more expensive.

BISYNC   (Binary Synchronous)   A synchronous, serial data communications method developed by IBM.

BIT   A Binary digit that can have either of two values, 1 or 0.

BIT RATE   The number of bits per second moved between two points in a computer or communications system.

BLOB   An area or region in a video picture at a particular intensity level.

BLOCK DIAGRAM   A diagram containing several blocks connected by lines to indicate the flow of materials, signals, or information.

BOARD/CARD   Interchangeable terms representing one printed circuit board capable of being inserted into the backplane and containing a collection of electronic components.

BRANCHING   In a computer program, a change in the sequential flow of operations to allow a different set of operations to take place.

BUFFER   Memory area or electronic register used for temporary storage of information until needed.

BUS   A circuit or group of circuits that provide a communication path between two or more devices, such as a central processing unit, memory, and peripherals.

BYTE   Originally, a unit of information containing 8 bits. Now refers to a collection of bits used together that may be 6 to 9 bits in length.

CABLE   A group of wires twisted together. Steel wires are used mechanically to support or move a load. Insulated copper or steel wires are used to carry electrical current or electrical signals.

CAD/CAM   (Computer-Aided Design/Computer-Aided Manufacturing.) It describes the use of computers for these major tasks in a factory or automated system.

CALIBRATION   Recording of a device's measured characteristics at known values of input. Used to determine the actual value when a particular value is indicated so that errors can be corrected.

CAM   (1) Mechanical device with projections or lobes used to cause mechanical movement of other devices when the cam moves. (2) Computer-Aided Manufacturing.

CARTESIAN COORDINATES   A coordinate system with three mutually perpendicular directions. A Cartesian coordinate robot can move in three directions, usually called X, Y, and Z. See section 1.2.3.

CASSETTE RECORDER   A magnetic tape-recording device used to record the movements of a robot during teaching. This tape can then be used, in the playback mode, to control the movements of the robot. The magnetic tape is permanently enclosed in a protective housing called a "cassette."

CCD CAMERA   A solid-state camera using Charge-Coupled Devices for light sensing.

CENTER OF GRAVITY   That point in a rigid body at which the entire mass of the body could be concentrated and produce the same external effect as the body itself.

CENTRAL PROCESSING UNIT (CPU)   In a computer, the unit that provides the primary mathematical, control, and timing functions. It contains the arithmetic and logic unit, the control unit, and the instruction sequencer.

CHAIN DRIVE A drive to transfer mechanical motion from one point to another by the use of a flexible chain driven by or driving toothed sprocket wheels.

CHARACTER A group of bits that have a unique meaning such as a letter, number, or control identifier.

CHARACTER SET The total set of characters available in a computer system. Limited by the number of bits available in the code used. ASCII can have 2^7 or 128 characters.

CHECKSUM The sum of digits resulting from an error-checking scheme. It is stored as generated and then compared after an operation is completed to detect errors.

CHIP An integrated circuit. An electronic solid-state circuit on one piece or chip of silicon (or other material). Also used to mean the packaged unit.

CIRCULAR INTERPOLATION Fitting a circular curve through three points. Used to generate a circular control path for a robot when three points are specified by teaching or by programming.

CLOCK A periodic signal used for timing and synchronizing all devices in a system.

CLOSED LOOP A control method in which the status or position of the output or moving member is fed back to control the input.

CMOS (Complementary Metal Oxide Semiconductor) A family of integrated curcuits with low power requirements, medium speed, and high resistance to noise.

COMPATIBLE Capable of operating on the same software and/or hardware.

COMPILER A computer program that accepts a high-level language (such as FORTRAN) and produces machine code (object code) as output.

COMPLIANCE The ability to move or bend without breaking in response to applied forces. In robots it allows the robot end effector to align with a hole or another object.

COMPUTE-BOUND A descriptive adjective used for a program that is limited by computational speed rather than by I/O operations.

COMPUTER-AIDED DESIGN (CAD) See CAD/CAM.

COMPUTER-AIDED MANUFACTURING (CAM) See CAD/CAM.

COMPUTED PATH CONTROL A control method in which the desired path of the robot wrist center specifies and controls the movement of joints and links under computer control as required.

COMPUTER NUMERICAL CONTROL (CNC) Use of a computer to provide numerical control of a machine tool or robot. Numerical control was formerly provided by punched paper tape, which provided digital signals to control the motion of the machine.

CONTACT SENSOR A device capable of detecting mechanical contact and providing an electrical signal as a result.

CONTINUOUS-PATH CONTROL A control method that provides so many controlled points that the path is continuous within the desired accuracy.

CONTROL CHARACTER A special character used to control computer functions. Any character can be used as a control character if it is preceded by a designated control character.

CONTROLLER The device or person that directs the operation of a robot or system. A computer controller specifies the operation of each part of a robot.

CONTROL SYSTEM The equipment and procedures used to control an operation to achieve a desired result. Sensors, logic units, computers, and actuators are some components of such a system.

CPU (Central Processing Unit.) The unit of a computer system that accesses, decodes, and executes programmed instructions.

CRT (Cathode Ray Tube) A vacuum tube with a phosphor screen used for display of information. The familiar TV tube is a CRT.

CURSOR An intensified point on the CRT screen of a computer terminal that identifies the current location of the operation being performed. It can be moved by control signals from the keyboard.

CYCLE A sequence of events that is repeated frequently or continuously.

CYCLE TIME The time required to carry out one complete cycle of operations and return to the starting point.

CYLINDRICAL COORDINATE SYSTEM A coordinate system that has two linear movements, vertical and horizontal, and a circular movement around the vertical axis. A robot arm that can extend radially, move vertically, and rotate about the vertical axis sweeps out a cylindrical volume with a hollow center determined by the minimum length of the arm.

DAMPING Absorption of energy from a moving body due to frictional or other effects. May be used intentionally to reduce oscillations or vibration. It is important in the control of closed-loop systems.

DATA Signals, command, or information used or stored in a system.

DATABASE A collection of data used in a system. Also refers specifically to data stored in a computer under the control of a data base management program.

DATA BUS A set of signal wires used to carry data between parts of a computer or computer system.

DATA COMMUNICATIONS Any transfer of data between digital devices. Usually serial but may be parallel.

DATA SET Bell Telephone's name for a modem used to convert digital information for transmission over telephone lines and vice versa.

DATA TERMINAL Devices used to input and receive data from computers and communication lines. They usually have a keyboard and a video display or printer.

DEAD ZONE A movement or signal range within which a change in input does not cause a change in output in a sensor or mechanical device.

DEBUG To find and correct errors in hardware or software.

DECIMAL Based on the base-10 number system.

DEFAULT A condition or state that is assumed to be true when no other information is provided. Default characteristics are those that are built into the system but that can be altered by program control or hardware operation.

DEGREE OF FREEDOM A possible direction of motion in a robot. Each joint has at least one degree of freedom.

DIGITAL Use of discrete or individual numbers to represent information.

DIGITAL-TO-ANALOG Conversion of digital information to analog form, usually for control of analog devices.

DISK A rotating circular plate coated with a magnetic material that can be written on by a magnetic head.

DISK DRIVE A mechanism for rotating a disk and moving one or more magnetic heads to a selected concentric track on the disk surface.

DMA (Direct Memory Access) Providing access from a peripheral device directly to the computer memory and bypassing the computer. Provides high-speed data transfer.

DOWNTIME The period during which a system is not available for productive work.

DUTY CYCLE The fraction of time during which a device or system operates.

DYNAMIC MEMORY A solid-state memory that loses information rapidly and must be rewritten ("refreshed") every few milliseconds. It has a lower cost than static memory.

EBCDIC (Extended Binary-Coded Decimal Interchange Code) A special IBM character set using 8 bits to obtain 256 characters. Not used often on microcomputers.

EDIT To change the form or format of data, especially in a computer memory.

EEPROM (Electrically Erasable Programmable Read-Only Memory) Similar to EPROM, but can be written and erased electrically in a short time.

ELECTRO-OPTICAL ISOLATOR A circuit that converts an electrical signal to a light signal and back again in order to isolate two circuits electrically for protection against voltage differences or noise.

ELEVATION The distance above a horizontal plane or the angle between a plane and a straight line in a vertical plane intersecting the horizontal plane.

EMULATOR A program or hardware circuit that imitates another program or circuit in real time. Usually used temporarily when the real device or program is unavailable.

ENCODE To put information into a digital code.

ENCODER A device used to convert a linear or angular measurement into a digital code.

END EFFECTOR A gripper, vacuum pickup, drill, or other device used at the end of a robot arm to perform useful work.

EPROM (Erasable Programmable Read-Only Memory) A slow-write, fast read memory that can be erased by ultraviolet light. Used as a temporary ROM, especially in development.

ERROR SIGNAL The difference between the desired signal and the measured signal.

EVEN PARITY A code check method in which the sum of the 1 bits in a binary word is an even number. An odd number of bits signifies an error.

EXECUTE To carry out an instruction or run a program.

EXECUTION TIME The time required to execute an instruction or program.

EXOSKELETON An articulated framework worn by a human operator and containing sensors at each joint to convert human motions into control signals to operate a remote manipulator.

EXPONENT A number specifying the power to which another number, the base, is to be raised. If 10 is the base, the value of 10^2 is 100 where the number 2 is the exponent.

FAIL SAFE A device that can fail only in such a way that no damage to personnel or facilities results.

FAIL SOFT A type of failure in which there is no immediate total interruption of function, although performance may be degraded.

FAN-IN The number of input signals to a logic circuit or gate.

FAN-OUT The number of output signals that a particular logic circuit is able to produce; limited by the power capability in a circuit.

FAULT A condition or malfunction of equipment that prevents normal operation.

FEEDBACK Taking a signal from one part of a process and connecting it back to affect an earlier operation. An essential function in servo-controlled robots and other systems.

FETTLING Removing unwanted material such as sprues or flash from a casting by mechanical or other means.

FIRMWARE A program (software) stored in ROM and used for system control. Often used for operating systems and interpreters that change seldom.

FLAG LINE A signal line used to signal the status of a device to which it is attached.

FLIPFLOP A device, usually electronic, that can go from one output state to another. Electronic flipflops can store 1 bit.

FREQUENCY The number or repetitions of a cyclic event in 1 second. Usually used to describe alternating current or radio waves. The unit of measure is the hertz, defined as one cycle per second.

FRICTION A resistive force due to molecular attraction between two bodies moving in contact with one another.

FULL DUPLEX A type of communication channel capable of simultaneous transmission in both directions. Compare to half-duplex.

GANTRY A robot type with an overhead beam supported by rail-guided wheels at each end, which in turn supports the robot manipulator. It is named after the gantry-type crane and provides the ability to cover a large floor area.

GATE A minimal logic circuit with one output and more than one input such as the logical elements AND and OR.

GEAR The gear with the greater number of teeth in a gear pair; the other gear is called a "pinion."

GIGO A programmer's word meaning "garbage in—garbage out" to express the fact that a program is no better than its input.

GRAPHICS In computer systems, the generation of line drawings and graphs on a CRT terminal. Graphics systems include software and hardware for making drawings of many types.

GRAY-SCALE PICTURE A picture with more than two levels of light intensity. In robot vision systems, the pictures are digitized and may have as many as 256 levels of intensity.

GROUND A circuit at ground potential, either deliberately or due to a faulty circuit.

GROUND POTENTIAL In electrical circuits, the side of the circuit that is at the minimum zero reference voltage or potential. It may be connected to a copper stake driven into the ground.

GROUP TECHNOLOGY A method and practice of grouping parts into families based on their similarities, so that parts with similar characteristics are processed together.

GUARD In a factory environment, a safety device such as a screen or fence used to protect personnel and equipment.

HALF-DUPLEX A communication channel capable of transmitting in only one direction at a time. Compare to full duplex.

HANDSHAKE A signaling method or protocol used for controlling the transfer of data between devices at a rate acceptable to both devices. Either hardware or software may be used.

HARD COPY A printed document produced as output from a computer or other data processing device.

HARDWARE All the mechanical and electrical parts of a system. Sometimes used to differentiate between actual devices and programs and other software.

HARDWARE DRIVER A circuit used to impress a signal on a conductor or transmission line or to supply power to a device.

HARDWARE INTERRUPT An electrical signal generated by an external device to force the computer to recognize a task of higher priority than the one being executed.

HARDWIRED Permanently wired connections between pieces of electrical equipment.

HEAD The magnetic head used to write and read information to and from a magnetic disk or tape. It is essentially a small electromagnet with precise characteristics.

HERTZ One cycle per second in an electrical circuit or electromagnetic wave.

HIGHER-LEVEL LANGUAGE Computer languages (e.g., BASIC, FORTRAN, PASCAL) that provide powerful instructions and ease of programming. They must be interpreted or compiled before use.

HPIB (Hewlett-Packard Interface Bus) Also known as the GPIB and IEEE—488 bus. See Section 7.3.5.

HYDRAULIC MOTOR An actuator powered by the flow of oil or other fluid to produce linear or rotary mechanical motion.

HYSTERESIS A condition of a mechanical or electrical device in which there is a difference between its characteristics in the forward and reverse directions due to friction, stored energy, and other factors. It is a source of error in many systems.

IC Integrated Circuit. A solid-state microcircuit fabricated on a chip of semiconductor material. It may contain a few or several thousand logical or control circuits.

I/O Input and output operations or equipment.

I/O BOUND The state of a program whose operating speed is limited by the need for information interchange between devices rather than by computational requirements.

IEEE (Institute of Electrical and Electronic Engineers) The major professional organization for these engineers. It designs and approves standards and publishes many journals.

INDUSTRIAL ROBOT A reprogrammable, multifunctional manipulator designed to move material, parts, tools, or specialized devices through variable programmed motions for the performance of a variety of tasks. (Official definition by the Robotic Industries Association).

INERTIA The tendency of a mass at rest to remain at rest and of a mass in motion to remain in motion. A force is required to change the velocity of a mass. This force is proportional to the product of the mass multiplied by the acceleration imparted.

INITIALIZATION The process of setting the starting values in a computer system. Often the monitor does this automatically, using information stored in ROM.

INPUT Transferring information to the computer from an external device.

INPUT/OUTPUT (I/O) The processes that transfer information between external devices or peripherals and the computer.

INSTRUCTION SET A list of the machine language instructions that a computer can perform.

INTEGRAL CONTROL A type of servo control in which the input signal is modified by the time integral of the error signal.

INTEGRATED CIRCUIT A semiconductor chip with multiple circuits fabricated on it.

INTELLIGENT ROBOT A robot that can sense its environment, contains stored information (a knowledge base), and a set of production rules and programs so that it can perform some tasks autonomously. It is one level more capable than an adaptive robot.

INTERFACE A shared boundary between two devices, programs, or human to device interaction over which some kind of operation or communication is transferred. Standardization of codes or operations or connections is necessary to permit operation or communication.

INTERFACE CARD A device or circuit that converts signals from a computer bus into signals required by a peripheral device. Voltages, codes, signal speeds, and so on may be converted.

INTERLOCK A control connection between devices to ensure that devices requiring coordination do not operate independently. Often used as a safety device to protect people or equipment in case of failure of a device or process.

INTERPRETER A program that executes a higher-level language or converts it to machine language. A compiler is usually considered to be a form of interpreter.

INTERRUPT A signal calling for attention to a peripheral device. Also used to signal disruption of a process's normal flow.

INTERRUPT HANDLER A functional module capable of detecting interrupt requests and initiating appropriate responses.

INVERTER A logic element that outputs a 1 for a 0 input and 0 for a 1 input. Also called a "NOT gate."

JOINT In a robot arm, a mechanical device providing for motion of the arm in one rotational or translational degree of freedom.

JOYSTICK A movable handle connected to control circuits so that movement of the stick will generate signals for the control of robots and other devices. Typical joysticks have three degrees of freedom to control three axes of motion.

JOB SHOP A manufacturing facility that makes parts or equipment in small batches or lots but may make many different products.

JUMPER A short length of conductor used to make connections on an electrical circuit. Used to make semipermanent modifications.

K Computer notation for 1,024. This number is 2^{10} and is the closest binary value to 1,000.

KLUGE A term of disparagement. It refers to a poorly designed concoction of hardware and software that may not work well.

LAG A delay in motion or time between an input and an output.

LAN (Local Area Network) A group of computers and devices connected to serve a region. Ethernet is one of the better known LANs.

LANGUAGE In robots and computers, a set of symbols and rules for representing and communicating information.

LATCH A logic device that transfers input data to output during a clock signal transition and holds or "latches" the data after the clock transition.

LATCHING RELAY A relay that stays in position when initially operated until released by an external control action.

LCD (liquid crystals display) Tiny crystals are aligned by electrical fields and reflect ambient light. Used in slow-speed displays. Not visible in the dark. Provides an inexpensive display.

LEAD SCREW A precision machine screw that drives a precision nut so that rotation of the screw moves the nut to a desired location.

LEAD THROUGH In robot control, a means for programming or teaching the robot to perform a task by manually moving it through a series of operations and causing it to record the motion coordinates at several points on the path.

LED (light-emitting diode.) A solid-state device that produces light under electrical stimulation. Used primarily as a signal indicator.

LIMIT SWITCH An electrical switch that is actuated by mechanical motion when some part of a mechanism reaches a preset point.

LINEAR An input/output relationship in which the output is directly proportional to the input.

LINEAR INTERPOLATION The capability, in a robot control, to generate a straight-line path through two given points.

LOGIC The function performed by logical elements. Collectively, the circuits in a computer that perform logical and arithmetic functions.

LOGIC GROUND The connection used as a reference for the logical circuits in a system. May be at a different potential than earth ground.

LOOP In a computer program, a sequence of instructions that are executed repeatedly until some specified condition is met.

LSI (Large-scale integration.) An integrated circuit with 1,000 to 100,000 circuits on one chip.

MACHINE CODE Instructions in binary code that are directly executable by the processor.

MAGNETIC CORE MEMORY A type of memory that stores information in magnetic cores (small beads). This is an older type of memory, but is valuable because it holds information even when power is turned off.

MAINFRAME A large computer such as the IBM 370 series and its equivalents. These computers are enclosed in large cabinets or mainframes.

MANIPULATOR A mechanism made up of links and joints and capable of grasping and moving objects a short distance away under manual control. A robot arm is a powered manipulator.

MASS STORAGE A magnetic or optical storage device capable of storing large amounts of information in machine-readable, non-volatile form. Capacities in billions of bytes are available.

MASTER A functional module capable of initiating data bus transfers.

MATRIX A general term to describe the placing of numbers, objects, or symbols in a two-dimensional array of rows and columns like a checkerboard. It is possible to extend the idea of a matrix to an infinite number of dimensions in mathematical work.

MEAN TIME BETWEEN FAILURES (MTBF) The average time that a device will operate before it fails.

MEAN TIME TO REPAIR (MTTR) The average time required to repair a device after failure.

MEMORY In computers, the central high-speed device in which information may be stored and retrieved. Also called "primary storage."

MICROPROCESSOR A computer processor with all major central processing functions on one printed circuit board.

MICROSECOND One millionth of a second (10^{-6} second)

MILLISECOND One thousandth of a second (10^{-3} second).

MODEM (Modulator-Demodulator) A device to convert digital signals to tones for transmission over telephone lines and vice versa. Called a "data set" by the Bell Telephone System.

MODULAR Composed of subunits that can be combined in several ways.

MODULE A collection of electronic components with a single functional purpose. More than one module may exist on the same card.

MONITOR (1) A display or terminal used to present information to a human. (2) A program used to control the hardware of a computer system.

MOS (Metal Oxide Semiconductor) An integrated circuit technology that provides low cost, high density, and low power requirements.

MULTIPLEX (1) To switch computer signals between multiple inputs or outputs. (2) To interleave operations or messages on a single channel.

MULTIPROCESSING Use of two or more central processing units to cooperate in running one or more programs at the same time.

NANOSECOND One billionth of a second.

NBS National Bureau of Standards.

NEGATIVE LOGIC Also called "negative-true logic." A logic system in which a low voltage represents a logic 1 and a higher voltage represents a logic 0.

NEMA (National Electric Manufacturers Association) This organization sets standards for electrical equipment.

NEMA TYPE 12 A category of industrial enclosures frequently used in factories. It is intended for indoor use to provide protection against dust, dirt, and corrosive liquids. It does not provide protection against internal condensation.

NETWORK Several devices or computers connected together in a communication system. See also LAN.

NOISE Any type of unwanted and extraneous signal in electrical or electronic equipment.

NONVOLATILE A characteristic of those memory devices that maintain information when power is turned off. Magnetic disks, ROMs and magnetic cores are nonvolatile.

OBJECT CODE A program in machine code resulting from the compilation or interpretation of a higher-level language program. Also results from an assembler operating on assembly language.

OCTAL Operation using a base-8 number system with values 0 through 7. Can be converted to binary numbers using 3 binary bits.

ODD PARITY An error-detecting method for binary information that adds an extra 1 if necessary to make the total number of 1s in a binary word odd.

OFF-LINE Describes equipment or devices not directly connected to a computer or communications line.

ON-LINE Describes equipment or devices directly connected to a computer or communication line.

1s COMPLEMENT Obtained by inverting each bit of a binary number. 1s become 0's and 0's become 1s.

OPEN COLLECTOR A transistor output structure using an external resistor and voltage source to pull the device's output to a high voltage level. Used extensively to drive bus lines.

OPEN LOOP A method of control in which commands are given to a device or robot but no means is provided to determine whether the command was carried out. No feedback signal is provided.

OPERATING SYSTEM The software that controls and operates the system hardware in a computing system. Also performs supervisory and input/output functions.

OSHA (Occupational Safety and Health Act). Also refers to the enforcing agency, the Occupational Safety and Health Administration of the U.S. government. This act and agency set and enforce work rules and safety practices throughout business and industry.

OUTPUT Transfer of information from the computer to an external device.

PAN Describes rotary motion of a camera around a vertical axis to enable a wide area to be scanned.

PARALLEL I/O The fastest input/output method, since several bits are transferred at one time over multiple wires.

PARITY A method of detecting errors in a data word by adding an extra bit. In even parity, the total 1 bits are made to be even. Odd parity results in an odd number of 1 bits.

PARITY BIT An extra bit position added to a binary word to provide for parity checking.

PASCAL A structured programming language with controlled data types that is becoming increasingly popular.

PATTERN RECOGNITION Description and identification of pictures or other data structures(including voice patterns) by analyzing and categorizing their characteristics.

PAYLOAD The maximum weight or mass that can be handled by a robot or other device within speed and accuracy specifications.

PEAK The maximum or minimum value of a varying quantity.

PERIPHERAL A general term for all devices connected to a computer to provide input/output functions of various kinds.

PERIPHERAL PROCESSOR A processor whose main function is to control the operation of peripheral devices and bus operations.

PHOTO-ISOLATOR A solid-state device used to transmit signals between electrical circuits but allowing no actual electrical contact.

PICK-AND-PLACE ROBOT A simple robot used to transfer items from one place to another without the ability for trajectory control between points specified. It often has only two or three degrees of freedom and works between fixed stops.

PITCH The angle of rotation of a body above a horizontal reference plane and around an axis perpendicular to its direction of motion.

PIXEL (Picture Element.) One element of a matrix sensor array.

POINT-TO-POINT CONTROL Robot control in which the input commands specify only a few points along the desired path and the robot arm endpoint can move off the path between points. A less expensive but less accurate control method.

POLLING Periodically interrogating input/output devices to determine whether they have information to transfer to the computer. Compare to the INTERRUPT operation.

POSITIVE-TRUE LOGIC A logic system using a low voltage to represent 0 and a higher voltage to represent 1. TTL uses a positive-true logic.

POTENTIOMETER A position sensor or control device that depends on tapping the mechanical variation of resistance to control the signal amplitude.

POWER SUPPLY (1) The source of energy for a piece of equipment or device, such as a battery, generator, or power line. (2) The equipment used to convert power from an available voltage or type to another voltage or type.

PRIORITY The order of importance or precedence among competing events.

PRIORITY INTERRUPT An interrupt structure in which devices are assigned numerical values. Devices with higher values or higher priority may interrupt devices with lower values or lower priority.

PROCEDURE A sequence of actions taken to produce a desired result. In a computer, a subroutine or program with specific characteristics.

PROCESSOR (1) A computer unit that does mathematical calculations and control tasks. See Central Processing Unit. (2) A device that accepts a variety of inputs and produces different outputs based on its internal characteristics.

PROGRAM (1) A sequence of instructions to be executed by a computer. (2) Writing or generating the instructions for a computer. (3) Teaching a robot to carry out a task by leading it through a series of operations.

PROGRAM A series of computer instructions or statements that define a procedure or algorithm for carrying out a task.

PROGRAMMABLE CONTROLLER A control system providing computer capability to control industrial processes. It is designed to appear as a relay network using "ladder" diagrams to set up a sequence of operations so that electricians and others with no computer knowledge can use it for control.

PROGRAMMER One who prepares programs or teaches a robot to carry out a series of tasks. See definitions (2) and (3) of Program.

PROGRAMMING The act of preparing a program or teaching a robot.

PROM (Programmable Read-only Memory) A form of nonvolatile memory that can be written only once and then becomes permanent. Used for the development of new devices.

PROPORTIONAL CONTROL A control method in which the drive signal to the actuator increases monotonically and proportionally with the increased difference between the actual output and the desired output.

PROPORTIONAL INTEGRAL DERIVATIVE CONTROL (PID) A control method in which the actuator drive signal is generated by a weighted sum of: the difference between the actual output and the desired output; the time integral of the difference and the time derivative of the difference. In most situations, PID control provides faster response and greater accuracy than proportional control.

PROTOCOL A set of rules or procedures controlling the transfer of information in communication links. Protocols may be complex and multi layered. Timing, data formats, error detection, and so on may be specified.

PROXIMAL Used to describe a position close to the base and away from the end effector of an arm.

PROXIMITY SENSOR A device that senses objects only a short distance away. These sensors may measure magnetic field changes, reflected sound, or other phenomena.

PC BOARD (printed circuit board) A rigid, flat board used to support and interconnect semiconductor circuit chips.

PULSE A brief surge in voltage, current, or electromagnetic wave intensity. Used for signaling, synchronizing, time measurement, and so on.

PUNCHED CARD A standardized card that has been punched in specific positions to represent information. Read by optical readers or wires making contact through the holes.

PUNCHED TAPE Paper or plastic tape punched with machine-readable holes at specific locations to store information. Still used for control of certain numerically controlled machines.

QUEUE A list of operations or processes awaiting execution by a particular device in sequential order.

RAM (Random Access Memory) See Random Access Memory.

RANDOM ACCESS MEMORY (RAM) A memory that can be written into and read from repeatedly, usually at a high speed. All words in memory are equally accessible. Most RAM's are semiconductor devices.

RATE CONTROL Use of the desired velocity of the controlled member as the control input to a servo system.

READ-ONLY MEMORY (ROM) Memory that can be read rapidly , but whose content is permanently fixed.

REAL-TIME OPERATION Computer processing that operates fast enough to control the attached equipment in a timely manner.

RECORD-PLAYBACK ROBOT A robot that can be taught by leading it through a sequence of operations that are recorded on magnetic tape, magnetic disk, or other means so that the recorded path can be played back to control the robot action. Actual values of all or several joint encoders on the robot are stored in digital form during the recording and played back as desired to reproduce the operation sequence.

RECTANGULAR COORDINATE SYSTEM A coordinate system with two or three mutually perpendicular axes. See Cartesian Coordinates.

REDUNDANCY Provision of duplicate units or information in such a way that operation of a system will continue even though a device or unit fails.

REGISTER Electronic devices, usually made up of flipflops, that can store multiple bits of information and process them in several ways.

RELIABILITY The probability that a device will function without failure over a specified time period or amount of usage.

REMOTE CENTER COMPLIANCE (RCC) The capability to give or distort when a force is applied to allow alignment of a robot end effector with a hole or surface. RCC devices are usually mounted on a robot wrist to support and provide compliance to an end effector.

REPEATABILITY A measure of the ability of a robot arm or other mechanism to repeat a position motion at the same location under the same conditions as before. Repeatability is usually better than accuracy in a robot.

RESOLUTION The smallest measureable increment of change in a positioning device. Also used to specify the number of possible separable positions in a range of positions.

RESOLVER A transducer that converts mechanical rotation or translation to an electrical signal by the use of interacting electromagnetic fields between fixed and movable parts of the transducer.

ROBOT A device capable of mechanical motion under the control of a computer or control system and capable of operating without human control. See Industrial Robot for another definition.

ROBOTIC Related to or descriptive of robots or their actions.

ROBOTICS The field of activities of designing, building and applying robots.

ROLL Angular displacement of a moving body about an axis along its principal direction of motion.

ROM (Read-Only Memory.) See Read-Only Memory.

RS-232 STANDARD One of several interfaces used in computers and robotics. This is the most common standard for terminals. See Section 7.5.2 for a detailed description of this and other standards.

SCHEMATIC A drawing of the circuitry and interconnections in an electrical or logical device.

SDLC (Synchronous Data-Link Control) A procedure or protocol specifying a serial data control system using multiple levels of control.

SEGMENTATION In computer vision, dividing a picture into regions according to some property of the region. Also used to describe the action of joining together line segments by computer processing to assemble a complete line to delineate a region.

SEMICONDUCTOR   A class of solid-state devices made up of silicon, germanium, or other elements that have the property of conducting electrons under controlled conditions.

SENSORS   A device whose input is some physical phenomenon and whose output is an electrical signal reproducibly related to the input.

SERIAL I/O   Transferral of data 1 bit at a time (bit serial). Commonly implemented for terminals with the RS-232 interface.

SERVO   A control system that measures an output quantity and compares it to an input quantity through feedback to improve the control capability.

SETTLING TIME   The time required for a damped oscillatory response to decay down to a given limit value.

SHAFT ENCODER   An encoder used to convert the mechanical position of a shaft to an electrical signal.

SIGNAL   An electrical quantity or other phenomenon that conveys information from one point to another.

SIMULATOR   A device or program that imitates another device or program. It may operate either faster or slower than the original.

SLAVE   A functional module capable of responding to data transfer operations generated by a Master.

SLEW RATE   The maximum rate of motion that a system can follow.

SLOT   A single position at which a card may be inserted into the backplane. One slot may consist of more than one connector.

SOFTWARE   All the procedures, programs, and supporting documentation in a system.

SOLENOID   An electromagnet with a movable core element used to provide mechanical motion when the solenoid is energized.

SPHERICAL COORDINATE SYSTEM   A coordinate system specified by two angles and a distance from a referenced point called the "origin." The tip of a radius vector from the origin sweeps out a sphere with a radius equal to the length of the radius vector.

STATIC FRICTION   The minimal force required to initiate movement between two contacting bodies. Sometimes this is called "STICTION."

STATIC MEMORY   Solid-state memory that holds information without refreshing as long as power is supplied.

STATUS   Information about the current condition of a device.

STEADY STATE   A value that does not change with time. Also, the condition of a system after a transient response has died out.

STOP   A mechanical device or movable clamp used to stop mechanical motion at a designated point.

STRAIN GAGE   A sensing method that measures small amounts of strain or stretching by the change in electrical resistance of a wire cemented to the structure. Used to measure forces and torques.

STRING   A set of alphanumeric characters considered as a unit.

STROBE   A timing control signal to control information acceptance and transfer.

STRUCTURED LIGHT   A light pattern projected on an object to allow its dimensions and geometry to be measured with a vision system.

SUBROUTINE   A portion of a computer program that performs a specific task or set of tasks and can be called from other programs or a main program to perform its function.

TACHOMETER   A small generator used to generate a voltage proportional to velocity for use in servo control.

TACTILE   Pertaining to the sense of touch.

TEACH   In robots, programming a sequence of operations by moving the robot arm through the desired sequence. See Record—Play-Back Robot.

TEACH PENDANT   A hand-held control unit used for controlling a robot during the Teach operation.

TELEOPERATOR   A manipulator that can be operated at a distance using local manual control and a communication link.

TERMINAL   (1) Any fitting attached to the end of a wire or circuit to aid in making a good electrical connection. (2) A keyboard and display unit used to communicate with a computer or robot controller.

THRESHOLD   The transition level between two logic states in a logic circuit. In TTL circuits where 0 is represented by 0.8 volt and 1 by 2.0 volts, the low threshold range is −0.3 to 0.8 volt and the high range is 2.0 to 4.5 volts.

TOLERANCE   A specified allowance for error from a desired or measured quantity.

TRANSDUCER   A device used to convert physical phenomena into electrical signals.

TRANSLATION   Movement of a body between two locations such that there is no rotation of the body about any axis.

TRI-STATE   (Trademark of the National Semiconductor Corporation) A logic circuit with three levels: High level and low level the same as TTL, and an isolation state of high impedance that is effectively disconnected. Used for bus connections.

TTL   (Transistor-Transistor Logic) The most widely used logic family with microprocessors. It is low cost, fast and uses medium power.

TTY   A teletypewriter or similar device capable of printing.

2s COMPLEMENT   A 1s complement binary number to which one has been added.

UART   (Universal Asynchronous Receiver/Transmitter) An interface circuit that converts asynchronous serial pulses to parallel and parallel to serial upon command.

UNDERWRITER'S LABORATORIES (UL) An organization chartered to establish, maintain, and operate laboratories for the examination and testing of devices, systems, and materials. UL approval is required by law or by insurance companies for many types of equipment.

USART   (Universal Synchronous/Asynchronous Receiver/Transmitter) A UART that will handle convert either synchronous or asynchronous pulses on command.

VARIABLE   A quantity that can be varied, measured, or controlled.

VECTORED INTERRUPT   An interrupt system in which the interrupting device supplies a code or address to identify the location of the program used to process the interrupt. Eliminates any need for polling.

VELOCITY   The time rate of change in position. It is related to speed but is a vector quantity, so that it also specifies direction in engineering usage.

VOICE CHANNEL   A telephone channel originally designed for voice transmission. Digital signals converted to audio tones by modems can be transmitted over such channels.

VOLATILE MEMORY   A memory that loses information when its power supply is turned off.

WINDUP A term describing the twisting of a shaft when a torque is applied to it.

WORD The unit of information in a computer or computer memory that can be separately addressed and read or written. Usually 8, 16, or 32 bits in microprocessors.

WORK CELL A manufacturing unit containing one or more work stations.

WORK COORDINATES The coordinate system with its center and axes in a workpiece.

WORK ENVELOPE The volume reachable by the maximum and minimum movements of a robot arm. Defined as the motion of the end effector mounting plate center without an end effector attached.

WORK STATION A manufacturing unit consisting of one robot and the necessary auxiliary equipment to perform a task.

WORLD COORDINATES A coordinate system referenced to the base of a robot or to the earth.

WRIST A set of rotary joints mounted on a robot arm and capable of moving with at least two degrees of freedom in yaw, pitch, or roll and supporting an end effector.

YAW Angular displacement of a moving body about an axis perpendicular to the line of motion and through the apparent vertical of the body.

ZERO POINT The origin of a coordinate system.

Index